# Biomaterials, Bioengineering and Sustainability

## Volume 8

There is an urgent need to address the current paradigm shifts in bioengineering for Human Health aiming the creation of breakthrough tissue engineered products, manufacturing technologies and effective regenerative treatments for tackling different diseases/disorders in a personalized manner. Yet, the excessive costs and wastes related to the development and production of biomaterials and advanced therapy medicinal products, and the increasing use of plastics in cell culture methods and animal derived reagents has recognized the importance of decreasing the direct carbon footprint and thus, implement sustainable principles and solutions in the innovation ecosystem. The main goal of the volumes in Biomaterials, Bioengineering and Sustainability series is to catapult and consolidate new concepts and solutions towards the development of the next-generation of sustainable and eco-friendly biomaterials and tissue engineering and regenerative medicine approaches. Each volume will focus on the latest developments dealing with the identification of new sources of sustainable or recycled biomaterials, providing ideas for green technologies and methods that can be applied for biomaterials advanced processing and scaffolding strategies, and applications in biofabrication, tissue engineering, regenerative medicine, and drug delivery systems. It also aims to include the exploitation of renewable and sustainable source of human cells applied for cell therapies or in combination with sustainable biomaterials. The develop complex in vitro 3D/4D models and dynamic cell culture systems will be other subjects to be further explored from a sustainable perspective. This series aims to attract the contributions of leading experts in bioengineering, cell biology, materials engineering, and environmental sciences.

Joaquim Miguel Oliveira
Joana Silva-Correia • Rui Luís Reis
Editors

# Sustainable Scaffolds-based Strategies in Tissue Engineering and Regenerative Medicine

Springer

*Editors*
Joaquim Miguel Oliveira
3B's Research Group, I3Bs – Research Institute on Biomaterials, Biodegradables and Biomimetics
University of Minho
Barco, Guimarães, Portugal

Joana Silva-Correia
3B's Research Group, I3Bs – Research Institute on Biomaterials, Biodegradables and Biomimetics
University of Minho
Barco, Guimarães, Portugal

Rui Luís Reis
3B's Research Group, I3Bs – Research Institute on Biomaterials, Biodegradables and Biomimetics
University of Minho
Barco, Guimarães, Portugal

ISSN 2731-751X ISSN 2731-7528 (electronic)
Biomaterials, Bioengineering and Sustainability
ISBN 978-3-031-96273-8 ISBN 978-3-031-96274-5 (eBook)
https://doi.org/10.1007/978-3-031-96274-5



This Springer imprint is published by the registered company Springer Nature Switzerland AG
The registered company address is: Gewerbestrasse 11, 6330 Cham, Switzerland

# Preface

Scaffold-based technologies and approaches play a pivotal role in modulation of cellular functions, extracellular matrix production, and in tissue regeneration and remodeling. The increasing use of scaffolds (3D) as alternative substrates to 2D cell culture methods may pose different challenges in respect to their sustainable application in *in vitro* models, tissue engineering and regenerative medicine (TERM), in a near future.

In this book, we will provide a concise overview on the relevant research dealing with the design and fabrication methods of sustainable scaffolds (conventional and advanced), considering not only the technological hurdles but also the increasingly important sustainability issues. In this context, the vast options of sustainable biomaterials and eco-friendly methods for biomaterials synthesis and scaffold's processing techniques are discussed. Importantly, a thorough discussion on the latest developments in the field of scaffolds-based strategies for TERM applications is provided. This book is divided into three main sections comprising 15 chapters, as follows: (i) Part I: Sustainable Scaffold Techniques and Designs in Tissue Engineering (five chapters); (ii) Part II: Natural Materials and Eco-Wastes Used as Sacrificial Templates (four chapters); and (iii) Part III: Biomedical Applications (six chapters). In brief, relevant topics on the types of biomaterials and its sustainable sources, and functionalization and processing methods, including green technologies and additive manufacturing, and different applications of sustainable scaffolds that promise to revolutionize preclinical research and clinics, are overviewed in depth.

This updated book includes the contribution of leading and multidisciplinary researchers, which provide an expert discussion on the significant achievements dealing with sustainable scaffolds for biomedical applications, thus serving as a core reference for a new generation of "Hybrid" students and established TERM researchers who aim to consider the sustainability dimension in their research activities.

Barco, Guimarães, Portugal

Joaquim Miguel Oliveira
Joana Silva-Correia
Rui Luís Reis

# Acknowledgments

We acknowledge the European Commission for the financial support provided by EngVIPO project (ID: 101183041) funded under the Marie Skłodowska-Curie Actions—Staff Exchanges 2023 (HORIZON-MSCA-2023-SE-01-01).

# Contents

# Part I

# Sustainable Scaffold Techniques and Designs in Tissue Engineering

# Recent Advancement of Sustainable Scaffolds in Regenerative Medicine

1

Swati Sambita Mohanty, Sanghamitra Pati, and Sangram Keshari Samal

**Abstract**

Sustainable scaffolds offer efficient temporary support for cellular growth that can continue tissue regeneration for the development of desired organs without any toxic impact on the system in the foreseeable future. Regenerative medicine offers a transformative approach to tissue repair, but conventional scaffolds often raise environmental concerns. In recent years, several approaches have been adopted to incorporate sustainability into the designing of tissue engineering scaffolds during the fabrication process, which is achieved by using biodegradable/biocompatible, natural/synthetic polymers, energy-efficient and solvent-free processes, etc. Recent advancements in designing sustainable scaffolds and their potential use in tissue engineering applications are discussed. This sustainable approach paves the way for a future where promoting human health by reducing morbidity and mortality, goes hand-in-hand with fostering a healthier planet.

**Keywords**

Sustainable Biomaterials · Biodegradable Scaffolds · Cell Adhesion & Differentiation · Extracellular Matrix · Targeted Tissue Regeneration

S. S. Mohanty · S. Pati · S. K. Samal (✉)
Laboratory of Biomaterials and Regenerative Medicine for Advanced Therapies, ICMR-Regional Medical Research Centre, Bhubaneswar, Odisha, India



J. M. Oliveira et al. (eds.), *Sustainable Scaffolds-based Strategies in Tissue Engineering and Regenerative Medicine*, Biomaterials, Bioengineering and Sustainability 8, https://doi.org/10.1007/978-3-031-96274-5_1

## 1.1 Introduction

Sustainable scaffolds offer efficient temporary support for cellular growth that can continue tissue regeneration to develop desired organs without any toxic impact on the system in the foreseeable future. Regenerative medicine offers a transformative approach to tissue repair. However, conventional fabrication techniques for designing scaffolds often rely on non-renewable resources or involve harsh production processes, raising concerns for biocompatibility and biodegradability in human health. Sustainable scaffolds, crafted from eco-friendly materials and employing greener fabrication methods, are emerging as a critical advancement in this field. The health benefits of sustainable scaffolds cannot be avoided as it offers the potential to regenerate damaged tissues and restore functionality. The sustainable scaffolds have the potential to degrade into non-toxic components, creating a safer environment for cell growth during regeneration. This enhanced biocompatibility is crucial for successful tissue regeneration (Trucillo 2024). However, translating this potential into reality presents significant engineering challenges. Mimicking the complexity and functionality of natural tissues requires a multifaceted approach. Engineering researchers strive to develop biocompatible scaffolds that provide structural support and cues for cell growth, while selecting the right cell types and delivering essential growth factors to promote tissue formation. Additionally, vascularization of engineered tissues, crucial for oxygen and nutrient delivery, remains a hurdle. Overcoming these challenges will be essential to ensure the long-term survival and function of engineered tissues in the body (Pogorielov et al. 2017). The sustainable scaffold paradigm is a cornerstone of tissue engineering, providing a crucial temporary structure for cell growth and guiding tissue formation. These sustainable scaffolds act as three-dimensional matrices, typically porous and biodegradable, that mimic the natural extracellular matrix (ECM) found within tissues. The ECM plays a vital role not only in providing structural support but also in directing cell adhesion, migration, proliferation, and differentiation (Schmidt et al. 2021). By mimicking the ECM's properties, scaffolds offer a platform for cells to organize and function similarly to how they would in native tissues.

Developing ideal sustainable scaffolds for tissue engineering requires careful consideration of various factors. Biocompatibility is paramount, ensuring the scaffold does not elicit an adverse immune response within the body (Lee et al. 2014). Additionally, the scaffold's degradation rate needs to be tailored to the specific tissue being regenerated. For instance, bone scaffolds require a slower degradation rate to provide long-term mechanical support for new bone formation, while scaffolds for skin regeneration can degrade more rapidly as the new tissue takes over (Chan and Leong 2008). Designing scaffolds with controlled biodegradability is another important consideration. Ideally, the scaffold should degrade at a rate that coincides with tissue formation, providing temporary support while gradually being replaced by functional tissue (Chan and Leong 2008). Natural polymers like chitosan or polycaprolactone offer biodegradable alternatives to traditional synthetic materials, allowing for a more controlled degradation process (Bolívar-Monsalve et al. 2021).

Despite its success, the sustainable scaffold paradigm is not without limitations. Engineering scaffolds with the intricate architecture and biochemical cues found in natural ECM remains a challenge. Additionally, vascularization of engineered tissues, essential for nutrient and oxygen delivery, can be hindered by the presence of scaffolds (De Pieri et al. 2021). Researchers are actively exploring strategies to address these limitations, such as incorporating bioactive molecules within scaffolds or developing techniques for scaffold-free tissue engineering approaches. Figure 1.1 illustrates the overcoming limitations in designing polymeric scaffolds for tissue engineering, such as mimicking the complexity of natural tissues or achieving optimal biocompatibility and biodegradability. The scaffold provides support and guidance for cell growth and new tissue formation. The pores allow for cell infiltration, nutrient transport, and waste removal, mimicking the natural ECM (Tarun 2012).

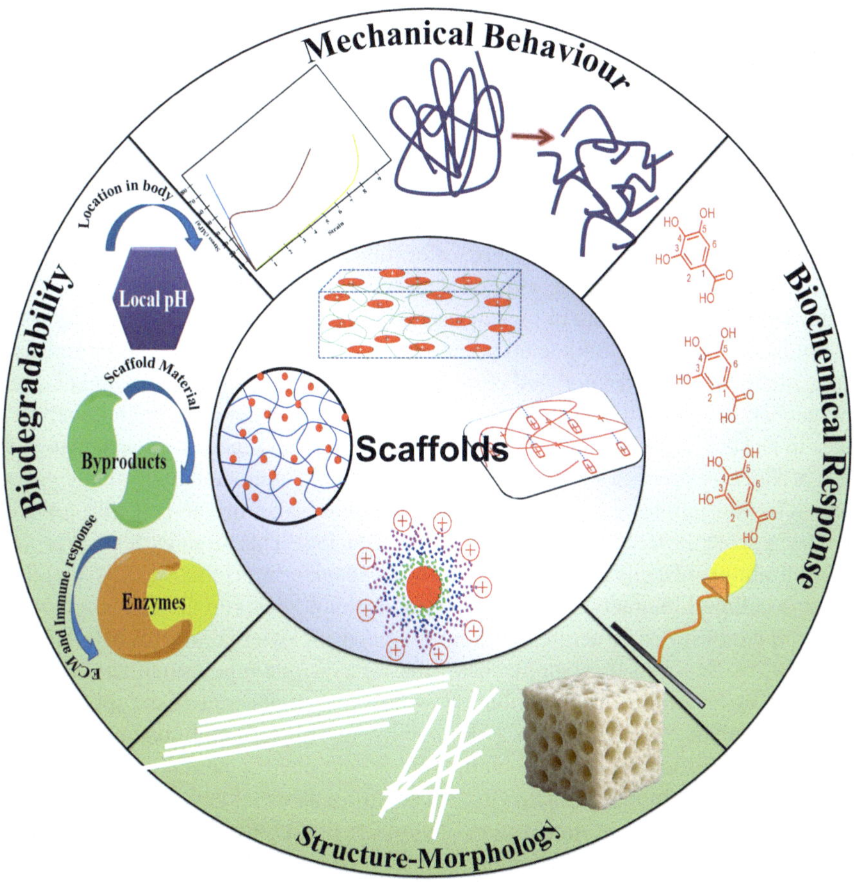


**Fig. 1.1** Design challenges in scaffolds for Tissue Engineering (Garg Tarun et al. 2012)

### 1.1.1 Limitations of Conventional Scaffolds and the Need for Sustainability

While the sustainable scaffold paradigm has been instrumental in tissue engineering advancements, conventional scaffolds present limitations that hinder their effectiveness and raise sustainability concerns. A major challenge lies in mimicking the complex architecture and heterogeneity of natural tissues. Most scaffolds lack the intricate network of pores and channels found within the ECM, potentially hindering cell migration, nutrient diffusion, and waste removal within the engineered tissue (Ikada 2006). Additionally, the materials used to fabricate conventional scaffolds often rely on non-degradable synthetic polymers. These materials can persist in the body long after tissue regeneration is complete, potentially leading to chronic inflammation and foreign body reactions (Park 2015). Furthermore, the production of these synthetic polymers often involves energy-intensive processes and raises concerns about their environmental impact.

### 1.1.2 Seeking Sustainable Alternatives: Biodegradable and Biocompatible Scaffolds

The limitations of conventional scaffolds highlight the need for developing more sustainable and biocompatible alternatives. Researchers are actively exploring the use of naturally-derived materials for scaffold fabrication. These materials, such as collagen, chitosan, and silk fibroin, offer inherent biocompatibility and often degrade at a rate more closely aligned with tissue formation (Gomes et al. 2008). Additionally, advancements in bio-fabrication techniques allow for the creation of scaffolds with tailored architectures that better mimic the structure of natural tissues, potentially improving cell function and tissue integration (Henkel and Hutmacher 2013). Furthermore, these naturally-derived materials often hold advantages in terms of sustainability. They can be sourced from renewable resources, and their production processes typically require less energy compared to synthetic polymers. The limitations associated with scaffolds have also driven research toward exploring scaffold-free tissue engineering approaches. This strategy leverages the inherent ability of cells to self-assemble and organize into functional tissues. By utilizing cell-cell interactions and bioprinting technologies, researchers aim to create three-dimensional structures without the need for external scaffolds (De Pieri et al. 2021). Scaffold-free approaches offer several potential benefits, including improved cell viability and distribution within the engineered tissue, potentially leading to enhanced functionality. Additionally, this approach eliminates concerns associated with the long-term biocompatibility and environmental impact of synthetic scaffolds. Ultimately, achieving successful and sustainable tissue regeneration will likely require a multifaceted approach that combines advancements in scaffold design, material selection, and scaffold-free techniques. By fostering collaboration between engineers, material scientists, and biologists, researchers can develop innovative solutions that address the limitations of conventional scaffolds while prioritizing biocompatibility and environmental sustainability. This combined effort holds the potential to unlock the full potential of tissue engineering for improving human health and well-being in an environmentally responsible manner.

## 1.2 Sustainable Natural Polymers Scaffolds: Abundance and Biocompatibility

### 1.2.1 Natural Polymers Scaffolds

Natural polymers, unlike their synthetic counterparts, boast a unique combination of abundance and biocompatibility. Cellulose, for instance, is the most abundant biopolymer on Earth, with plants being the major producer (Balaji et al. 2018). These readily available materials are also attractive due to their minimal negative impact on the human body. Their biocompatible nature stems from their structural similarity to naturally occurring components found in living organisms (Mallik et al. 2019). This makes them ideal candidates for a variety of biomedical applications, ranging from drug delivery systems to tissue engineering scaffolds (Rajeswari 2017). Beyond just abundance and general biocompatibility, natural polymers offer a spectrum of advantages due to their diverse origins and chemical structures. Some, like silk fibroin, possess excellent mechanical strength, making them suitable for sutures, ligaments, and other implantable devices (Balaji et al. 2018). Others, like chitosan derived from crustacean shells, boast inherent antibacterial properties, reducing the risk of infection associated with medical implants (Troy et al. 2021). This biocompatibility extends beyond human applications. The natural degradation of these materials by microorganisms minimizes environmental impact, making them attractive for sustainable packaging materials and agricultural products (Ponnusamy and Mani 2022).

However, natural polymers also have limitations. Their inherent variability, compared to precisely controlled synthetic polymers, can present challenges in maintaining consistent material properties for large-scale applications (Ponnusamy and Mani 2022). Additionally, some natural polymers may trigger allergic reactions in a small percentage of the population, requiring careful selection and potential prescreening for specific uses (Balaji et al. 2018). Despite these limitations, research continues to refine processing techniques and explore modifications to natural polymers, enhancing their performance and expanding their potential applications. The inherent advantages of abundance, biocompatibility, and environmental friendliness make natural polymers a promising class of materials for the future of biomedicine and sustainable technologies.

Natural polymers, like collagen, chitosan, and silk, offer a unique set of advantages over synthetic alternatives with their excellent biocompatibility and biodegradability properties. These materials are readily accepted by the human body, minimizing the risk of rejection when used in medical devices or implants (Jiang and Loos 2016). Collagen, for instance, is the main structural protein in our skin and bones, making it an ideal candidate for tissue engineering applications

(Parenteau-Bareil et al. 2010). Similarly, chitosan, derived from shellfish, shows promise in wound dressings and other tissue engineering applications due to its biocompatibility, antibacterial properties, and versatility in scaffold fabrication (Dash et al. 2017; Rajinikanth et al. 2024; Sultankulov et al. 2019). Unlike some synthetic polymers that persist for extended periods in the environment, these materials can break down naturally by microorganisms. This characteristic makes them environment friendly and reduces the risk of long-term pollution (Jiang and Loos 2016). For example, silk-based sutures dissolve after the wound heals, eliminating the need for a second surgery for removal (Altman et al. 2003). The several possible applications of chitosan scaffolds in tissue engineering are depicted in Fig. 1.2.

Finally, natural polymers often possess inherent biofunctional properties that make them well-suited for specific applications. Collagen's structure provides excellent strength and elasticity, while chitosan's positive charge allows it to interact with negatively charged cells and tissues (Rajinikanth et al. 2024; Parenteau-Bareil et al. 2010). Silk fibers boast remarkable tensile strength, making them a valuable asset for textiles and biomaterials field (Altman et al. 2003). These inherent properties can be further tailored through processing techniques, expanding the potential uses of natural polymers. In conclusion, natural polymers offer a compelling combination of biocompatibility, degradability, and functional properties. As research continues to explore their potential, these materials hold promise for a wide range of applications in medicine, engineering, and other fields.

Biomaterials form the fundamental building blocks of scaffolds in regenerative medicine. These scaffolds are crucial for repairing, replacing, or improving tissue

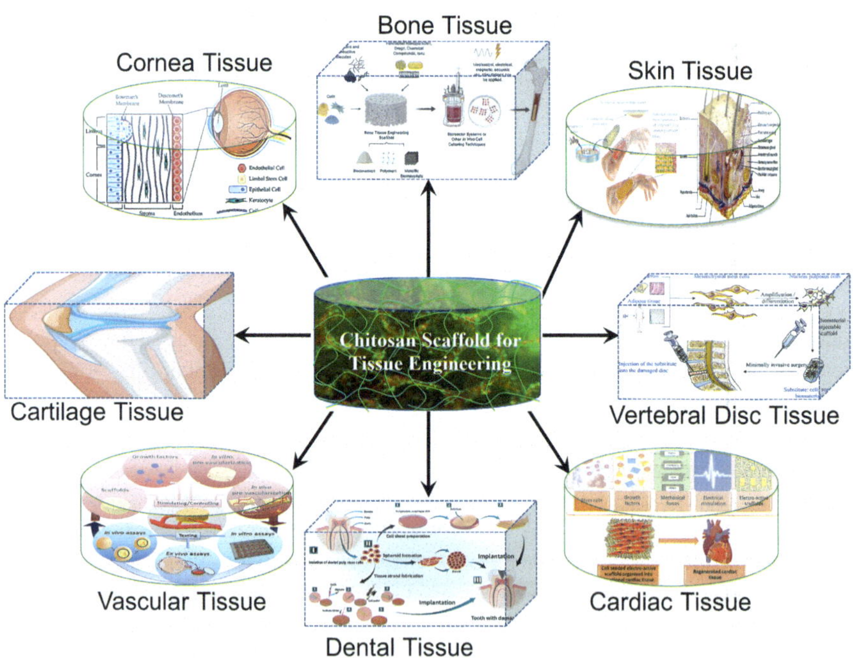


**Fig. 1.2** Applications of chitosan scaffolds in tissue engineering (Liu et al. 2023)

function. Williams has extensively detailed the nature and function of these biomaterials (Williams 2009). Despite their numerous benefits, achieving true sustainability with natural polymers remains a significant challenge with major concern in sourcing. Large-scale production of some natural polymers, like chitosan obtained from shrimp shells, can put pressure on marine ecosystems if not managed responsibly (Rinaudo 2006). Sustainable practices for harvesting raw materials and utilizing by-products from existing industries are crucial. Another challenge lies in processing. Traditional methods for extracting and processing natural polymers can be energy-intensive and involve harsh chemicals, impacting the overall environmental footprint (Chemat et al. 2012). Research on greener extraction techniques using enzymes or less harmful solvents is ongoing. Additionally, optimizing processing steps to minimize waste and energy consumption is essential. Life cycle assessment (LCA) plays a vital role in developing sustainable strategies. LCA allows for a comprehensive evaluation of the environmental impact of a natural polymer throughout its entire life cycle, from sourcing to disposal (Fonseca et al. 2023). By identifying hotspots within the production chain, researchers and manufacturers can target areas for improvement. Beyond the challenges of sourcing and processing, ensuring the overall sustainability of natural polymers requires addressing issues like land use and potential competition with food production. For instance, large-scale cultivation of plants for biopolymers could lead to deforestation or depletion of arable land (La Rosa 2016). Strategies such as utilizing waste biomass or non-food crops can help mitigate this concern. Furthermore, promoting the circular economy principles for natural polymers is crucial. This involves designing products with end-of-life in mind, facilitating efficient collection and recycling of bio-based materials (Brandão et al. 2021). Additionally, exploring possibilities for biodegradation in industrial composting facilities can further enhance the sustainability of these materials. Finally, fostering consumer awareness plays a significant role. Educating consumers about the benefits and limitations of natural polymers can encourage responsible choices and support the development of a sustainable market for these materials (Anquez et al. 2022). Transparency throughout the supply chain, including clear labelling and certification programs, can build trust and encourage consumer adoption.

### 1.2.2 Synthetic Biodegradable Polymers: Tailoring Properties

Synthetic biodegradable polymers offer a powerful advantage—the ability to precisely control their properties for specific applications. Unlike their naturally derived counterparts, these materials can be meticulously designed at the molecular level to achieve desired characteristics. This tailored design approach unlocks a vast potential for various fields, including medicine, tissue engineering, and environmental solutions. One key aspect of tailoring synthetic biodegradable polymers is manipulating their degradation rate. By altering the chemical structure of the polymer backbone or incorporating specific additives, scientists can control how quickly the material breaks down in the body or environment (Rejinold et al. 2021). For instance, polylactic acid (PLA) have relatively fast degradation, making it ideal for short-term implants or drug delivery systems. Conversely, polycaprolactone (PCL) degrades at a slower pace, finding use in long-term implants or controlled release applications (Thakur and Thakur 2015).

Another crucial property under control is the mechanical strength of the polymer. Depending on the intended use, the material can be engineered to be rigid and supportive, like bone scaffolds, or flexible and elastic, mimicking soft tissues (Hutmacher 2001). This control over mechanical properties is achieved through various techniques, including copolymerization with different monomers or modifying the polymer architecture through branching or crosslinking (Colquhoun and Tanner 2015). By harnessing the power of synthetic chemistry, scientists can create a diverse range of biodegradable polymers with precisely tailored properties. This ability to fine-tune material characteristics paves the way for the development of innovative solutions that address specific needs across various sectors. The ability to tailor properties extends far beyond just the degradation rate in synthetic biodegradable polymers. Scientists can introduce various functionalities to these materials, creating a new level of control over their behavior within the body or environment. One exciting area is the incorporation of bioactive molecules. These molecules can be physically blended, chemically attached, or encapsulated within the polymer matrix (Hutmacher et al. 2023). For instance, antibiotics loaded onto synthetic scaffolds can provide localized infection control for bone implants, while growth factors can promote cell adhesion and proliferation for tissue engineering applications (Yao et al. 2020). Another approach involves tailoring the surface properties of the polymer. By introducing specific chemical groups, scientists can design hydrophilic (water-loving) or hydrophobic (water-repelling) surfaces (Katti et al. 2008). This control over surface chemistry influences factors like cell attachment, protein adsorption, and drug release kinetics. For example, hydrophilic surfaces can promote cell adhesion, while hydrophobic surfaces may be better suited for controlled drug delivery applications. The concept of biocompatibility also comes into play when tailoring synthetic biodegradable polymers. By incorporating specific functionalities, researchers can modulate the immune response to the implanted material (Ma et al. 2007). This can be crucial for minimizing inflammation and rejection, especially for long-term implants or devices in contact with sensitive tissues.

In brief, tailoring properties in synthetic biodegradable polymers goes beyond just basic biodegradation. The ability to introduce functionalities and control surface chemistry unlocks a new realm of possibilities. This allows scientists to design biomaterials that actively interact with the biological environment, leading to the development of more advanced and targeted solutions in areas like medicine, drug delivery, and tissue engineering.

Synthetic polymers also offer greater flexibility in mechanical property design. Depending on the intended use, the material can be engineered to possess specific strength, elasticity, or stiffness. This control over mechanical properties is achieved through techniques like copolymerization or modifying the polymer architecture. For example, PLA can be blended with other polymers to create stronger and more rigid materials, while PCL can be modified to be more flexible for applications mimicking soft tissues (Capuana et al. 2021; Zheng and Pan 2020). Furthermore, synthetic polymers can be readily functionalized with specific molecules to enhance their performance. These functionalities can range from bioactive molecules like growth factors for tissue engineering to antimicrobial agents for medical devices. This ability to introduce desired functionalities allows for the creation of targeted biomaterials with specific therapeutic or diagnostic capabilities (Martinez-Robinson 2020).

Finally, synthetic polymers offer the advantage of being readily functionalized with specific molecules to enhance their performance. These functionalities can range from bioactive molecules like growth factors for tissue engineering to antimicrobial agents for medical devices. This ability to introduce desired functionalities allows for the creation of targeted biomaterials with specific therapeutic or diagnostic capabilities (Martinez-Robinson 2020). Figure 1.3 represents the synthetic polymers for tissue scaffolds. In conclusion, synthetic biodegradable polymers offer a powerful advantage for controlled design due to their consistency, tunable degradation rates, tailorable mechanical properties, and ease of functionalization. These features empower scientists and engineers to create innovative solutions across various fields, from medicine and drug delivery to tissue engineering and environmental applications.

Synthetic polymers, a cornerstone of modern life, present a complex challenge in terms of sustainability. Their undeniable benefits in applications ranging from medical devices to food packaging are often overshadowed by their environmental impact. Herein, we explore two key considerations for a more sustainable future of synthetic polymers: biodegradability and resource utilization. Biodegradability refers to a polymer's ability to decompose naturally by microorganisms into harmless substances. Conventional synthetic polymers, derived from fossil fuels, are notoriously resistant to degradation, leading to massive plastic pollution in landfills and ecosystems (Martinez-Robinson 2020). Research on biodegradable polymers

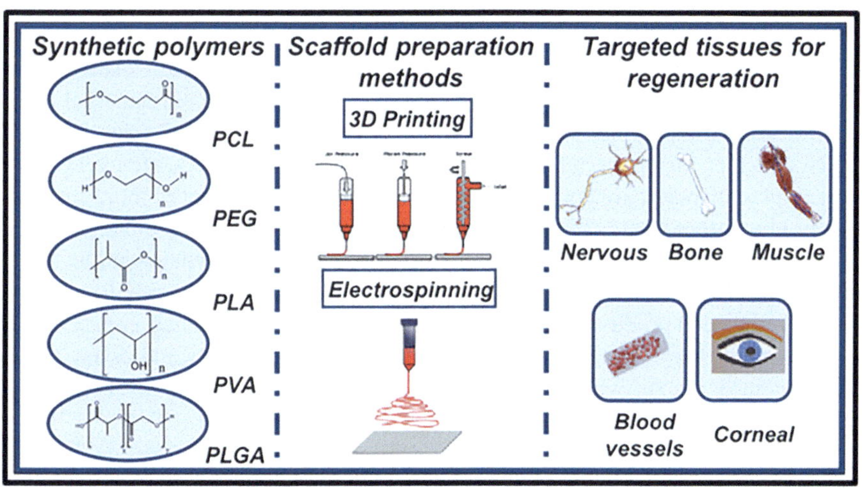


**Fig. 1.3** Biocompatible synthetic polymers for tissue engineering purposes (Perez-Puyana et al. 2020)

derived from renewable resources like plant starches or microorganisms offers a promising solution (Park 2015). These materials can significantly reduce plastic waste accumulation, mitigating environmental harm (Zhu et al. 2016b).

While biodegradability is a crucial aspect of sustainable polymers, it's not the only factor. Polymers are commonly engineered for strength and durability, but not for efficient breakdown. Developing polymers that can be easily broken down into their parts at the end of their useful life facilitates efficient recycling and reduces reliance on virgin materials (Kuhl et al. 2015). Not all biodegradable polymers are suitable for composting. Some require industrial composting facilities with specific temperature and pressure conditions, limiting their environmental benefit. Developing polymers that can biodegrade in home or backyard compost piles offers a more accessible and sustainable end-of-life solution (Ghosh and Jones 2021). Biodegradable polymers themselves, or the additives used in their production, can potentially introduce new toxins into the environment during degradation. Careful selection of materials and thorough toxicity assessments are essential to ensure biodegradability doesn't come at the expense of environmental or human health (Adekomaya et al. 2021).

Sustainability goes beyond just environmental impact. For synthetic polymers to be truly sustainable, social and economic factors must also be considered. Sustainable alternatives often have higher upfront costs. Life cycle cost analysis, which considers the total cost of a polymer throughout its lifetime, including disposal or recycling, can provide a more complete picture and incentivize the adoption of sustainable options (Kneifel and Webb 2020). Effective recycling and composting infrastructure is crucial for maximizing the sustainability of synthetic polymers. Investment in building and maintaining this infrastructure is essential to support the transition towards a circular economy for these materials (Epps et al. 2022).

Resource utilization during polymer production is another critical factor. The traditional reliance on non-renewable resources like petroleum for polymer feedstocks is unsustainable. Developing methods for utilizing recycled materials or renewable feedstocks like biomass can significantly reduce the environmental footprint of synthetic polymer production (Epps et al. 2022). Life cycle assessments, which consider the environmental impact throughout a polymer's life cycle from extraction to disposal, are crucial tools in guiding the development of more sustainable production processes (Amarakoon et al. 2022). By focusing on biodegradability and efficient resource utilization, the future of synthetic polymers can be more sustainable. Continuous research and development efforts are essential to create new materials and production processes that minimize environmental impact while retaining the valuable properties that synthetic polymers offer. By addressing these diverse considerations, we can create a future for synthetic polymers that benefit the environment, society, and the economy.

### 1.2.3 Decellularized Extracellular Matrix (dECM): Mimicking the Native Niche

The human body is a complex ecosystem, with each tissue harboring a unique microenvironment essential for proper cell function. This microenvironment, known as the ECM, provides structural support, biochemical cues, and mechanical signals that guide cellular behavior. In regenerative medicine, replicating this intricate niche is crucial for promoting successful cell transplantation and tissue engineering. dECM emerges as a promising solution, offering a natural scaffold that mimics the native tissue environment and has also shown promise in hemostatic applications (Cai and Weng 2023). dECM is derived from various tissues by removing the cellular component while preserving the intricate three-dimensional structure and biochemical composition of the ECM. This process retains vital components like collagen, glycosaminoglycans, and growth factors, which orchestrate cellular processes like adhesion, proliferation, and differentiation (Gilpin and Yang 2017). By providing a familiar biological blueprint, dECM scaffolds can significantly enhance cell function compared to traditional synthetic biomaterials (Yue 2014).

The ability of dECM to mimic the native niche extends beyond structural similarity. Studies have shown that dECM derived from specific tissues can promote the growth and differentiation of corresponding cell types. For instance, cardiac dECM effectively supports the growth and maturation of cardiomyocytes, while lung dECM promotes the development of lung epithelial cells (Golebiowska et al. 2024; Stabler et al. 2015). This tissue-specific biomimicry holds immense potential for regenerative medicine applications targeting damaged or diseased tissues.

However, challenges remain in optimizing dECM technology. Decellularization techniques need to be carefully tailored to ensure complete cell removal while preserving the delicate ECM structure and composition. Additionally, sourcing dECM from various tissues and ensuring consistent quality across batches requires further development (Yi et al. 2017). Despite these challenges, dECM offers a revolutionary approach for mimicking the native cellular niche, paving the way for advancements in regenerative medicine. The utilization of dECM scaffolds for bone regeneration: a cell-based approach is illustrated in Fig. 1.4.

The tissue-specific advantage of dECM goes beyond structural similarity. Studies have shown remarkable success in directing cell fate using dECM. For example, cardiac dECM scaffolds effectively promote the growth and maturation of cardiomyocytes, the muscle cells of the heart (Barbulescu et al. 2022). Similarly, lung dECM has been shown to support the development of functional lung epithelial cells, which are crucial for gas exchange in the lungs (Gilpin and Wagner 2018). This inherent ability of dECM to guide cell differentiation towards specific lineages holds immense potential for regenerative medicine applications targeting damaged or diseased tissues. Furthermore, dECM scaffolds offer inherent biological cues that synthetic materials often lack. The retained growth factors and signaling molecules within the dECM can directly influence cellular behavior. These signals guide

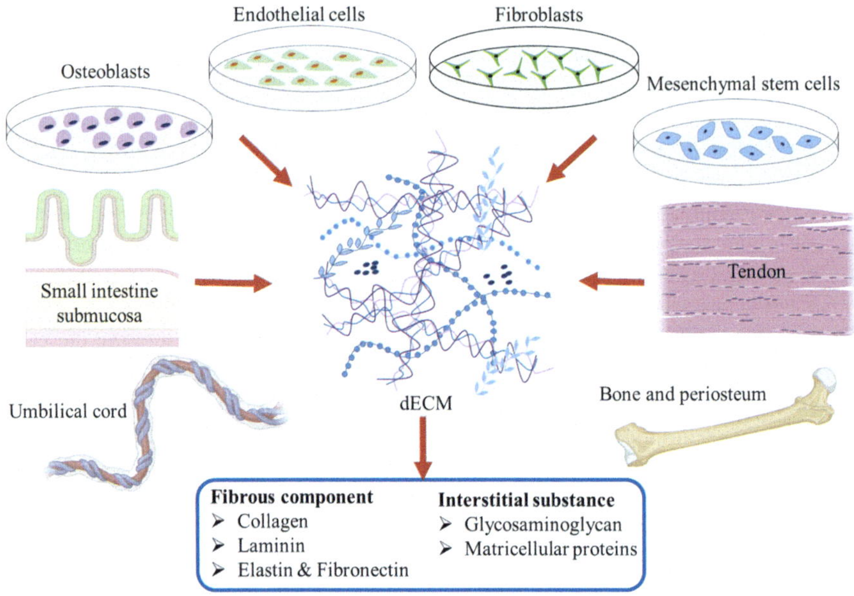


**Fig. 1.4** Decellularized extracellular matrices in bone tissue engineering: from cells to tissues (Gilpin and Yang 2017)

processes like migration, proliferation, and differentiation, creating a microenvironment conducive to tissue regeneration (Zhang et al. 2022b). Additionally, dECM scaffolds possess inherent vascularization potential, promoting the formation of blood vessels within the scaffold, which is essential for nutrient delivery and waste removal in newly formed tissues (McInnes et al. 2022). In conclusion, dECM scaffolds offer a revolutionary approach for tissue-specific applications in regenerative medicine. By mimicking the native tissue environment and providing essential biological cues, dECM promotes targeted cell growth and differentiation. As research continues to optimize decellularization techniques and explore dECM applications for various tissues, these scaffolds hold promising potential for the regeneration of damaged or diseased tissues across the body.

dECM holds immense promise for regenerative medicine due to its ability to mimic the native tissue environment. However, the processing of dECM presents several sustainability challenges that need to be addressed for its widespread adoption. Herein, we explore some key challenges and potential strategies for optimizing dECM processing towards a more sustainable future. One major challenge lies in the resource intensity of decellularization techniques. Conventional methods often rely on harsh chemicals and detergents to remove cells, raising concerns about environmental impact and potential toxicity. Disposing of these chemicals after use can pollute waterways and require significant energy for wastewater treatment

(Neishabouri et al. 2022). Additionally, the high volume of detergents needed can be resource-intensive and contribute to the depletion of non-renewable resources. Strategies to overcome this challenge involve exploring alternative decellularization methods. Research is ongoing to develop techniques using enzymes, sonication, or even supercritical fluids, which offer a more environment friendly approach (Allu et al. 2023; Duarte et al. 2009). Enzymes, for instance, can specifically target cellular components while leaving the ECM intact, reducing the need for harsh chemicals. Similarly, sonication and supercritical fluids utilize physical forces to disrupt cellular membranes, offering a potentially more sustainable alternative. Another challenge is the variability associated with dECM derived from different sources. Decellularization efficiency can be impacted by factors like tissue type, donor health, and processing protocols. This variability can lead to inconsistencies in the resulting dECM scaffolds, affecting their performance and ultimately requiring the use of more tissue samples to achieve consistent results (Badylak 2002). Strategies mitigating this challenge involve standardization of decellularization protocols. Developing well-defined protocols for specific tissue types, considering factors like decellularization time, detergent concentration, and temperature control, can help ensure consistent quality across dECM batches. Additionally, implementing quality control measures throughout the processing pipeline can further minimize variability in the final product.

Finally, the scalability of dECM production for broader clinical applications needs to be addressed. Currently, obtaining sufficient tissue for dECM production can be challenging, especially for tissues with limited availability. Strategies for overcoming this hurdle include exploring the use of xenogeneic (animal-derived) dECM sources or utilizing smaller tissue biopsies with optimized decellularization protocols (Massaro et al. 2021). Furthermore, research into bioprinting dECM-like structures using biocompatible inks holds promise for large-scale production of tissue-specific scaffolds in the future. By addressing these challenges and adopting sustainable processing strategies, dECM can truly fulfill its potential as a revolutionary tool for regenerative medicine. Continuous research and development efforts focused on environment friendly decellularization methods, standardized protocols, and scalable production techniques are crucial for realizing the full potential of dECM in a sustainable manner.

## 1.3 Fabrication Techniques for Sustainable Scaffolds

Tissue engineering relies heavily on scaffolds to provide structural support and biochemical cues for cell growth and differentiation. However, traditional scaffold materials often raise concerns about sustainability due to factors like reliance on non-renewable resources or challenges with biodegradability. To address these issues, research is focusing on developing sustainable fabrication techniques for scaffolds used in tissue engineering. One promising approach involves utilizing natural biomaterials, derived from sources like collagen, chitosan, or cellulose that are often biodegradable and can be processed using environment friendly methods. Techniques like solvent casting, freeze-drying, or electrospinning can be employed to create scaffolds with desired pore sizes and architectures from these natural polymers (Brovold et al. 2018; Zhao et al. 2023). Additionally, advancements in 3D printing allow for precise control over scaffold design and the incorporation of bioactive molecules for enhanced cell function (Tajik et al. 2023). Another strategy for sustainable scaffold fabrication focuses on utilizing waste materials. For instance, recycled plastics or discarded agricultural byproducts can be processed into biocompatible scaffolds. Techniques like melt extrusion or solvent casting can be adapted to create scaffolds from these recycled materials, offering a solution for waste management while creating valuable tools for tissue engineering applications (Balla et al. 2021; Šafarič et al. 2020).

Beyond the materials themselves, sustainable fabrication techniques consider the environmental impact of the processing methods. Techniques with lower energy consumption or reduced reliance on harsh chemicals are preferred. For example, supercritical fluid processing offers an environment friendly alternative to traditional solvent-based methods for scaffold fabrication (Manjare and Dhingra 2019). In addition, research is ongoing to develop biofabrication techniques that utilize living cells to create scaffolds, potentially eliminating the need for harsh processing conditions (Vesvoranan et al. 2022). There are various conventional methods of sustainable scaffold fabrication (Fig. 1.5) that have been developed for drug delivery, 3D cell culture and tissue engineering applications. The choice of fabrication technique ultimately depends on the desired properties of the scaffold and the specific tissue engineering application. However, by focusing on natural biomaterials, recycled materials, and environment friendly processing methods, researchers are paving the way for a more sustainable future for scaffold fabrication in tissue engineering.

Tissue engineering relies on the development of scaffolds that mimic the natural ECM to promote cell growth and differentiation. Solvent casting/particulate leaching (SCPL) emerges as a simple and versatile technique for creating porous scaffolds for various tissue engineering applications. This approach offers several advantages, making it a valuable tool for researchers and scientists.

The core principle of SCPL is straightforward. A polymer is first dissolved in a suitable solvent to create a homogeneous solution. Next, well-defined particles, often salts like sodium chloride (table salt), are incorporated into the polymer solution. This mixture is then cast into a mold with the desired shape for the final scaffold. As the solvent evaporates, the polymer forms a solid matrix around the embedded particles (Hutmacher 2001). Finally, the scaffold is immersed in a solvent that dissolves the sacrificial particles, leaving behind a network of interconnected pores within the polymer matrix. The pore size and distribution can be controlled by the size and shape of the leached particles (Lyons et al. 2020).

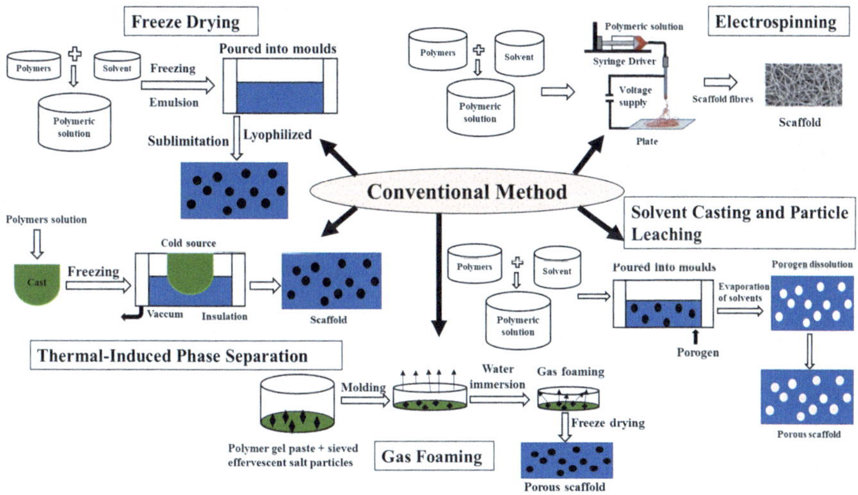


**Fig. 1.5** Sustainable techniques in scaffold fabrication process for tissue engineering applications (Song et al. 2018; Zeng et al. 2017)

The simplicity of SCPL is one of its key advantages. This technique requires minimal specialized equipment and can be readily implemented in various laboratory settings. Additionally, SCPL offers a high degree of versatility. A wide range of polymers can be utilized, allowing researchers to tailor the scaffold properties to specific tissue engineering needs. For instance, biodegradable polymers like PLA can be used for scaffolds that degrade over time, mimicking the natural remodeling process in tissues (Langer and Vacanti 1993).

Despite its advantages, SCPL also has limitations to consider. The choice of solvent can impact the final properties of the scaffold, and residual solvent traces may require additional processing steps. Additionally, the selection of suitable porogens (leachable particles) is crucial for achieving the desired pore morphology. Finally, SCPL may not be suitable for creating highly complex scaffold architectures due to limitations in mold design (Kaliaraj et al. 2023). SCPL offers a valuable approach for fabricating porous scaffolds for tissue engineering applications. Its simplicity, versatility, and compatibility with various materials make it a widely used technique. While limitations exist regarding solvent selection, porogen choice, and scaffold complexity, ongoing research is focused on optimizing SCPL protocols for diverse tissue engineering needs.

### 1.3.1 Electrospinning: Nanofiber Scaffolds for Enhanced Cell-Material Interactions

Tissue engineering strives to create functional tissues by mimicking the natural cellular environment. A key element in achieving this is providing the cells with a scaffold that offers not only structural support but also crucial cues for growth and differentiation. Electrospinning emerges as a powerful fabrication technique for generating nanofiber scaffolds that promote enhanced cell-material interactions. Electrospinning utilizes an electric field to draw a polymer solution into ultrathin fibers, typically ranging in diameter from tens to hundreds of nanometers. These nanofibers mimic the ECM, the natural network of proteins and polysaccharides that surrounds cells in tissues (Flores-Rojas et al. 2023). The high surface area-to-volume ratio inherent to nanofibers provides a multitude of attachment sites for cells, promoting cell adhesion and proliferation (Bhardwaj and Kundu 2010). Additionally, the interconnected porosity of electro-spun scaffolds allows for efficient nutrient and waste transport, which is crucial for maintaining cell viability within the scaffold. Beyond mimicking the ECM structure, electrospinning offers versatility in tailoring the scaffold properties for specific cell types. By manipulating factors like polymer choice, fiber diameter, and alignment, researchers can create scaffolds with specific mechanical properties, degradation rates, and biological functionalities (Hu et al. 2014). For instance, incorporating bioactive molecules like growth factors or adhesion peptides into the nanofibers can further enhance cell-material interactions and promote specific cellular responses (Flores-Rojas et al. 2023). The ability of electrospinning to create complex architectures further expands its potential in tissue engineering. By controlling the electric field and collector design, researchers can fabricate scaffolds with specific pore sizes, gradients, and even multilayered structures. These intricate architectures can mimic the native tissue microenvironment in greater detail, providing even more sophisticated cues for cell behavior (Xing et al. 2023).

Hence, electrospinning offers a robust and versatile technique for creating nanofiber scaffolds that enhance cell-material interactions. By mimicking the ECM structure, offering a high surface area for cell attachment, and allowing for tailored properties and architectures, electro-spun scaffolds provide a valuable tool for researchers in tissue engineering. Figure 1.6 outlines the nanofiber scaffolds fabricated by electrospinning for wound healing. As research continues to refine electrospinning techniques and explore novel material combinations, its impact on developing functional tissue replacements is likely to grow even stronger.

### 1.3.2 3D Printing: Designing Complex Architectures for Functional Tissues

A critical challenge in tissue engineeris still lies in replicating the intricate three-dimensional (3D) structures of native tissues. 3D printing emerges as a revolutionary technology, allowing researchers to design and fabricate scaffolds with complex architectures that mimic the natural environment of various tissues. Unlike traditional scaffold fabrication techniques with limited control over structure, 3D

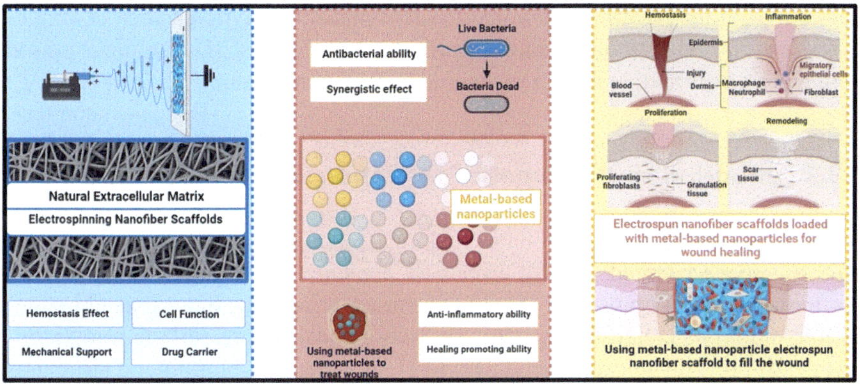


**Fig. 1.6** Electrospun sustainable nanofiber scaffolds for wound healing applications (Dang et al. 2023)

printing offers unparalleled freedom in designing. Using computer-aided design (CAD) software, researchers can create digital models of scaffolds with intricate internal features, channels, and gradients. These models can then be translated into instructions for the 3D printer, enabling the precise deposition of biocompatible materials layer-by-layer to create the desired scaffold architecture (Fang et al. 2023). This level of control allows for the creation of scaffolds that closely resemble the complex geometries of natural tissues, such as the branching networks of blood vessels or the porous structure of bone.

The ability to design complex architectures with 3D printing extends beyond mere shape replication. By incorporating channels and gradients within the scaffold, researchers can control factors like nutrient and oxygen diffusion, crucial for cell survival and functionality within the scaffold (Bose et al. 2012). Additionally, 3D printing allows for the precise patterning of bioactive molecules or cells within the scaffold. For instance, specific growth factors can be deposited in designated regions to promote the differentiation of desired cell types, fostering the development of functional tissues (Gu et al. 2016). The versatility of 3D printing extends to the materials used. A wide range of biocompatible materials, including natural polymers, synthetic biomaterials, and even bioinks containing living cells, can be utilized for 3D printing. This allows researchers to tailor the scaffold properties to match the specific needs of different tissues (Zieliński et al. 2023). For example, soft hydrogels can be used to mimic brain tissue, while stronger polymers may be suitable for bone or cartilage regeneration. 3D printing offers immense potential for tissue engineering, however, there are some specific challenges remain. Optimizing printing parameters for various biomaterials and ensuring the fidelity of complex structures during printing are ongoing areas of research. Additionally, the development of robust vascularization strategies within 3D-printed scaffolds is crucial for the survival of cells within thick tissues (Chan and Leong 2008). 3D printing revolutionizes tissue engineering by enabling the design and fabrication of scaffolds with intricate architectures that mimic native tissues. The ability to control structure, incorporate functional cues, and utilize diverse biomaterials makes 3D printing a powerful tool for creating functional tissue replacements. As research continues to address current challenges and explore new possibilities, 3D printing is poised to play a transformative role in regenerative medicine.

### 1.3.3 Bioprinting: Biomimetic Fabrication with Cellular Control

Bioprinting emerges as a ground-breaking technology in tissue engineering, offering unprecedented control over cell placement and organization within scaffolds. It builds upon traditional 3D printing principles, but utilizes biocompatible materials called bio inks to precisely deposit living cells in a layer-by-layer fashion. This allows researchers to create intricate, biomimetic structures that closely resemble natural tissues, fostering enhanced cell function and tissue regeneration. A key strength of bioprinting lies in its ability to mimic the cellular microenvironment. Bio inks can be formulated using natural biomaterials like collagen, alginate, or hyaluronic acid, which provide essential structural support and biochemical cues for cells (Waidi et al. 2024). These biomaterials mimic the natural ECM, the complex network of proteins and polysaccharides that surrounds cells in tissues. Furthermore, bio inks can be enriched with bioactive molecules like growth factors or adhesion peptides (Wang et al. 2021). These act as signals that guide cellular behaviour and promote differentiation towards specific lineages, such as muscle cells or bone cells. This biomimetic approach allows for the creation of scaffolds that not only offer structural support but also actively influence cell fate, leading to the development of functional tissues. Figure 1.7 illustrates the "Bioprinting Functional Scaffolds for Tissue Regeneration" (Bioprinting emphasizes the 3D printing aspect specifically for biological applications).

3D printing revolutionizes tissue engineering by enabling the design and fabrication of scaffolds with intricate architectures that mimic native tissues. The ability to control structure, incorporate functional cues, and utilize diverse biomaterials makes 3D printing a powerful tool for creating functional tissue replacements. As research continues to address current challenges and explore new possibilities, 3D printing is poised to play a transformative role in regenerative medicine.

### 1.3.4 Bioprinting: Biomimetic Fabrication with Cellular Control

Bioprinting emerges as a ground-breaking technology in tissue engineering, offering unprecedented control over cell placement and organization within scaffolds. It builds upon traditional 3D printing principles, but utilizes biocompatible materials called bio inks to precisely deposit living cells in a layer-by-layer fashion. This allows researchers to create intricate, biomimetic structures that closely resemble natural tissues, fostering enhanced cell function and tissue regeneration. A key strength of bioprinting lies in its ability to mimic the cellular microenvironment. Bio inks can be formulated

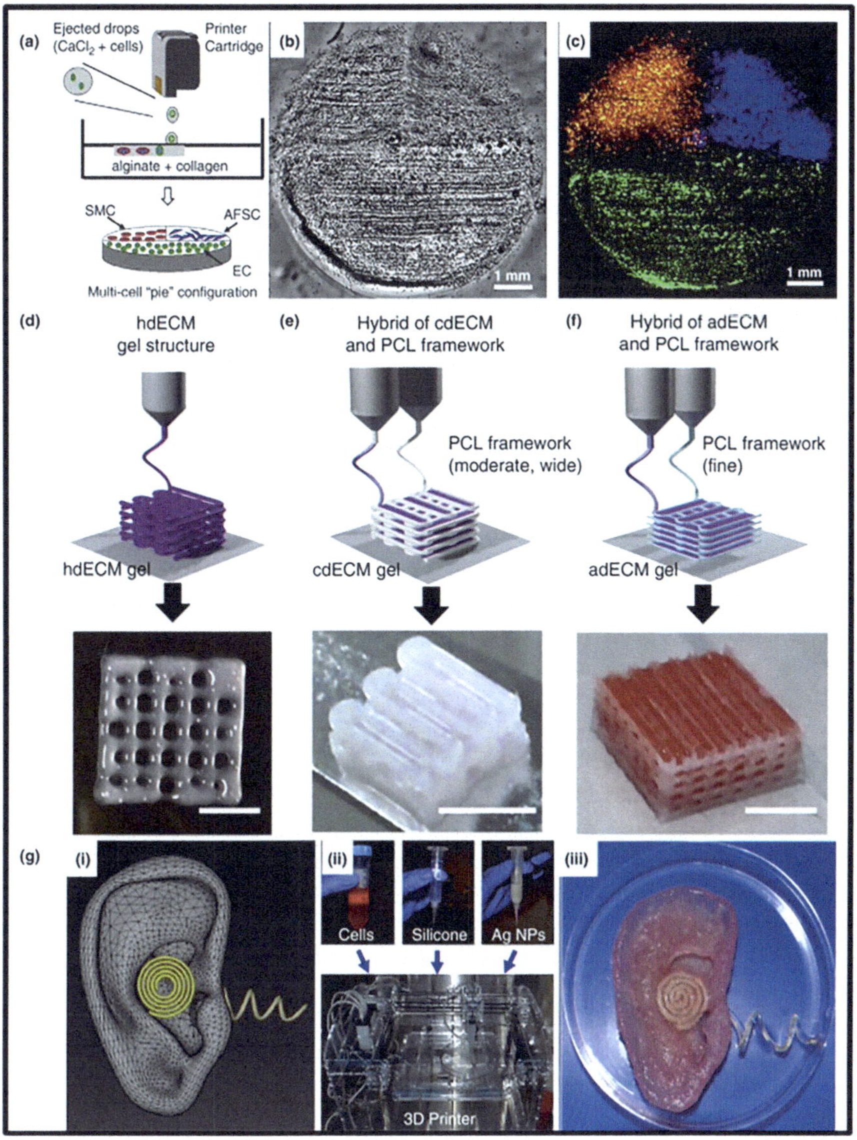


**Fig. 1.7** 3D printing of functional biomaterials for tissue engineering (Zhu et al. 2016a): (**a**) Schematic showing multi-cellular "pie"-like configuration created by bioprinting $CaCl_2$ + cell droplets onto an alginate + collagen layer, organizing different cell types—smooth muscle cells (SMC), amniotic fluid stem cells (AFSC), and endothelial cells (EC)—into distinct regions. (**b**) Optical image of the printed construct showing the layered deposition pattern. (**c**) Fluorescently labeled image illustrating distinct regions corresponding to different cell types in the printed structure (orange, blue, green), demonstrating spatial organization. (**d**) Printing of human-derived ECM (hdECM) gel using extrusion-based bioprinting, yielding a soft grid-like gel structure. (**e**) Hybrid structure combining cartilage-derived ECM (cdECM) gel and polycaprolactone (PCL) framework, with moderate and wide spacing to provide mechanical support and biological compatibility. (**f**) Hybrid structure with adipose-derived ECM (adECM) gel and finely spaced PCL framework to optimize tissue integration and stability. (**g**) 3D bioprinting of a functional bionic ear: (i) Computer model of the ear integrated with an electronic coil antenna, (ii) Multimaterial bioprinting process using cells, silicone, and silver nanoparticles (Ag NPs), (iii) Final printed bionic ear construct exhibiting integrated electronics and biomaterials

using natural biomaterials like collagen, alginate, or hyaluronic acid, which provide essential structural support and biochemical cues for cells (Waidi et al. 2024). These biomaterials mimic the natural ECM, the complex network of proteins and polysaccharides that surrounds cells in tissues. Furthermore, bio inks can be enriched with bioactive molecules like growth factors or adhesion peptides (Wang et al. 2021). These act as signals that guide cellular behaviour and promote differentiation towards specific lineages, such as muscle cells or bone cells. This biomimetic approach allows for the creation of scaffolds that not only offer structural support but also actively influence cell fate, leading to the development of functional tissues.

However, traditional bio inks often rely on synthetic polymers or harsh processing techniques, raising concerns about sustainability. Research efforts are now focused on developing sustainable alternatives. One promising approach involves utilizing dECM as a bio ink source (Zhang and Zhang 2015). dECM is derived from natural tissues and retains the inherent biological cues needed for cell function. Additionally, researchers are exploring the use of minimally processed biomaterials like plant-derived hydrogels (Nikolova and Chavali 2019). These sustainable bio inks offer comparable performance to their synthetic counterparts while minimizing environmental impact. Sustainability considerations extend beyond the bio ink materials themselves. Bioprinting techniques that utilize cell spheroids or self-assembling biomaterials hold promise for reducing the overall bio ink volume required (Mandrycky et al. 2016). Cell spheroids are clusters of cells that can self-organize and mimic some aspects of tissue structure. Self-assembling biomaterials are designed to form specific structures when exposed to certain cues, reducing the need for precise bio ink deposition during the printing process. By minimizing bio ink usage and exploring sustainable materials, researchers are ensuring that bioprinting remains an environmentally responsible technology.

Bioprinting with cellular control offers a powerful tool for regenerative medicine. By combining biomimetic scaffold design, precise cell placement, and the use of sustainable bio inks, researchers are paving the way for creating functional tissue replacements for a variety of diseases and injuries. Examples include bioprinting skin grafts for burn victims, heart patches for damaged cardiac tissue, or even complex structures like bones or ears. As bioprinting technology continues to advance and integrate sustainability principles, its impact on regenerative medicine is poised to be transformative.

## 1.4 Strategies for Optimizing Scaffold Sustainability

Tissue engineering relies heavily on scaffolds to provide structural support and biochemical cues for cell growth and differentiation. However, traditional scaffold materials often raise sustainability concerns due to factors like reliance on non-renewable resources or challenges with biodegradability. To address these issues, researchers are actively exploring various strategies for optimizing scaffold sustainability. One key approach focuses on utilizing natural biomaterials. These materials, derived from sources like collagen, chitosan, or cellulose, are often biodegradable and can be processed using environmentally friendly methods. Techniques like solvent casting, freeze-drying, or electrospinning can be employed to create scaffolds with desired pore sizes and architectures from these natural polymers (Bhat and Kumar 2012; Xu et al. 2019). Additionally, advancements in 3D printing allow for precise control over scaffold design and the incorporation of bioactive molecules for enhanced cell function, all while utilizing biocompatible and potentially renewable materials (Barcena et al. 2023).

Another strategy for sustainable scaffold fabrication focuses on utilizing waste materials. For instance, recycled plastics or discarded agricultural byproducts can be processed into biocompatible scaffolds. Techniques like melt extrusion or solvent casting can be adapted to create scaffolds from these recycled materials, offering a solution for waste management while creating valuable tools for tissue engineering applications (Alavi et al. 2023; Saravanan et al. 2013). This approach not only reduces reliance on virgin resources but also provides a valuable pathway for upcycling waste materials. Beyond the materials themselves, sustainable fabrication techniques consider the environmental impact of the processing methods. Techniques with lower energy consumption or reduced reliance on harsh chemicals are preferred. For example, supercritical fluid processing offers an environmentally friendly alternative to traditional solvent-based methods for scaffold fabrication (Kankala et al. 2017). Additionally, research is ongoing to develop bio fabrication techniques that utilize living cells to create scaffolds, potentially eliminating the need for harsh processing conditions altogether (De Pieri et al. 2021). These bio fabrication techniques often involve utilizing cell spheroids or self-assembling biomaterials to create desired tissue structures in a more natural and sustainable manner. Optimizing scaffold design for efficient material usage also contributes to sustainability. By employing computer modeling and optimization techniques, researchers can design scaffolds with minimal material waste while maintaining the necessary structural integrity and functional properties (Hollister 2005). Additionally, exploring techniques like bioprinting that allow for precise deposition of bio inks can further minimize material usage compared to traditional scaffold fabrication methods. Finally, promoting the development of biodegradable scaffolds that degrade at a controlled rate after implantation is crucial for long-term sustainability. Biodegradable scaffolds eventually break down into harmless byproducts by the body, eliminating the need for a secondary surgical procedure for removal and minimizing waste associated with non-degradable implants (De Pieri et al. 2021).

Hence, optimizing scaffold sustainability requires a multifaceted approach. Utilizing natural biomaterials, waste materials, and environmentally friendly processing techniques alongside efficient design and biodegradability are all crucial strategies. As research in these areas continues to progress, tissue engineering can achieve its goals of promoting regeneration while minimizing its environmental impact.

### 1.4.1 Biocompatibility and Biodegradability: Essential Considerations

At the heart of tissue engineering field lie scaffolds, which provide essential structural support and biochemical cues for cell growth and differentiation. However, for a scaffold to be successful, it must possess two key properties: biocompatibility and biodegradability. Biocompatibility refers to the ability of a material to interact harmoniously with the biological system without causing adverse effects. An ideal biocompatible scaffold should not trigger an immune response, inflammation, or cytotoxicity (cell death) within the body (Ratner et al. 2013). It should also allow for proper cell adhesion, proliferation, and differentiation, ultimately fostering tissue regeneration. Testing biocompatibility involves a battery of *in vitro* and *in vivo* studies to assess the scaffold's interaction with cells and tissues at the implantation site (Qin 2016). Biodegradability refers to the ability of a material to break down over time within the body. Biodegradable scaffolds are designed to gradually degrade at a controlled rate as the new tissue forms. This allows for the transfer of load-bearing function from the scaffold to the regenerated tissue and prevents the need for a second surgery to remove the implant (Hutmacher 2001). The degradation products should also be non-toxic and easily eliminated by the body's natural processes. The ideal degradation rate of a scaffold depends on the specific tissue being engineered, as different tissues regenerate at varying paces (Langer and Vacanti 1993).

The importance of both biocompatibility and biodegradability cannot be overstated. A non-biocompatible scaffold, even if structurally sound, can trigger rejection or inflammation, hindering the healing process. Conversely, a scaffold that degrades too quickly may not provide sufficient support for proper tissue regeneration, leading to implant failure. Therefore, striking a balance between these two properties is crucial for successful tissue engineering applications. Researchers are constantly developing new materials and optimizing existing ones to achieve optimal biocompatibility and biodegradability. Natural materials like collagen or chitosan often exhibit good biocompatibility, while synthetic polymers like PLA offer controlled biodegradability (Qasim et al. 2018). Advanced techniques like surface modification or drug delivery can further enhance the biocompatibility and functionality of scaffolds.

In conclusion, biocompatibility and biodegradability are fundamental considerations for designing successful scaffolds in tissue engineering. By carefully selecting materials and optimizing their properties, researchers can create scaffolds that effectively promote tissue regeneration while ensuring safety and minimal long-term impact on the body.

### 1.4.2 Tailoring Degradation Rates: Matching Tissue Regeneration Needs

Scaffolds play a crucial role in providing structural support and biochemical cues for cell growth and differentiation. However, a critical aspect of scaffold design often goes beyond initial functionality: tailoring the degradation rate to match the specific regeneration timeline of the target tissue. The ideal degradation rate for a scaffold depends on the tissue it aims to regenerate. Rapidly regenerating tissues like skin or muscle require scaffolds that degrade within weeks or months (Dhandayuthapani et al. 2011). These scaffolds provide initial support for cell attachment and proliferation while gradually being replaced by the newly formed tissue. Conversely, tissues with slower regeneration rates, such as bone or cartilage, necessitate scaffolds that degrade over a much longer timeframe, often years (Langer 2000b). These slower-degrading scaffolds offer extended support for cell differentiation and matrix deposition, allowing for the formation of robust and mechanically stable tissues.

Tailoring degradation rates can be achieved through various strategies. The inherent properties of the scaffold material play a significant role. Synthetic polymers like PLA or polyglycolic acid (PGA) degrade at predictable rates depending on their molecular weight and composition (Makadia and Siegel 2011). By altering these factors, researchers can create scaffolds with degradation profiles suited for different tissues. Additionally, blending synthetic polymers with natural materials like collagen or chitosan can offer a balance between mechanical strength and biodegradability (Liu et al. 2023). Surface modification techniques further enhance control over degradation rates. For instance, by incorporating enzymes that break down the scaffold material onto its surface, researchers can trigger a more localized and controlled degradation process (Rossi et al. 2015). Conversely, crosslinking agents can be used to create more stable scaffolds with slower degradation rates, suitable for applications in bone or ligament regeneration (Xue et al. 2019). Optimizing degradation rates is not only crucial for proper tissue formation but also for long-term functionality of the implant. A scaffold that degrades too rapidly in a slow-regenerating tissue may not provide sufficient support, leading to implant failure. On the other hand, a scaffold that degrades too slowly in a fast-regenerating tissue can hinder complete tissue remodeling and potentially lead to encapsulation by fibrous tissue.

In brief, tailoring degradation rates is an essential aspect of designing effective scaffolds for tissue engineering. By understanding the regeneration timelines of different tissues and utilizing various material properties, surface modification techniques, and design strategies, researchers can create scaffolds that degrade in concert with tissue formation, ultimately promoting successful regeneration and long-term functionality.

### 1.4.3 Incorporation of Bioactive Molecules: Growth Factors and Signalling Cues for Enhanced Tissue Formation

In order scaffolds can provide the structural foundation for tissue regeneration, they often require additional cues to guide cell behaviour. Incorporation of bioactive molecules, such as growth factors and signalling peptides, emerges as a powerful strategy to enhance tissue formation within scaffolds. Growth factors are naturally occurring proteins that play a critical role in cell proliferation, differentiation, and migration. By incorporating specific growth factors into scaffolds during fabrication, researchers can create a localized environment that promotes the desired cell behaviour for tissue regeneration (Atala et al. 2012). For instance, bone morphogenetic proteins (BMPs) can be incorporated into scaffolds for bone tissue engineering to stimulate the differentiation of stem cells into bone-forming osteoblasts (Oliveira et al. 2021). Similarly, vascular endothelial growth factor (VEGF) can be used to promote blood vessel formation within scaffolds, crucial for nutrient and oxygen supply to support tissue growth (Filippi et al. 2020). The incorporation of various growth factors, a class of bioactive molecules, is a widely employed technique to induce chondrogenesis is given in Fig. 1.8.

Signalling peptides are short sequences of amino acids that can bind to cell surface receptors and trigger specific cellular responses. These peptides can mimic the natural signals that guide cell behaviour during tissue development. For instance, the RGD (arginine-glycine-aspartic acid) peptide sequence is a common motif found in many ECM proteins that promotes cell adhesion (Dhavalikar et al. 2020). By incorporating RGD or other relevant signalling peptides into scaffolds, researchers can enhance cell attachment and spreading, facilitating tissue formation. The delivery of bioactive molecules can be achieved through various techniques. One approach involves directly encapsulating the molecules within the scaffold material during fabrication. Another strategy utilizes controlled release mechanisms, where the molecules are gradually released from the scaffold over time, mimicking the natural signalling cascades that occur during tissue development (Langer 2000a). This controlled release ensures sustained delivery of cues for cell activity throughout the regeneration process.

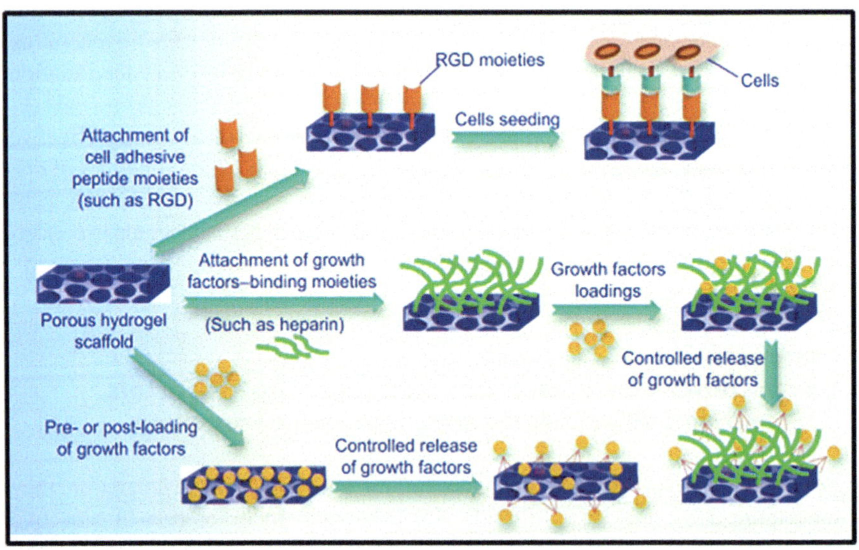


**Fig. 1.8** Bioactive factors in the form of various growth factors with sustainable scaffolds (Langer 2000a)

Incorporating bioactive molecules offers significant advantages for tissue engineering. By providing localized and controlled delivery of growth factors and signalling peptides, researchers can promote specific cellular responses and accelerate tissue formation. This approach can potentially improve the efficacy of scaffold-based therapies for a wide range of tissue regeneration applications. However, certain challenges remain such as optimization of the type, dosage, and release profile of bioactive molecules which is crucial for ensuring their effectiveness and safety within the body. Additionally, research is ongoing to develop strategies for co-delivery of multiple bioactive molecules to mimic the complex signalling environment found in native tissues. Thus, the incorporation of bioactive molecules represents a significant advancement in scaffold design for tissue engineering. By providing essential cues for cell behaviour, this approach holds promise for enhancing tissue formation and promoting successful regeneration of damaged or diseased tissues.

### 1.4.4 Sustainable Cell Seeding and Delivery Techniques: Minimizing Waste and Maximizing Cell Viability

Cell seeding, the process of introducing viable cells onto a scaffold, is a crucial step in tissue engineering. However, traditional cell seeding techniques can be wasteful and contribute to reduced cell viability. Sustainable practices in cell seeding and delivery are becoming increasingly important, aiming to minimize cell loss while reducing environmental impact. One key approach focuses on minimizing cell waste during the seeding process. High cell densities are often required for successful tissue formation, but traditional techniques like cell suspension can lead to cell aggregation and clumping, reducing the number of viable cells effectively seeded onto the scaffold (Echeverria Molina et al. 2021). Advanced techniques like microfluidic cell seeding offer improved control over cell distribution and minimize cell-cell interactions that can lead to cell death during seeding (Tien 2014). Additionally, utilizing cell spheroids, clusters of pre-aggregated cells, can enhance cell survival and function within the scaffold compared to individual cells (Echeverria Molina et al. 2021). Another aspect of sustainable cell seeding involves utilizing biocompatible and potentially renewable materials during the cell delivery process. Hydrogels derived from natural materials like alginate or collagen can be used as cell carriers, offering a biomimetic environment that promotes cell attachment and survival (Asa'ad et al. 2016). These materials can be formulated into injectable formats, allowing for minimally invasive delivery of cells directly to the site of implantation, minimizing waste associated with additional surgical procedures (Nii and Katayama 2021).

Sustainability also extends to optimizing cell culture conditions prior to seeding. Traditional cell culture often relies on fetal bovine serum (FBS) as a growth supplement. However, FBS is a non-renewable resource and can introduce variability into experiments. Research is ongoing to develop synthetic serum replacements or feeder-free culture systems, reducing reliance on animal products and promoting a more sustainable cell culture process (Tan and Barker 2013). Beyond minimizing waste, maximizing cell viability during and after seeding is crucial for successful tissue engineering. Techniques like pre-conditioning cells with specific growth factors or mechanical stimulation can enhance their survival and engraftment within the scaffold (Sun et al. 2022). Additionally, developing bioreactors that mimic the physiological environment, including factors like fluid flow and oxygen tension, can further improve cell viability and promote tissue formation (Langer and Vacanti 1993). Integrating these sustainable practices can have a significant impact on the field of tissue engineering. By minimizing cell waste, utilizing biocompatible delivery materials, and optimizing cell culture conditions, researchers can achieve successful cell seeding while reducing the environmental footprint of these procedures. This approach ultimately paves the way for more sustainable and ethically sourced cellular therapies.

## 1.5 Translational Applications of Sustainable Scaffolds

Sustainable scaffolds, designed with both efficacy and environmental responsibility in mind, hold immense promise for various translational applications in tissue engineering. These scaffolds offer a compelling alternative to traditional materials, paving the way for more eco-friendly regenerative therapies.

One key translational application lies in bone regeneration. Musculoskeletal injuries and diseases often necessitate bone grafting procedures. Sustainable scaffolds derived from natural biomaterials like decellularized bone matrix or cellulose can provide structural support for bone growth while minimizing reliance on synthetic polymers or donor tissues (Castañeda-Rodríguez et al. 2023; Gu et al. 2016). Additionally, these scaffolds can be incorporated with bioactive molecules like growth factors to stimulate bone cell differentiation and promote faster healing. Another promising area for sustainable scaffolds is in skin regeneration. Burn victims or patients with chronic wounds often require skin grafts for functional recovery. Scaffolds fabricated from biocompatible and biodegradable materials like collagen or chitosan offer a temporary matrix for cell migration and proliferation, facilitating the formation of new skin tissue (Khumalo et al. 2022; Zhang et al. 2022a). These sustainable alternatives can potentially overcome limitations associated with traditional skin graft harvesting procedures.

Sustainable scaffolds also hold potential for applications in cartilage repair. Cartilage injuries in joints can be debilitating and have limited repair capabilities. Hydrogels derived from natural materials like alginate or hyaluronic acid can be utilized as injectable scaffolds for cartilage regeneration (Xu et al. 2022). These biocompatible materials provide a supportive environment for chondrocyte growth and differentiation, promoting the formation of new cartilage tissue. Beyond specific tissue types, sustainable scaffolds can be instrumental in developing novel drug delivery platforms. These scaffolds can be loaded with therapeutic agents and implanted at the site of disease. The controlled degradation of the scaffold material allows for sustained release of the drug, minimizing systemic side effects and improving treatment efficacy (Dhandayuthapani et al. 2011). Utilizing sustainable materials for such delivery systems further promotes an environmentally conscious approach to therapeutic intervention. The successful translation of sustainable scaffolds to clinical applications necessitates addressing certain challenges. One key issue involves ensuring the mechanical properties of these scaffolds are sufficient to withstand the specific physiological demands of the target tissue (Han et al. 2020). Additionally, rigorous *in vivo* testing and regulatory approval processes are crucial for demonstrating safety and efficacy before widespread clinical use.

### 1.5.1 Bone Regeneration: Restoring Skeletal Integrity

The human skeleton provides a rigid framework for support, movement, and protection of vital organs. However, fractures, diseases, and congenital malformations can compromise bone integrity, leading to pain, functional limitations, and decreased quality of life. Bone regeneration, a branch of tissue engineering, offers promising solutions for restoring skeletal integrity and functionality.

Bone is a dynamic tissue with a remarkable ability to self-heal under certain conditions. However, this process can be slow and incomplete for large defects caused by trauma, tumours, or congenital disorders. Bone regeneration strategies aim to accelerate and enhance this natural healing process by providing a suitable environment for bone formation. These strategies can involve three key components: cells, signalling molecules, and scaffolds. Mesenchymal stem cells (MSCs) are a promising cell source for bone regeneration. These pluripotent stem cells not only differentiate into bone-forming osteoblasts but also respond to mechanical cues from their microenvironment. MSCs can be harvested from bone marrow or other tissues and expanded in culture before being incorporated into scaffolds for implantation at the defect site (Jin and Lee 2018). Signalling molecules, such as growth factors, play a crucial role in stimulating bone cell activity. Bone morphogenetic proteins (BMPs) are potent growth factors that can induce osteoblast differentiation and bone formation. These molecules can be incorporated into scaffolds or delivered locally to promote bone regeneration (Ball et al. 2023). Scaffolds provide a three-dimensional (3D) structure that supports cell attachment, proliferation, and differentiation. Ideal bone regeneration scaffolds should be biocompatible, biodegradable, and possess suitable mechanical properties to withstand physiological stresses. Natural materials like collagen or synthetic polymers like PLA are commonly used for scaffold fabrication (Raeisdasteh Hokmabad et al. 2017). Scaffolds are implemented in a damaged tissue niche and bone tissue engineering (BTE) principles are applied to trigger specific biological responses from the local environment, to induce the healing process (Fig. 1.9). The diagram highlights these criteria through color-coded sections: *Biomaterials (Yellow):* The critical factors for choosing the appropriate material for the scaffold matrix are summarized. *Cell Type (Green):* Selection of the optimal cell type to facilitate bone healing and formation within the scaffold. *Inductive Signals (Purple):* Incorporation of signaling molecules within the scaffold to stimulate bone growth and regeneration.

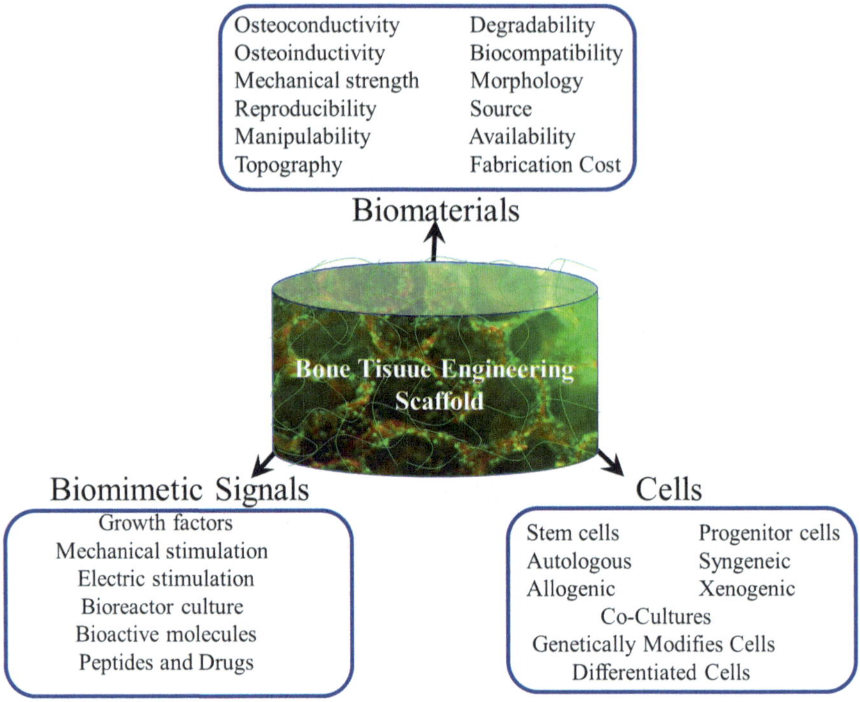


**Fig. 1.9** Principles of bone tissue engineering (Filippi et al. 2020)

By carefully considering these elements, researchers can design targeted and efficacious scaffolds for bone tissue engineering applications (Filippi et al. 2020). Current advancements in bone regeneration extend beyond these traditional approaches. Bioprinting, a computer-controlled additive manufacturing technique, allows for the creation of customized scaffolds with complex geometries that mimic the architecture of natural bone (Xie et al. 2020). Additionally, research is ongoing on incorporating vascularization networks within scaffolds to enhance oxygen and nutrient delivery, crucial for promoting bone formation within larger defects (Klimek and Ginalska 2020). The field of bone regeneration offers significant potential for improving patient outcomes and enhancing quality of life. By combining cell therapies, growth factors, and advanced scaffold designs, researchers are paving the way for more effective strategies to restore skeletal integrity and function after bone injuries or diseases. However, different scientific challenges still remain unsolved. Optimizing the delivery and efficacy of growth factors within the body, ensuring long-term functionality of implanted scaffolds, and controlling the differentiation of stem cells into mature bone tissue are ongoing areas of research. Thus, bone regeneration represents a transformative approach to treating bone defects and restoring skeletal function. By harnessing the body's natural healing potential and utilizing innovative technologies, researchers continuously strive to develop more effective and personalized therapies for patients suffering from bone-related conditions.

### 1.5.2 Cartilage Repair: Addressing Joint Degeneration

Cartilage, a smooth, avascular connective tissue, plays a critical role in facilitating pain-free movement within joints. Unfortunately, cartilage lacks a robust self-healing capacity, and injuries or degeneration due to aging can lead to significant functional limitations. Cartilage repair strategies aim to address joint degeneration and restore joint function in individuals suffering from conditions like osteoarthritis. The primary challenge in cartilage repair lies in the limited regenerative potential of chondrocytes, the specialized cells that form cartilage. While attempts have been made to stimulate cartilage repair through microfracture surgery, which creates small holes in the underlying bone to encourage blood-derived stem cells to migrate and form fibrocartilage, this approach often leads to the formation of a less durable tissue compared to hyaline cartilage (Song and Park 2019).

Current cartilage repair strategies focus on two main approaches: cell-based therapies and scaffold-based interventions. Cell-based therapies involve transplanting healthy chondrocytes, either autologous (from the patient) or allogeneic (from a donor), onto the defect site. This approach can promote the formation of new hyaline cartilage, but challenges include limited availability of healthy autologous cells and potential immune rejection with allogeneic cells (Harris et al. 2010). Scaffold-based interventions utilize three-dimensional structures to provide a supportive microenvironment for cell attachment, proliferation, and differentiation. These scaffolds can be seeded with chondrocytes or stem cells before implantation, or they can be implanted alone to encourage native cell recruitment. Biocompatible and biodegradable materials like collagen or synthetic polymers are often used for scaffold fabrication (Weigel et al. 2006).

Advanced techniques are constantly evolving in the field of cartilage repair. Bioprinting allows for the creation of customized scaffolds with intricate structures that mimic the natural architecture of articular cartilage (Bartolo et al. 2022). Additionally, research is ongoing on incorporating bioactive molecules like growth factors or stimulating electrical signals within scaffolds to further enhance cartilage repair processes (Fahy et al. 2018). The successful development of long-term, durable cartilage repair strategies holds immense potential for improving patient outcomes and quality of life. By restoring joint function and reducing pain, these strategies can enable individuals to resume daily activities and participate in sports or physical hobbies. However, other significant challenges remain to be fully addressed, including optimizing cell sources, designing highly functional scaffolds, and ensuring long-term integration of the repaired tissue with surrounding healthy cartilage. Therefore, cartilage repair research offers promising solutions for addressing joint degeneration. By exploring innovative cell-based therapies, advanced scaffold technologies, and growth factor delivery methods, researchers are paving the way for a future where pain-free joint mobility can be restored for individuals suffering from cartilage injuries or osteoarthritis.

### 1.5.3 Skin Regeneration: Promoting Wound Healing and Functional Recovery

Skin, the largest organ of the human body, serves as a critical barrier against external threats, regulates body temperature, and plays a vital role in sensory perception. However, various factors like burns, injuries, and chronic wounds can disrupt the natural healing process, leading to functional impairment and increased risk of infection. Skin regeneration strategies aim to accelerate wound closure, promote tissue restoration, and improve functional recovery for patients suffering from these conditions. The human body possesses a remarkable ability to heal skin wounds through a complex cascade of cellular events. This process involves inflammation, tissue granulation, re-epithelialization, and remodelling (Kamel et al. 2013). However, this natural healing can be overwhelmed in cases of large or deep wounds, leading to chronic wounds that fail to close completely. Skin regeneration interventions aim to bridge this gap by providing additional support to expedite the healing process.

One common approach involves utilizing skin grafts. These grafts, composed of healthy skin tissue, can be harvested from an uninjured area of the patient's body (autograft) or from a donor (allograft). Autograft transplantation remains the gold standard for treating large wounds, but it can be limited by donor site morbidity and insufficient tissue availability (Bhatia 2010). Advanced bioengineered skin substitutes are emerging as promising alternatives to traditional skin grafts. These substitutes often utilize scaffolds fabricated from biocompatible materials like collagen or hyaluronic acid. These scaffolds can be seeded with cells, such as fibroblasts or keratinocytes, to promote cellular growth and differentiation within the wound bed, facilitating faster closure and improved tissue regeneration (Kaliaraj et al. 2023). Figure 1.10 represents the process of skin regeneration after wound healing.

Beyond promoting wound closure, skin regeneration strategies aim to restore the functional integrity of the healed tissue. This includes regaining barrier function, sensation, and a cosmetically acceptable appearance. Advanced biomaterials are being developed to mimic the natural structure and properties of skin, ensuring the regenerated tissue provides optimal protection and functionality (De Chiara et al. 2024). Researchers are also exploring the potential of stem cell therapies for skin regeneration. Mesenchymal stem cells hold promise for promoting wound healing due to their immunomodulatory properties and ability to differentiate into various skin cell types (Jo et al. 2021). Additionally, research is ongoing on incorporating growth factors and other bioactive molecules within scaffolds to further enhance the regenerative potential of these therapies.

The successful development of skin regeneration techniques can significantly improve patient outcomes and quality of life. By accelerating wound closure, minimizing scarring, and restoring functional skin integrity, these strategies can help individuals regain their health and return to daily activities with renewed confidence. However, challenges remain, including optimizing biomaterial design, ensuring long-term graft survival, and controlling the cost associated with these

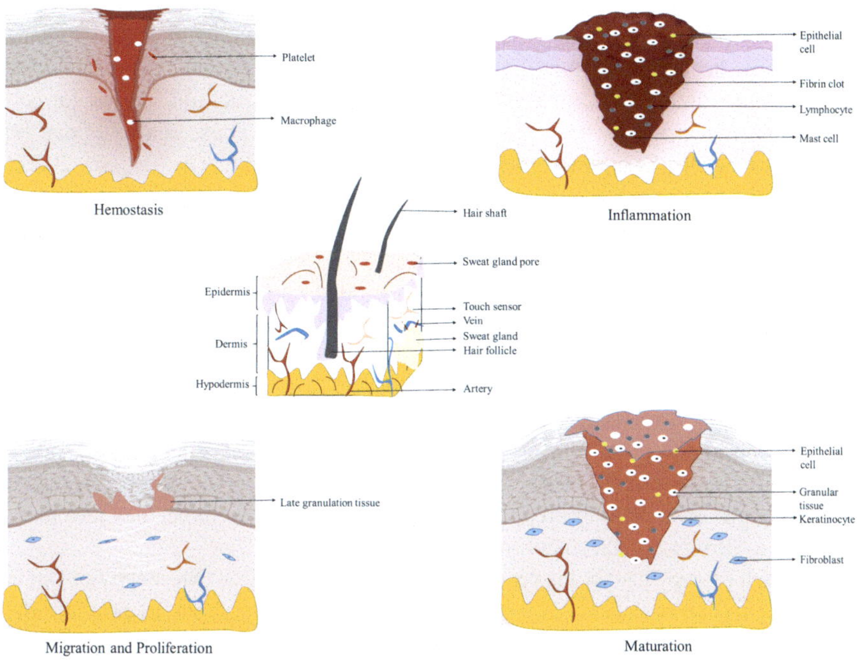


**Fig. 1.10** Skin regeneration and wound healing phase (Kaliaraj et al. 2023)

advanced therapies. Therefore, skin regeneration research offers a beacon of hope for patients suffering from extensive wounds and chronic skin conditions. By harnessing the body's healing potential and utilizing innovative bioengineering approaches, researchers are continuously working towards creating more effective therapies that promote faster wound closure, improved functionality, and optimal patient recovery.

### 1.5.4 Exploring Applications in Other Tissue Engineering Areas

The principles of tissue engineering extend far beyond bone, cartilage, and skin regeneration. Researchers are actively exploring innovative strategies for repairing and regenerating a wide variety of tissues in the body. Herein, we delve into two promising areas: myocardial repair and neural regeneration. Heart disease, often caused by coronary artery blockage, can lead to heart muscle damage and decreased pumping efficiency. Myocardial repair strategies aim to regenerate damaged heart tissue and improve cardiac function after a heart attack or other injuries. Cell-based therapies are a promising approach for myocardial repair. Studies are ongoing to investigate the potential of transplanting mesenchymal stem cells or cardiac progenitor cells directly into the injured area. These cells have the potential to differentiate into cardiomyocytes, the muscle cells of the heart, or to promote tissue regeneration through paracrine signalling (Csöbönyeiová et al. 2022). Another strategy involves utilizing biocompatible scaffolds to provide structural support for cell attachment and growth within the damaged heart tissue. These scaffolds can be seeded with cells or growth factors before implantation (Guo et al. 2020). Additionally, researchers are exploring the potential of bioengineered heart patches, fabricated from a combination of cells and scaffolds, to replace damaged sections of the heart wall.

The central nervous system (CNS) has a limited capacity for self-repair after injury. Neural regeneration strategies aim to promote the regrowth of damaged neurons and the reestablishment of neural connections, potentially leading to functional recovery from conditions like spinal cord injuries or neurodegenerative diseases. One approach involves utilizing biomaterials to create scaffolds that can bridge the gap in the injured spinal cord and provide a supportive environment for axonal regeneration, the regrowth of nerve fibres (Ren et al. 2019). Additionally, research is ongoing on incorporating growth factors or other bioactive molecules into these scaffolds to further enhance neuronal regeneration.

Stem cell therapy also holds promise for neural regeneration. Mesenchymal stem cells or neural stem cells can be transplanted into the injured site to promote the growth and differentiation of new neurons or to support the survival of existing neurons (Shammaa et al. 2020). Furthermore, researchers are exploring the use of biomaterials to deliver neurotrophic factors, which promote neuronal survival and function, directly to the site of injury. These are just a few examples of how tissue engineering is being explored for the regeneration of various tissues. The field holds immense potential for improving patient outcomes and quality of life across a wide range of medical conditions.

## 1.6 Challenges and Future Directions

Despite the immense promise of sustainable scaffolds for tissue engineering, several challenges remain. Optimizing the mechanical properties of these biomaterials to match the specific needs of different tissues while maintaining biocompatibility and biodegradability is a crucial area of focus (Prem Ananth and Jayram 2024). Additionally, ensuring long-term *in vivo* functionality and minimizing potential immunogenic responses require further investigation (McInnes et al. 2022). Future directions include the development of biomimetic scaffolds that not only provide structural support but also actively promote cell differentiation and tissue formation through controlled release of growth factors or mimicking the native ECM (Deb et al. 2018). Furthermore, integrating vascularization networks within scaffolds is essential for promoting nutrient and oxygen delivery, particularly for larger tissue constructs (Jafari et al. 2017). Finally, the development of cost-effective and scalable production methods for sustainable scaffolds is crucial for broader clinical translation and accessibility (Chia and Wu 2015). By addressing these challenges and pursuing innovative strategies, sustainable scaffolds have the potential to revolutionize the field of tissue engineering and regenerative medicine, offering eco-friendly and effective solutions for promoting tissue repair and improving patient outcomes.

The successful development of novel tissue engineering therapies hinges on a robust pre-clinical testing pipeline. While *in vitro* testing offers valuable insights into cellular behaviour and biomaterial properties, it often fails to fully capture the complex physiological environment present within a living organism. Bridging this gap between *in vitro* studies and clinically relevant pre-clinical research is crucial for translating promising therapies into effective clinical applications. One key approach involves utilizing more sophisticated *in vitro* models that better mimic the *in vivo* environment. This can include 3D cell culture systems that incorporate multiple cell types and recapitulate the tissue architecture more accurately than traditional monolayer cultures (Perez-Puyana et al. 2020). Additionally, microfluidic devices can be used to create dynamic environments with controlled flow conditions, providing a more realistic test bed for evaluating cell behaviour and therapeutic efficacy (Tien 2014). Moving beyond *in vitro* studies, well-designed animal models play a vital role in pre-clinical testing. Choosing the appropriate animal model with a high degree of homology to the targeted disease or injury in humans is critical. For example, using immunocompromised animals for testing human cell therapies avoids issues related to immune rejection, allowing for a focused evaluation of the therapeutic potential of the cells themselves (Lin et al. 2017).

However, animal models have limitations including species-specific differences in anatomy, physiology, and disease processes that can lead to discrepancies between pre-clinical findings and clinical outcomes in humans. Therefore, it is essential to carefully interpret data from animal studies and acknowledge these limitations when designing clinical trials. Advanced pre-clinical testing methodologies are also emerging. Large animal models like pigs offer greater anatomical and physiological similarity to humans compared to smaller rodents. Additionally, the use of genetically modified animals with specific disease pathologies can provide valuable insights into the effectiveness of novel therapies for targeted conditions (Yang et al. 2021).

Ultimately, a successful pre-clinical testing strategy should strive for a balance between scientific rigor and ethical considerations. Refined *in vitro* models, carefully chosen animal models, and the exploration of advanced pre-clinical methodologies all contribute to a more robust evaluation process, ultimately paving the way for safer and more effective tissue engineering therapies to reach patients in need.

The successful application of tissue engineering therapies hinges on the long-term *in vivo* performance of implanted scaffolds. While scaffolds play a crucial role in supporting cell attachment, proliferation, and differentiation, ensuring their functionality and biocompatibility within the complex biological environment remains a significant challenge. One key challenge lies in maintaining the mechanical integrity of the scaffold over time. Biodegradation is a desired feature for many scaffolds, as it allows for gradual replacement by newly formed tissue. However, the degradation rate needs to be carefully controlled to ensure the scaffold provides adequate support throughout the critical period of tissue regeneration (Jafari et al. 2017). Furthermore, the degradation products should be non-toxic and readily cleared by the body. Another challenge involves minimizing the potential for inflammatory responses. The introduction of any foreign material into the body can trigger an immune reaction. Optimizing scaffold design and material properties to minimize foreign body reactions and promote biocompatibility is crucial for long-term success (Golebiowska et al. 2024).

Future directions in scaffold design focus on creating biomimetic environments that closely resemble the native ECM of the target tissue. These scaffolds can provide essential cues for cell adhesion, migration, and differentiation, potentially leading to improved tissue integration and functionality (Ebhodaghe 2021). Additionally, strategies are being explored to incorporate bioactive molecules within scaffolds, such as growth factors or anti-inflammatory agents, to further enhance tissue regeneration and minimize immune responses (Perez-Puyana et al. 2020). The development of *in vitro* and *in vivo* models that more accurately predict long-term scaffold performance is also crucial. These models should consider factors like degradation kinetics, cellular interactions with the scaffold material, and potential immune responses within the host organism (Huzum et al. 2021). Finally, close collaboration between material scientists, engineers, biologists, and clinicians is essential for translating promising scaffold designs into effective clinical applications. By addressing the challenges associated with long-term *in vivo* performance, researchers can ensure the development of safe and efficacious tissue engineering therapies that provide lasting benefits for patients. While sustainable scaffolds offer an environmentally friendly approach to tissue engineering, ensuring cost-effectiveness and scalability remains a challenge. Current production methods for some biomaterials derived from natural sources can be labor-intensive or require specialized equipment, leading to higher costs compared to traditional synthetic materials (Bikiaris et al. 2023). Additionally, upscaling production methods for large-scale clinical applications needs further development to meet potential patient demand. Future directions include exploring cost-effective, scalable harvesting and processing techniques for naturally derived biomaterials. Furthermore, research on developing synthetic alternatives with comparable sustainability profiles and improved processability is ongoing (Bartolo et al. 2022). Additionally, investigating strategies to improve the reusability or recyclability of scaffolds could potentially reduce overall costs associated with these therapies. Ultimately, achieving a balance between sustainability, cost-effectiveness, and scalability is crucial for the successful clinical translation of sustainable scaffold-based tissue engineering strategies. The path from promising tissue engineering strategies to widespread clinical use is paved with regulatory hurdles. Demonstrating the safety and efficacy of novel scaffolds and cell therapies within the complex regulatory landscape remains a significant challenge (Dzobo et al. 2018). Regulatory agencies require extensive pre-clinical testing to ensure patient safety and establish a clear therapeutic benefit. Additionally, obtaining regulatory approval can be a lengthy and costly process, potentially hindering the timely translation of these innovative therapies to patients in need (Cossu et al. 2018). Future directions involve fostering greater collaboration between researchers, regulatory agencies, and industry stakeholders. Developing clear and streamlined regulatory pathways specifically tailored for tissue engineering products could expedite the approval process while maintaining rigorous safety standards. Additionally, utilizing established manufacturing practices and robust quality control systems can further ensure the safety and efficacy of these novel therapies, facilitating their successful clinical translation.

## 1.7 Conclusion

Sustainable scaffolds hold immense promise for revolutionizing the field of regenerative medicine. By harnessing the power of biocompatible and environment friendly materials, researchers are paving the way for developing innovative therapies that promote tissue repair and regeneration. Addressing the challenges related to long-term *in vivo* performance, cost-effectiveness, scalability, and navigating regulatory pathways will be crucial for broader clinical translation. Sustainable scaffolds offer a ground-breaking approach to tissue engineering, by utilizing environment friendly materials while maintaining functionality, holds a promise for wide range of translational applications. From bone regeneration to skin grafts and cartilage repair, sustainable scaffolds pave the way for eco-friendly and effective regenerative therapies, ultimately improving patient outcomes and contributing to a more sustainable healthcare system. Sustainable cell seeding and delivery techniques are crucial for responsibly advancing tissue engineering. By minimizing cell waste, utilizing renewable materials, and optimizing cell culture practices, researchers can ensure a high degree of cell viability while promoting environment friendly approaches to tissue regeneration. However, with continued advancements in material science, bioengineering, and clinical research, sustainable scaffolds have the potential to become a cornerstone of regenerative medicine, offering patients a wealth of eco-friendly treatment options for a wide range of tissue injuries and diseases.

**Acknowledgement** Dr. Sangram Keshari Samal highly acknowledges the Ramanujan fellowship (SB/S2/RJN-038/2016) of the Department of Science and Technology, and Ramalingaswami Re-entry fellowship (Ref: D.O. No. BT/HRD/35/02/2006) of the Department of Biotechnology, Government of India. The authors also thank the Indian Council of Medical Research-Regional Medical Research Centre, Bhubaneswar, for providing a scientific platform.

## References

Adekomaya O, Majozi T, Adedoyin S (2021) Bio-based and biodegradable plastic materials: life cycle assessment. In: Handbook of nanomaterials and nanocomposites for energy and environmental applications. Springer International Publishing, pp 3689–3706. https://doi.org/10.1007/978-3-030-36268-3_180

Alavi MS, Memarpour S, Pazhohan-Nezhad H, Salimi Asl A, Moghbeli M, Shadmanfar S, Saburi E (2023) Applications of poly(lactic acid) in bone tissue engineering: a review article. Artif Organs 47(9):1423–1430. https://doi.org/10.1111/aor.14612

Allu I, Sahi AK, Koppadi M, Gundu S, Sionkowska A (2023) Decellularization techniques for tissue engineering: towards replicating native extracellular matrix architecture in liver regeneration. J Funct Biomater 14(10):518. https://doi.org/10.3390/jfb14100518

Altman GH, Diaz F, Jakuba C, Calabro T, Horan RL, Chen J, Lu H, Richmond J, Kaplan DL (2003) Silk-based biomaterials. Biomaterials 24(3):401–416. https://doi.org/10.1016/S0142-9612(02)00353-8
Amarakoon M, Alenezi H, Homer-Vanniasinkam S, Edirisinghe M (2022) Environmental impact of polymer fiber manufacture. Macromol Mater Eng 307(11). https://doi.org/10.1002/mame.202200356
Anquez E, Raab K, Cechella FS, Wagner R (2022) Consumers' perception of sustainable packaging in the food industry. Rev Direitos Cult 17(41):251–265. https://doi.org/10.20912/rdc.v17i41.728
Asa'ad F, Pagni G, Pilipchuk SP, Giannì AB, Giannobile WV, Rasperini G (2016) 3D-printed scaffolds and biomaterials: review of alveolar bone augmentation and periodontal regeneration applications. Int J Dent 2016:1–15. https://doi.org/10.1155/2016/1239842
Atala A, Kasper FK, Mikos AG (2012) Engineering complex tissues. Sci Transl Med 4(160). https://doi.org/10.1126/scitranslmed.3004890
Badylak SF (2002) The extracellular matrix as a scaffold for tissue reconstruction. Semin Cell Dev Biol 13(5):377–383. https://doi.org/10.1016/S1084952102000940
Balaji AB, Pakalapati H, Khalid M, Walvekar R, Siddiqui H (2018) Natural and synthetic biocompatible and biodegradable polymers. In: Biodegradable and biocompatible polymer composites. Elsevier, pp 3–32. https://doi.org/10.1016/B978-0-08-100970-3.00001-8
Ball JR, Shelby T, Hernandez F, Mayfield CK, Lieberman JR (2023) Delivery of growth factors to enhance bone repair. Bioengineering 10(11):1252. https://doi.org/10.3390/bioengineering10111252
Balla E, Daniilidis V, Karlioti G, Kalamas T, Stefanidou M, Bikiaris ND, Vlachopoulos A, Koumentakou I, Bikiaris DN (2021) Poly(lactic acid): a versatile biobased polymer for the future with multifunctional properties—from monomer synthesis, polymerization techniques and molecular weight increase to PLA applications. Polymers 13(11):1822. https://doi.org/10.3390/polym13111822
Barbulescu GI, Bojin FM, Ordodi VL, Goje ID, Barbulescu AS, Paunescu V (2022) Decellularized extracellular matrix scaffolds for cardiovascular tissue engineering: current techniques and challenges. Int J Mol Sci 23(21):13040. https://doi.org/10.3390/ijms232113040
Barcena AJR, Dhal K, Patel P, Ravi P, Kundu S, Tappa K (2023) Current biomedical applications of 3D-printed hydrogels. Gels 10(1):8. https://doi.org/10.3390/gels10010008
Bartolo P, Malshe A, Ferraris E, Koc B (2022) 3D bioprinting: materials, processes, and applications. CIRP Ann 71(2):577–597. https://doi.org/10.1016/j.cirp.2022.06.001
Bhardwaj N, Kundu SC (2010) Electrospinning: a fascinating fiber fabrication technique. Biotechnol Adv 28(3):325–347. https://doi.org/10.1016/j.biotechadv.2010.01.004
Bhat S, Kumar A (2012) Biomaterials in regenerative medicine. J Postgrad Med Educ Res 46(2):81–89. https://doi.org/10.5005/jp-journals-10028-1018
Bhatia SK (2010) Tissue engineering for clinical applications. Biotechnol J 5(12):1309–1323. https://doi.org/10.1002/biot.201000230
Bikiaris ND, Koumentakou I, Samiotaki C, Meimaroglou D, Varytimidou D, Karatza A, Kalantzis Z, Roussou M, Bikiaris RD, Papageorgiou GZ (2023) Recent advances in the investigation of poly(lactic acid) (PLA) nanocomposites: incorporation of various nanofillers and their properties and applications. Polymers 15(5):1196. https://doi.org/10.3390/polym15051196
Bolívar-Monsalve EJ, Alvarez MM, Hosseini S, Espinosa-Hernandez MA, Ceballos-González CF, Sanchez-Dominguez M, Shin SR, Cecen B, Hassan S, Di Maio E, Trujillo-de Santiago G (2021) Engineering bioactive synthetic polymers for biomedical applications: a review with emphasis on tissue engineering and controlled release. Mater Adv 2(14):4447–4478. https://doi.org/10.1039/D1MA00092F
Bose S, Roy M, Bandyopadhyay A (2012) Recent advances in bone tissue engineering scaffolds. Trends Biotechnol 30(10):546–554. https://doi.org/10.1016/j.tibtech.2012.07.005
Brandão AS, Gonçalves A, Santos JMRCA (2021) Circular bioeconomy strategies: from scientific research to commercially viable products. J Clean Prod 295:126407. https://doi.org/10.1016/j.jclepro.2021.126407

Brovold M, Almeida JI, Pla-Palacín I, Sainz-Arnal P, Sánchez-Romero N, Rivas JJ, Almeida H, Dachary PR, Serrano-Aulló T, Soker S, Baptista PM (2018) Naturally-derived biomaterials for tissue engineering applications. Adv Exp Med Biol:421–449. https://doi.org/10.1007/978-981-13-0947-2_23

Cai D, Weng W (2023) Development potential of extracellular matrix hydrogels as hemostatic materials. Front Bioeng Biotechnol 11. https://doi.org/10.3389/fbioe.2023.1187474

Capuana E, Lopresti F, Carfì Pavia F, Brucato V, La Carrubba V (2021) Solution-based processing for scaffold fabrication in tissue engineering applications: a brief review. Polymers 13(13):2041. https://doi.org/10.3390/polym13132041

Castañeda-Rodríguez S, González-Torres M, Ribas-Aparicio RM, Del Prado-Audelo ML, Leyva-Gómez G, Gürer ES, Sharifi-Rad J (2023) Recent advances in modified poly (lactic acid) as tissue engineering materials. J Biol Eng 17(1):21. https://doi.org/10.1186/s13036-023-00338-8

Chan BP, Leong KW (2008) Scaffolding in tissue engineering: general approaches and tissue-specific considerations. Eur Spine J 17(S4):467–479. https://doi.org/10.1007/s00586-008-0745-3

Chemat F, Vian MA, Cravotto G (2012) Green extraction of natural products: concept and principles. Int J Mol Sci 13(7):8615–8627. https://doi.org/10.3390/ijms13078615

Chia HN, Wu BM (2015) Recent advances in 3D printing of biomaterials. J Biol Eng 9(1):4. https://doi.org/10.1186/s13036-015-0001-4

Colquhoun R, Tanner KE (2015) Mechanical behaviour of degradable phosphate glass fibres and composites—a review. Biomed Mater 11(1):014105. https://doi.org/10.1088/1748-6041/11/1/014105

Cossu G, Birchall M, Brown T, De Coppi P, Culme-Seymour E, Gibbon S, Hitchcock J, Mason C, Montgomery J, Morris S, Muntoni F, Napier D, Owji N, Prasad A, Round J, Saprai P, Stilgoe J, Thrasher A, Wilson J (2018) Lancet Commission: stem cells and regenerative medicine. Lancet 391(10123):883–910. https://doi.org/10.1016/S0140-6736(17)31366-1

Csöbönyeiová M, Beerová N, Klein M, Debreová-Čeháková M, Danišovič Ľ (2022) Cell-based and selected cell-free therapies for myocardial infarction: how do they compare to the current treatment options? Int J Mol Sci 23(18):10314. https://doi.org/10.3390/ijms231810314

Dang Z, Ma X, Yang Z, Wen X, Zhao P (2023) Electrospun nanofiber scaffolds loaded with metal-based nanoparticles for wound healing. Polymers 16(1):24. https://doi.org/10.3390/polym16010024

Dash M, Samal SK, Douglas TEL, Schaubroeck D, Leeuwenburgh SC, Van Der Voort P, Declercq HA, Dubruel P (2017) Enzymatically biomineralized chitosan scaffolds for tissue-engineering applications. J Tissue Eng Regen Med 11(5):1500–1513. https://doi.org/10.1002/TERM.2048

De Chiara F, Ferret-Miñana A, Fernández-Costa JM, Ramón-Azcón J (2024) The tissue engineering revolution: from bench research to clinical reality. Biomedicines 12(2):453. https://doi.org/10.3390/biomedicines12020453

De Pieri A, Rochev Y, Zeugolis DI (2021) Scaffold-free cell-based tissue engineering therapies: advances, shortfalls and forecast. NPJ Regen Med 6(1):18. https://doi.org/10.1038/s41536-021-00133-3

Deb P, Deoghare AB, Borah A, Barua E, Das Lala S (2018) Scaffold development using biomaterials: a review. Mater Today Proc 5(5):12909–12919. https://doi.org/10.1016/j.matpr.2018.02.276

Dhandayuthapani B, Yoshida Y, Maekawa T, Kumar DS (2011) Polymeric scaffolds in tissue engineering application: a review. Int J Polym Sci 2011:1–19. https://doi.org/10.1155/2011/290602

Dhavalikar P, Robinson A, Lan Z, Jenkins D, Chwatko M, Salhadar K, Jose A, Kar R, Shoga E, Kannapiran A, Cosgriff-Hernandez E (2020) Review of integrin-targeting biomaterials in tissue engineering. Adv Healthc Mater 9(23). https://doi.org/10.1002/adhm.202000795

Duarte ARC, Mano JF, Reis RL (2009) Supercritical fluids in biomedical and tissue engineering applications: a review. Int Mater Rev 54(4):214–222. https://doi.org/10.1179/174328009X411181

Dzobo K, Thomford NE, Senthebane DA, Shipanga H, Rowe A, Dandara C, Pillay M, Motaung KSCM (2018) Advances in regenerative medicine and tissue engineering: innovation and transformation of medicine. Stem Cells Int 2018:1–24. https://doi.org/10.1155/2018/2495848

Ebhodaghe SO (2021) Natural polymeric scaffolds for tissue engineering applications. J Biomater Sci Polym Ed 32(16):2144–2194. https://doi.org/10.1080/09205063.2021.1958185
Echeverria Molina MI, Malollari KG, Komvopoulos K (2021) Design challenges in polymeric scaffolds for tissue engineering. Front Bioeng Biotechnol 9. https://doi.org/10.3389/fbioe.2021.617141
Epps TH, Korley LTJ, Yan T, Beers KL, Burt TM (2022) Sustainability of synthetic plastics: considerations in materials life-cycle management. JACS Au 2(1):3–11. https://doi.org/10.1021/jacsau.1c00191
Fahy N, Alini M, Stoddart MJ (2018) Mechanical stimulation of mesenchymal stem cells: implications for cartilage tissue engineering. J Orthop Res 36(1):52–63. https://doi.org/10.1002/jor.23670
Fang W, Yang M, Wang L, Li W, Liu M, Jin Y, Wang Y, Yang R, Wang Y, Zhang K, Fu Q (2023) Hydrogels for 3D bioprinting in tissue engineering and regenerative medicine: current progress and challenges. Int J Bioprinting 9(5):759. https://doi.org/10.18063/ijb.759
Filippi M, Born G, Chaaban M, Scherberich A (2020) Natural polymeric scaffolds in bone regeneration. Front Bioeng Biotechnol 8. https://doi.org/10.3389/fbioe.2020.00474
Flores-Rojas GG, Gómez-Lazaro B, López-Saucedo F, Vera-Graziano R, Bucio E, Mendizábal E (2023) Electrospun scaffolds for tissue engineering: a review. Macromolecules 3(3):524–553. https://doi.org/10.3390/macromol3030031
Fonseca A, Ramalho E, Gouveia A, Figueiredo F, Nunes J (2023) Life cycle assessment of PLA products: a systematic literature review. Sustain For 15(16):12470. https://doi.org/10.3390/su151612470
Ghosh K, Jones BH (2021) Roadmap to biodegradable plastics—current state and research needs. ACS Sustain Chem Eng 9(18):6170–6187. https://doi.org/10.1021/acssuschemeng.1c00801
Gilpin SE, Wagner DE (2018) Acellular human lung scaffolds to model lung disease and tissue regeneration. Eur Respir Rev 27(148):180021. https://doi.org/10.1183/16000617.0021-2018
Gilpin A, Yang Y (2017) Decellularization strategies for regenerative medicine: from processing techniques to applications. Biomed Res Int 2017:1–13. https://doi.org/10.1155/2017/9831534
Golebiowska AA, Intravaia JT, Sathe VM, Kumbar SG, Nukavarapu SP (2024) Decellularized extracellular matrix biomaterials for regenerative therapies: advances, challenges and clinical prospects. Bioact Mater 32:98–123. https://doi.org/10.1016/j.bioactmat.2023.09.017
Gomes M, Azevedo H, Malafaya P, Silva S, Oliveira J, Silva G, Sousa R, Mano J, Reis R (2008) Natural polymers in tissue engineering applications. In: Tissue engineering. Elsevier, pp 145–192. https://doi.org/10.1016/B978-0-12-370869-4.00006-9
Gu BK, Choi DJ, Park SJ, Kim MS, Kang CM, Kim C-H (2016) 3-dimensional bioprinting for tissue engineering applications. Biomater Res 20(1). https://doi.org/10.1186/s40824-016-0058-2
Guo R, Morimatsu M, Feng T, Lan F, Chang D, Wan F, Ling Y (2020) Stem cell-derived cell sheet transplantation for heart tissue repair in myocardial infarction. Stem Cell Res Ther 11(1):19. https://doi.org/10.1186/s13287-019-1536-y
Han F, Wang J, Ding L, Hu Y, Li W, Yuan Z, Guo Q, Zhu C, Yu L, Wang H, Zhao Z, Jia L, Li J, Yu Y, Zhang W, Chu G, Chen S, Li B (2020) Tissue engineering and regenerative medicine: achievements, future, and sustainability in Asia. Front Bioeng Biotechnol 8. https://doi.org/10.3389/fbioe.2020.00083
Harris JD, Siston RA, Pan X, Flanigan DC (2010) Autologous chondrocyte implantation. J Bone Joint Surg Am 92(12):2220–2233. https://doi.org/10.2106/JBJS.J.00049
Henkel J, Hutmacher DW (2013) Design and fabrication of scaffold-based tissue engineering. BioNanoMaterials 14(3–4). https://doi.org/10.1515/bnm-2013-0021
Hollister SJ (2005) Porous scaffold design for tissue engineering. Nat Mater 4(7):518–524. https://doi.org/10.1038/nmat1421
Hu X, Liu S, Zhou G, Huang Y, Xie Z, Jing X (2014) Electrospinning of polymeric nanofibers for drug delivery applications. J Control Release 185:12–21. https://doi.org/10.1016/j.jconrel.2014.04.018

Hutmacher DW (2001) Scaffold design and fabrication technologies for engineering tissues — state of the art and future perspectives. J Biomater Sci Polym Ed 12(1):107–124. https://doi.org/10.1163/156856201744489

Hutmacher DW, Tandon B, Dalton PD (2023) Scaffold design and fabrication. In: Tissue engineering. Elsevier, pp 355–385. https://doi.org/10.1016/B978-0-12-824459-3.00011-1

Huzum B, Puha B, Necoara R, Gheorghevici S, Puha G, Filip A, Sirbu P, Alexa O (2021) Biocompatibility assessment of biomaterials used in orthopedic devices: an overview (Review). Exp Ther Med 22(5):1315. https://doi.org/10.3892/etm.2021.10750

Ikada Y (2006) Challenges in tissue engineering. J R Soc Interface 3(10):589–601. https://doi.org/10.1098/rsif.2006.0124

Jafari M, Paknejad Z, Rad MR, Motamedian SR, Eghbal MJ, Nadjmi N, Khojasteh A (2017) Polymeric scaffolds in tissue engineering: a literature review. J Biomed Mater Res B Appl Biomater 105(2):431–459. https://doi.org/10.1002/jbm.b.33547

Jiang Y, Loos K (2016) Enzymatic synthesis of biobased polyesters and polyamides. Polymers 8(7):243. https://doi.org/10.3390/polym8070243

Jin Y-Z, Lee JH (2018) Mesenchymal stem cell therapy for bone regeneration. Clin Orthop Surg 10(3):271. https://doi.org/10.4055/cios.2018.10.3.271

Jo H, Brito S, Kwak BM, Park S, Lee M-G, Bin B-H (2021) Applications of mesenchymal stem cells in skin regeneration and rejuvenation. Int J Mol Sci 22(5):2410. https://doi.org/10.3390/ijms22052410

Kaliaraj G, Shanmugam D, Dasan A, Mosas K (2023) Hydrogels—a promising materials for 3D printing technology. Gels 9(3):260. https://doi.org/10.3390/gels9030260

Kamel RA, Ong JF, Eriksson E, Junker JPE, Caterson EJ (2013) Tissue engineering of skin. J Am Coll Surg 217(3):533–555. https://doi.org/10.1016/j.jamcollsurg.2013.03.027

Kankala RK, Zhang YS, Wang S, Lee C, Chen A (2017) Supercritical fluid technology: an emphasis on drug delivery and related biomedical applications. Adv Healthc Mater 6(16). https://doi.org/10.1002/adhm.201700433

Katti D, Vasita R, Shanmugam K (2008) Improved biomaterials for tissue engineering applications: surface modification of polymers. Curr Top Med Chem 8(4):341–353. https://doi.org/10.2174/156802608783790893

Khumalo M, Sithole B, Tesfaye T, Lekha P (2022) Valorization of waste chicken feathers: fabrication and characterization of novel keratin nanofiber conduits for potential application in peripheral nerve regeneration. J Nanomater 2022:1–10. https://doi.org/10.1155/2022/7080278

Klimek K, Ginalska G (2020) Proteins and peptides as important modifiers of the polymer scaffolds for tissue engineering applications—a review. Polymers 12(4):844. https://doi.org/10.3390/polym12040844

Kneifel J, Webb D (2020) Life cycle cost manual for the federal energy management program. https://doi.org/10.6028/NIST.HB.135-2020

Kuhl N, Bode S, Hager MD, Schubert US (2015) Self-healing polymers based on reversible covalent bonds, pp 1–58. https://doi.org/10.1007/12_2015_336

La Rosa AD (2016) Life cycle assessment of biopolymers. In: Biopolymers and biotech admixtures for eco-efficient construction materials. Elsevier, pp 57–78. https://doi.org/10.1016/B978-0-08-100214-8.00004-X

Langer R (2000a) Biomaterials in drug delivery and tissue engineering: one laboratory's experience. Acc Chem Res 33(2):94–101. https://doi.org/10.1021/ar9800993

Langer R (2000b) Tissue engineering. Mol Ther 1(1):12–15. https://doi.org/10.1006/mthe.1999.0003

Langer R, Vacanti JP (1993) Tissue engineering. Science 260(5110):920–926. https://doi.org/10.1126/science.8493529

Lee EJ, Kasper FK, Mikos AG (2014) Biomaterials for tissue engineering. Ann Biomed Eng 42(2):323–337. https://doi.org/10.1007/s10439-013-0859-6

Lin W, Xu L, Zwingenberger S, Gibon E, Goodman SB, Li G (2017) Mesenchymal stem cells homing to improve bone healing. J Orthop Translat 9:19–27. https://doi.org/10.1016/j.jot.2017.03.002

Liu S, Yu J-M, Gan Y-C, Qiu X-Z, Gao Z-C, Wang H, Chen S-X, Xiong Y, Liu G-H, Lin S-E, McCarthy A, John JV, Wei D-X, Hou H-H (2023) Biomimetic natural biomaterials for tissue engineering and regenerative medicine: new biosynthesis methods, recent advances, and emerging applications. Mil Med Res 10(1):16. https://doi.org/10.1186/s40779-023-00448-w
Lyons JG, Plantz MA, Hsu WK, Hsu EL, Minardi S (2020) Nanostructured biomaterials for bone regeneration. Front Bioeng Biotechnol 8. https://doi.org/10.3389/fbioe.2020.00922
Ma Z, Mao Z, Gao C (2007) Surface modification and property analysis of biomedical polymers used for tissue engineering. Colloids Surf B: Biointerfaces 60(2):137–157. https://doi.org/10.1016/j.colsurfb.2007.06.019
Makadia HK, Siegel SJ (2011) Poly lactic-co-glycolic acid (PLGA) as biodegradable controlled drug delivery carrier. Polymers 3(3):1377–1397. https://doi.org/10.3390/polym3031377
Mallik A, Shahruzzaman M, Sazedul Islam Md, Haque P, Rahman MM (2019) In: Nayak AK, Hasnain MS, Pal D (eds) Natural polymers for pharmaceutical applications. Apple Academic Press. https://doi.org/10.1201/9780429328350
Mandrycky C, Wang Z, Kim K, Kim D-H (2016) 3D bioprinting for engineering complex tissues. Biotechnol Adv 34(4):422–434. https://doi.org/10.1016/j.biotechadv.2015.12.011
Manjare SD, Dhingra K (2019) Supercritical fluids in separation and purification: a review. Mater Sci Energy Technol 2(3):463–484. https://doi.org/10.1016/j.mset.2019.04.005
Martinez-Robinson KG (2020) Polymer conjugates in biomedical applications. In: Tailor-made polysaccharides in biomedical applications. Elsevier, pp 175–196. https://doi.org/10.1016/B978-0-12-821344-5.00008-4
Massaro MS, Pálek R, Rosendorf J, Červenková L, Liška V, Moulisová V (2021) Decellularized xenogeneic scaffolds in transplantation and tissue engineering: immunogenicity versus positive cell stimulation. Mater Sci Eng C 127:112203. https://doi.org/10.1016/j.msec.2021.112203
McInnes AD, Moser MAJ, Chen X (2022) Preparation and use of decellularized extracellular matrix for tissue engineering. J Funct Biomater 13(4):240. https://doi.org/10.3390/jfb13040240
Neishabouri A, Soltani Khaboushan A, Daghigh F, Kajbafzadeh A-M, Majidi Zolbin M (2022) Decellularization in tissue engineering and regenerative medicine: evaluation, modification, and application methods. Front Bioeng Biotechnol 10. https://doi.org/10.3389/fbioe.2022.805299
Nii T, Katayama Y (2021) Biomaterial-assisted regenerative medicine. Int J Mol Sci 22(16):8657. https://doi.org/10.3390/ijms22168657
Nikolova MP, Chavali MS (2019) Recent advances in biomaterials for 3D scaffolds: a review. Bioact Mater 4:271–292. https://doi.org/10.1016/j.bioactmat.2019.10.005
Oliveira ÉR, Nie L, Podstawczyk D, Allahbakhsh A, Ratnayake J, Brasil DL, Shavandi A (2021) Advances in growth factor delivery for bone tissue engineering. Int J Mol Sci 22(2):903. https://doi.org/10.3390/ijms22020903
Parenteau-Bareil R, Gauvin R, Berthod F (2010) Collagen-based biomaterials for tissue engineering applications. Materials 3(3):1863–1887. https://doi.org/10.3390/ma3031863
Park BK (2015) Biodegradable polymers for tissue engineering: review article. J Biomed Eng Res 36(6):251–263. https://doi.org/10.9718/JBER.2015.36.6.251
Perez-Puyana V, Jiménez-Rosado M, Romero A, Guerrero A (2020) Polymer-based scaffolds for soft-tissue engineering. Polymers 12(7):1566. https://doi.org/10.3390/polym12071566
Pogorielov M, Oleshko O, Hapchenko A (2017) Tissue engineering: challenges and selected application. Adv Tissue Eng Regen Med Open Access 3(2):330–334. https://doi.org/10.15406/atroa.2017.03.00058
Ponnusamy PG, Mani S (2022) Material and environmental properties of natural polymers and their composites for packaging applications—a review. Polymers 14(19):4033. https://doi.org/10.3390/polym14194033
Prem Ananth K, Jayram ND (2024) A comprehensive review of 3D printing techniques for biomaterial-based scaffold fabrication in bone tissue engineering. Ann 3D Printed Med 13:100141. https://doi.org/10.1016/j.stlm.2023.100141
Qasim S, Zafar M, Najeeb S, Khurshid Z, Shah A, Husain S, Rehman I (2018) Electrospinning of chitosan-based solutions for tissue engineering and regenerative medicine. Int J Mol Sci 19(2):407. https://doi.org/10.3390/ijms19020407

Qin Y (2016) Biocompatibility testing for medical textile products. In: Medical textile materials. Elsevier, pp 191–201. https://doi.org/10.1016/B978-0-08-100618-4.00014-5

Raeisdasteh Hokmabad V, Davaran S, Ramazani A, Salehi R (2017) Design and fabrication of porous biodegradable scaffolds: a strategy for tissue engineering. J Biomater Sci Polym Ed 28(16):1797–1825. https://doi.org/10.1080/09205063.2017.1354674

Rajeswari S (2017) Natural polymers: a recent review. World J Pharm Pharm Sci:472–494. https://doi.org/10.20959/wjpps20178-9762

Rajinikanth BS, Rajkumar DSR, Keerthika K, Vijayaragavan V (2024) Chitosan-based biomaterial in wound healing: a review. Cureus. https://doi.org/10.7759/cureus.55193

Ratner BD, Hoffman AS, Schoen FJ, Lemons JE (2013) Biomaterials science (Introduction – biomaterials science: an evolving, multidisciplinary endeavor a history of biomaterials), 3rd edn. Elsevier. https://doi.org/10.1016/C2009-0-02433-7

Rejinold NS, Choi G, Choy J-H (2021) Recent developments on semiconducting polymer nanoparticles as smart photo-therapeutic agents for cancer treatments—a review. Polymers 13(6):981. https://doi.org/10.3390/polym13060981

Ren X, Kim C-Y, Canavero S (2019) Bridging the gap: spinal cord fusion as a treatment of chronic spinal cord injury. Surg Neurol Int 10:51. https://doi.org/10.25259/SNI-19-2019

Rinaudo M (2006) Chitin and chitosan: properties and applications. Prog Polym Sci 31(7):603–632. https://doi.org/10.1016/j.progpolymsci.2006.06.001

Rossi F, Santoro M, Perale G (2015) Polymeric scaffolds as stem cell carriers in bone repair. J Tissue Eng Regen Med 9(10):1093–1119. https://doi.org/10.1002/term.1827

Šafarič R, Fras Zemljič L, Novak M, Dugonik B, Bratina B, Gubeljak N, Bolka S, Strnad S (2020) Preparation and characterisation of waste poultry feathers composite fibreboards. Materials 13(21):4964. https://doi.org/10.3390/ma13214964

Saravanan S, Sameera DK, Moorthi A, Selvamurugan N (2013) Chitosan scaffolds containing chicken feather keratin nanoparticles for bone tissue engineering. Int J Biol Macromol 62:481–486. https://doi.org/10.1016/j.ijbiomac.2013.09.034

Schmidt T, Xiang Y, Bao X, Sun T (2021) A paradigm shift in tissue engineering: from a top–down to a bottom–up strategy. PRO 9(6):935. https://doi.org/10.3390/pr9060935

Shammaa R, El-Kadiry AE-H, Abusarah J, Rafei M (2020) Mesenchymal stem cells beyond regenerative medicine. Front Cell Dev Biol 8. https://doi.org/10.3389/fcell.2020.00072

Song SJ, Park CH (2019) Microfracture for cartilage repair in the knee: current concepts and limitations of systematic reviews. Ann Transl Med 7(S3):S108–S108. https://doi.org/10.21037/atm.2019.05.11

Song P, Zhou C, Fan H, Zhang B, Pei X, Fan Y, Jiang Q, Bao R, Yang Q, Dong Z, Zhang X (2018) Novel 3D porous biocomposite scaffolds fabricated by fused deposition modeling and gas foaming combined technology. Compos Part B 152:151–159. https://doi.org/10.1016/j.compositesb.2018.06.029

Stabler CT, Lecht S, Mondrinos MJ, Goulart E, Lazarovici P, Lelkes PI (2015) Revascularization of decellularized lung scaffolds: principles and progress. Am J Phys Lung Cell Mol Phys 309(11):L1273–L1285. https://doi.org/10.1152/ajplung.00237.2015

Sultankulov B, Berillo D, Sultankulova K, Tokay T, Saparov A (2019) Progress in the development of chitosan-based biomaterials for tissue engineering and regenerative medicine. Biomol Ther 9(9):470. https://doi.org/10.3390/biom9090470

Sun Y, Wan B, Wang R, Zhang B, Luo P, Wang D, Nie J-J, Chen D, Wu X (2022) Mechanical stimulation on mesenchymal stem cells and surrounding microenvironments in bone regeneration: regulations and applications. Front Cell Dev Biol 10. https://doi.org/10.3389/fcell.2022.808303

Tajik S, Garcia CN, Gillooley S, Tayebi L (2023) 3D printing of hybrid-hydrogel materials for tissue engineering: a critical review. Regen Eng Transl Med 9(1):29–41. https://doi.org/10.1007/s40883-022-00267-w

Tan S, Barker N (2013) Engineering the niche for stem cells. Growth Factors 31(6):175–184. https://doi.org/10.3109/08977194.2013.859683

Tarun G (2012) Scaffold: tissue engineering and regenerative medicine. Int Res J Pharm 2(12):37–42.

Thakur VK, Thakur MK (eds) (2015) Handbook of polymers for pharmaceutical technologies. Wiley. https://doi.org/10.1002/9781119041412
Tien J (2014) Microfluidic approaches for engineering vasculature. Curr Opin Chem Eng 3:36–41. https://doi.org/10.1016/j.coche.2013.10.006
Troy E, Tilbury MA, Power AM, Wall JG (2021) Nature-based biomaterials and their application in biomedicine. Polymers 13(19):3321. https://doi.org/10.3390/polym13193321
Trucillo P (2024) Biomaterials for drug delivery and human applications. Materials 17(2):456. https://doi.org/10.3390/ma17020456
Vesvoranan O, Anup A, Hixon KR (2022) Current concepts and methods in tissue interface scaffold fabrication. Biomimetics 7(4):151. https://doi.org/10.3390/biomimetics7040151
Waidi YO, Kariim I, Datta S (2024) Bioprinting of gelatin-based materials for orthopedic application. Front Bioeng Biotechnol 12. https://doi.org/10.3389/fbioe.2024.1357460
Wang X, Wang Z, Zhai W, Wang F, Ge Z, Yu H, Yang W (2021) Engineering biological tissues from the bottom-up: recent advances and future prospects. Micromachines 13(1):75. https://doi.org/10.3390/mi13010075
Weigel T, Schinkel G, Lendlein A (2006) Design and preparation of polymeric scaffolds for tissue engineering. Expert Rev Med Devices 3(6):835–851. https://doi.org/10.1586/17434440.3.6.835
Williams DF (2009) On the nature of biomaterials. Biomaterials 30(30):5897–5909. https://doi.org/10.1016/j.biomaterials.2009.07.027
Xie Z, Gao M, Lobo AO, Webster TJ (2020) 3D bioprinting in tissue engineering for medical applications: the classic and the hybrid. Polymers 12(8):1717. https://doi.org/10.3390/polym12081717
Xing J, Zhang M, Liu X, Wang C, Xu N, Xing D (2023) Multi-material electrospinning: from methods to biomedical applications. Mater Today Bio 21:100710. https://doi.org/10.1016/j.mtbio.2023.100710
Xu Y, Chen C, Hellwarth PB, Bao X (2019) Biomaterials for stem cell engineering and biomanufacturing. Bioact Mater 4:366–379. https://doi.org/10.1016/j.bioactmat.2019.11.002
Xu Y, Zhang F, Zhai W, Cheng S, Li J, Wang Y (2022) Unraveling of advances in 3D-printed polymer-based bone scaffolds. Polymers 14(3):566. https://doi.org/10.3390/polym14030566
Xue J, Wu T, Dai Y, Xia Y (2019) Electrospinning and electrospun nanofibers: methods, materials, and applications. Chem Rev 119(8):5298–5415. https://doi.org/10.1021/acs.chemrev.8b00593
Yang W, Chen X, Li S, Li X-J (2021) Genetically modified large animal models for investigating neurodegenerative diseases. Cell Biosci 11(1):218. https://doi.org/10.1186/s13578-021-00729-8
Yao T, Baker MB, Moroni L (2020) Strategies to improve nanofibrous scaffolds for vascular tissue engineering. Nanomaterials 10(5):887. https://doi.org/10.3390/nano10050887
Yi S, Ding F, Gong L, Gu X (2017) Extracellular matrix scaffolds for tissue engineering and regenerative medicine. Curr Stem Cell Res Ther 12(3):233–246. https://doi.org/10.2174/1574888X11666160905092513
Yue B (2014) Biology of the extracellular matrix. J Glaucoma 23:S20–S23. https://doi.org/10.1097/IJG.0000000000000108
Zeng S, Ye J, Cui Z, Si J, Wang Q, Wang X, Peng K, Chen W (2017) Surface biofunctionalization of three-dimensional porous poly(lactic acid) scaffold using chitosan/OGP coating for bone tissue engineering. Mater Sci Eng C 77:92–101. https://doi.org/10.1016/j.msec.2017.03.220
Zhang X, Zhang Y (2015) Tissue engineering applications of three-dimensional bioprinting. Cell Biochem Biophys 72(3):777–782. https://doi.org/10.1007/s12013-015-0531-x
Zhang L, Lee W, Li X, Jiang Y, Fang NX, Dai G, Liu Y (2022a) 3D direct printing of mechanical and biocompatible hydrogel meta-structures. Bioact Mater 10:48–55. https://doi.org/10.1016/j.bioactmat.2021.08.015
Zhang X, Chen X, Hong H, Hu R, Liu J, Liu C (2022b) Decellularized extracellular matrix scaffolds: recent trends and emerging strategies in tissue engineering. Bioact Mater 10:15–31. https://doi.org/10.1016/j.bioactmat.2021.09.014

Zhao R, Mao F, Qian H, Yang X, Zhu X, Zhang X (2023) Micro-/nano-structured biomaterials for bone regeneration: new progress. J Inorg Mater 38(7):750. https://doi.org/10.15541/jim20220580

Zheng Y, Pan P (2020) Crystallization of biodegradable and biobased polyesters: polymorphism, cocrystallization, and structure-property relationship. Prog Polym Sci 109:101291. https://doi.org/10.1016/j.progpolymsci.2020.101291

Zhu W, Ma X, Gou M, Mei D, Zhang K, Chen S (2016a) 3D printing of functional biomaterials for tissue engineering. Curr Opin Biotechnol 40:103–112. https://doi.org/10.1016/j.copbio.2016.03.014

Zhu Y, Romain C, Williams CK (2016b) Sustainable polymers from renewable resources. Nature 540(7633):354–362. https://doi.org/10.1038/nature21001

Zieliński PS, Gudeti PKR, Rikmanspoel T, Włodarczyk-Biegun MK (2023) 3D printing of bio-instructive materials: toward directing the cell. Bioact Mater 19:292–327. https://doi.org/10.1016/j.bioactmat.2022.04.008

# Green Manufacturing Techniques in Scaffold Fabrication

2

Mohamed Aboughaly

**Abstract**

Sustainable scaffolds have become crucial in tissue engineering and repair as well as regenerative steam cell therapy, offering eco-friendly solutions for tissue repair and regeneration. Despite advancements in biomaterials, current technologies still face challenges in achieving desired stability and efficiency in tissue defect repair. The advancement in regenerative tissue engineering and scaffold synthesis is evolving toward the development of durable scaffolds that provide structural support for tissue, promoting healing while minimizing waste through sustainable practices. The global sustainable scaffold market is projected to reach €7 billion by 2025, with a 15% annual growth rate. The key innovations highlighted include the synthesis of three-dimensional porous and hydrogel scaffolds that mimic the synthesis of extracellular matrix using natural vibers that enable cell infiltration and tissue regeneration. The porosity, pore size, and interconnectivity of scaffolds are critical for effective cell growth. Green manufacturing processes, such as the use of biodegradable materials (chitosan, cellulose, alginate), solvent-free methods (freeze drying, electrospinning, supercritical $CO_2$ foaming), and 3D printing, enhance sustainability by minimizing energy consumption, chemical waste, and environmental impact. These processes improve biocompatibility while reducing material waste and carbon footprints. Overall, sustainable scaffold fabrication supports regenerative medicine's goals while aligning with broader environmental sustainability objectives, making it a key area for future biomedical advances.

M. Aboughaly (✉)
Chemical Engineering Department, Mcmaster University, Hamilton, ON, Canada
e-mail: mohamed.aboughaly@ontariotechu.net; abougm1@mcmaster.ca



J. M. Oliveira et al. (eds.), *Sustainable Scaffolds-based Strategies in Tissue Engineering and Regenerative Medicine*, Biomaterials, Bioengineering and Sustainability 8, https://doi.org/10.1007/978-3-031-96274-5_2

**Keywords**
Sustainable scaffolds · Tissue engineering · Regenerative medicine · Green manufacturing · 3D printing · Biodegradable materials · Solvent-free methods

## 2.1 Introduction

Sustainable materials and scaffolds have risen as new biomedical fields for synthesis of materials and scaffolds used in tissue engineering and steam cell therapy (Ribeiro et al. 2015; Godara et al. 2008; Pryadko et al. 2021). The inherit complexity of biomaterials and its manufacturing technologies haven't yet reached the desired stability and satisfactory effects of engineered tissue defects repair. New generative medicine refers to the development of biomaterials that are long lasting and act as support structures and scaffolds to help group new tissues and repair damaged ones that provide framework for cells to attach and grow and develop new functional tissues that promotes healing and faster regeneration and minimizes waste and rely on sustainable practice use of materials. The sustainable scaffold market is expected to exceed €7 billion by 2025 with expected annual growth of 15% which is considered the fastest growing sector in regenerative medical industry.

This field focuses on synthesis of safe, ecofriendly and effective solutions for medical treatments. The design and processing of new biomaterials that are functionalized and used in synthesis of cell material and scaffold interactions and imaging to enhance biocompatibility and new biomaterials interactions. The surface characterization and functionalization used in cell material interactions and imaging to characterize biomaterials are used in synthesis of three dimensional porous and hydrogel scaffolds are used for cell growth forming 3D porous and hydrogel scaffold which are used in cell synthesis to form natural extracellular matrix (ECM) used in synthesis of porous and hydrogel scaffolds in regenerative processes. The pore size, interconnectivity and porosity of tissue structures that hold direct influence over the functionality of tissue engineered structures. The scaffold based engineered tissue, high porosity is crucial in waste removal and cell infiltration. The pore size plays a vital role in the ECM production.

Green manufacturing techniques for scaffold fabrication processes aims to reduce the environmental impact, enhance process sustainability and ensure safe processes in tissue engineering. These green processes focus on eco-friendly materials, reduction of energy consumption, minimization of energy usage and usage of sustainable processes throughout various production processes. Green manufacturing techniques include usage of biodegradable materials such as chitosan, cellulose, alginate that are non-toxic and biodegradable. Biomaterials enhances the sustainability and reduce environmental impact and ensure safety in tissue engineering. Usage of biomaterials eliminate the need for extraction and disposal. Also, green and sustainable techniques allow usage of organic solvents and usage of solvent free methods such as freeze drying, electrospinning with water-based solutions and carbon dioxide foaming (Nga et al. 2023; Vavouranakis et al. 2017; Adolph et al. 2014).

Green processes avoid harmful chemicals and reduce potential for environmental contamination and enhances biocompatibility. Also, green techniques such as supercritical $CO_2$ foaming and hydrogel-based 3D printing offers energy efficient pathways that fabricate scaffolds by reducing process treatment processing times and high temperature requirements (Li and Tuan 2009; Chen et al. 2013; Hwang et al. 2018). Also, green processes minimize chemical waste and reduce carbon foot print. Green fabrication processes encourage the recycling of raw materials and scaffolds such as polylactic acids, and reduction of material waste. Additionally, reusage of scaffold materials ensure cost efficiency and sustainability in production as well as increase cost efficiency.

In terms of additive manufacturing, 3D printing is a successful strategy for reduction of material waste through usage of green materials for creation of scaffolds layer by layer and avoid usage of excessive waste in manufacturing processes. These methods allow precision in design, reduction for post processing and lowering energy consumption. Also, water-based processes such as water jet cutting, freeze casting and usage of aqueous based inks in 3D bioprinting minimizes environmental impact and green technologies through usage of water replacing harmful solvents (Prem Ananth and Jayram 2024; Liang et al. 2023; Misra et al. 2017).

Green manufacturing emphasizes the reduction of waste by optimizing scaffold designs to maximize material efficiency, avoiding overproduction, and using closed-loop systems to capture and reuse waste materials. By implementing these techniques, scaffold fabrication in tissue engineering not only supports the development of regenerative medicine but also aligns with broader environmental sustainability goals. These methods ensure biocompatibility, reduce toxic chemical use, and minimize energy and resource consumption, making them key to future advances in sustainable biomedical manufacturing. The elevated level of porosity facilitates cellular penetration and ECM development, which is regulated by the dimensions of the pores. When designed with open, interconnected pores, scaffolds significantly enhance cellular proliferation and movement, supporting substantial natural ECM production. Researchers typically employ two main strategies: developing acellular scaffolding biomaterials or creating cell-containing scaffolding structures that replicate the tissue's extracellular environment. These approaches result in three-dimensional porous structures or hydrogel scaffolds that support cellular growth, enabling the formation of natural extracellular matrices essential for regenerative processes (Ribeiro et al. 2015; Godara et al. 2008; Yari et al. 2022; Pörtner et al. 2005). Natural polymers such as polysaccharides are used for synthesis of three-dimensional artificial ECM such as hydrogels using various processing methods and conditions. Natural polymers have shown excellent biological response and able to be modified into various forms of scaffolds including nanofibrous, 3D printed structures and used in synthesis of various tissue regenerations such as bones and cartilages, retinal, neural and other synthesized tissue parts (Chan and Leong 2008). Figure 2.1 depicts a schematic diagram of scaffold-based tissue engineered approach.

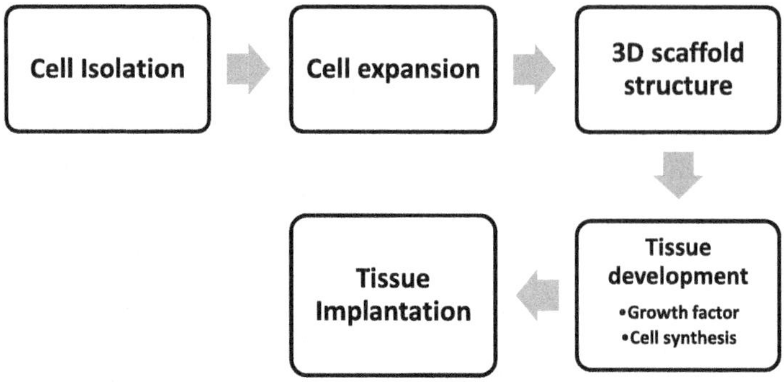


**Fig. 2.1** Schematic of the scaffold-based tissue engineering approach

## 2.2 Biomaterials for Tissue Engineering and Regenerative Medicine

Various tissue engineering applications utilize calcium phosphate ceramics as biomaterials because of their rigidity and gradual decomposition rate, making them appropriate for combination with biodegradable polymers. This pairing creates enhanced structures like bioactive glasses, metallic-based substances, and polymer composites for diverse tissue engineering needs (Berdimurodov et al. 2023; Das et al. 2023; Nurchi et al. 2023). In creating skin and soft tissue structures, researchers frequently employ biodegradable polymers. These include naturally derived polymers such as collagen, gelatin, alginate, and chitosan, as well as manufactured polymers like polylactic acid (PLLA), poly lactic co-glycolic acid, and polycaprolactone (PCL)—which are partially crystalline biodegradable polyesters (Harris et al. 2021; Farag 2023). Biodegradable polymers fall into two categories based on their source: natural or synthetic. For tissue engineering applications, keratin-based materials or natural biomaterials containing genetically modified cells are utilized in developing viable implants. Their appeal stems from their interconnected microstructure, bioactive properties, clearly defined molecular arrangements, and functional sites that allow for additional modifications with other materials. These engineered substances enable the creation of a vast array of materials with specific characteristics and precisely calibrated functionalities (Tavakoli 2017; Stie et al. 2022; Purewal et al. 2024; Mokhtarinia et al. 2023).

Natural materials currently face several challenges, including batch-to-batch inconsistency, quick breakdown, and insufficient mechanical strength, which hinder their integration. Custom-designed biomimetic natural biomaterials (BNBMs) offer enhanced physicochemical, mechanical, and biological characteristics that facilitate adhesion, penetration, and movement of oxygen and nutrients throughout cells and tissues. The widespread use of BNBMs in tissue engineering stems from their nontoxic, non-genotoxic, and non-teratogenic properties in relation to healthy native tissues, along with high bioactivity that encourages cellular growth, attachment, movement, and differentiation processes (Sommer et al. 2018; Gao et al. 2024; Park et al. 2024; Zhang et al. 2023). The primary sustainable biomimetic natural biomaterials employed in tissue engineering and regenerative medicine include bio polyesters such as polylactic acid (PLA), poly-hydroxy-alkenoates (PHA), and their derivatives. These also encompass polypeptides and proteins like collagen, gelatin, fibroin, and antimicrobial peptides (AMPs), which can be customized to deliver superior physicochemical, mechanical, and biological properties supporting cellular infiltration, adhesion, differentiation, and nutrient transport. The following sections present various examples and applications of biomaterials, along with strategies and purposes as illustrated in Fig. 2.2.

Calcium phosphate ceramics serve as valuable biomaterials in tissue engineering applications owing to their inherent rigidity and prolonged degradation characteristics, making them compatible companions for biodegradable polymers. This combination produces enhanced structures including bioactive glasses, metal-incorporated materials, and polymer composites suitable for diverse tissue engineering implementations (Berdimurodov et al. 2023; Das et al. 2023; Nurchi et al. 2023). For skin and soft tissue synthesis, researchers frequently employ biodegradable polymers, including naturally derived options (collagen, gelatin, alginate, and chitosan) and laboratory-synthesized alternatives such as polylactic acid (PLLA), poly lactic co-glycolic acid, and polycaprolactone (PCL), which function as semi-crystalline biodegradable polyesters (Harris et al. 2021; Farag 2023).

The classification of biodegradable polymers follows their origin, dividing them into natural or synthetic categories. In designing living implants for tissue engineering applications, researchers utilize keratin-based biomaterials or naturally derived materials containing genetically modified cells. These materials offer advantages through their interconnected microstructure, bioactive properties, well-defined molecular organization, and functional sites that permit additional modifications with complementary materials. Such engineered substances facilitate the development of an extensive range of materials with precise characteristics and carefully calibrated functionalities (Tavakoli 2017; Stie et al. 2022; Purewal et al. 2024; Mokhtarinia et al. 2023).

Current challenges with natural materials include inconsistency between production batches, accelerated degradation rates, and insufficient mechanical strength, which collectively impede their integration processes. Custom-engineered biomimetic natural biomaterials (BNBMs) deliver enhanced physicochemical, mechanical, and biological properties that facilitate cellular adhesion, penetration, and efficient transport of oxygen and nutrients throughout cellular structures and tissues. The widespread adoption of BNBMs in tissue engineering stems from their non-toxic, non-genotoxic, and non-teratogenic profile in relation to healthy native tissues, coupled with high bioactivity that stimulates cell proliferation, attachment, migration, and differentiation regulation.

Primary sustainable biomimetic natural biomaterials employed in tissue engineering and regenerative medicine encompass bio polyesters including polylactic

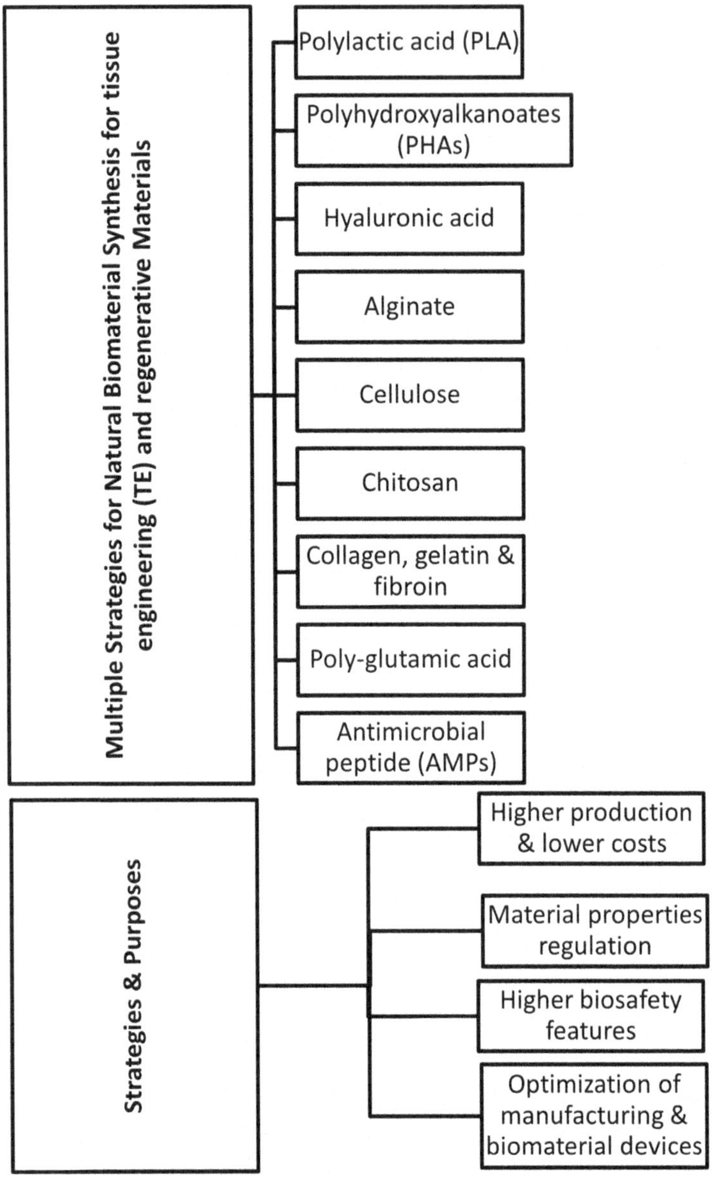


**Fig. 2.2** Different strategies and purposes for natural biomaterial synthesis for tissue engineering and regenerative medicine using novel techniques and tools

acid (PLA), poly-hydroxy-alkenoates (PHA), and their derivatives. These also include polypeptides and proteins such as collagen, gelatin, fibroin, and antimicrobial peptides (AMPs), which can be customized to deliver superior physicochemical, mechanical, and biological characteristics supporting cellular infiltration, adhesion, differentiation, and nutrient transport (Mokhtarinia et al. 2023; Brizuela et al. 2024; Srinivasan et al. 2022). Various examples and applications of these biomaterials, along with implementation strategies and functional purposes, are illustrated in Fig. 2.2.

PLA and PLGA demonstrate reduced biodegradation rates and generate acidic biodegradable monomers and oligomers during breakdown (Wasyłeczko et al. 2020; Subramanian et al. 2009; Rodríguez-Vázquez et al. 2015; Carotenuto et al. 2022). Biopolymer composites are extensively utilized to minimize cellular and tissue reactions that might otherwise lead to restricted local inflammatory responses (Tavakoli 2017; Stie et al. 2022; Liu et al. 2023a). Additionally, the controlled release characteristics of acidic biodegradable polymers serve to decrease acid burden and minimize cellular and tissue reactions that could trigger localized inflammation.

As a linear polysaccharide, cellulose comprises extended chains containing hundreds to thousands of β-glucopyranose units with cellobiose functioning as the fundamental dimeric repeating structure. Recognized as Earth's most abundant natural material, cellulose exhibits no odor, demonstrates hydrophilic properties, biodegrades naturally, shows biocompatibility, and can effectively regulate degradation rates in tissue-engineered scaffold structures. Bacterial cellulose specifically demonstrates exceptional water retention capabilities and elasticity, making it valuable for wound treatment applications. Similarly, animal-derived cellulose variants, such as those from sea squirts, offer advantages through their molecular weight characteristics, mechanical resilience, superior water retention capacity, air permeability, and thermal stability (Liu et al. 2023a).

Both natural and synthetic polymers—including proteins, polysaccharides, polyglycolic acid, polylactic acid (PLA)—and inorganic biomaterials—including titanium and its alloys, alongside ceramics such as alumina, zirconia, and calcium phosphate compounds—demonstrate enhanced rigidity when combined with natural polymers due to their structural similarities with extracellular matrix. These inorganic biomaterials exhibit strong biocompatibility, osteoconductive properties, and bioresorbability (Wasyłeczko et al. 2020; Niu et al. 2023; Fu et al. 2022).

Various advancements have emerged in osteochondral tissue engineering focused on regenerating biomimetic tissues using diverse materials that deliver essential components within engineered tissues, scaffolds providing structural and functional support for cellular growth, and decellularized matrices facilitating tissue regeneration. Nevertheless, current synthesized tissues continue to face challenges stemming from inherent complexity, insufficient mechanical performance, and limited chemical stability (Niu et al. 2023; Fu et al. 2022). Beyond medical applications, these materials contribute to biosensor development for detecting biological and chemical threats, and engineered tissues serve in toxicity testing for various biomedical devices.

Fabrication techniques now enable the creation of highly porous, biodegradable materials with customized dimensions and porosity through parameter adjustment and functional 3D tissue engineering scaffold polymers. These approaches have applications in drug delivery systems and cell therapy. Incorporating physical and biological features—such as modified surface chemistry and bioactive ligands—onto scaffolds enhances machinery interaction and regulates growth and vascularization processes essential for tissue regeneration.

Biomaterials incorporated into engineered tissues can be synthesized through multiple techniques: selective laser sintering, stereolithography, solvent-based extrusion, fused deposition modeling, and bioprinting methodologies (Zare and Shabani 2016; Li et al. 2021). Natural polymer scaffolds—including collagen, alginate, hyaluronic acid, chitosan, and fibrin—are frequently integrated into various polymer scaffold systems. Despite their potential, biomaterials and natural polymers face limitations in tissue engineering applications due to inadequate mechanical properties, potential disease transmission risks, and immunogenicity concerns. Consequently, synthetic polymers such as PEG (polyethylene glycol), PLA (polylactic acid), GGA (glycolic acid), OLGA (oligolactic acid), and PCL (polycaprolactone) find extensive application across numerous engineered tissue constructs (Sultana 2018).

## 2.3 Green Scaffolding Techniques for Regeneration and Engineering of Tissues

Various scaffolding approaches for tissue synthesis have been developed to enhance scalability, sustainability, and spatial control in engineered tissues. These include freeze casting, gas foaming, sol-gel techniques, solvent casting, compression molding, leaching, and phase separation processes (Fu et al. 2022). In implant fabrication, numerous biopolymers are utilized, such as silk fibroin, collagen, gelatin, keratin, elastin, and polysaccharides including chitosan, chitin, alginate, and their derivatives. Additionally, structural proteins like elastic, fibrin, and silk serve as sutures for scaffold fabrication and drug delivery systems (Niu et al. 2023; Fu et al. 2022; Suamte et al. 2023). Scaffold fabrication employs diverse design methodologies: solvent casting, particle leaching, freeze drying, thermal-induced phase separation, gas foaming, powder foaming, sol-gel processes, electrospinning, fused deposition modeling (FDM), selective laser sintering (SLS), binder jetting techniques, laser-assisted bioprinting, and metal-based additive manufacturing (Suamte et al. 2023). For perspective, creating a 50 μm thick cell sheet requires approximately 100,000 human endometrial gland-derived cells per square centimeter.

Exocrine glands such as salivary and lacrimal glands, which contribute to oral health maintenance and are utilized in surgical procedures, represent important engineered tissue applications. The developmental processes are replicated using engineered tissues to mimic gland formation, which originates from ectoderm through reciprocal epithelial-mesenchymal interactions during embryogenesis

(Suamte et al. 2023; Contessi Negrini et al. 2021). Successful engineered tissue development necessitates precise selection of cell sources, appropriate scaffolds, and cell culture configurations that facilitate replication of morphogenetic signals. These engineered tissues find applications in developing teeth, hair follicles, and salivary and lacrimal glands, with particular emphasis on biomaterials and cell culture arrangements (Suamte et al. 2023).

The freeze casting technique enables fabrication of graded scaffolds with enhanced mechanical properties as compared to foam replication methods. This advantage stems from avoiding defects typically generated during polymer foam template pyrolysis, resulting in scaffolds with regulated porosity and pore dimensions that exhibit suitable compressive properties for various engineered tissue applications (Fu et al. 2022). Composite scaffolds incorporating silica-rich bioactive glasses prepared through solid-liquid separation methods demonstrate different characteristics compared to those created using solvent casting particulate leaching or phase inversion techniques. Variations in microstructure, crystallinity, degradation kinetics, and bioactivity are observed across composite materials depending on fabrication methodologies and biomaterial compositions. These fabrication approaches help regulate pore architecture, polymer crystallinity, and biodegradation rates, which subsequently influence cell adhesion and proliferation (Loh and Choong 2013).

Lyophilization involves cooling a mixture through freeze drying to remove solvent and water, creating macro and micropores within the scaffold structure. This method offers the advantage of modifiable pore size and porosity through solution characteristic adjustments, including concentration, viscosity, quenching rate, and freezing temperature. However, limitations include high organic solvent requirements and potential emulsion instability (Fu et al. 2022). Freeze casting represents a common manufacturing technique involving controlled solidification processes, including solvent sublimation under reduced pressure followed by densification. This method applies to various materials, producing scaffolds with adjustable micro- and macro-structures.

Gas foaming subjects raw materials to elevated carbon dioxide pressure to create porous structures, offering benefits of high uniformity in cell infiltration, though improvements remain necessary. Alternatively, microfluidic foaming generates foam through microfluidics under precisely controlled conditions, producing homogeneous pores with monodispersity and interconnections that facilitate abundant cell infiltration and versatility (Fu et al. 2022).

The microstructure, crystallinity, degradation rates, and bioactivity of sustainable scaffolds made of poly-ε-caprolactone and silica-rich bioactive materials made by solid-liquid phase separation techniques using solvent casting particles vary depending on the production process. Cell adhesion and proliferation are impacted by these changes in pore architecture, polymer crystallinity, and biodegradation properties. Interfacial tension during solvent evaporation causes surface skin formation, which is related to freeze drying techniques and solid-liquid separation fabrication procedures. It is possible to modify these fabrication methods to regulate

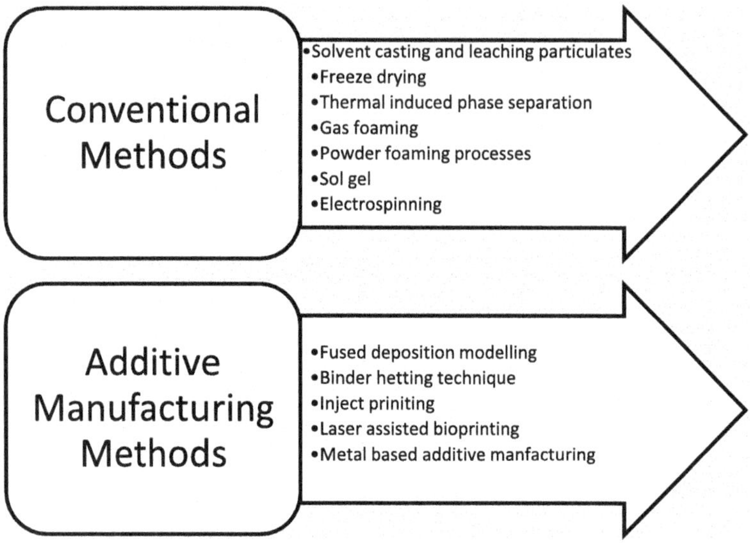


**Fig. 2.3** Creating the perfect scaffolds for a range of biomedical applications using additive and conventional manufacturing techniques

porosity and produce scaffolds that are connected. Furthermore, it has been documented that ethanol and dry ice combinations can create scaffolds with open architecture that feature interconnected pores and structures (Loh and Choong 2013; Turnbull et al. 2018). The numerous types of conventional and additive manufacturing processes utilized to create appropriate scaffolds for a range of biomedical applications are depicted in Fig. 2.3.

As shown below, there are four primary scaffolding methodologies that are employed in different production techniques:

Confluent cells with secret extracellular matrix;

- Hydrogel-encapsulated cells;
- Prefabricated porous scaffolds for cell seeding; and
- Decellularized extracellular matrix.

In the first way, readymade porous scaffolds are constructed utilizing a cell seeding methodology. Pyrogens such calcium carbonate, ammonium bicarbonate, and ammonium carbonate are incorporated, along with alternate manufacturing processes that use woven or non-woven fibers. However, scaffolds made with the phase separation approach are appropriate for applications involving synthetic tissue. Scaffolds with clearly defined hierarchical structures that offer optimal porosity and topographical features are produced using this technique. This technique is well-known for producing synthetic tissues with effective cell adhesion and long-term vitality. The physical characteristics of the scaffold are altered by the size, shape, and structural arrangement of the material's pores. Making scaffolds from different polymer composites that are used to increase the scaffold's bioactivity improves its mechanical strength (Loh and Choong 2013; Turnbull et al. 2018; Liu et al. 2023b; Yousefiasl et al. 2023; Zhu et al. 2024).

Another scaffolding approach is confluent cells with secreted extracellular matrix which processes the extracellular matrix using confluent cells. The main advantage of this method is its high biocompatibility; however, it requires multiple laminations. The main disadvantage includes high cellularity, epithelial tissues and development of thin layer tissues (Chan and Leong 2008; Liu et al. 2023b). The mechanical strength could be further enhanced by manufactured scaffolds from various polymer composites which are utilized to improve the bioactivity of scaffolds. Phase separation improves the manufacturing processes for producing scaffolds designed specifically for engineered tissue applications. Liquid to liquid phase separation is used for development of porous polymeric scaffolds that consist of biodegradable polymers resembling extracellular matrix. These types of polymer composites that use porous polymeric scaffolds have tiny fiber diameters that range between 50 and 350 nm. The fiber diameter, porosity and surface area are controlled by adjusting the concentration of the polymer solution.

The most common technique for synthesis of porous scaffolds is cell seeding which is a common method of tissue engineering. Seeded therapeutic cells are used to synthesize pre-made porous scaffolds synthesized of biodegradable materials used for synthesis of well-established scaffolds (Chan and Leong 2008; Pooyan et al. 2018; Ghassemi et al. 2018).

In engineered tissues, cells are seeded on scaffolds and act as a template for cell growth and used to guide the formation of functional tissues and organs. Synthesized tissue should be biocompatible, non-toxic or immunogenic, biodegradable and capable of regenerated a repaired and able to provide porous structure and a 3D environment for cell and tissue penetration as well as media for transportation of waste and nutrients and surface conductive to support the diverse cellular activities in newly formed tissues. Engineered scaffolds should be biocompatible and non-toxic to living cells and have high mechanical stability for surgical handling and ease of manufacturing and sterilization (Chan and Leong 2008; Liu et al. 2023b; Yousefiasl et al. 2023).

Also, biodegradable scaffolds should be able to be remodeled in tune with regeneration and repair process. Porous sustainable scaffolds are used to provide suitable 3D environment for cell and tissue penetration and waste transportation and a surface media for cell functions. Porous engineered tissue should have high mechanical stability for surgical handling and ease of manufacturing and sterilization. The scaffolds are also able to carry biological signals such as growth factors, cytokines in a controlled manner and serve as drug delivery systems with high mechanical stability for surgical handling (Sultana 2018; Pooyan et al. 2018; Thangavel and Elsen Selvam 2022). The porous 3D nanofibrous scaffolds possess a rough surface, lose porous structure and excellent water absorption characteristics. Nanofibrous scaffolds are designed effectively to enhance cell adhesion, proliferation and migration compared to its 2D structures. The nanofibrous scaffolds are used to promote wound healing with effective cell adhesion, proliferation and migration behaviors compared to alternative 2D materials (Thangavel and Elsen Selvam 2022; Yu et al. 2019). Cloned renal cells are seeded in scaffolds consisted of three collages coated cylindrical silastic catheters forming three membranes connected to catheters used to create renal neo-organ with mechanism for collecting excreted urinary fluids.

Animal derived biomaterials have shown promising performance for efficient scalable production of sustainable, inexpensive and biodegradable scaffolds. They are inexpensive, nontoxic, edible, degradable biomaterials used in scaffold engineered tissues. Cutting edge techniques such as electrospinning, modified spinning and 3D bio printed materials for directional cell growth and structured product production. Extended focus on engineered tissue materials that are biodegradable, eco-friendly and safer for green electrospinning. Nanofibrous scaffolds can mimic the structural features of natural ECM which have shown excellent performance for tissue engineering. The surface modification of nanofibrous scaffolds include blending and coating, plasma treatment, wet chemical methods and surface graft polymerization. The applications of nanofibrous scaffolds in tissue engineering. Finally, the applications of nanofibrous scaffolds in tissue engineering are used in various applications owing to their high surface to volume ratio, nanofibrous scaffolds which increase the cell adhesion and differentiation (Garbayo et al. 2021; Breuer et al. 2023; Jana et al. 2022; Tay 2021). Surface modification of nanofibrous scaffolds include coating and blending, thermal plasma treatment, wet chemical methods and surface polymerization.

In designing scaffolds for bone tissue engineering, a biomimetic approach is developed as 3D surface modification technique used to grow partially carbonated hydroxyapatite in porous, polymer scaffolds using simulated fluids. For the rational design of scaffolding materials and optimization of biometric processes, various materials and processing parameters in relation to apatite formation on 3D polymer scaffolds are synthesized. The growth in internal pores of poly(L-lactide) and poly (D-lactide) scaffolds are noticed to be faster than those in poly lactide-co-glycolide scaffolds (Blanco et al. 2021).

Also, biodegradable scaffolds should be able to be remodeled in tune with regeneration and repair process. Porous sustainable scaffolds are used to provide suitable 3D environment for cell and tissue penetration and waste transportation and a surface media for cell functions. Porous engineered tissue should have high mechanical stability for surgical handling and ease of manufacturing and sterilization. The scaffolds are also able to carry biological signals such as growth factors, cytokines in a controlled manner and serve as drug delivery systems with high mechanical stability for surgical handling (Sultana 2018; Pooyan et al. 2018; Thangavel and Elsen Selvam 2022). The porous 3D nanofibrous scaffolds possess a rough surface, lose porous structure and excellent water absorption characteristics. Nanofibrous scaffolds are designed effectively to enhance cell adhesion, proliferation and migration compared to its 2D structures. The nanofibrous scaffolds are used to promote wound healing with effective cell adhesion, proliferation and migration behaviors compared to alternative 2D materials (Thangavel and Elsen Selvam 2022; Yu et al. 2019). Cloned renal cells are seeded in scaffolds consisted of three collages coated cylindrical silastic catheters forming three membranes connected to catheters used to create renal neo-organ with mechanism for collecting excreted urinary fluids.

Various fabrication techniques are classified into two categories, either conventional fabrication or rapid prototyping methods. Include electrospinning, freeze drying, gas foaming, solvent casting and leaching as well as thermally induced phase separation. While, rapid prototyping method includes selective laser sintering, stereolithography, solvent based extrusion, fused deposition modelling and bioprinting (Adolph et al. 2014; Breuer et al. 2023; Kamat et al. 2020). 3D printing plays a crucial role in various scaffolding base applications. Polymeric scaffolds have a critical role in the implantation and proliferation of cells and new 3D regenerated tissues for various organs. Natural polymeric scaffolds such as collagen, alginate, hyaluronic acid, chitosan and fibrin have been applied in tissues in recent years. Despite the several advantages of natural polymers, they endure high degradation rates and unstable mechanical performance as well as potential disease transmission and immunogenicity. Polymers such as PEG. PLA, PGA, PLGA and PCL are utilized as engineered tissue scaffolds and have shown highly controlled degradation rate. The ability of sustainable scaffolds to be developed into complex shapes improves the cell attachment and susceptibility to deliver soluble molecules. Synthetic polymers could be fabricated in large quantities at low cost and have high rate of durability (Chan and Leong 2008; Niu et al. 2023; Liu et al. 2023b; Huang et al. 2023).

The construction of scaffolds used in tissue engineering applications, material properties, porosity, surface area, morphology and other mechanical properties plays a vital role in influencing the cell behavior, ensures structural integrity during the healing process. Different types of natural and synthetic materials are used to fabricate the cartilage tissue scaffolds with various solubility, cross-linking and mechanical properties as shown in Table 2.1.

## 2.4 Future Prospective for Engineered Tissues Using Various Sustainable Scaffold Strategies

Despite the major advances in designing scaffolds to be used in tissue engineering technologies scaffold tissue engineering technologies, limitations such as the high number of cells required to produce 3D engineered tissues limits its applications. Engineered tissues such as temperature responsive surface derived single layer scaffolds and the electrochemical desorption for cell transfer requires high numbers of cells to produce 3D scaffolds. Future prospective include synthesis of emerging sustainable materials such as plant-based polymers, recycled bioplastics and biofabricated materials. Bioprinting and decellularized organs are used newly in scaffold construction. Engineered tissues are used in development of medical devices

**Table 2.1** Different scaffolds used for engineered cartilage tissues and their properties

| Scaffolding material | Advantage | Disadvantage | References |
|---|---|---|---|
| Chitosan | High biocompatibility and biodegradability<br>Similar molecular structure to Glycosaminoglycans (GAGs) | Low solubility and requires cross linking and low filtration rate due to lower pore size | Sultana (2018) |
| Collagen | Abundant material and supports chondrogenesis. Found abundant in cartilage tissue. | Rapid degradation and poor mechanical stability | Huang et al. (2023) and Tan et al. (2024) |
| Hyaluronic acid | Main glycosaminoglycans molecule in extracellular matrix | Negative charge leads to lower cell adhesion. Low mechanical strength and high-water adsorption rate | Vavouranakis et al. (2017) |
| Fibrin | It is a natural component that exist in intravascular cells and has spontaneous repair activities with abundant collage and sulfate GAGs. This material is common in biomedical and biomaterial synthesis due to its high biocompatibility and cell promotion, cell adhesion and high tissue repair | Mechanical weakness including low tensile strength and elasticity and high degradation under mechanical stress.<br>Fibrin also could trigger immune response and allergic reactions in some patients. | Vavouranakis et al. (2017), Adolph et al. (2014), and Jankauskaite et al. (2022) |
| Cellulose | This material has excellent mechanical strength superior to natural cartilage | Requires coating with other biopolymers as well as low biocompatibility | Zare and Shabani (2016), Breuer et al. (2023), Jana et al. (2022), and Jankauskaite et al. (2022) |
| Synthesized Polylactic acid (PLA) and Polylactic-co-glycolic acid (PLGA) | A major advantage of this biopolymer is its non-toxicity since it contains lactic acid and glycolic acid that are metabolized by the body and eventually eliminate. This eliminates the need for surgical removal of scaffolds after regeneration of tissues. | High acid degradation into lactic acid and glycolic acid that leads to local acidosis that affects surrounding tissues and cells through inhibition of cell growth. | Nga et al. (2023), Vavouranakis et al. (2017), Cordeiro et al. (2022), and Gentile et al. (2014) |

such as personalized environmentally friendly devices with minimal environmental impact. Also, to overcome the dimensionality limitations of scaffold structures, biomaterials combined with various scaffold conformations such as hydrogels and tissue grafts are used in 3D printing and electrospinning. Engineered tissues have also been used with urethral bladders, blood vessels and tracheal replacement and organ donor replacements which encourages future efforts for engineered organs. Engineered tissues have also been used for engineered bladders and used in growth of neural structures into neo bladders restored after implantation. Engineered tissues have been used in synthesis of small intestine and neuro organs. The orchestrated delivery of neurotrophic factors aids in musk nerve recovery and synaptic interface development with engineered tissues. Vascular development in engineered tissues is supplemented using growth factors like VEGF (vascular endothelial growth factors) and FGFs (fibroblast growth factors). The addition of antigenic compounds into the scaffolds like organ specific matrix from liver facilitates the regeneration of sinusoidal endothelial cells. Tissue engineering synthetic biomaterials designed to give signals formerly provided by the ECM suggests replacement of the native molecules. Novel scaffold materials' synthesis not only sustains large sophisticated organs, but also allows for post implantation performance evaluation. The expensive clinical trials that involve bioengineered products could benefit from alternative approaches, such as using bioengineered scaffolds. With the advancement of synthesis technology for bioengineered tissues, the expense reduces significantly.

The future of engineered tissues and advancement of sustainable scaffold structures using emerging biodegradable, bioresorbable and bioactive materials offer more efficient 3D bioprinting and smart scaffold structures with enhanced mechanical and biochemical properties that revolutionize tissue regeneration and usage of biomaterials such as plant derived polymers that improve sustainability without compromising the functionality. The development of self-healing stimulus responsive scaffolds is able to elevate tissue longevity and functionality. The future of scaffold technologies focuses on sustainability, biocompatibility and scalability creating novel opportunities for effective tissue engineering applications. These advancements contribute to significantly reduction of environmental impact and enhance engineered tissues.

## 2.5 Conclusion

To sum up, the approach to development of tissue engineering and regenerative medicine is greatly aided by chronic supportive biomaterials and scaffolds due to their critical role in frameworks for cell attachment, growth, matrix formation, and healing. The construction of the 3D porous and hydrogel scaffolds is of utmost importance together with parameters such as biomaterials pore size and pore interconnectivity which affects cell as well as tissue functioning. Overcoming the challenges pertaining to biocompatibility, biodegradability as well as promoting cell adhesion and proliferation is possible with a combination of natural and synthetic biomaterials. Scaffolds for diverse biomedical applications are manufactured, utilizing different fabrication methods like solvent casting, electrospinning, and 3D printing. These innovations tend to refine accomplishing organ renovation, and stimulate natural like processes such as regeneration of salivary glands, synthesis of bones, and skin tissues. Engineered tissues for various biomedical applications are made possible through the combination of sustainable materials and diverse fabrication methods. New engineered tissues are developed with the advancement of scaffolding strategies and fracturing techniques. The impact of advancement in various scaffold strategies and fabrication techniques are used to develop newly engineering tissues and scaffolds.

## References


Adolph EJ, Pollins AC, Cardwell NL, Davidson JM, Guelcher SA, Nanney LB (2014) Biodegradable lysine-derived polyurethane scaffolds promote healing in a porcine full-thickness excisional wound model. J Biomater Sci Polym Ed 25(17). https://doi.org/10.1080/09205063.2014.965997

Berdimurodov E et al (2023) Green electrospun nanofibers for biomedicine and biotechnology. https://doi.org/10.3390/technologies11050150

Blanco FG et al (2021) From residues to added-value bacterial biopolymers as nanomaterials for biomedical applications. https://doi.org/10.3390/nano11061492

Breuer T, Jimenez M, Humphrey JD, Shinoka T, Breuer CK (2023) Tissue engineering of vascular grafts: a case report from bench to bedside and back. https://doi.org/10.1161/ATVBAHA.122.318236

Brizuela C, Meza G, Khoury M (2024) Revolutionizing endodontics: innovative approaches for treating mature teeth with closed apices and apical lesions: a report of two cases. J Endodont 50(5). https://doi.org/10.1016/j.joen.2024.02.012

Carotenuto F et al (2022) From soft to hard biomimetic materials: tuning micro/nano-architecture of scaffolds for tissue regeneration. Micromachines (Basel) 13(5). https://doi.org/10.3390/mi13050780

Chan BP, Leong KW (2008) Scaffolding in tissue engineering: general approaches and tissue-specific considerations. Eur Spine J. https://doi.org/10.1007/s00586-008-0745-3

Chen H, Truckenmüller R, Van Blitterswijk C, Moroni L (2013) Fabrication of nanofibrous scaffolds for tissue engineering applications. In: Nanomaterials in tissue engineering: fabrication and applications. https://doi.org/10.1533/9780857097231.1.158

Contessi Negrini N, Angelova Volponi A, Higgins CA, Sharpe PT, Celiz AD (2021) Scaffold-based developmental tissue engineering strategies for ectodermal organ regeneration. https://doi.org/10.1016/j.mtbio.2021.100107

Cordeiro R, Henriques M, Silva JC, Antunes F, Alves N, Moura C (2022) Corncob cellulose scaffolds: a new sustainable temporary implant for cartilage replacement. J Funct Biomater 13(2). https://doi.org/10.3390/jfb13020063

Das A et al (2023) Nanoparticles in bone tissue engineering. In: Nanostructured materials for tissue engineering. https://doi.org/10.1016/B978-0-323-95134-0.00012-2

Farag MM (2023) Recent trends on biomaterials for tissue regeneration applications: review. https://doi.org/10.1007/s10853-022-08102-x

Fu JN et al (2022) Scaffold-based tissue engineering strategies for osteochondral repair. https://doi.org/10.3389/fbioe.2021.812383

Gao Y et al (2024) Injectable and photocurable gene scaffold facilitates efficient repair of spinal cord injury. ACS Appl Mater Interfaces 16(4). https://doi.org/10.1021/acsami.3c14902

Garbayo E et al (2021) Delivery of cardiovascular progenitors with biomimetic microcarriers reduces adverse ventricular remodeling in a rat model of chronic myocardial infarction. Acta Biomater 126. https://doi.org/10.1016/j.actbio.2021.03.017

Gentile P, Chiono V, Carmagnola I, Hatton PV (2014) An overview of poly(lactic-co-glycolic) acid (PLGA)-based biomaterials for bone tissue engineering. https://doi.org/10.3390/ijms15033640
Ghassemi T, Shahroodi A, Ebrahimzadeh MH, Mousavian A, Movaffagh J, Moradi A (2018) Current concepts in scaffolding for bone tissue engineering. https://doi.org/10.22038/abjs.2018.26340.1713
Godara P, McFarland CD, Nordon RE (2008) Design of bioreactors for mesenchymal stem cell tissue engineering. https://doi.org/10.1002/jctb.1918
Harris AF, Lacombe J, Zenhausern F (2021) The emerging role of decellularized plant-based scaffolds as a new biomaterial. https://doi.org/10.3390/ijms222212347
Huang WH et al (2023) Collagen for neural tissue engineering: materials, strategies, and challenges. https://doi.org/10.1016/j.mtbio.2023.100639
Hwang TI et al (2018) Facile fabrication of spongy nanofibrous scaffold for tissue engineering applications. Mater Lett 219. https://doi.org/10.1016/j.matlet.2018.02.040
Jana S et al (2022) Waste-derived biomaterials as building blocks in the biomedical field. J Mater Chem B 10(4). https://doi.org/10.1039/d1tb02125g
Jankauskaite L et al (2022) Functionalized electrospun scaffold–human-muscle-derived stem cell construct promotes in vivo neocartilage formation. Polymers (Basel) 14(12). https://doi.org/10.3390/polym14122498
Kamat P et al (2020) Adipose tissue and the vascularization of biomaterials: stem cells, microvascular fragments and nanofat—a review. https://doi.org/10.1016/j.jcyt.2020.03.433
Li WJ, Tuan RS (2009) Fabrication and application of nanofibrous scaffolds in tissue engineering. https://doi.org/10.1002/0471143030.cb2502s42
Li X et al (2021) Supramolecular polymer nanocomposites for biomedical applications. https://doi.org/10.3390/polym13040513
Liang W et al (2023) Recent advances in 3D printing of biodegradable metals for orthopaedic applications. https://doi.org/10.1186/s13036-023-00371-7
Liu S et al (2023a) Biomimetic natural biomaterials for tissue engineering and regenerative medicine: new biosynthesis methods, recent advances, and emerging applications. https://doi.org/10.1186/s40779-023-00448-w
Liu H et al (2023b) Recent advancement in vascularized tissue-engineered bone based on materials design and modification. https://doi.org/10.1016/j.mtbio.2023.100858
Loh QL, Choong C (2013) Three-dimensional scaffolds for tissue engineering applications: role of porosity and pore size. https://doi.org/10.1089/ten.teb.2012.0437
Misra SK et al (2017) 3D-printed multidrug-eluting stent from graphene-nanoplatelet-doped biodegradable polymer composite. Adv Healthc Mater 6(11). https://doi.org/10.1002/adhm.201700008
Mokhtarinia K, Rezvanian P, Masaeli E (2023) Sustainable hydrogel-based cell therapy. In: Sustainable hydrogels: synthesis, properties, and applications. https://doi.org/10.1016/B978-0-323-91753-7.00009-0
Nga NK, Huyen TTT, Dung TN (2023) Solvent casting-particulate leaching synthesis of a nano-SiO2/chitosan composite scaffold for potential use in bone tissue engineering. Vietnam J Chem 61(5). https://doi.org/10.1002/vjch.202300020
Niu X, Li N, Du Z, Li X (2023) Integrated gradient tissue-engineered osteochondral scaffolds: challenges, current efforts and future perspectives. https://doi.org/10.1016/j.bioactmat.2022.06.011
Nurchi C, Buonvino S, Arciero I, Melino S (2023) Sustainable vegetable oil-based biomaterials: synthesis and biomedical applications. https://doi.org/10.3390/ijms24032153
Park S et al (2024) Rice grains integrated with animal cells: a shortcut to a sustainable food system. Matter 7(3). https://doi.org/10.1016/j.matt.2024.01.015
Pooyan P, Brewster LP, Tannenbaum R, Garmestani H (2018) Biomimetic synthesis of two different types of renewable cellulosic nanomaterials for scaffolding in tissue engineering. Green Process Synth 7(3). https://doi.org/10.1515/gps-2016-0196

Pörtner R, Nagel-Heyer S, Goepfert C, Adamietz P, Meenen NM (2005) Bioreactor design for tissue engineering. https://doi.org/10.1263/jbb.100.235
Prem Ananth K, Jayram ND (2024) A comprehensive review of 3D printing techniques for biomaterial-based scaffold fabrication in bone tissue engineering. https://doi.org/10.1016/j.stlm.2023.100141
Pryadko A, Surmeneva MA, Surmenev RA (2021) Review of hybrid materials based on polyhydroxyalkanoates for tissue engineering applications. https://doi.org/10.3390/polym13111738
Purewal SS, Kaur A, Bangar SP, Singh P, Singh H (2024) Protein-based films and coatings: an innovative approach. https://doi.org/10.3390/coatings14010032
Ribeiro C, Sencadas V, Correia DM, Lanceros-Méndez S (2015) Piezoelectric polymers as biomaterials for tissue engineering applications. https://doi.org/10.1016/j.colsurfb.2015.08.043
Rodríguez-Vázquez M, Vega-Ruiz B, Ramos-Zúñiga R, Saldaña-Koppel DA, Quiñones-Olvera LF (2015) Chitosan and its potential use as a scaffold for tissue engineering in regenerative medicine. https://doi.org/10.1155/2015/821279
Sommer H, Juliá-Hernández F, Martin R, Marek I (2018) Walking metals for remote functionalization. ACS Cent Sci 4(2). https://doi.org/10.1021/acscentsci.8b00005
Srinivasan A, Murugesan S, Mahendran AR (2022) Progress in polymer research for biomedical, energy and specialty applications. https://doi.org/10.1201/9781003200710
Stie MB, Kalouta K, da Cunha CFB, Feroze HM, Vetri V, Foderà V (2022) Sustainable strategies for waterborne electrospinning of biocompatible nanofibers based on soy protein isolate. Sustain Mater Technol 34. https://doi.org/10.1016/j.susmat.2022.e00519
Suamte L, Tirkey A, Barman J, Jayasekhar Babu P (2023) Various manufacturing methods and ideal properties of scaffolds for tissue engineering applications. https://doi.org/10.1016/j.smmf.2022.100011
Subramanian A, Krishnan UM, Sethuraman S (2009) Development of biomaterial scaffold for nerve tissue engineering: biomaterial mediated neural regeneration. https://doi.org/10.1186/1423-0127-16-108
Sultana N (2018) Mechanical and biological properties of scaffold materials. In: Functional 3D tissue engineering scaffolds: materials, technologies, and applications. https://doi.org/10.1016/B978-0-08-100979-6.00001-X
Tan SH, Liu S, Teoh SH, Bonnard C, Leavesley D, Liang K (2024) A sustainable strategy for generating highly stable human skin equivalents based on fish collagen. Biomater Adv 158. https://doi.org/10.1016/j.bioadv.2024.213780
Tavakoli J (2017) Tissue engineering of the intervertebral disc's annulus fibrosus: a scaffold-based review study. https://doi.org/10.1007/s13770-017-0024-7
Tay A (2021) Scientific challenges and solutions for cultured meat manufacturing. https://doi.org/10.1016/j.jfutfo.2023.06.002
Thangavel M, Elsen Selvam R (2022) Review of physical, mechanical, and biological characteristics of 3D-printed bioceramic scaffolds for bone tissue engineering applications. https://doi.org/10.1021/acsbiomaterials.2c00793
Turnbull G et al (2018) 3D bioactive composite scaffolds for bone tissue engineering. https://doi.org/10.1016/j.bioactmat.2017.10.001
Vavouranakis E, Kalogeras K, Tousoulis D (2017) Novel developments in interventional strategies. In: Coronary artery disease: from biology to clinical practice. https://doi.org/10.1016/B978-0-12-811908-2.00022-2
Wasyłeczko M, Sikorska W, Chwojnowski A (2020) Review of synthetic and hybrid scaffolds in cartilage tissue engineering. https://doi.org/10.3390/membranes10110348
Yari D et al (2022) Biochemical aspects of scaffolds for cartilage tissue engineering; from basic science to regenerative medicine. https://doi.org/10.22038/ABJS.2022.55549.2766
Yousefiasl S, Sharifi E, Salahinejad E, Makvandi P, Irani S (2023) Bioactive 3D-printed chitosan-based scaffolds for personalized craniofacial bone tissue engineering. Eng Regen 4(1). https://doi.org/10.1016/j.engreg.2022.09.005

Yu H et al (2019) Novel porous three-dimensional nanofibrous scaffolds for accelerating wound healing. Chem Eng J 369. https://doi.org/10.1016/j.cej.2019.03.091

Zare Y, Shabani I (2016) Polymer/metal nanocomposites for biomedical applications. https://doi.org/10.1016/j.msec.2015.11.023

Zhang Q et al (2023) Review on strategies toward efficient piezocatalysis of BaTiO3 nanomaterials for wastewater treatment through harvesting vibration energy. https://doi.org/10.1016/j.nanoen.2023.108507

Zhu P et al (2024) 3D printed cellulose nanofiber aerogel scaffold with hierarchical porous structures for fast solar-driven atmospheric water harvesting. Adv Mater 36(1). https://doi.org/10.1002/adma.202306653

# Additive Manufacturing as an Advanced Sustainable Technique for Scaffold Production

3

Hossein Rayat Pisheh and Mahsa Sani

**Abstract**

Tissue engineering, as one of the most innovative fields in regenerative medicine, focuses on developing biological substitutes to repair and regenerate damaged tissues. Scaffolds, as three-dimensional matrices, play a pivotal role in this process. These porous structures provide a suitable environment for cell adhesion, proliferation, and differentiation, thereby facilitating tissue regeneration. In recent decades, significant advancements in additive manufacturing technology have revolutionized the design and fabrication of scaffolds. Additive manufacturing, with its ability to create complex and customized structures, has paved the way for producing scaffolds with unique properties. This technology, by enabling precise control over the micro and nano structures of scaffolds, allows for the mimicry of natural tissue environments, consequently significantly enhancing the efficacy of these scaffolds.

**Keywords**

Additive manufacturing · Scaffold · Tissue engineering · Biomaterials · 3D printing

H. R. Pisheh
Department of Tissue Engineering and Applied Cell Sciences, School of Advanced Medical Sciences and Technologies, Shiraz University of Medical Sciences, Shiraz, Iran
e-mail: h_rayatpisheh@sums.ac.ir

M. Sani (✉)
Department of Tissue Engineering and Applied Cell Sciences, School of Advanced Medical Sciences and Technologies, Shiraz University of Medical Sciences, Shiraz, Iran

Shiraz Institute for Stem Cell & Regenerative Medicine, Shiraz University of Medical Sciences, Shiraz, Iran
e-mail: sani_mahsa@sums.ac.ir



J. M. Oliveira et al. (eds.), *Sustainable Scaffolds-based Strategies in Tissue Engineering and Regenerative Medicine*, Biomaterials, Bioengineering and Sustainability 8, https://doi.org/10.1007/978-3-031-96274-5_3

## 3.1 Introduction

The landscape of manufacturing is constantly evolving. Additive manufacturing (AM), as an emerging technology, has revolutionized the field of tissue engineering. This method, with its ability to create complex and customized structures, has paved the way for producing scaffolds with unique properties. Scaffolds, serving as three-dimensional (3D) matrices, play a pivotal role in tissue engineering, providing a suitable environment for cell adhesion, proliferation, and differentiation. 3D printing, as an emerging technology with immense potential for societal and environmental transformation, has emerged as the optimal choice for low-volume production, frequent design changes, and highly complex designs. Over time, this technology has evolved to become more user-friendly, accurate, and capable of creating more intricate models (Mikaeeli Kangarshahi et al. 2024). With its innovative approach, 3D printing is gradually replacing traditional manufacturing methods due to its greater flexibility. Current technologies require materials with a unique combination of properties that surpass traditional materials. The development of such composites has replaced traditional materials. 3D printing is also known as AM. This technology is rapidly growing in the manufacturing sector. The significant reduction in the cost of 3D printers has increased market demand. 3D printing produces components using 3D computer data containing the geometric information of the object. This technology is the best option for low-volume production, frequent design changes, and highly complex designs. AM methods typically utilize computer data and STL files containing the geometry of the object (Vasvari et al. 2021).

AM and traditional manufacturing are two distinct approaches to fabrication and production, each with its own advantages and limitations. From the perspective of tissue engineering, which aims to construct scaffolds and biological structures for repairing and regenerating damaged tissues, the differences between these two methods become particularly significant. One key difference between the two methods is design flexibility. AM allows tissue engineers to create highly complex and customized structures with great precision. In contrast, traditional manufacturing, due to the limitations of molds and tools, does not readily facilitate the creation of such complex structures. This design flexibility is crucial for tissue engineering, which requires the fabrication of scaffolds with precise geometries that match various tissues. Another difference lies in the types of materials that can be used (Shashikumar et al. 2024). A variety of materials can be used in the AM process. This allows tissue engineers to select materials with mechanical and biological properties suitable for the intended application. In contrast, traditional manufacturing is often limited to specific materials. Production speed is another notable difference. Due to its layer-by-layer nature, AM can accelerate the production process and reduce the time required to fabricate scaffolds. This is particularly important in tissue engineering research and development, where there is a need to rapidly produce prototypes for testing and evaluation. Production cost is another difference that must be considered. While the initial cost of additive manufacturing equipment may

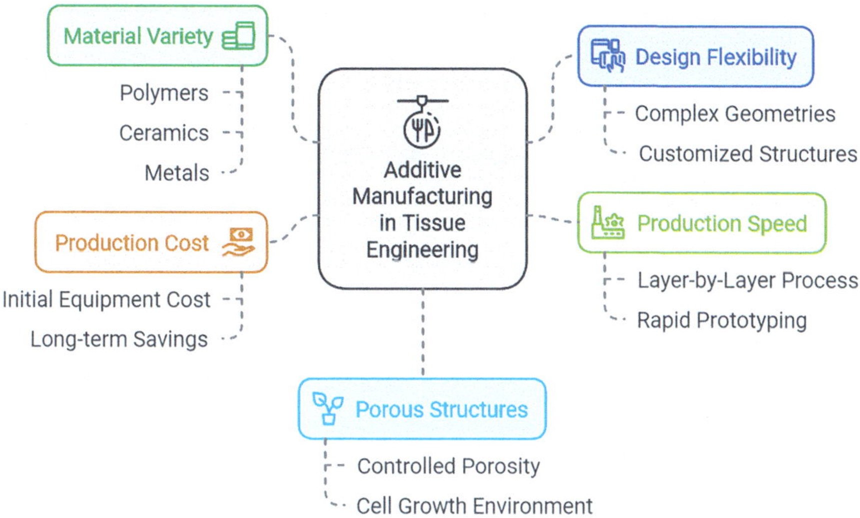


**Fig. 3.1** Advantage of additive manufacturing in tissue engineering

be high, it can reduce costs in the long run due to decreased material consumption and increased production speed. Additionally, the ability to produce large numbers of small, customized parts without the need for additional tooling reduces production costs. AM, due to its ability to create porous structures with controlled porosity, enables the creation of an environment similar to natural tissue. This feature is crucial for cell growth, proliferation, and the formation of new tissue (Fig. 3.1) (Deja et al. 2021; Isaacson et al. 2022).

Herein, we will conduct a comprehensive review of AM as an advanced and sustainable technique for producing scaffolds. First, we will examine the various AM methods used in scaffold fabrication, followed by the suitable materials for this purpose and the parameters that affect the quality of the produced scaffold. Finally, we will address the existing challenges and the future outlook of this technology.

## 3.2 3D Bioprinting Methods

3D bioprinting is a novel technology inspired by conventional 3D printers that enables the creation of living tissues and artificial organs. In this method, living cells, biomaterials, and growth factors are printed layer by layer onto a 3D scaffold to create complex tissue-like structures. This technology promises a significant revolution in the field of regenerative medicine and tissue engineering. There are various types of 3D bioprinting (Fig. 3.2), each suitable for specific applications, which will be discussed further in Table 3.1. One of the main advantages of 3D bioprinting is the ability to create customized tissues (Liu et al. 2023a). Using this technology, tissues with the desired structure and function can be produced individually for each

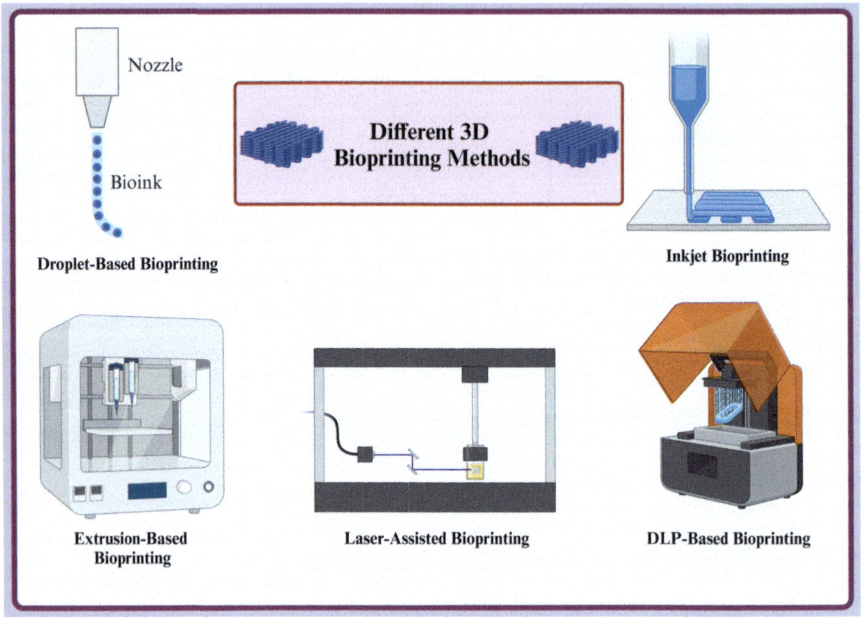


**Fig. 3.2** The main methods of 3D bioprinting

patient. This is particularly promising for the treatment of diseases such as diabetes, heart disease, and spinal cord injuries. In addition, 3D bioprinting enables the creation of accurate models of patient tissues, which is very useful for developing new drugs and testing new treatments.

### 3.2.1 Droplet-Based Bioprinting (DBB)

Droplet-based bioprinting (DBB) is a widely used method for producing three-dimensional biological structures. It has gained significant attention from researchers due to its high precision, speed, and adaptability. In this method, very small droplets containing biomaterials, cells, and growth factors are deposited with high precision and control onto a substrate. These droplets can have dimensions of about one picoliter and are layered on top of each other to create the desired 3D structure. One of the main advantages of droplet-based bioprinting is the high precision in droplet placement, resulting in the creation of high-resolution structures. This feature is possible due to the lack of direct nozzle contact with the substrate and the use of very small droplets. Additionally, this method has a high printing speed and can create complex structures in a short time. Furthermore, droplet-based bioprinting is highly adaptable and can be used to print a wide range of biomaterials with different viscosities (Arjoca et al. 2024).

**Table 3.1** Comparison of different 3D bioprinting methods

| Bioprinting technique | Working mechanism | Advantages | Disadvantages |
|---|---|---|---|
| Drop-on-Demand (DoD) bioprinting | Dispensing small droplets of bioink | High precision, moderate speed, compatibility with various bioinks | Limited to low-viscosity bioinks, high cost, nozzle clogging |
| Inkjet bioprinting | Similar to DoD, using inkjet technology for droplet formation | High precision, high speed, compatibility with various bioinks | Limited to low-viscosity bioinks, high cost, nozzle clogging |
| Stereolithography (SLA) bioprinting | Polymerization of resin using a focused laser beam | Very high precision, high speed, complex structures | High equipment cost, limited biomaterial choices, requires photopolymerizable resins |
| Digital Light Processing (DLP) bioprinting | Polymerization of resin using a digital projector to create patterns | High precision, high speed, complex structures, simultaneous build of multiple parts | High equipment cost, limited biomaterial choices, requires photopolymerizable resins |
| Extrusion-based bioprinting | Extruding bioink through a nozzle under pressure | High material flexibility, complex structures, relatively low equipment cost | Low speed, lower precision compared to some methods, nozzle clogging |
| Acoustic bioprinting | Using sound waves to generate and deposit droplets | High precision, compatibility with various bioinks, non-contact nozzle | Requires specialized equipment, limited build size |
| Magnetic 3D bioprinting (M3DB) | Manipulating cells and biomaterials using magnetic fields | Precise spatial control, complex structures without scaffolds | Requires magnetic materials, limited build size |

However, droplet-based bioprinting also faces some limitations. One of these limitations is the need for greater droplet compatibility. Droplets must be designed to adhere to each other quickly after deposition on the substrate and form a cohesive structure. Additionally, the structural and mechanical integrity of printed products in this method is often lower than other methods. This is due to the droplet-based nature of the process and the lack of strong bonds between droplets. Moreover, controlling porosity in printed products is challenging with this method. When multiple types of ink are printed simultaneously, the size and distribution of the voids created in the final structure are unpredictable (Brunel et al. 2024; Ng et al. 2022).

### 3.2.2 Inkjet Bioprinting

Inkjet bioprinting is one of the most common and well-known droplet-based bioprinting methods. This method has been developed based on the technology of conventional two-dimensional inkjet printers and therefore benefits from some of the advantages of this technology. In this method, a mixture of biomaterials (such as hydrogels) and cells, called bioink, is placed in a special cartridge. Then, using a thermal or piezoelectric mechanism, very small droplets of this ink are deposited with high precision and control onto a substrate. One of the most important advantages of inkjet bioprinting is its low cost. Due to its structural similarity to commercial inkjet printers, the production and maintenance costs of this type of printer are lower than other bioprinting methods. Also, high printing speed is another prominent feature of this method. Using multiple printheads, a large number of droplets can be deposited on the substrate per unit time, thus reducing the production time of complex structures. Additionally, studies have shown that inkjet bioprinting has high cell viability. Typically, more than 80% of the printed cells remain alive after the printing process (Azizi Machekposhti et al. 2020; Waasdorp et al. 2018).

Despite the mentioned advantages, inkjet bioprinting also faces some limitations. One of these limitations is changes in ink viscosity during the printing process. As the volume of ink in the cartridge decreases, its viscosity may increase, resulting in changes in the size and shape of the droplets. This can affect the accuracy and uniformity of the printed structure. Additionally, precise control over the distribution of cells within the structure is another challenge of this method. Cells may aggregate or separate during the printing process, which affects the function of the engineered tissue (Zheng et al. 2024).

### 3.2.3 Laser-Assisted Bioprinting

Laser-assisted bioprinting is a novel and precise method for producing three-dimensional biological structures. In this method, a high-precision laser beam is used to induce physical changes in biomaterials to form the desired structures. This method is generally performed in two ways: direct laser writing and laser-induced forward transfer (LIFT). In direct laser writing, the laser beam is directly irradiated onto the biomaterial layer, causing local changes and the formation of the structure. In LIFT, the laser is irradiated onto an energy-absorbing layer, and the resulting energy creates bubbles that propel the biomaterial upward and onto the underlying substrate (Chang and Sun 2023; Grosfeld et al. 2022).

One of the main advantages of laser-assisted bioprinting is its very high precision in creating micro and nanometer structures. Additionally, since this method does not use a nozzle, it allows the use of bioinks with high viscosity and density. This feature is particularly important for printing cells while maintaining their viability. Furthermore, laser-assisted bioprinting enables the precise arrangement of cells and the creation of complex structures (Wu et al. 2023). However, laser-assisted bioprinting also faces some challenges. One of these challenges is the low speed of this method. Due to the need for rapid gelation of the bioink to maintain the shape of the structure, the printing speed in this method is lower than other methods. Additionally, precise control of laser parameters to achieve the desired results is very important and requires expertise and specific equipment. Moreover, the high cost of laser equipment is another limitation of this method (Wang et al. 2023).

### 3.2.4 Extrusion-Based Bioprinting

Extrusion-based bioprinting is one of the most widely used methods for producing three-dimensional biological structures. Due to its simplicity, affordability, and ability to produce large structures, it has received widespread attention. In this method, biomaterials (such as cells, hydrogels, and growth factors) are deposited layer by layer onto a substrate in the form of a paste or gel through a very small (micro) diameter nozzle. This process is very similar to the extrusion of polymeric materials, except that biomaterials are used here. The operating mechanism in extrusion-based bioprinting involves applying mechanical or pneumatic pressure to the biomaterial inside the cartridge, causing the material to exit the nozzle in the form of continuous filaments and deposit on the substrate. The precise movement of the nozzle in three spatial directions and precise control of the output flow rate enable the creation of three-dimensional structures with high complexity (Arjoca et al. 2024).

Despite being inexpensive and affordable, this method enables the use of various types of biomaterials with high viscosity, which is very important for creating scaffolds with suitable mechanical properties. Also, the cells placed in the biomaterial are well preserved due to the mild printing conditions and lack of direct contact with hot surfaces or corrosive chemicals. However, this method has limitations such as lower precision compared to other methods, difficulty in printing very small structures, and challenges in creating uniform porosity (Putra et al. 2021; Zhang et al. 2020).

### 3.2.5 Acoustic Bioprinting

Acoustic bioprinting is a novel technology that utilizes sound waves to enable the creation of three-dimensional biological structures with high precision and speed. In this method, cells and biomaterials are encapsulated in very small droplets within a liquid medium and are then deposited layer by layer onto each other using sound waves to form the desired three-dimensional structure. One of the distinguishing features of acoustic bioprinting is the elimination of the need for a nozzle. In traditional bioprinting methods, the use of a nozzle often leads to clogging and damage to cells (Castro et al. 2018; Jentsch et al. 2021). However, in acoustic bioprinting, cells are non-contactly ejected from the liquid medium using sound waves and deposited onto the substrate. This protects cells from damage caused by heat, pressure, and shear stresses, significantly increasing their viability (Ning et al. 2021).

The mechanism of acoustic bioprinting works by a piezoelectric substrate (such as quartz or lithium niobate) generating surface acoustic waves through vibrations. These sound waves induce vibrational forces in the liquid medium, resulting in the detachment of very small droplets of bioink and their ejection upwards. To adjust the speed and size of the droplets, the frequency and amplitude of the sound waves must be controlled and changed. The advantages of acoustic bioprinting include the ability to print three-dimensional structures with very high resolution and precise control over cell positioning, the ability to use various types of cells and biomaterials, high printing speed, and no cell damage due to direct contact with the nozzle (Sabzevari et al. 2023).

Despite all the advantages of acoustic bioprinting, this technology is still in the early stages of development and faces some challenges. These challenges include the complexity of printing systems, limitations in the types of biomaterials that can be used, and the need to optimize printing parameters for each type of cell and biomaterial. However, with recent advances in tissue engineering and 3D printing technologies, it is expected that acoustic bioprinting will be used in the near future as a powerful tool for constructing engineered tissues and treating diseases (Amukarimi et al. 2022; Rahimnejad et al. 2021).

### 3.2.6 Magnetic Bioprinting

Magnetic 3D bioprinting (M3DB) is an innovative technique in the field of tissue engineering that enables the creation of complex and customized biological structures using magnetic forces. This method allows researchers to organize cells and biomaterials with high precision and control in a three-dimensional structure. One of the main advantages of magnetic bioprinting is the precise spatial control over cells and biomaterials. In this method, magnetic forces are used to manipulate and guide cells to specific locations. This enables the creation of complex structures with precise spatial patterns (Wu et al. 2023; Szychlinska et al. 2022). Additionally, magnetic bioprinting allows researchers to facilitate the synthesis of endogenous extracellular matrix without the need for artificial protein substrates. This feature allows the created biological structures to provide a more natural environment for cell growth and proliferation. In magnetic bioprinting, two main methods are used to manipulate cells. In the first method, the diamagnetic phenomenon is used. In this method, the cell environment is combined with a paramagnetic buffer and then exposed to a magnetic field. This magnetic field causes the cells to form aggregates that can be easily manipulated. In the second method, cells are coated with magnetic nanoparticles such as magnetic iron oxide. These nanoparticles allow cells to respond to the magnetic field and be manipulated three-dimensionally. After creating cell aggregates, these aggregates are placed next to each other in a specific order using gentle magnetic forces generated by a magnetic plate. The spatial pattern of these aggregates can be easily adjusted by changing the shape of the magnetic plate. Despite all the advantages of magnetic bioprinting, this technology still faces some challenges, including the need to develop more biocompatible magnetic materials and more precise control of magnetic forces (Gahl and Kunze 2018; Hu et al. 2023; Liu et al. 2023b).

### 3.2.7 DLP-Based Bioprinting

Digital Light Processing (DLP)-based bioprinting is a novel and widely used method for producing three-dimensional biological structures. This method, utilizing ultraviolet (UV) light and light-sensitive resins, enables the creation of highly precise and complex structures at high speed (Sultan et al. 2022). In DLP bioprinting, a thin layer of liquid resin containing cells and biomaterials is deposited on a substrate. Then, a digital projector projects the desired pattern onto the resin. The parts of the resin exposed to UV light polymerize and solidify. By repeating this process for subsequent layers, the desired 3D structure is built. Due to the use of UV light and light-sensitive resins, this method is capable of producing structures with very high resolution and fine details (Rayat Pisheh et al. 2024a, b). In that method, the resin polymerization process occurs quickly, leading to increased printing speed. Moreover, the ability to use various types of light-sensitive resins and cells makes this method very flexible. Despite the mentioned advantages, this method also has limitations. The resin viscosity must be such that it can be spread as a uniform thin layer on the substrate. Polymerized resins may have low oxygen permeability, which can affect cell viability (Garciamendez-Mijares et al. 2024).

## 3.3 Biopolymers Used in Scaffold Production

Biopolymers are derived from renewable biological sources such as animals, plants, and microorganisms, possessing tunable biocompatibility, chemical adaptability, non-toxicity, bioabsorbability, bioactivity, and biodegradability. The utilization of these stable materials in medical applications is steadily growing. This includes areas such as bone, heart, and liver regeneration, wound healing, and drug delivery systems, driven by the potential for more precise and effective treatments (Das et al. 2023; Nath et al. 2024; Rekhi et al. 2022). The functional properties of additively manufactured tissues and organs are significantly influenced by the biocompatibility of the constituent biomaterials. Furthermore, successful integration of these constructs relies on the adhesion and viability of natural cells, which is crucial for maintaining proper tissue function and intercellular interaction (Liang et al. 2024). A key challenge in the development of synthetic biopolymers for biomedical applications is mitigating the potential for inflammatory responses. Addressing the safety and efficacy concerns associated with these materials is crucial for their successful implementation. Chemical modifications offer a route to synthesizing composite scaffolds with tailored characteristics (Gherasim et al. 2023). Natural living organisms, including algae, fungi, and bacteria, degrade biopolymers into small molecules through anaerobic or aerobic techniques, forming organic products $H_2O$ and $CO_2$ (Premnath et al. 2021). In addition, these materials exhibit very compatible behavior due to their similarity to the extracellular matrix (ECM). The ECM is a complex network of interconnected tissue layers, primarily composed of structural proteins and adhesive polysaccharides. These components work together to provide structural support and facilitate cell adhesion and communication. The ECM not only provides a framework for tissues but also plays a critical role in regulating essential cellular activities such as adhesion, interaction, proliferation, and differentiation, all of which are crucial for tissue development and homeostasis (Mitchell and Lo 2023).

The use of biopolymers in the production of biological scaffolds enables the design and fabrication of scaffolds with tunable mechanical and biological properties. However, synthetic biopolymers often face challenges such as inducing inflammatory responses. To overcome this limitation, researchers have turned to the synthesis of composite scaffolds through chemical modifications. This approach enables the combination of the desired properties of natural and synthetic biopolymers, resulting in scaffolds with optimal performance. One of the significant advantages of biopolymers is their biodegradability. Living organisms, including microorganisms, can degrade biopolymers into simpler components. This feature prevents the accumulation of waste material in the body after scaffold implantation, thereby increasing the biocompatibility of the scaffold. In addition, the structural similarity of biopolymers to the ECM leads to strong interactions between the scaffold and cells, thereby accelerating the tissue repair process (Capuana et al. 2022; Reddy et al. 2021; Shokrani et al. 2022).

### 3.3.1 Natural Biopolymers

Nature provides a rich source of biodegradable polymers, such as polysaccharides and proteins, which are increasingly utilized in biomedical applications. These materials, sourced from plants, animals, or microbes, offer unique advantages for applications ranging from tissue engineering to controlled drug release. Natural biopolymers, such as polysaccharides (cellulose, chitin, alginate), proteins (collagen, fibroin, gelatin), and polypeptides, provide a suitable environment for cell growth and proliferation due to their structural and functional similarities to the ECM (Alsalhi 2024; Dabholkar et al. 2021). Scaffolds made from these materials, by mimicking the three-dimensional structure of the ECM, provide the necessary biological signals to guide cellular processes and, consequently, play a significant role in repairing damaged tissues and engineering new tissues. In general, natural biopolymers have a high molecular weight, resulting in viscous polymer solutions that enable their use in 3D printing. Consequently, the processability and printability of these polymers remain challenging (Nashchekina et al. 2024). Chemical modification strategies can be employed to enhance the properties of certain biopolymers, leading to improvements in non-toxicity, biocompatibility, and biodegradability. Some natural biopolymers include chitosan, silk fibroin (SF), collagen, cellulose, gelatin, hemicellulose, alginate, hyaluronic acid (HA), lignin, and starch. While natural biopolymers offer attractive biological properties, including excellent bioactivity, biocompatibility, and biodegradability, they often exhibit limitations such as suboptimal mechanical strength, high water solubility, inconsistent sourcing, potential immunogenicity, and susceptibility to denaturation during processing (Table 3.2) (Cao et al. 2021; Sergi et al. 2020; Sreena and Nathanael 2023).

Various natural biopolymers are used to produce 3D printed scaffolds. Some of the most important of these biopolymers include:

- **Polysaccharides:** Cellulose, chitin, alginate, hyaluronan, and agar are among the most commonly used polysaccharides in this field. These materials, due to their hydrophilic properties, biocompatibility, and ability to form gels, are used as base materials for the construction of 3D scaffolds. Cellulose and chitin are two natural polysaccharides that have received attention in the production of

**Table 3.2** Comparison table of synthetic and natural biomaterials in bio 3D printing and characteristics of bioinks

| Biomaterial | Crosslinking method | Suitable cell types | Target tissue | Advantages | Disadvantages | Additional notes |
|---|---|---|---|---|---|---|
| Natural biomaterials | | | | | | |
| Hyaluronic acid (HA) | Physical (temperature, pH), chemical (crosslinking) | Various, especially skin and cartilage cells | Skin, cartilage, connective tissue | High biocompatibility, lubricity, promotes tissue repair | High cost | Major component of extracellular matrix. |
| Alginate | Physical (calcium ions), chemical (crosslinking) | Various | Soft tissue, tissue engineering | High biocompatibility, rapid gel formation, biodegradability | Limited mechanical properties, sensitive to calcium ions | Widely used for cell encapsulation. |
| Collagen | Physical (temperature, pH), chemical (crosslinking) | Various, especially fibroblasts and epithelial cells | Skin, bone, cartilage | High biocompatibility, mimics ECM | High cost, low stability | Various types of collagen (I, II, III) exist with different properties. |
| Silk fibroin | Physical (temperature), chemical (crosslinking) | Various, especially neural cells | Skin, soft tissue, neural engineering | High strength, biocompatibility, biodegradability | Difficult to process | Excellent mechanical properties and biocompatibility. |
| Synthetic biomaterials | | | | | | |
| Polylactic acid (PLA) | Chemical (crosslinking) | Various | Bone, cartilage, soft tissue | Good biocompatibility, tunable mechanical properties, good printability | Slow degradation rate | Can be used pure or in combination with other polymers. |

(continued)

**Table 3.2** (continued)

| Biomaterial | Crosslinking method | Suitable cell types | Target tissue | Advantages | Disadvantages | Additional notes |
|---|---|---|---|---|---|---|
| Polyglycolic acid (PGA) | Chemical (crosslinking) | Various | Bone, cartilage | Good biocompatibility, high degradation rate | Brittleness | Often combined with PLA to improve mechanical properties. |
| Polycaprolactone (PCL) | Chemical (crosslinking) | Various | Soft tissue, tissue engineering | High flexibility, good biocompatibility, good printability | Slow degradation rate | Can be used as a drug carrier. |
| Polyethylene glycol (PEG) | Chemical (crosslinking) | Various | Soft tissue, tissue engineering | High biocompatibility, hydrophilicity, ability to be chemically modified | Weak mechanical properties | Often used as a crosslinking agent for other polymers. |
| Polyethylene fumarate (PEF) | Chemical (crosslinking) | Various | Soft tissue, tissue engineering | Good biocompatibility, high mechanical properties, biodegradability | High cost | Produced from renewable resources. |

biological scaffolds due to their unique properties. As the most abundant polysaccharide on Earth, cellulose forms the primary structural component of plant cell walls, providing rigidity and support. This polysaccharide has a linear and crystalline structure that gives it high strength and resistance. Due to its biocompatibility, biodegradability, and good mechanical properties, cellulose is used as a reinforcing material in the production of composite scaffolds (Murizan et al. 2020). As the second most prevalent polysaccharide on Earth, chitin plays a crucial role in providing structural support to the exoskeletons of arthropods, a diverse group that includes insects and crustaceans. Chitosan, a derivative of chitin, is widely used in medicine due to its antibacterial and biocompatible properties. Both chitin and chitosan can be used as self-supporting scaffolds or as reinforcements for other biopolymers. Alginate is a natural polysaccharide extracted from brown algae. This polysaccharide has strong gel-forming properties and easily forms gels with calcium ions (Kulka and Sionkowska 2023; Wijesekara and Xu 2024). Alginate gels are used as base materials for the construction of 3D scaffolds due to their biocompatibility, biodegradability, and ease of manipulation. In addition, alginate can be used as a carrier for cells and growth factors (Jahani et al. 2024). Hyaluronan is a natural polysaccharide widely found in connective tissues. This polysaccharide has high lubricating and water-absorbing properties and acts as a major component of the extracellular matrix. Hyaluronan is widely used in the production of biological scaffolds due to its biocompatibility, biodegradability, and ability to promote tissue repair. Hyaluronan can be used as a self-supporting scaffold or as a reinforcement for other biopolymers. In addition, hyaluronan can be used as a carrier for drugs and genes (Abdou et al. 2024).

- **Proteins:** Collagen, silk fibroin, and gelatin are among the proteins that, due to their structure similar to the ECM, are widely used in the production of biological scaffolds. These proteins can be used as self-supporting scaffolds or as reinforcements for other biopolymers. Collagen is the most abundant protein in vertebrates and plays a key role in the structure of skin, bones, tendons, and cartilage. Due to its high biocompatibility, good tensile strength, and ability to facilitate cell adhesion and proliferation, collagen is one of the most popular biomaterials used in tissue engineering. Collagen scaffolds can be produced in the form of hydrogels, sponges, or fibers and can be used for the repair of soft and hard tissues (Han et al. 2024; Mneimneh and Mehanna 2021). Silk fibroin is the main protein that makes up silk. This protein, due to its high strength, biocompatibility, biodegradability, and tunable mechanical properties, has been introduced as an attractive option for the production of biological scaffolds. Silk fibroin can be produced in the form of fibers, films, and hydrogels and is used for the repair of soft tissues such as skin and nerves (Lyu et al. 2023).
- **Polyhydroxyalkanoates (PHAs):** PHAs are natural polyesters produced by some bacteria. These materials, due to their good mechanical properties, biodegradability, and biocompatibility, have been considered as alternatives to synthetic polymers in the production of biological scaffolds (de Melo et al. 2023; Duran-Lara et al. 2024).

### 3.3.2 Synthetic Biopolymers

Synthetic biopolymers are polymers that are artificially designed and synthesized to possess properties similar to natural biopolymers. These materials have gained attention as alternatives to natural biopolymers in tissue engineering due to their tunable properties, mass production capabilities, and potential for chemical modification. One of the important applications of synthetic biopolymers is the production of 3D printed scaffolds, which play a significant role in repairing damaged tissues (Capuana et al. 2022; Randhawa et al. 2023). Various synthetic biopolymers are used to produce 3D printed scaffolds. Some of the most important of these biopolymers include:

- **Polylactic Acid (PLA):** Renewable resources, including corn starch and sugarcane, serve as the feedstock for the production of PLA, a biodegradable polymer. Due to its high biocompatibility, good mechanical properties, and 3D printability, PLA is one of the most popular materials used in the production of biological scaffolds. In 3D bioprinting, PLA is used to create 3D scaffolds with complex structures that can serve as replacements for damaged tissues. These scaffolds provide a suitable environment for cells to grow and proliferate, ultimately aiding in tissue repair. PLA, due to its good mechanical properties, biocompatibility, and 3D printability, is widely used in the production of biological scaffolds (Singhvi et al. 2019; Vach Agocsova et al. 2023).
- **Polyglycolic Acid (PGA):** PGA is a biodegradable linear polyester composed of repeating units of glycolic acid. This polymer, due to its high degradation rate, biocompatibility, and tunable mechanical properties, is widely used in tissue engineering and especially in the production of 3D printed scaffolds. PGA has a higher water absorption and degradation rate compared to polylactic acid (PLA), making it suitable for applications that require faster scaffold degradation. However, the high brittleness of PGA is one of its limitations, which is often overcome by combining it with other polymers such as PLA to improve mechanical properties. PGA is commonly combined with PLA to adjust the mechanical properties and degradation rate of the scaffold (Sitthisang et al. 2024).
- **Polycaprolactone (PCL):** PCL is a biodegradable polyester that, due to its good mechanical properties, biocompatibility, and high flexibility, is widely used in tissue engineering and especially in the production of 3D printed scaffolds. PCL has a relatively low melting point, which facilitates the 3D printing process (Feng et al. 2020). Also, this polymer has a slower degradation rate compared to polyglycolic acid (PGA), making it suitable for applications that require longer-lasting scaffolds. Other advantages of PCL include the ability to combine with other polymers and biomaterials to improve scaffold properties (Gharibshahian et al. 2023).
- **Polyethylene Glycol (PEG):** PEG is a synthetic polymer characterized by repeating units of ethylene glycol, giving it a variety of properties depending on its molecular weight. This polymer, due to its unique properties such as biocompatibility, high hydrophilicity, flexibility, and ability to undergo chemical modification, is widely used in various fields including pharmaceuticals and tissue engineering.

In 3D bioprinting, PEG is used as an important component in the formulation of bioinks. PEG, due to its hydrophilic properties, helps create a moist environment for cells and prevents them from drying out. Also, PEG can be used as a cross-linking agent to create polymeric networks with tunable mechanical properties (Shin et al. 2021; Zermeno-Perez et al. 2024; Rayat Pisheh et al. 2024c).

The use of synthetic biopolymers in the production of 3D printed scaffolds has several advantages (Table 3.2) (Rahimnejad et al. 2021; Leonovich et al. 2023):

- The mechanical, chemical, and biological properties of synthetic biopolymers can be easily adjusted by changing their structure and composition.
- Synthetic biopolymers can be produced in large quantities and at lower costs compared to natural biopolymers.
- Many synthetic biopolymers are resistant to various environments and maintain their properties.
- The surface of synthetic biopolymers can be modified with various functional groups to improve cell adhesion and biological properties of the scaffold.

Despite the numerous advantages mentioned, the use of synthetic biopolymers in 3D printing also faces some challenges (Table 3.2) (Rahimnejad et al. 2021; Leonovich et al. 2023):

- Some synthetic biopolymers may induce an immune response in the body.
- Some synthetic biopolymers may not have sufficient mechanical properties for specific applications.
- The production of some synthetic biopolymers with high purity is costly.

To overcome these challenges, researchers are working to develop new synthetic biopolymers with higher biocompatibility and better mechanical properties. Also, the use of surface modification methods and the combination of synthetic biopolymers with natural biopolymers can help improve the properties of biological scaffolds.

## 3.4 Future Perspective

Despite significant advancements in 3D bioprinting, this technology still faces numerous challenges that must be overcome for widespread commercialization. One of the primary challenges is developing bioinks with precise and predictable properties. Bioinks must possess the right viscosity, adhesiveness, and biocompatibility to enable the printing of complex and stable structures. Moreover, maintaining cell viability and function during and after the printing process is a major hurdle. Additionally, the high cost of equipment and materials, the complexity of the printing process, and regulatory hurdles are among the obstacles hindering the commercialization of this technology.

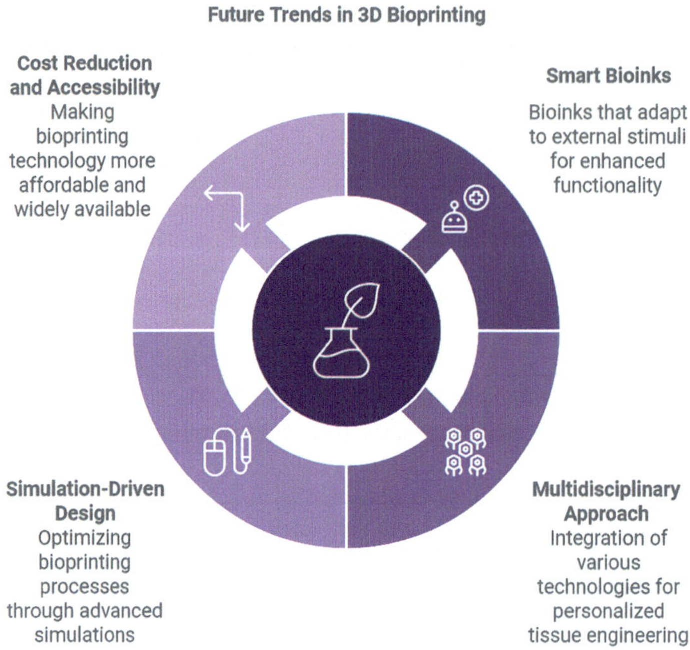


**Fig. 3.3** Future trends in 3D bioprinting

However, the future outlook for 3D bioprinting is very promising (Fig. 3.3). Recent advancements in tissue engineering, materials science, and printing technologies have made it possible to produce bio-scaffolds with extremely high complexity and precision. In the future, we can expect 3D bioprinting to become a standard method for producing engineered tissues and artificial organs. Some of the most important future trends in this field include:

- Future bioinks will be able to respond to external stimuli and dynamically change their properties.
- Combining 3D bioprinting with other technologies such as genetic engineering and nanotechnology will enable the production of complex and personalized tissues.
- By using simulation models, the 3D bioprinting process can be optimized, and the quality of the final products can be improved.
- With technological advancements and increased production, the costs of producing bio-scaffolds will decrease, making this technology more accessible to patients.

## 3.5 Conclusion

3D bioprinting, an emerging technology, holds immense potential to revolutionize the fields of medicine and tissue engineering. By enabling the creation of intricate, personalized biological scaffolds, this technology paves the way for engineered tissues and artificial organs. Despite considerable progress, several key challenges remain. Developing bioinks with precise and predictable properties, ensuring cell viability, reducing costs, and simplifying the printing process remain pivotal hurdles. Nevertheless, the future of 3D bioprinting appears exceptionally promising. With advancements in related technologies and intensified research, 3D bioprinting is poised to become a standard method for producing engineered tissues and artificial organs. In the future, it may address a multitude of medical challenges, such as organ transplantation, tissue repair, and chronic disease treatment. Thus, 3D bioprinting is a multidisciplinary field demanding collaboration among engineers, materials scientists, biologists, and clinicians. Through collective efforts, a bright future can be envisioned for this transformative technology.

## References

Abdou SM, Moustafa A, Allam NK (2024) Patterned PCL/PGS nanofibrous hyaluronic acid-coated scaffolds promote cellular response and modulate gene expression profiles. ACS Appl Bio Mater 7(4):2569–2581

Alsalhi A (2024) Applications of selected polysaccharides and proteins in dentistry: a review. Int J Biol Macromol 260(Pt 1):129215

Amukarimi S et al (2022) Smart biomaterials: from 3D printing to 4D bioprinting. Methods 205:191–199

Arjoca S et al (2024) Hydrogel extrusion speed measurements for the optimization of bioprinting parameters. Gels 10(2)

Azizi Machekposhti S, Movahed S, Narayan RJ (2020) Physicochemical parameters that underlie inkjet printing for medical applications. Biophys Rev (Melville) 1(1):011301

Brunel LG et al (2024) Embedded 3D bioprinting of collagen inks into microgel baths to control hydrogel microstructure and cell spreading. Adv Healthc Mater 13(25):e2303325

Cao L et al (2021) Polysaccharides and proteins-based nanogenerator for energy harvesting and sensing: a review. Int J Biol Macromol 173:225–243

Capuana E et al (2022) Poly-l-lactic acid (PLLA)-based biomaterials for regenerative medicine: a review on processing and applications. Polymers (Basel) 14(6)

Castro JO et al (2018) Continuous tuneable droplet ejection via pulsed surface acoustic wave jetting. Soft Matter 14(28):5721–5727

Chang J, Sun X (2023) Laser-induced forward transfer based laser bioprinting in biomedical applications. Front Bioeng Biotechnol 11:1255782

Dabholkar N et al (2021) Biodegradable microneedles fabricated with carbohydrates and proteins: revolutionary approach for transdermal drug delivery. Int J Biol Macromol 170:602–621

Das A et al (2023) A comprehensive review on recent advances in preparation, physicochemical characterization, and bioengineering applications of biopolymers. Polym Bull (Berl) 80(7):7247–7312

de Melo RN et al (2023) Recent updates to microbial production and recovery of polyhydroxyalkanoates. 3 Biotech 13(6):204

Deja M et al (2021) Applications of additively manufactured tools in abrasive machining-a literature review. Materials (Basel) 14(5)

Duran-Lara EF et al (2024) Bacterial polyhydroxyalkanoates-based therapeutics-delivery nanosystems. Curr Med Chem 31(36):5884–5897
Feng Z et al (2020) A novel waterborne polyurethane with biodegradability and high flexibility for 3D printing. Biofabrication 12(3):035015
Gahl TJ, Kunze A (2018) Force-mediating magnetic nanoparticles to engineer neuronal cell function. Front Neurosci 12:299
Garciamendez-Mijares CE et al (2024) Design considerations for digital light processing bioprinters. Appl Phys Rev 11(3):031314
Gharibshahian M et al (2023) Recent advances on 3D-printed PCL-based composite scaffolds for bone tissue engineering. Front Bioeng Biotechnol 11:1168504
Gherasim O, Grumezescu V, Irimiciuc SA (2023) Overview of antimicrobial biodegradable polyester-based formulations. Int J Mol Sci 24(3)
Grosfeld EV et al (2022) Theoretical and experimental assay of shock experienced by yeast cells during laser bioprinting. Int J Mol Sci 23(17)
Han Q et al (2024) Injectable bioadhesive photocrosslinkable hydrogels with sustained release of kartogenin to promote chondrogenic differentiation and partial-thickness cartilage defects repair. Adv Healthc Mater 13(9):e2303255
Hu H, Krishaa L, Fong ELS (2023) Magnetic force-based cell manipulation for in vitro tissue engineering. APL Bioeng 7(3):031504
Isaacson N et al (2022) Compressive properties and failure behavior of photocast hydroxyapatite gyroid scaffolds vary with porosity. J Biomater Appl 37(1):55–76
Jahani A et al (2024) Biomolecules-loading of 3D-printed alginate-based scaffolds for cartilage tissue engineering applications: a review on current status and future prospective. Arch Bone Joint Surg 12(2):92–101
Jentsch S et al (2021) Multiscale 3D bioprinting by nozzle-free acoustic droplet ejection. Small Methods 5(6):e2000971
Kulka K, Sionkowska A (2023) Chitosan based materials in cosmetic applications: a review. Molecules 28(4)
Leonovich M et al (2023) Poly(lactic acid) and nanocrystalline cellulose methacrylated particles for preparation of cryogelated and 3D-printed scaffolds for tissue engineering. Polymers (Basel) 15(3)
Liang W et al (2024) Prospective applications of bioactive materials in orthopedic therapies: a review. Heliyon 10(16):e36152
Liu S et al (2023a) 3D Bioprinting tissue analogs: current development and translational implications. J Tissue Eng 14:20417314231187113
Liu J et al (2023b) Multifunctional droplets formed by interfacially self-assembled fluorinated magnetic nanoparticles for biocompatible single cell culture and magnet-driven manipulation. ACS Appl Mater Interfaces 15(13):17324–17334
Lyu Y et al (2023) Application of silk-fibroin-based hydrogels in tissue engineering. Gels 9(5)
Mikaeeli Kangarshahi B, Naghib SM, Rabiee N (2024) 3D printing and computer-aided design techniques for drug delivery scaffolds in tissue engineering. Expert Opin Drug Deliv:1–22
Mitchell J, Lo KW (2023) The use of small-molecule compounds for cell adhesion and migration in regenerative medicine. Biomedicines 11(9)
Mneimneh AT, Mehanna MM (2021) Collagen-based scaffolds: an auspicious tool to support repair, recovery, and regeneration post spinal cord injury. Int J Pharm 601:120559
Murizan NIS et al (2020) Review on nanocrystalline cellulose in bone tissue engineering applications. Polymers (Basel) 12(12)
Nashchekina Y et al (2024) Precisely printable silk fibroin/carboxymethyl cellulose/alginate bioink for 3D printing. Polymers (Basel) 16(8)
Nath PC et al (2024) Recent advances in production of sustainable and biodegradable polymers from agro-food waste: applications in tissue engineering and regenerative medicines. Int J Biol Macromol 259(Pt 1):129129
Ng WL et al (2022) Controlling droplet impact velocity and droplet volume: key factors to achieving high cell viability in sub-nanoliter droplet-based bioprinting. Int J Bioprint 8(1):424

Ning Y et al (2021) Mechanism and stability investigation of a nozzle-free droplet-on-demand acoustic ejector. Analyst 146(18):5650–5657
Premnath N et al (2021) A crucial review on polycyclic aromatic hydrocarbons – environmental occurrence and strategies for microbial degradation. Chemosphere 280:130608
Putra NE et al (2021) Extrusion-based 3D printing of ex situ-alloyed highly biodegradable MRI-friendly porous iron-manganese scaffolds. Acta Biomater 134:774–790
Rahimnejad M et al (2021) Biomaterials in bone and mineralized tissue engineering using 3D printing and bioprinting technologies. Biomed Phys Eng Express 7(6)
Randhawa A et al (2023) Recent advances in 3D printing of photocurable polymers: types, mechanism, and tissue engineering application. Macromol Biosci 23(1):e2200278
Rayat Pisheh H et al (2024a) Cardiac tissue engineering: an emerging approach to the treatment of heart failure. Front Bioeng Biotechnol 12:1441933
Rayat Pisheh H et al (2024b) Effective and new technologies in kidney tissue engineering. Front Bioeng Biotechnol 12
Rayat Pisheh H et al (2024c) Development of HEMA-succinic acid-PEG bio-based monomers for high-performance hydrogels in regenerative medicine. Biopolymers:e23631
Reddy MSB et al (2021) A comparative review of natural and synthetic biopolymer composite scaffolds. Polymers (Basel) 13(7)
Rekhi P et al (2022) Polyhydroxyalkanoates biopolymers toward decarbonizing economy and sustainable future. Crit Rev Biotechnol 42(5):668–692
Sabzevari A et al (2023) Progress in bioprinting technology for tissue regeneration. J Artif Organs 26(4):255–274
Sergi R, Bellucci D, Cannillo V (2020) A review of bioactive glass/natural polymer composites: state of the art. Materials (Basel) 13(23)
Shashikumar U et al (2024) Innovative technologies for the fabrication of 3D/4D smart hydrogels and its biomedical applications – a comprehensive review. Adv Colloid Interf Sci 328:103163
Shin YJ et al (2021) 3D bioprinting of mechanically tuned bioinks derived from cardiac decellularized extracellular matrix. Acta Biomater 119:75–88
Shokrani H et al (2022) Green polymer nanocomposites for skin tissue engineering. ACS Appl Bio Mater 5(5):2107–2121
Singhvi MS, Zinjarde SS, Gokhale DV (2019) Polylactic acid: synthesis and biomedical applications. J Appl Microbiol 127(6):1612–1626
Sitthisang S et al (2024) Nanomechanical mapping of PLA hydroxyapatite composite scaffolds links surface homogeneity to stem cell differentiation. Sci Rep 14(1):21097
Sreena R, Nathanael AJ (2023) Biodegradable biopolymeric nanoparticles for biomedical applications-challenges and future outlook. Materials (Basel) 16(6)
Sultan MT et al (2022) Three-dimensional digital light-processing bioprinting using silk fibroin-based bio-ink: recent advancements in biomedical applications. Biomedicines 10(12)
Szychlinska MA et al (2022) Three-dimensional bioprinting for cartilage tissue engineering: insights into naturally-derived bioinks from land and marine sources. J Funct Biomater 13(3)
Vach Agocsova S et al (2023) Resorbable biomaterials used for 3D scaffolds in tissue engineering: a review. Materials (Basel) 16(12)
Vasvari GF et al (2021) CMT additive manufacturing parameters defining aluminium alloy object geometry and mechanical properties. Materials (Basel) 14(6)
Waasdorp R et al (2018) Accessing individual 75-micron diameter nozzles of a desktop inkjet printer to dispense picoliter droplets on demand. RSC Adv 8(27):14765–14774
Wang J, Cui Z, Maniruzzaman M (2023) Bioprinting: a focus on improving bioink printability and cell performance based on different process parameters. Int J Pharm 640:123020
Wijesekara T, Xu B (2024) New insights into sources, bioavailability, health-promoting effects, and applications of chitin and chitosan. J Agric Food Chem 72(31):17138–17152
Wu BX et al (2023) Application of three-dimensional (3D) bioprinting in anti-cancer therapy. Heliyon 9(10):e20475

Zermeno-Perez D et al (2024) Bioresorbable polyester coatings with antifouling and antimicrobial properties for prevention of biofilm formation in early stage infections on Ti6Al4V hard-tissue implants. ACS Appl Bio Mater 7(8):5728–5739
Zhang B et al (2020) Solvent-based extrusion 3D printing for the fabrication of tissue engineering scaffolds. Int J Bioprint 6(1):211
Zheng Y et al (2024) Effect of polymethyl methacrylate on in situ patterning of perovskite quantum dots by inkjet printing. Luminescence 39(2):e4691

# Smart Natural-Based Membranes as Delivery Systems for Biomedical Applications

4

Simone S. Silva, Luísa C. Rodrigues, Emanuel M. Fernandes, and Rui L. Reis

**Abstract**

This chapter embarks on a journey of discovery, exploring the use of smart membranes, particularly those derived from natural-origin polymers and proteins, in tandem with cutting-edge processing methods. These smart membranes are engineered by integrating stimuli-responsive materials, enabling them to autonomously modulate their physical and chemical characteristics in response to environmental cues, including temperature, pH, light, magnetic or electric fields, redox conditions, and specific ions/molecules. Smart membranes exhibit considerable potential in biomedical applications, encompassing controlled drug release, bioseparation, and tissue engineering due to their adjustable permeability, selectivity, and distinctive biomimetic properties. The chapter further discusses the formulations and processing strategies employed to fabricate these smart polymeric membranes, as well as methodologies to enhance their performance and functionality. Beyond their conventional roles, smart membranes can deliver drugs or other therapeutic agents, creating optimal environments for cellular growth and differentiation for tissue regeneration, skin repair, guided bone regeneration, and other disease treatments. Therefore, with their potential to advance medical treatments and improve patient outcomes, these membranes bring hope and optimism to the biomedical field, promising a brighter future for healthcare.

S. S. Silva (✉) · L. C. Rodrigues · E. M. Fernandes · R. L. Reis
3B's Research Group, I3Bs – Research Institute on Biomaterials, Biodegradables and Biomimetics, University of Minho, Headquarters of the European Institute of Excellence on Tissue Engineering and Regenerative Medicine, AvePark - Parque de Ciência e Tecnologia, Rua Ave 1, Edifício 1 (Sede), Barco, Guimarães, Portugal

ICVS/3B's – PT Government Associate Laboratory, Braga, Guimarães, Portugal
e-mail: simonesilva@i3bs.uminho.pt



J. M. Oliveira et al. (eds.), *Sustainable Scaffolds-based Strategies in Tissue Engineering and Regenerative Medicine*, Biomaterials, Bioengineering and Sustainability 8, https://doi.org/10.1007/978-3-031-96274-5_4

**Keywords**

Biomaterials · Membrane · Natural polymers · Proteins · Skin repair · Tissue regeneration · Wound dressing · Drug delivery · Guided bone regeneration

## Abbreviations

| | |
|---|---|
| ALG | Alginate |
| AI | Artificial intelligence |
| AMPs | Antimicrobial peptides |
| ANG | Angiogenin |
| BC | Bacterial cellulose |
| BiPEP | Two peptide drugs |
| Bio-ILs | Biocompatible ionic liquid |
| CMCs | Carboxymethyl chitosan |
| BV/TV | Bone volume to tissue volume |
| CNCs | Cellulose nanocrystals |
| CNC-g-PDMAEMA | Cellulose nanocrystal-poly[2-(dimethylamino) ethyl methacrylate] |
| CHA | Carbonated hydroxyapatite |
| CHT | Chitosan |
| CC | Chitosan carboxylate |
| CsA | Cyclosporine |
| CIP | Ciprofloxacin |
| CMC | Carboxymethylcellulose |
| CMCHT | Carboxymethyl chitosan |
| CPX | Ciprofloxacin hydrochloride monohydrate |
| Cur | Curcumin |
| DDS | Drug delivery system |
| DTZ | Diltiazem |
| DS | Diclofenac sodium |
| EDC | 1-ethyl-3-(3-dimethylaminopropyl) carbodiimide |
| EGF | Epidermal growth factor |
| FDA | Food and Drug Administration |
| FGF | Fibroblast growth factor |
| FU | Fluorouracil |
| GAGAGS | Glycine-alanine-glycine-alanine-glycine-serine |
| GBR | Guided bone regeneration |
| G-blocks | α-L-guluronic acid |
| GO | Graphene oxide |
| GQDs@PSi | Graphene quantum dots-decorated luminescent porous silicon |
| HA | Hyaluronic acid |
| HPMC | Hydroxypropyl methylcellulose |

| | |
|---|---|
| ILs | Ionic liquids |
| LbL | Layer-by-layer |
| M-blocks | β-(1-4)-D-mannuronic acid |
| MGRS | Modified glutinous rice starch |
| NPTDD | Natural polymer-based transdermal drug delivery systems |
| PAH | Poly(allylamine hydrochloride) |
| PBS | Phosphate buffer solution |
| PCL | Polycaprolactone |
| pDNA | Plasmid DNA |
| PHBV | Poly(3-hydroxybutyrate-co-3-hydroxy valerate) |
| PLGA | Poly(lactic-co-glycolic acid) |
| PMGly | Poly(N-methacryloyl glycine) |
| PVA | Polyvinyl alcohol |
| PVP | Polyvinyl pyrrolidone |
| PEDOT:PSS | Poly(3,4-ethylenedioxythiophene):polystyrene sulfonate |
| QSG | Quince Seed Gum |
| ROS | Reactive oxygen species |
| SF | Silk fibroin |
| SGT | Simulated gastric fluid |
| SIF | Simulated intestinal fluid |
| SS | Silk sericin |
| TDS | Transdermal delivery system |
| TERM | Tissue engineering and regenerative medicine |
| TEOS | Tetraethyl orthosilicate |
| TH | Tetracycline hydrochloride |
| TK | Thioketal |
| TPP | Tripolyphosphate |
| VCO | Virgin coconut oil |
| [Cho]Ac | Choline acetate |
| 3-PtTPyP | Metalloporphyrin |

## 4.1 Introduction

Natural polymers derived from renewable resources have the potential to create membranes for a wide range of applications, including water treatment, controlled drug release, and tissue engineering. These materials play a pivotal role in developing sustainable technologies (Silva et al. 2017, 2020; Gentile et al. 2011; Zou et al. 2022). Membranes made from renewable natural resources, such as polysaccharides from marine, plant, and microbial sources, e.g., chitosan, alginate, cellulose, hyaluronic acid, and silk fibroin, offer numerous benefits. These include cost-effectiveness, lack of toxicity, ease of modification, biodegradability, biocompatibility, anti-bacterial properties, and film-forming ability (Li et al. 2022). These properties open up numerous potential applications for these membranes, including their use as barriers and for the controlled release of drugs or active substances.

In the era of intelligent materials, smart membranes are engineered to respond to various stimuli, including pH, temperature, light, electric field, magnetic fields, and specific ions/molecules. For instance, they can sustain release in response to changes in pH, temperature, or light exposure, making them useful for controlled drug release and targeted therapy with reduced side effects. Efforts have been made to promote/improve the functionality of smart membranes, such as incorporating nanoparticles or stimuli-responsive polymers to improve membrane sensitivity and responsiveness. These strategies would advance the potential benefits and reliability of membranes in tissue engineering, drug delivery, wound healing, cancer therapy, and other medical fields (Zou et al. 2022).

The design of delivery systems using smart membranes has been stimulated to provide efficient transport of therapeutic substances to the cells or tissue target through improved aqueous solubility, chemical stability, and higher pharmacological activity (Li et al. 2019; Alavi et al. 2024). In this process, the selection of drug carriers and additives, such as plasticizers and cross-linking agents, and their processing are crucial for the safe and effective transport of the active molecules. These additives can improve the flexibility, mechanical strength, and stability of the membranes, enhancing their performance and functionality.

Smart membranes have significant potential and commercial prospects in biomedical applications, particularly in controlled drug release, biological separation, and tissue engineering (Zou et al. 2022). In controlled drug release, conventional drug delivery systems face challenges in avoiding toxicity and drug wastage. By employing rational structural design and selecting responsive materials, smart membranes can precisely regulate the timing and location of drug release, thereby effectively addressing these issues (Zou et al. 2022; Alavi et al. 2024; Van Gheluwe et al. 2021). In the context of biological separation, smart membranes offer advantages such as low-pressure drop, short processing time, high flow rate, accessible equipment, and excellent antifouling properties, contributing to reducing energy consumption and equipment costs while enhancing production efficiency and yield of biomolecules, crucial demands in the healthcare sector. Non-invasive administration methods like transdermal delivery systems (TDS) provide a stable, efficient, and safe approach to disease therapy, offering low rejection rates, ease of administration, and high patient compliance (Sabbagh and Kim 2022). TDS is widely used in pharmaceuticals, skincare, and biomedicine, enabling localized drug delivery to maximize therapeutic efficacy and minimize side effects. This advanced technology employs membranes or patches with specific layers that can administer various drugs through effective design and materials. Despite progress, there is ongoing research into enhancers like sterols, ceramides, fatty acids, terpenes, terpenoids, and urea to improve drug absorption through the stratum corneum, offering safety, low skin irritation, and biodegradability, increasing accepted by consumers (Schafer et al. 2023). Aligning with this tendency, ionic liquids (ILs) can also be applied as penetration enhancers. Their function is to weaken the barrier of the stratum corneum, allowing poorly soluble drugs to be delivered through the skin. Additionally, they can help design transdermal preparations to improve formulations (Zhang et al. 2023; Lu et al. 2022; Chen et al. 2024).

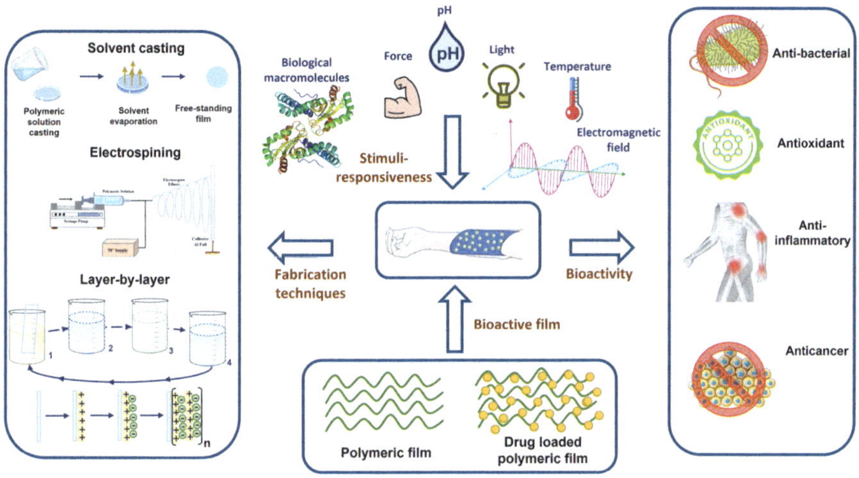


**Fig. 4.1** Production methodologies, stimuli-responsiveness, and bioactive functions of smart polymeric-based films

Herein, we will explore the creation of smart membranes using natural sources as vehicles for drug delivery. These membranes are designed for targeted delivery and effective penetration. We will also cover diverse biomedical applications of these delivery system-based membranes, such as skin repair, tissue regeneration, cancer treatment, guided bone regeneration, and treatment of inflammatory diseases (Fig. 4.1).

## 4.2 Materials and Processing Techniques for the Production of Sustainable Membranes

Over the last few years, the use of membranes in various medical applications has continued to advance. They are useful in creating controlled-release systems for active substances and improving the delivery properties of topical and transdermal formulations. These membrane-based systems consist of an active molecule, a film-forming polymer from marine, plant, or microbial sources, either combined with other natural or synthetic materials or used alone and with a solvent to develop enhanced membrane-based materials (Touitou et al. 2023). Fabrication techniques such as solvent casting, electrospinning, and Layer-by-Layer (LbL) assembly are employed to create membranes with tailored structures and functionalities. Intelligent membranes can also be produced by incorporating active molecules or nanoparticles to enhance their responsiveness to external stimuli, such as changes in temperature or pH. These molecules are incorporated into membranes through encapsulation or physical entrapment, offering sustained drug delivery and enhanced therapeutic efficacy. Once prepared, smart membranes can work as transdermal delivery systems by modulating drug release across the skin barrier in response to specific stimuli. They offer several advantages, including biocompatibility with tissues, biodegradability at rates adjusted to match tissue regeneration, non-toxicity, non-immunogenicity, appropriate mechanical properties, and morphology conducive to moving cells, gases, metabolites, nutrients, and signaling molecules. To serve this purpose, many natural polymers such as chitosan, collagen, alginate, gelatin, and bacterial cellulose, alone or in combination with other natural materials, have been used to prepare membrane-based systems, ensuring biocompatibility, biodegradability at an adjusted rate, and adequate mechanical properties, among other features according to the desired application or function.

### 4.2.1 Natural Biomacromolecules

The integration of diverse stimuli-responsive materials into membranes has allowed for the creation of intelligent membranes inspired by the capability of cell membranes to react to external cues. These membranes are capable of sensing environmental triggers, including pH, light, magnetic fields, temperature, and specific molecules. As a result, they can adjust their physical and chemical properties, such as pore dimensions and surface properties, in response to these stimuli. This adaptative behavior enables the membranes to achieve adjustable permeability and selectivity, similar to their biological counterparts.

Smart membranes show great promise and are commercially viable avenues in biomedical applications, particularly in controlled drug release, biological separation, and tissue engineering (Zou et al. 2022; Abedini et al. 2023). In controlled drug release, smart membranes offer precise control over drug release timing and localization, addressing challenges such as toxic side effects and inefficient drug utilization associated with conventional delivery systems (Zou et al. 2022). Smart membranes offer numerous advantages in biological separation, such as low-pressure drop, short processing times, high flow rates, easy equipment accessibility, and excellent antifouling properties (Zou et al. 2022). These benefits result in reduced energy consumption and equipment costs, improved production efficiency, and increased biomolecule yields, which have significant implications for healthcare applications. In tissue engineering, intelligent membranes facilitate the rapid and non-destructive retrieval of cell sheets from culture media, thereby minimizing the harmful effects of enzymatic hydrolysis and mechanical scraping associated with traditional cell and tissue culture methods. That system enhances cell viability and protection, simplifying the overall tissue engineering process.

Smart membranes are often used in capsule form in controlled drug release due to their easy design and implementation (Zou et al. 2022). However, this topic will not be further discussed herein.

Biopolymeric membranes offer distinct advantages over their synthetic counterparts in the tissue engineering field. They show favorable biocompatibility, biodegradability, intrinsic bioactivity, cell-attractive properties, and susceptibility to degradation through cell-mediated proteolysis, facilitating natural remodeling processes (Shiohara et al. 2021). Among natural biopolymers, collagen, and chitosan are the most widely investigated materials for tissue regeneration. Additionally, other natural biomacromolecules are employed to develop natural-based smart delivery systems. These biomacromolecules and their sources, processing methodologies, and possible applications are presented in Fig. 4.2.

**Alginate (ALG)**, derived from brown algae, is a linear anionic polysaccharide polymer. Its structure comprises varying proportions of β-(1-4)-D-mannuronic acid (M-blocks) and α-L-guluronic acid (G-blocks). The alginate source determines the arrangement of monomer blocks and the molecular weight of the polymer and significantly impacts its physiochemical properties. Alginate materials exhibit excellent water absorption capacity, facilitating exudate management and creating a moist wound environment conducive to wound healing (Manna et al. 2023). Consequently, alginate-based wound dressings are widely utilized in commercial and research settings (Manna et al. 2023).

Numerous commercially available alginate wound dressings are utilized to treat cutaneous wounds, including burns, donor sites, surgical wounds, and particularly chronic wounds. These dressings predominantly rely on sodium or calcium-mediated ionic crosslinking of alginate into hydrogels, which can be further processed into porous or fibrous membranes. Examples of commercially available alginate wound dressings include AlgiSite M, Comfeel Plus, Algivon, Kaltostat, and hydrogels such as Purilon Gel and Nu-Gel (Manna et al. 2023).

Alginate has a high loading capacity for bioactive molecules, with silver-containing alginate wound dressings like Algicell Ag and AlgiSite M being commonly used. In research, alginate has been studied for loading various bioactive compounds, including antimicrobial agents like chitosan, vancomycin, silver, and zinc, as well as oxygen-releasing microspheres to enhance neovascularization and cell proliferation, and for delivering anti-inflammatory drugs such as ibuprofen and even plasmids.

Alginate is a widely utilized biopolymer for developing transdermal films in drug delivery and biomedical applications (Manna et al. 2023) (Fig. 4.1), especially when combined with other polymeric matrices. Table 4.1 includes several examples of its applications, a few of which are presented below.

A chitosan/halloysite nanocomposite film sandwiched between two alginate layers was proposed as a pH-controlled carrier approach (Lisuzzo et al. 2019). The alginate layers slowed down the diclofenac release, allowing it to be controlled under variable physiological pH conditions. The diclofenac release was negligible ($<1\%$), at pH = 3 (stomach), and a slight increase was detected (*ca.* 5%) at pH = 5.7 (duodenum/ileum), and an exponential increase was noted for alkaline conditions (pH = 7.8, typical in the colon), reaching 70% after 30 hours (Lisuzzo et al. 2019).

An electrochemically active hydrogel film, prepared by gelling a mixture of poly(3,4-ethylenedioxythiophene): polystyrene sulfonate (PEDOT: PSS) and ALG, has been loaded with curcumin (CUR), a hydrophobic drug with a wide spectrum of clinical applications (Puiggalí-Jou et al. 2020). The hydrogel system demonstrates a controlled drug release upon application of a negative electrical voltage, which affects the intermolecular interactions. Such a stimuli-responsive drug delivery system may be used as a drug delivery implant, where the release of the drug can be regulated according to the patient's requirements by applying an electrical stimulus.

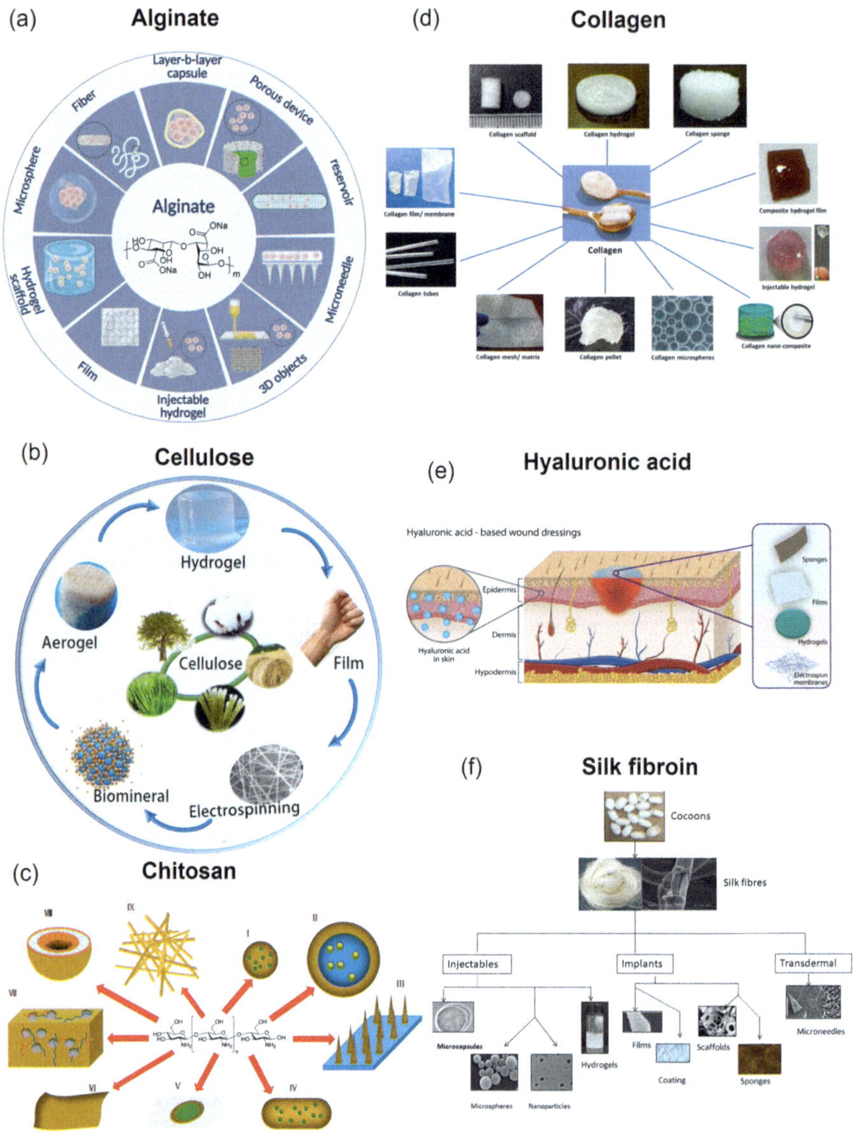


**Fig. 4.2** (**a**) Alginate-based possible products for drug delivery applications. (Reprinted with Copyright permission from Creative Commons Corporation (Kavand et al. 2024)); (**b**) Cellulose sources and different application approaches in drug carrier systems. (Reprinted with permission from Elsevier (Yan et al. 2020)); (**c**) Various forms of application of chitosan and its derivatives in transdermal drug delivery systems (yellow represents chitosan and derivatives thereof). I: nanoparticles; II: emulsions; III: transdermal microneedles; IV: nanocapsules; V: transdermal patches; VI: transdermal membranes; VII: hydrogels; VIII: liposomes; IX: nano-scaffolds. (Reprinted with Copyright permission from Creative Commons Corporation (Ma et al. 2022)); (**d**) Collagen-based products achieved using different processing strategies. (Reprinted with Copyright permission from Creative Commons Corporation (Amit Kumar 2022)); (**e**) Hyaluronic acid- based wound dressing different possible approaches. (Reprinted with Copyright permission from Creative Commons Corporation (Juhaščik et al. 2022)); and (**f**) Silk fibroin source and different viable application strategies for drug delivery design (Pandey et al. 2020). (Reprinted with permission from Elsevier (Yan et al. 2020))

**Table 4.1** Smart natural-based membranes for delivery of various bioactive molecules or drugs

| Natural polymer | Membrane composition | Bioactive molecule/ drug | Major outcomes | Potential application | Reference |
|---|---|---|---|---|---|
| Alginate | Alginate-QSG | Curcumin | Stearic acid-modified curcumin films (Q25-Sa-Cur) with sustained drug release, higher collagen synthesis, and β TGF-expression. | Transdermal delivery and wound healing | Abedini et al. (2023) |
| | Alginate/ poly(4-vinylpyridine) | Ciprofloxacin HCL | Multilayer membrane: hydrophilic and stable in acidic buffer up to pH 4.2;<br>Optimum swelling capacity, allowing higher drug loading;<br>The drug is well dispersed and released via a t1/2 kinetics diffusion mechanism. | Transdermal delivery | Alshhab and Yilmaz (2020) |
| | CHT/ alginate | Diclofenac | Slow release at pH 3 (<1%), increases at pH 5.7 (ca. 5%), and exponentially rises at pH 7.8, reaching 70% after 30 hours. | Drug carrier | Lisuzzo et al. (2019) |
| | Alginate/3-PtTPyP | Hydrocortisone | When exposed to visible light, the film degraded due to ROS generated by the photo-excited porphyrin, enabling controlled release of HC. | Chronic skin diseases. | Quadrado et al. (2023) |
| Cellulose | CMC/CHT | 5-Fluorouracil (5-FU) Tetracycline Hydrochloride (TH) | Redox-responsive hydrogel films *via* using CHT microspheres into CMC hydrogel film;<br>CMC/CHT-loaded dual drugs have both pH and redox-responses drug delivery features;<br>Exhibited inhibition to HepG2 cells and antimicrobial efficacy. | Cancer therapy | Wang et al. (2019) |
| | CNC-g-PDMAEMA + PHBV | Tetracycline hydrochloride (TH) | Accelerated drug release at specific pathological sites stimulated by internal temperature and pH variations;<br>Enhanced therapeutic efficacy while reducing potential toxic side effects. | Drug delivery | Chen et al. (2023) |
| | CNC/CHT | Doxorubicin | LBL assembled film could load a substantial amount of doxorubicin;<br>Doxorubicin release in a sustained manner under physiological conditions. | Drug delivery | Mohanta et al. (2014) |
| | BC/poly(N-methacryloyl glycine) (PMGly) | Diclofenac | In situ polymerization films of methacryloyl glycine monomer within the BC network;<br>Modulated drug release at different pH over 24 hours, indicating controlled and pH-sensitive drug delivery. | Transdermal and oral drug delivery | Saïdi et al. (2017) |

(continued)

**Table 10.1** (continued)

| Natural polymer | Membrane composition | Bioactive molecule/drug | Major outcomes | Potential application | Reference |
|---|---|---|---|---|---|
| Chitosan | CHT | Naringenin | Consistent drug release across the skin over 72 hours;<br>Skin permeation tests showed the ability to pass through the epidermis and dermis while minimizing drug retention in skin layers. | Treating inflammatory conditions | Quintão et al. (2022) |
| | CHT carboxylate (CC) | Cyclosporine (CsA) | *In vitro* permeation studies showed that membranes could release the cyclosporin-A in internal layers of skin.<br>Membranes containing CsACC and free CsA showed a significant decrease in human breast cancer MDA-MB-231 cell viability. | Implantable subcutaneous device and Treatment of breast cancer | Trombino et al. (2021) |
| | CHT/Virgin Coconut Oil hollow capsules | Curcumin | LbL to produce biocompatible photo-responsive multilayer film system;<br>Drug release profile is not significantly influenced by the medium pH reaching a maximum of 37% ± 1% after 48 hours. | Topical applications | Rodrigues et al. (2024) |
| | CHT—Tamarind Seed | Tamarind gum | Drug release duration for T50% ranged from 200 to 180 hours, while drug permeability for T50% varied between 200 and 201 hours;<br>Appropriate penetration enhancer can improve the bioavailability of peptide molecules in the formulation. | Transdermal film | Malviya et al. (2021) |
| | CHT/HPMC and polyvinylpyrrolidone/HPMC | Gentamycin | High ratio of CHT or HPMC in different formulas induces the longest drug release, with <70% drug released in 24 hours;<br>Gentamycin loaded in HPMC/CHT film induces a stronger activity than the drug loaded in HPMC/PVP film after 24 hours. | Target skin infection | Jaber (2023) |
| | CHT/starch | Lidocaine hydrochloride | Mucoadhesiveness of MGRS-based films was not affected by the presence of CHT;<br>A higher CHT ratio resulted in stronger but less flexible films compared to MGRS films;<br>Drug permeation across porcine mucosa increased in line with the CHT ratio. | Buccal delivery | Soe et al. (2020) |

| | | | | | |
|---|---|---|---|---|---|
| | CHT/guar gum/PVP | Ciprofloxacin hydrochloride | In SIF and PBS, almost 100% drug was delivered out from the hydrogel film in 70 minutes in SIF, and in 90 minutes in PBS;<br>All films showed bacteriostatic behaviour against both *E. coli* and *S. aureus* bacteria. | Intravenous drug release administration | Ghauri et al. (2021) |
| | CHT/PAH | CPX | Sustained drug release was observed in PBS, as compared to SGF. | Tissue regeneration | Sarwar et al. (2020) |
| | CHT/gelatin | Methylene blue | Increasing the degree of crosslinking within the CHT, leads to concomitant increase in film swelling, and decreases the drug release;<br>The proportion of genipin within the crosslinked CHT exerted a significant influence on the controlled drug release. | Drug delivery | Koc and Altıncekic (2021) |
| | BiPEP-GQDs@PSi/CHT matrix | EGF and insulin | In combination with CHT degradation in the acidic and oxidative microenvironment, the film showed stimuli-responsive drug release toward slightly acid and highly oxidative conditions;<br>*In vitro* and *in vivo* results showed the smart system enhanced the proliferation and migration of cells as well as significantly healed diabetic wounds; | Diabetic wounds | Cui et al. (2021) |
| | CHT, BC, and PVA | Ibuprofen | *In vitro* tests showed that variations in pH, the presence of BC, and the film composition modulate the release rate;<br>High concentrations of BC, and increased hydrogen ion concentrations led to a decrease in ibuprofen release. | Drug delivery | Pavaloiu et al. (2014) |
| | CHT/PVA/graphene oxide | Allicin | Nanofibrous membranes have strong antibacterial activity and sustained-release properties;<br>Release rate and amount of Alli could be regulated by the content of GO in the membrane. | Wound dressing and tissue engineering | Liu et al. (2020) |
| | CHT and poly(allylamine hydrochloride) (PAH) | CPX | Homogeneous and surface morphology of films with increased PAH concentration;<br>pH-responsive behavior, releasing CPX more rapidly in PBS (pH = 7.4) than SGT (pH = 1.2). | Wound dressing, oral administration | Sarwar et al. (2020) |
| | Imine-functionalized CHT | Cinnamaldehyde, citronellal, hydroxy cinnamaldehyde, and citral | pH-responsive film based on CHT to release various aldehydes;<br>Films synthesized utilizing the reversible bonding capacity of aldehydes with CHT. | Drug release | Heras-Mozos et al. (2022) |

(continued)

**Table 10.1** (continued)

| Natural polymer | Membrane composition | Bioactive molecule/drug | Major outcomes | Potential application | Reference |
|---|---|---|---|---|---|
| | PEG-coated vinyl trimethoxy silane-g-CHT (PEG@VTMS-g-CS) | DTZ | Copolymer with low swelling capacity and sustained release;<br>Biocompatibility of the film confirmed in viability assays on HaCaT and PBMC cell lines;<br>Enhanced DTZ therapy by improving drug encapsulation, sustained release, and physiological acceptance. | Drug delivery | Anirudhan et al. (2018) |
| | PVA–CHT | Ciprofloxacin and rutin hydrate | Synthesis of dual drug-loaded nanofibers using the electrospinning method;<br>Dressing material with efficacious antibacterial activity. | Infected skin wounds | Alotaibi et al. (2024) |
| | CHT/pectin | Antimicrobial peptides (D-GL13K and IDR-1018) | Mucoadhesion was enhanced by coating the membranes with oxidized sugars/ polysaccharide (pectin);<br>Absorption of AMPs (D-GL13K or IDR-1018) conferred simultaneous antimicrobial potency against relevant oral streptococci and cytocompatibility with both soft tissue and hard tissue-associated cells;<br>The membranes can be used for therapeutic delivery in the oral cavity in a pH-controlled manner. | Adhesive gingival grafts and GBR membranes | Boda et al. (2020) |
| Collagen | Collagen/PCL | ZIF-8 nanoparticles ($Zn^{2+}$ ions) | Double-layer fabricated by electrospinning has tunable mechanical properties, flexibility, and controlled degradation rate;<br>pH-responsive release of $Zn^{2+}$ ions, hence triggering rBMSC recruitment and coordination of subsequent osteogenesis and angiogenesis processes. | Bone tissue engineering | Xue et al. (2021) |
| | Collagen/PVA/choline acetate ([Cho]Ac) ionic liquid (IL). | CIP | CIP was controlled released due to the incorporation of [Cho]Ac into the film;<br>Films loaded with CIP showed excellent antibacterial activity against both *E. coli* and *S. aureus* bacteria. | Wound dressing and other | Ahmed et al. (2023) |
| | Collagen/CHT | Lidocaine, tetracaine, and benzocaine | Uniform dispersion of anesthetics, with a gradual release over 6–24 hours;<br>*In vitro* non-cytotoxicity promoted cell proliferation and biocompatibility with human dermal fibroblast cells, making them for drug delivery and tissue proliferation. | Local anesthetics delivery | Di Martino et al. (2019) |

| | | | | | |
|---|---|---|---|---|---|
| | Collagen/PVA | Ciprofloxacin hydrochloride | Anionic films with sustained drug delivery, inhibiting growth of *S. aureus* and *E. coli*;<br>Suitable mechanical properties, water content, hydrophilicity, permeability, and pH for corneal interaction. | Corneal ulcer treatment | Daza et al. (2020) |
| | Collagen/polyurethane/cellulose | Ketokonazole | Incorporation of lignin or lignin-metal complexes enhances the mechanical strength and imparts controlled drug release along with biocidal activity. | Transcutaneous drug delivery | Anghel et al. (2021) |
| Gelatin | Gelatin/CHT/CHA/TEOS | Metronidazole | Composition 0.3/0.3/0.5/1.87w/w/w/w has superior tensile strength, controlled drug release, and degradation rates compared to collagen film, Dentium™;<br>Mathematical modeling indicated diffusion-controlled drug release;<br>Release profiles of metronidazole can be improved with the TEOS crosslinker. | Regeneration of infective bone defects | Mahmudi et al. (2024) |
| | Gelatin/CHT | Methylene blue | High controllable release, complying with the zero-order kinetic model at low pH (pH 1.5), except for the 1.5% genipin ratio. | pH-delivery system | Koc and Altıncekic (2021) |
| Hyaluronic acid | PVA and Hyaluronic acid | Dexamethasone and levofloxacin | Delivery platform for the concomitant release of lipophilic and hydrophilic drugs;<br>Easy and potentially scalable method;<br>Modified release platform for the reduction of administration frequency. | Ocular drug delivery | Ghezzi et al. (2023) |
| | CHT and hyaluronic acid | Aspirin derivative (ACS14) | LbL multilayer film composed of catechol group modified CHT and hyaluronic acid;<br>Reversible sensitive pH responsiveness. | pH-delivery system | Lu et al. (2019) |

(continued)

**Table 10.1** (continued)

| Natural polymer | Membrane composition | Bioactive molecule/drug | Major outcomes | Potential application | Reference |
|---|---|---|---|---|---|
| Silk fibroin | Silk thioketal | Camptothecin | Films showed significant ROS-responsive cleavage behavior;<br>The TK with carboxyl (–COOH) groups at both ends was successfully incorporated between SF macromolecules via amidation. | Treat tumors and chronic inflammation | Li et al. (2024) |
| | Silk/CHT | Diclofenac sodium | Short drug loading time of 2 hours;<br>The 70CHT/30RSF film showed prolonged drug release within therapeutic level up to 11 hours. | Disposable contact lenses—ophthalmic delivery | Jeencham et al. (2020) |

**Abbreviation:** *AMPs* antimicrobial peptides, *BC* bacterial cellulose, *CC* chitosan carboxylate, *CHT* chitosan, *CHA* carbonated hydroxyapatite, *[Cho]Ac* choline acetate, *CMC* carboxymethylcellulose, *CIP* ciprofloxacin, *CPX* ciprofloxacin hydrochloride monohydrate, *CsA* Cyclosporine, *Cur* curcumin, *DTZ* Diltiazem, *E. coli Escherichia coli*, *5-FU* 5-Fluorouracil, *EGF* epidermal growth factor, *HC* hydrocortisone, *HPMC* hydroxypropyl methylcellulose, *GBR* guided bone regeneration, *GO* graphene oxide, *LBL* layer-by-layer, *MGRS* modified glutinous rice starch, *PBS* phosphate buffer solution, *PAH* poly(allylamine hydrochloride), *PEG* polyethylene glycol, *PMGly* poly(N-methacryloyl glycine), *PVA* polyvinyl alcohol, *PVP* polyvinyl pyrrolidone, *TEOS* Tetraethyl orthosilicate, *ROS* reactive oxygen species, *SA* sodium alginate, *SIF* simulated intestinal fluid, *SGT* simulated gastric fluid, *QSG* Quince Seed Gum, *TH* Tetracycline Hydrochloride, *TK* thioketal, *SF* Silk fibroin, *DS* diclofenac sodium, *S. aureus Staphylococcus aureus*, *VCO* virgin coconut oil pH-sensitive chitosan/alginate/tripolyphosphate (CHT/ALG/TPP) hydrogel membranes were prepared based on the CHT/ALG polyelectrolyte complex cross-linked with TPP, which influenced its molecular and supermolecular structure, roughness and swelling behavior, without changing its surface morphology (Gierszewska et al. 2018). The swelling of the membranes was found to be pH-dependent; hydrogel membranes in buffered solutions of pH $\geq$ 7.4 showed an "overshooting effect", that is not visible in an acidic medium. This pH-sensitive CHT/ALG/TPP membrane seems to be suitable for drug delivery systems, but the initial "burst" release limits their application as a long-term delivery system.

Sodium alginate and polyvinyl alcohol were combined to produce an ionic cross-linked alginate porous film (Saraiva et al. 2023). Due to immiscibility, phase separation occurred between the alginate and polyvinyl alcohol. Calcium chloride was employed to dissolve the polyvinyl alcohol phase, leaving behind the alginate phase, which was subsequently crosslinked with calcium ions to create a porous alginate film. The pore size of this porous membrane could be regulated by changes in pH, indicating pH responsiveness. This property suggests the potential utility of this porous membrane for applications such as separation membranes or controlled drug release systems.

**Cellulose** is the most abundant biopolymer globally, composed of linear glucose rings linked by β(1→4) glycosidic bonds. It is sourced from wood fibers, tunicates, and certain bacteria (Sheikhi et al. 2019). Nanocrystals can be extracted from cellulose precursors through methods such as acid hydrolysis, sonochemical fragmentation, microbial or enzymatic digestion, and other top-down fabrication techniques. The source and extraction method used influence the precise morphology of the resulting cellulose nanocrystals (CNCs), which typically exhibit lateral dimensions of 3–5 nm and lengths of 100–300 nm. CNCs are valued for their impressive mechanical properties, including a high surface area (~250 $m^2/g$), substantial tensile strength (7500 MPa), and significant stiffness (Young's modulus up to 140 GPa) (Nasseri et al. 2020). Cellulose-based responsive biomaterials are summarized in Table 4.1. Still, some examples comprising CNCs and bacterial cellulose were also described.

CNCs can be electrospun with polymers, nanoparticles, and target drugs, such as poly(lactic-co-glycolic acid) (PLGA), curcumin (Cur), and PEI-carboxymethyl chitosan (CMCHT). The PLGA/CNC/pDNA-ANG composite demonstrated significantly higher ANG expression than controls containing only PLGA/CNCs, with a 1.6-fold increase in expression on day 3 and a 2.8-fold increase by day 7. *In vivo* studies revealed that rats treated with the pDNA-ANG nanocomposite experienced accelerated wound recovery and re-epithelialization. Chen et al. investigated drug delivery applications of double stimuli-responsive cellulose nanocrystal-poly[2-(dimethylamino)ethyl methacrylate] (CNC-g-PDMAEMA) reinforced poly(3-hydroxybutyrate-co-3-hydroxy valerate) (PHBV) electrospun composite membranes, employing tetracycline hydrochloride (TH) as a model drug (Chen et al. 2020). The electrospun nanofiber membranes developed by Chen et al. (2023) possess pH and temperature responsiveness, allowing for accelerated drug release at specific pathological sites when stimulated by internal temperature and pH variations. This property enhances therapeutic efficacy while reducing potential toxic side effects. Drug-loading ratios and release capabilities are critical factors in designing nanocellulose-based composite films (Huo et al. 2022). Saidi et al. (2017) developed bacterial cellulose (BC)/poly(N-methacryloyl glycine) (PMGly) composite films for controlled diclofenac delivery. These films were created through the *in situ* polymerization of methacryloyl glycine monomer within the BC network. Diclofenac was loaded into the BC/PMGly films by immersing the wet composite films in a drug solution. The films exhibited a slower drug release rate (9%) at pH 2.1 over 24 hours but a significantly higher release (85%) at pH 7.4 within the same period, indicating controlled and pH-sensitive drug delivery properties suitable for transdermal and oral delivery.

**Chitosan (CHT)**, derived from chitin, is the second most abundant naturally occurring cationic amino polysaccharide after cellulose and has been utilized in the pharmaceutical arena since 1990 due to the versatility of its active amino groups (Fazal et al. 2023).

Chitosan has recently gained attention as a promising and adaptable solution for biomedical applications due to its natural origin and structural similarity to glycosaminoglycans found in living tissues (Han et al. 2023). Its intrinsic characteristics, including biodegradability, biocompatibility, non-toxicity, antimicrobial properties, and capacity for functionalization, render it appealing in water treatment, biosensors, and medical fields. However, a significant limitation of chitosan is its poor solubility in aqueous and basic solutions. This obstacle can be mitigated by modifying its structure with diverse functional groups or processing it into various forms such as gels, microspheres, films, nanosheets, nanoparticles, and hydrogels (Ma et al. 2022). These modifications enhance chitosan's versatility and potential as a biomaterial in numerous biomedical and industrial applications. Free primary amino groups in its backbone contribute to its excellent physicochemical properties and favorable interactions with living tissues and cells.

Chitosan films can be synthesized via solvent casting, electrostatic spraying, and dry-wet phase separation (Quintão et al. 2022; Heras-Mozos et al. 2022; Sousa et al. 2023; Sarwar et al. 2020; Jaber 2023). The molecular weight, mucoadhesive properties, and degree of deacetylation are crucial factors influencing its effectiveness as a drug delivery vehicle. Chitosan films have been extensively explored for drug delivery across various epithelia, including nasal, ocular, buccal, intestinal, and transdermal routes. Still, the precise mechanism of chitosan's interaction with living tissues remains debated. Several combinations have been studied to achieve chitosan-based biomaterials responsive to various physico-chemical stimuli; examples are included in Table 4.1. For example, blend films comprising CHT and poly(allylamine hydrochloride) (PAH) for drug delivery applications were prepared using the solution casting method, exhibiting homogeneous architecture and improved surface morphology with increased PAH concentration. They demonstrated a pH-responsive behavior, releasing ciprofloxacin hydrochloride monohydrate (CPX) more rapidly in phosphate buffer saline (pH = 7.4) compared to simulated gastric fluid (pH = 1.2). These findings suggest their potential for drug delivery, tissue regeneration, and biomedical applications like wound dressing, but they still need to be improved in their suitability for oral administration (Sarwar et al. 2020). Another possible strategy is incorporating responsive capsules into films to induce a multifunctional performance, such as emulsion films composed of chitosan/virgin coconut oil doped with photosensitive loaded capsules (Rodrigues et al. 2024). The emulsion films with CUR-loaded capsules were designed for transdermal drug delivery of CUR under a UV-light stimulus. The permeability of CUR (bioactive agent) was easily adjusted by varying the radiation exposure, acting the multilayer shell as a barrier. Oil migration from the emulsion film favored CUR migration due to its lipophilic properties. The system demonstrated efficient release and adequate antioxidant potential.

**Collagen**, the predominant component of connective tissues, is the most plentiful protein in the human body. Its manifold attributes, encompassing mechanical robustness and biocompatible degradability, have identified and explored 29 distinct types of collagen for potential utilization in developing drug delivery systems (Fazal et al. 2023). Collagen-based biomaterials can be formulated into different forms according to the desired medical application of the drug delivery system, as can be observed in Fig. 4.2.

Collagen membranes exhibit remarkable biocompatibility, facilitating excellent tissue integration, osteoblastic adhesion, and rapid vascularization without eliciting a foreign body reaction (Shiohara et al. 2021). Additionally, they stimulate fibroblast chemotaxis and promote hemostasis, thereby expediting wound healing and prompting a reduced immune response compared to synthetic polymers like polystyrene and polyethylene. Despite these advantages, collagen membranes do possess certain limitations. They are susceptible to degradation by enzymes such as collagenase/proteases, macrophage/polymorphic-nuclear leukocyte-derived enzymes, and bacterial proteases, potentially leading to membrane resorption. Furthermore, they may lose their efficacy as space maintainers in humid conditions. Collagen membranes often suffer from inadequate mechanical strength and rapid biodegradation. Researchers have resorted to crosslinking techniques to address these issues to induce strength and degradation control. Glutaraldehyde has traditionally been the primary agent for collagen crosslinking, yet concerns regarding its cytotoxicity have been reported. Consequently, alternative crosslinking materials like genipin and oligomeric proanthocyanidins have been investigated as potential solutions (Amit Kumar 2022; Gu et al. 2019).

Collagen-chitosan-based films and membranes serve as biomaterials, finding applications in antibiotic release, wound healing, cancer treatment, transdermal delivery, cardiac illness, and tissue engineering (Amit Kumar 2022; Di Martino et al. 2019; Liu et al. 2017). Engineering collagen-chitosan-graphene oxide (GO) composite films crosslinked using 1-ethyl-3-(3-dimethylaminopropyl) carbodiimide (EDC), enhanced thermal endurance and mechanical strength due to GO, improving wound healing through controlled release of basic fibroblast growth factor (FGF) (Liu et al. 2017). Implanted on rats, these films facilitated wound remodeling and repair of full-thickness skin wounds, suggesting potential as wound dressing materials for drug delivery and healing. Daza et al. designed collagen/polyvinyl alcohol (PVA) anionic membranes as carriers for ciprofloxacin hydrochloride, with added antibacterial efficacy for ulcerative keratitis treatment (Daza et al. 2020). These membranes sustained drug delivery, inhibiting *S. aureus* and *E. coli* growth for up to 48 hours. They displayed suitable mechanical properties, water content, hydrophilicity, permeability, and pH for corneal interaction. Collagen fibrils in the membrane reduced stromal damage and improved epithelial regeneration. These cost-effective membranes offer secure biomaterials for corneal ulcer treatment.

**Hyaluronic acid (HA)**, is an anionic linear polysaccharide composed of N-acetyl-D-glucosamine and D-glucuronic acid units linked by β-1,4- and β-1,3-glycosidic bonds. Due to its widespread presence in the skin and extracellular tissues, HA is extensively used as a natural polymer matrix for drug delivery, owing to its exceptional biocompatibility, biodegradability, and non-immunogenicity (Han et al. 2023). In recent years, its high moisturizing effect and skin affinity have led to intensive research on its application in natural polymer-based transdermal drug delivery systems (NPTDD). The molecular structure of HA, rich in carboxyl and hydroxyl groups, offers opportunities for structural modifications and property adjustments. Moreover, HA-based drug delivery systems can achieve targeted drug delivery by binding to CD44 receptors, which are often overexpressed by tumor cells, thus demonstrating promising applications in localized precision tumor therapy.

Hyaluronic acid and chitosan were covalently functionalized with catechol groups to enable the assembly of polyelectrolyte layer-by-layer films, benefiting from intrinsic stabilization through dopamine interactions (Lu et al. 2019). This modification not only enhanced the solubility of chitosan but also improved the stability of the resulting films. The proposed film exhibited pH responsiveness, swelling in an acidic environment, and shrinking in an alkaline environment, suggesting its potential for pH-responsive drug release applications. Hydrophilic films, comprising hyaluronic acid and two types of PVA, were loaded with dexamethasone and levofloxacin for post-cataract surgery management and were considered potentially beneficial for treating eye infections accompanied by pain and inflammation (Ghezzi et al. 2023). These films, prepared using a straightforward and scalable method, exhibited a high loading capacity, controlled drug release, and the ability to deliver dexamethasone and levofloxacin to the cornea and across the sclera, potentially reaching the posterior eye segment. By adjusting the hydrolysis grade of the PVA, the swelling behavior of the films could be modified, for customization of film characteristics according to the specific application site within the eye. In summary, this device serves as a versatile delivery platform capable of simultaneously releasing lipophilic and hydrophilic drugs.

**Silk fibroin (SF)**, is a natural protein called the "Queen of fiber". It is obtained from various insects, including spiders, and different species of silkworms, such as mulberry (*Bombyx mori*) and nonmulberry (*Antheraea assama*, *Antheraea mylitta*, *Antheraea pernyi*, and *Philosamia ricini*). It can also be produced synthetically through bioengineering (Silva et al. 2024). Silkworms, especially the *Bombyx mori* silkworm species, spin silk cocoons as a low-density structural constituent to protect the egg, which is composed of two types of proteins: fibrous silk fibroin (SF) and globular silk sericin (SS) (Atila et al. 2024). SF is capable of self-assembling into an anti-parallel β-sheet structure due to repetitive glycine-alanine-glycine-alanine-glycine-serine (GAGAGS) amino acid sequences; hence, the structural rigidity of silk arises from the SF component. On the other hand, sericin (SS) is a rubbery and adhesive coverage surrounding the SF fibers, which is degummed and removed during the SF isolation process. SF exhibits excellent mechanical properties, tunable biodegradability, ability to induce tissue regeneration, and excellent biological compatibility (Silva et al. 2024). SF is a promising Food and Drug Administration (FDA) approved material for drug delivery or biomedical purposes (Pandey et al. 2020). SF is an optimal drug and gene delivery candidate since it can be processed into films, hydrogels, and nano-micro particles. Then, SF devices for drug delivery have often involved incorporating or dissolving the active agent or drug in its solution, which is processed into different shapes, where the drug and its interaction with SF plays a critical role in determining the release kinetics (Pandey et al. 2020; Dorishetty et al. 2020). An increasing number of applications for the protein-based delivery of drugs, genes, and other active molecules has stimulated clinical investigations on SF-based delivery systems. Jeencham et al. (2020) used the soaking method to produce SF films with chitosan loaded with diclofenac sodium, a hydrophilic anti-inflammatory drug. The drug loading capacity and drug release profile were evaluated with the film fibroin content and suggested the use as disposable therapeutic contact lenses-based ophthalmic delivery. The ability to regulate antibiotic delivery in an on/off manner presents a significant advantage by enabling patient-specific tuning of the antibiotic delivery profile. The successful fabrication of *Bombyx mori* SF films and their subsequent modification to create a conductive interpenetrating polymer network has been achieved (Mousavi et al. 2020). The conductive films were electrochemically doped with drugs from aqueous solutions. These films demonstrated low levels of passive release of Texas-Red Gentamicin, with significantly enhanced delivery upon the application of an electrical stimulus throughout the experiments. This capability highlights the potential of SF films for precise and controllable electrochemical drug delivery systems.

### 4.2.2 Fabrication Techniques

Many fabrication techniques, including solvent casting, electrospinning, electrostatic spraying, layer-by-layer assembly, and dry-wet phase separation, have produced membranes with customized structures and specialized functions (Fig. 4.3A(a–d)).

**Solvent casting** is a commonly used method for creating membranes or films. This process involves drying a solution that forms a film at room temperature or a specific temperature, which leads to forming films with a smooth surface and uniform thickness (Jaber 2023; Gu et al. 2019; He et al. 2021) (Fig. 4.3A–a). It is essential to consider the solubility or affinity of the polymer and other components (e.g., active substances or drugs) with the solvent when selecting a solvent system for preparing the film-forming solution. The inclusion of other components, such as

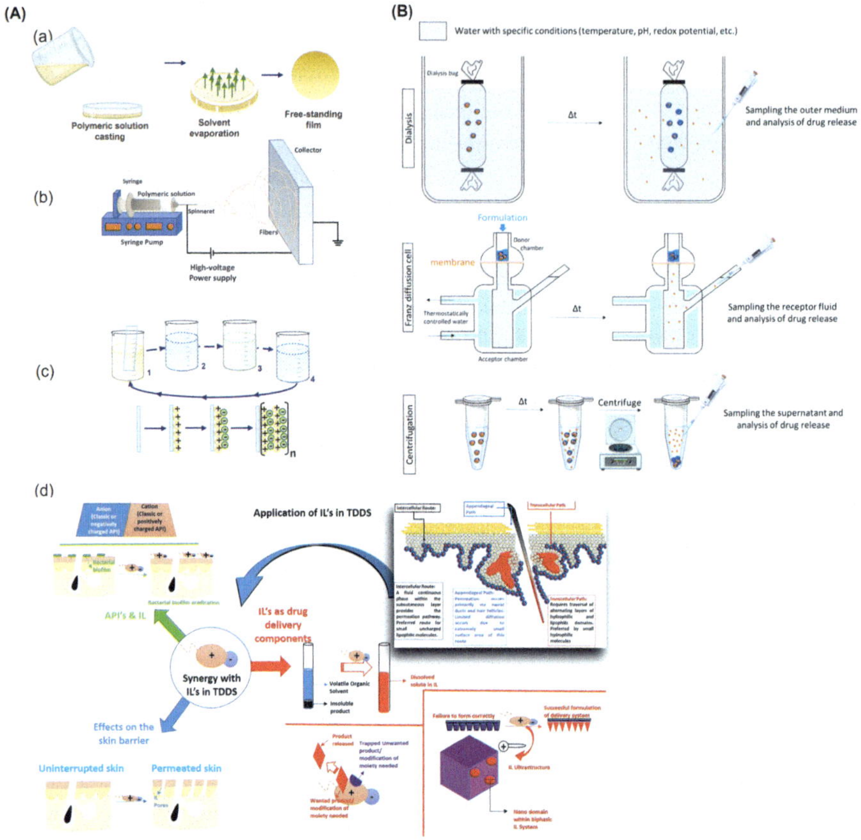


**Fig. 4.3** (**A**) Preparation methodologies: (a) solvent casting, (b) electrospinning, (c) Layer-by-layer, and (d) Transdermal permeation approach. (Reprinted with Copyright permission from Creative Commons Corporation (Sidat et al. 2019)). (**B**) *In vitro* triggered drug release studies: methods used to separate the released drug from its smart system. (Reprinted with Copyright permission from Creative Commons Corporation (Van Gheluwe et al. 2021))

plasticizers and stabilizers, into the solution ensures adequate dissolution and prevents the presence of particles or aggregates in the resulting film.

**Electrospinning** is a process that utilizes electrostatic forces to produce nanofibers from a polymeric solution. In this process, a jet of liquid creates nanofibers and deposits them onto a metallic collector (Fig. 4.3A–b). Various parameters such as concentration, molecular weight, viscosity, elasticity, electrical conductivity, surface tension, and polymer polarity directly affect nanofiber formation (He et al. 2021; Li et al. 2023). Compared to solvent casting, electrospinning can produce stronger films without plasticizers (Li et al. 2023). Electrospinning is preferred for its ability to produce materials with nanoscale features, allowing the processing of different polymers into polymeric nanofibers. The advantages of electrospun nanofibers in drug delivery systems include a high surface area-to-volume ratio, which enhances drug loading and absorption, and their porosity and tunable pore size enable controlled and sustained drug release (Cui et al. 2021). Additionally, the customization of the polymer selection allows the delivery of hydrophilic and hydrophobic drugs, proteins, and genes. At the same time, active substances within nanofibers protect them from degradation and enhance stability and shelf life.

**Layer-by-layer (LbL)** assembly technology is a simple, cost-effective, reproducible, and versatile methodology. It has been widely used to functionalize surfaces and create controlled and sustained drug-release LbL multilayer films (Sarwar et al. 2020; Ghezzi et al. 2023; Khansari et al. 2009; Budker et al. 2014). In the LbL assembly process, multilayer films are deposited onto the surface of a substrate via alternate adsorption of the oppositely charged polyelectrolytes, followed by rinsing and drying steps (Fig. 4.3A–c). The primary force in LbL self-assembly is electrostatic attraction, with other interactions, including van der Walls forces, ionic bonding, and hydrogen bonds, potentially contributing to the multilayer formation. Changing the immersion sequence and material type enables the construction of multilayer films with distinct structures and compositions, desired properties, and improved functions. Owing to the mild construction conditions, lack of restrictions on the size or shape of the substrate, and versatility, therapeutical molecules can be incorporated into them, enhancing their potential use as drug and/or gene delivery systems (Sarwar et al. 2020; Budker et al. 2014; Trombino et al. 2021). Despite that, the conventional LbL method produces ultrathin films, and reaching micro-scale thickness requires hundreds of layers. This necessitates specialized automated machines or adjustments to the solution pH or polyelectrolyte molecular weight to accelerate film growth.

The research focuses on developing environmentally friendly methods and preparation techniques for high-performance membranes to meet sustainability demands, resource efficiency, health safety, and performance. In membrane production, there is a growing interest in using less-toxic solvents, envisioning to preserve human health and the environment. Ionic liquids (ILs), which are organic salts, are being used in biomedicine, especially in TDS, due to their unique features, such as good chemical stability and biocompatibility (Lu et al. 2022). Indeed, ILs, especially the biocompatible ones (Bio-ILs), are considered the best option for replacing hazardous solvents without compromising membrane performance. This is due to their non-volatility, ability to customize specific properties, solubility in water, and biocompatibility. ILs offer new perspectives for sustainable material fabrication. ILs can serve as penetration enhancers to bypass the barrier of the stratum corneum (Zhang et al. 2023; Mahmudi et al. 2024) (Fig. 4.3A–d). For instance, hydrophilic ILs open tight junctions in the stratum corneum by enhancing fluidization in protein and lipid regions to promote paracellular transport. Hydrophobic ILs enhance their distribution into the epithelial membrane by providing channels, thereby promoting the transcellular transport of lipid regions.

### 4.2.3 *In Vitro* Triggered Drug Release Characterization

Characterizing drug release *in vitro* is crucial in developing and evaluating drug delivery systems. These systems are designed to release therapeutic or active molecules in response to contact with a liquid medium or specific environmental triggers such as pH, temperature, or enzyme activity. By accurately mimicking these triggers in a controlled laboratory setting, an evaluation of the efficiency, kinetics, and mechanism of drug release is required to predict *in vivo* performance and optimize the design of the systems. Typically, the release of the active substance or drug is evaluated by placing the system in contact with a liquid medium, such as phosphate buffer solution (PBS) or another buffer that mimics the intended stimulus. Samples of the release medium are collected at pre-determined intervals, and the amount of active substance released is quantified in the release medium using appropriate analytical methods, designing a release kinetics curve. Three major separation methods are commonly used: dialysis, Franz cells, and centrifugation

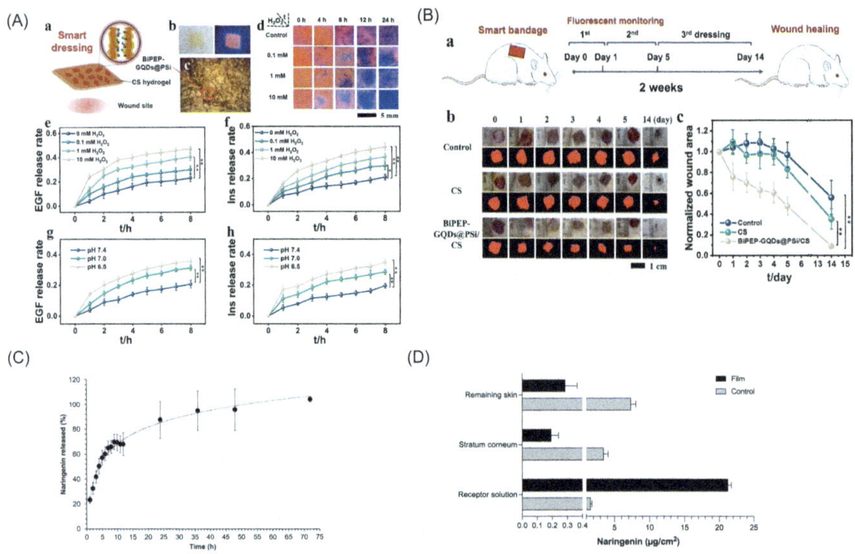


**Fig. 4.4** (**A**) (a) Schematic illustration for the structure of smart dressing based on BiPEP-GQDs@PSi/CS matrix. (b) Photographs of the as-prepared wound dressing with or without the UV excitation (377 nm). (c) Optical microscopic image of the dressing comprising BiPEP-GQDs@ PSi and CS matrix. (Scale bar length is 200 μm). (d) Fluorescence photos of the smart dressing on simulated skin with different concentration of $H_2O_2$. (e, f) Drug release curve of EGF (e) and insulin (f) form the smart dressing with different concentration of $H_2O_2$. (g, h) Drug release curve of EGF (g) and insulin (h) from smart dressing at different pH. ∗Represents $P < 0.05$, and ∗∗represents $P < 0.01$ when the data were compared with control, $n = 3$. (Reprinted with permission from Elsevier (Cui et al. 2021)). (**B**) Effect of different dressings (Control, CS, or BiPEP-GQDs@PSi/CS) on diabetic wounds healing. (a) Schematic illustration of the *in vivo* diabetic wound experiments. (b) Representative images and corresponding schematic diagram of the wounds in different groups for 14 days. (c) Statistical analysis of normalized wound area in different groups for 14 days. ∗∗Represents $P < 0.01$ when the data were compared with control, $n = 3$. (Reprinted with permission from Elsevier (Cui et al. 2021)). (**C**) Release profile of naringenin from the chitosan film in PBS with 2.5% (w/v) Pluronic® F127 (pH 7.4). (Reprinted with permission from Elsevier (Quintão et al. 2022)). (**D**) Naringenin recovered from the remaining skin (epidermis + dermis) and the stratum corneum and quantified in the receptor solution after 72 hours of skin treatment with the chitosan-based film and the naringenin control solution. (Reprinted with permission from Elsevier (Quintão et al. 2022))

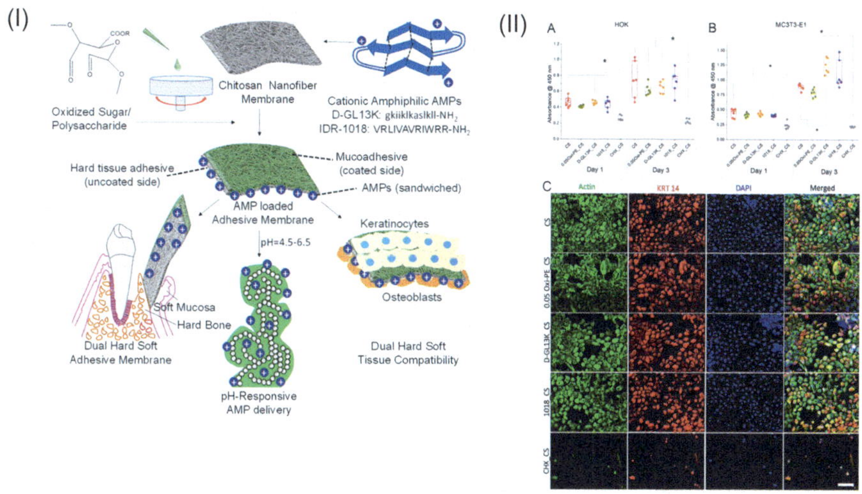


**Fig. 4.5** (I) Schematic illustration of the workflow in the current study: (i) fabrication of oral tissue adhesive nanofiber membranes loaded with cationic AMPs, (ii) pH-responsive delivery of AMPs to acidogenic oral biofilms, (iii) dual hard and soft tissue compatibility of the membranes for oral tissue regeneration. (Reprinted with permission from ACS Publications (Boda et al. 2020)). (II) Proliferation of (A) immortalized human oral keratinocytes (OKF-6/TERT2) and (B) murine pre-osteoblasts (MC3T3-E1) on surface modified chitosan nanofiber membranes. Each data point represents one replicate, and boxplots present statistical values for each group. *Indicates statistical significance at $p < 0.05$ between compared groups. (C) Triple immunofluorescent staining of adhered oral keratinocytes on surface modified chitosan nanofiber membranes after 3 days of culture. Green = actin, red = cytokeratin 14, blue = nuclei, scale bar = 75 μm. (Reprinted with permission from ACS Publications (Boda et al. 2020))

(Fig. 4.3B) (Van Gheluwe et al. 2021). These methods are essential for accurately determining the drug release profile and ensuring the reliability of the *in vitro* studies. Below is a summary of these methods.

**Dialysis method** This method involves placing the drug delivery system in a dialysis membrane of appropriate pore size containing release media (inner media). It is sealed and then placed in a larger vessel containing release media (outer media). The bag allows the drug to diffuse into the medium while retaining the system. At a pre-determined time, an appropriate volume of the outer medium is withdrawn and measured. This method is suitable for pH triggering as the outer medium.

**Franz Diffusion cells** These cells are used to study drug release with a synthetic membrane. Franz cells consist of two primary chambers (donor and acceptor compartment) separated by a membrane. The drug delivery system is placed in the donor compartment, and the drug diffuses through the membrane into the bottom chamber (acceptor compartment) containing fluid from which samples can be taken at regular analysis intervals (Van Gheluwe et al. 2021; He et al. 2021). For example, the acceptor compartment is filled with the buffer media of interest (specific pH) for pH-triggered drug release studies. For temperature-triggered drug release, the system is kept in a temperature-controlled water bath to maintain the temperature of interest in the donor compartment.

**Centrifugation** In this method, the drug delivery system is placed in a medium with specific conditions (pH, temperature), and at predetermined times, aliquots are withdrawn and centrifuged, and the supernatant is analyzed (Van Gheluwe et al. 2021).

## 4.3 Applications of Smart Membranes in the Biomedical Field

Smart-based membranes made of biomacromolecules seem attractive for biomedical applications due to their diversity and ease of processing. They have shown promise in diverse biomedical applications, including wound healing, cancer therapy, tissue regeneration, bone tissue regeneration, corneal ulcer treatment, implantable devices, and ocular drug delivery. Table 4.1 provides a summary of many examples of these systems. Recent findings have shown that successful drug delivery for wound healing involves using stimuli-responsive films to control drug release and regulate different factors of the pathological microenvironment (Li et al. 2023). For example, Cui et al. (2021) developed a stimuli-responsive smart dressing for theranostics of diabetic wounds using graphene quantum dots-decorated luminescent porous silicon (GQDs@PSi). This dressing, loaded with a peptide and embedded in a CHT film, displayed red fluorescence under UV light for easy wound monitoring (Fig. 4.4A-a, b). The GQDs can enhance peptide drug loading and form a stimulus-responsive visual ratiometric sensor with luminescent Psi, offering sustained peptide release for effective diabetic wound treatment. Embedded in a CHT film, the dressing shows $H_2O_2$-responsive fluorescence changes and smoother drug release kinetics than free particles (Fig. 4.4A-e–h). To evaluate the real wound-healing effect of BiPEP-GQDs@PSi/CS smart dressing, the wound-healing *in vivo* was performed *via* a rat model (Fig. 4.4B-a). Photographic imaging over 2 weeks showed that wounds treated with the dressing healed significantly faster than those in the control group, with almost complete healing after 14 days (Fig. 4.4B-b, c). Thus, the smart dressing significantly accelerates diabetic wound healing due to its efficient therapeutic peptide delivery.

Researchers are exploring the use of antioxidants to treat chronic inflammatory disorders, which are frequently linked to an oxidative stress process (Khansari et al. 2009). To ensure greater comfort and acceptance of long-term therapy of chronic inflammatory diseases, Quintão et al. (2022) proposed a transdermal delivery system for naringenin, a bioflavonoid found in citrus fruits, by incorporating it into a CHT-based film. This system enhances bioavailability and reduces toxicity by avoiding first-pass metabolism. It also allows treatment to be stopped easily by removing the film. The drug release profile (Fig. 4.4C) shows an initial burst effect, which creates a high gradient for drug diffusion into the skin. After a 72-hours treatment, naringenin retention in the skin's first layer increased 17-fold compared to the control (Fig. 4.4D), indicating that the CHT film effectively permeabilized the stratum corneum, enabling rapid drug diffusion. Thus, the CHT film acted as an efficient platform for the transdermal delivery of naringenin, ensuring consistent drug permeation with minimal retention in the skin layers.

Breast cancer treatment often involves surgery, chemotherapy, radiotherapy, and treatment with monoclonal antibodies, but these methods come with numerous side effects. One strategy to reduce these side effects is to target the drug directly to tumors and cancer cells (Budker et al. 2014). Trombino et al. (2021) designed implantable dense polymeric membranes made from carboxylated CHT and cyclosporine A (CsA), a lipophilic drug and a powerful immunosuppressant used in organ transplant treatment. Thus, this system aims to prolong drug release at the tumor site, ensuring even distribution and minimizing side effects. *In vitro* permeation studies using CHT membranes with coumarin-6, a lipophilic model drug, and analogy to CsA, showed that the drug released from the membrane was present on the skin surface and in deeper subcutaneous layers. Membranes containing CsA and free CsA significantly reduced the viability of human breast cancer MDA-MB-231 cells, an effect not seen with free CsA alone at the same concentration. Thus, these membranes could be a promising strategy for delivering CsA to breast cancer patients, reducing systemic toxicity.

During infection and inflammation of bone tissues, administering drugs through systemic delivery often fails to treat defects/wounds in bone tissues due to the lack of adequate tissue penetration of drugs and poor blood supply to the bone (Mahmudi et al. 2024). Boda et al. (2020) fabricated CHT-based nanofiber membranes. They were surface-coated with oxidized pectin to enhance the mucoadhesion properties (Fig. 4.5I). To demonstrate the potential application of the oral tissue dual adhesive nanofiber membranes for the delivery of antimicrobials in a pH-responsive manner, TAMRA-tagged antimicrobial GL13K peptides were incorporated into the CHT nanofiber membrane. The cumulative release profiles at pH 4.5, 5.5, and 6.5 from chitosan membranes for TAMRA-GL13K showed a pH-dependent release of antimicrobial peptide from the nanofiber matrixes. The pH dependence of antimicrobial peptide release can be attributed to two factors: (i) increasing peptide solubility with decreasing pH and/or (ii) greater chitosan degradation with decreasing pH, which would enable the release of the absorbed peptides. In addition, a good proliferation of human oral keratinocytes (HOK) and murine pre-osteoblasts (MC3T3-E1) on pristine and surface-modified CHT membranes was showed, and the immunofluorescent images of adhered HOK on all membranes displayed good cytoskeletal spreading, morphology and cytokeratin 14(KRT14) marker expression (Fig. 4.5II).

Bone tissue engineering with the help of guided bone biomaterials has emerged in response to substantial clinical demand for the ability to repair critical-sized bony defects. GBR is an established treatment that enhances bone regeneration through cellular exclusion and avoids the invasion of epithelial and connective tissues that grow at the defective site instead of bone tissue (Gentile et al. 2011). In line of this research field, polymeric membranes are frequently used for bone regeneration in oral and periodontal surgery (Toledano-Osorio et al. 2022; Alqahtani 2023; Xue et al. 2021). For example, Xue et al. (2021) designed collagen/polycaprolactone (PCL) composite membranes modified with zeolitic imidazolate framework-8 (ZIF-8) for GBR. In this study, the PCL/collagen/ZIF-8 membrane would serve as a barrier layer that prevents the infiltration of fibroblasts, while a ZIF-8 crystal layer provides a beneficial microenvironment for GBR. The findings indicated that incorporating ZIF-8 nanoparticles rendered the membrane with an enlarged surface area and pH-responsive release of $Zn^{2+}$ ions, hence triggering rBMSC recruitment and coordination of subsequent osteogenesis and angiogenesis processes. Once implanted in rats, the PCL/Col/ZIF-8 composite membrane adhered to the bone tissue surrounding the defect without additional grafting materials and fixation pins. The 3D reconstructed images displayed newly formed bone within the defects, while the bone volume to tissue volume (BV/TV) in the composite group was markedly higher than that in the other groups (blank, PCL/collagen and collagen group). Moreover, the histological observation based on hematoxylin–eosin indicated the new bone grown from the defect margins in the composite membrane and collage groups, suggesting the osteoconductive ability of the membranes.

## 4.4 Final Remarks and Future Trends

The development of novel drug delivery methods to address a variety of diseases has drawn a lot of attention in the medical field. This chapter summarizes recent academics on the design of smart systems using stimuli-responsive polymeric films. Natural biomacromolecules possess inherent advantages and can produce intelligent delivery systems. However, polymeric films often face challenges such as low mechanical properties, low flexibility, and difficulty in providing customized drug release. Routes on blends based on natural or in combination with synthetic polymers and chemical modification are used to reinforce the polymeric network and address these drawbacks, resulting in improved encapsulation efficiency and extended release compared to pure formulations. Furthermore, an optimal film design should consider essential properties, including compatibility between the drug and polymer, barrier properties (e.g., swelling capacity), mechanical strength, adhesion, drug release kinetics, and drug permeation. Still, despite the extensive background of studied systems, an infinite potential exists for combinations of drugs, polymers, excipients, manufacturing processes, and routes of administration to be explored.

Innovative formulations and multifunctional materials, serving as permeation enhancers and dispersion agents, can significantly simplify the design and development of transdermal drug delivery systems. Furthermore, incorporating green solvents such as ionic liquids into polymeric formulations can enhance skin penetration and contribute to the release of active molecules.

Multifunctional polymeric film systems are primarily used for topical and transdermal delivery but can also be administrated to other organs, including the oral cavity and eyes, presenting a valuable alternative for controlled drug delivery. The data collected emphasizes the versatility of these drug delivery systems and the pressing need for further research and collaboration, as it can be suitable for nearly all therapies requiring controlled release to maintain drug levels over longer time periods.

Despite extensive research, most studies have been limited to *in vitro* characterization, with only a small percentage addressing *in vivo* investigations or clinical trials. As such, successful market translation continues to be extremely difficult, but it remains essential to design appropriate scale-up strategies, including time and technological development, marketing resource allocation, and providing claim validation and regulatory compliance.

Artificial intelligence (AI) might bring significant advances in the development of stimuli-responsive polymeric films as drug-delivery systems. AI can assist in selecting and optimizing the polymeric system and simulating drug release profiles. It also has the potential to predict biocompatibility, drug behavior under several conditions, and therapeutic efficacy. This reduces the necessity for extensive testing and facilitates the creation of personalized drug delivery systems customized to meet individual patient needs. As AI technology advances, it can also help navigate regulatory pathways and optimize the scale-up process from laboratory to industrial production, ensuring consistency, quality, and cost-effectiveness. Furthermore, close collaboration among researchers, clinicians, stakeholders, politicians, and society is essential for translating advancements in drug delivery systems into clinical practice.

**Acknowledgments** This work is also financially supported by Portuguese FCT (CEEC Institutional FCT program- CEECINST/00018/2021/CP2806/CT0012 (S.S.S.) DOI https://doi.org/10.54499/CEECINST/00018/2021/CP2806/CT0012 (https://doi.org/10.54499/CEECINST/00018/2021/CP2806/CT0012); CEECINST/00018/2021/CP 2806/CT0013 (E.M.F), (DOI https://doi.org/10.54499/CEECINST/00018/2021/CP2806/CT0013) and DL 57/2016/CP1377/CT0054 (L.C.R) (https://doi.org/10.54499/DL57/2016/CP1377/CT0054). The authors would like to thank the contributions to this research from the project "TERM RES Hub—Scientific Infrastructure for Tissue Engineering and Regenerative Medicine", reference PINFRA/22190/2016 (Norte-01-0145-FEDER-022190), funded by the Portuguese National Science Foundation (FCT) in cooperation with the Northern Portugal Regional Coordination and Development Commission (CCDR-N), for providing relevant lab facilities, state-of-the-art equipment, and highly qualified human resources.

## References

Abedini A, Pircheraghi G, Kaviani A, Hosseini S (2023) Exploration of curcumin-incorporated dual anionic alginate-quince seed gum films for transdermal drug delivery. Int J Biol Macromol 248:125798. https://doi.org/10.1016/j.ijbiomac.2023.125798

Ahmed M, Bhat AR, Verma AK, Patel R (2023) Collagen–PVA films plasticized with choline acetate ionic liquid for sustained drug release: UV shielding, mechanical, antioxidant, and antibacterial properties. ACS Appl Bio Mater 6(2):663–673. https://doi.org/10.1021/acsabm.2c00918

Alavi SE, Alharthi S, Alavi SZ, Raza A, Ebrahimi Shahmabadi H (2024) Bioresponsive drug delivery systems. Drug Discov Today 29(1):103849. https://doi.org/10.1016/j.drudis.2023.103849

Alotaibi BS, Khan AK, Kharaba Z, Yasin H, Yasmin R, Ijaz M et al (2024) Development of poly(vinyl alcohol)–chitosan composite nanofibers for dual drug therapy of wounds. ACS Omega 9(11):12825–12834. https://doi.org/10.1021/acsomega.3c08856

Alqahtani AM (2023) Guided tissue and bone regeneration membranes: a review of biomaterials and techniques for periodontal treatments. Polymers 15(16):3355. https://doi.org/10.3390/polym15163355

Alshhab A, Yilmaz E (2020) Sodium alginate/poly(4-vinylpyridine) polyelectrolyte multilayer films: preparation, characterization and ciprofloxacin HCl release. Int J Biol Macromol 147:809–820. https://doi.org/10.1016/j.ijbiomac.2019.10.058

Amit Kumar V (2022) Ch. 5: Collagen-based biomaterial as drug delivery module. In: Nirmal M, Sanjiban C (eds) Collagen biomaterials. IntechOpen, Rijeka. https://doi.org/10.5772/intechopen.103063

Anghel N, Dinu VM, Verestiuc L, Spiridon IA (2021) Transcutaneous drug delivery systems based on collagen/polyurethane composites reinforced with cellulose. Polymers (Basel) 13(11). https://doi.org/10.3390/polym13111845

Anirudhan TS, Nair AS, Gopika SS (2018) The role of biopolymer matrix films derived from carboxymethyl cellulose, sodium alginate and polyvinyl alcohol on the sustained transdermal release of diltiazem. Int J Biol Macromol 107:779–789. https://doi.org/10.1016/j.ijbiomac.2017.09.053

Atila D, Gurses S, Tezcaner A (2024) Self-composite silk fibroin membranes for guided bone regeneration: 1 material 3 architectures with intrinsic integrity. Mater Today Commun 39:109215. https://doi.org/10.1016/j.mtcomm.2024.109215

Boda SK, Fischer NG, Ye Z, Aparicio C (2020) Dual oral tissue adhesive nanofiber membranes for pH-responsive delivery of antimicrobial peptides. Biomacromolecules 21(12):4945–4961. https://doi.org/10.1021/acs.biomac.0c01163

Budker VG, Monahan SD, Subbotin VM (2014) Loco-regional cancer drug therapy: present approaches and rapidly reversible hydrophobization (RRH) of therapeutic agents as the future direction. Drug Discov Today 19(12):1855–1870. https://doi.org/10.1016/j.drudis.2014.08.009

Chen Y, Abdalkarim SYH, Yu H-Y, Li Y, Xu J, Marek J et al (2020) Double stimuli-responsive cellulose nanocrystals reinforced electrospun PHBV composites membrane for intelligent drug release. Int J Biol Macromol 155:330–339. https://doi.org/10.1016/j.ijbiomac.2020.03.216

Chen K, Li Y, Li Y, Tan Y, Liu Y, Pan W et al (2023) Stimuli-responsive electrospun nanofibers for drug delivery, cancer therapy, wound dressing, and tissue engineering. J Nanobiotechnol 21(1):237. https://doi.org/10.1186/s12951-023-01987-z

Chen X, Li Z, Yang C, Yang D (2024) Ionic liquids as the effective technology for enhancing transdermal drug delivery: design principles, roles, mechanisms, and future challenges. Asian J Pharm Sci 19(2):100900. https://doi.org/10.1016/j.ajps.2024.100900

Cui Y, Duan W, Jin Y, Wo F, Xi F, Wu J (2021) Graphene quantum dot-decorated luminescent porous silicon dressing for theranostics of diabetic wounds. Acta Biomater 131:544–554. https://doi.org/10.1016/j.actbio.2021.07.018

Daza JHU, Righetto GM, Chaud MV, da Conceição Amaro Martins V, Lopes Baratella da Cunha Camargo I, Maria de Guzzi Plepis A (2020) PVA/anionic collagen membranes as drug carriers of ciprofloxacin hydrochloride with sustained antibacterial activity and potential use in the treatment of ulcerative keratitis. J Biomater Appl 35(3):301–312. https://doi.org/10.1177/0885328220931733

Di Martino A, Drannikov A, Surgutskaia NS, Ozaltin K, Postnikov PS, Marina TE et al (2019) Chitosan-collagen based film for controlled delivery of a combination of short life anesthetics. Int J Biol Macromol 140:1183–1193. https://doi.org/10.1016/j.ijbiomac.2019.08.228

Dorishetty P, Dutta NK, Choudhury NR (2020) Silk fibroins in multiscale dimensions for diverse applications. RSC Adv 10(55):33227–33247. https://doi.org/10.1039/d0ra03964k

Fazal T, Murtaza BN, Shah M, Iqbal S, Rehman M-u, Jaber F et al (2023) Recent developments in natural biopolymer based drug delivery systems. RSC Adv 13(33):23087–23121. https://doi.org/10.1039/D3RA03369D

Gentile P, Chiono V, Tonda-Turo C, Ferreira AM, Ciardelli G (2011) Polymeric membranes for guided bone regeneration. Biotechnol J 6(10):1187–1197. https://doi.org/10.1002/biot.201100294

Ghauri ZH, Islam A, Qadir MA, Gull N, Haider B, Khan RU et al (2021) Development and evaluation of pH-sensitive biodegradable ternary blended hydrogel films (chitosan/guar gum/PVP) for drug delivery application. Sci Rep 11(1):21255. https://doi.org/10.1038/s41598-021-00452-x
Ghezzi M, Ferraboschi I, Fantini A, Pescina S, Padula C, Santi P et al (2023) Hyaluronic acid – PVA films for the simultaneous delivery of dexamethasone and levofloxacin to ocular tissues. Int J Pharm 638:122911. https://doi.org/10.1016/j.ijpharm.2023.122911
Gierszewska M, Ostrowska-Czubenko J, Chrzanowska E (2018) pH-responsive chitosan/alginate polyelectrolyte complex membranes reinforced by tripolyphosphate. Eur Polym J 101:282–290. https://doi.org/10.1016/j.eurpolymj.2018.02.031
Gu L, Shan T, Ma Y x, Tay FR, Niu L (2019) Novel biomedical applications of crosslinked collagen. Trends Biotechnol 37(5):464–491. https://doi.org/10.1016/j.tibtech.2018.10.007
Han W, Liu F, Li Y, Liu G, Li H, Xu Y et al (2023) Advances in natural polymer-based transdermal drug delivery systems for tumor therapy. NanoMicro Small 19(35):2301670. https://doi.org/10.1002/smll.202301670
He S, Jacobsen J, Nielsen CU, Genina N, Østergaard J, Mu H (2021) Exploration of in vitro drug release testing methods for saquinavir microenvironmental pH modifying buccal films. Eur J Pharm Sci 163:105867. https://doi.org/10.1016/j.ejps.2021.105867
Heras-Mozos R, Gavara R, Hernández-Muñoz P (2022) Chitosan films as pH-responsive sustained release systems of naturally occurring antifungal volatile compounds. Carbohydr Polym 283:119137. https://doi.org/10.1016/j.carbpol.2022.119137
Huo Y, Liu Y, Xia M, Du H, Lin Z, Li B et al (2022) Nanocellulose-based composite materials used in drug delivery systems. Polymers 14(13):2648. https://doi.org/10.3390/polym14132648
Jaber SA (2023) Transdermal patches based on chitosan/hydroxypropyl methylcellulose and polyvinylpyrrolidone/hydroxypropyl methylcellulose polymer blends for gentamycin administration. J Adv Pharm Technol Res 14(3):202–207. https://doi.org/10.4103/japtr.japtr_130_23
Jeencham R, Sutheerawattananonda M, Rungchang S, Tiyaboonchai W (2020) Novel daily disposable therapeutic contact lenses based on chitosan and regenerated silk fibroin for the ophthalmic delivery of diclofenac sodium. Drug Deliv 27(1):782–790. https://doi.org/10.1080/10717544.2020.1765432
Juhaščik M, Kováčik A, Huerta-Ángeles G (2022) Recent advances of hyaluronan for skin delivery: from structure to fabrication strategies and applications. Polymers 14(22):4833. https://doi.org/10.3390/polym14224833
Kavand A, Noverraz F, Gerber-Lemaire S (2024) Recent advances in alginate-based hydrogels for cell transplantation applications. Pharmaceutics 16(4):469. https://doi.org/10.3390/pharmaceutics16040469
Khansari N, Shakiba Y, Mahmoudi M (2009) Chronic inflammation and oxidative stress as a major cause of age- related diseases and cancer. Recent Patents Inflamm Allergy Drug Discov 3(1):73–80. https://doi.org/10.2174/187221309787158371
Koc FE, Altıncekic TG (2021) Investigation of gelatin/chitosan as potential biodegradable polymer films on swelling behavior and methylene blue release kinetics. Polym Bull 78(6):3383–3398. https://doi.org/10.1007/s00289-020-03280-7
Li C, Wang J, Wang Y, Gao H, Wei G, Huang Y et al (2019) Recent progress in drug delivery. Acta Pharm Sin B 9(6):1145–1162. https://doi.org/10.1016/j.apsb.2019.08.003
Li J, Xiang H, Zhang Q, Miao X (2022) Polysaccharide-based transdermal drug delivery. Pharmaceuticals (Basel, Switzerland) 15(5). https://doi.org/10.3390/ph15050602
Li H, Li B, Lv D, Li W, Lu Y, Luo G (2023) Biomaterials releasing drug responsively to promote wound healing via regulation of pathological microenvironment. Adv Drug Deliv Rev 196:114778. https://doi.org/10.1016/j.addr.2023.114778
Li P, Yu Y, Song H, Dai M, Wang F, Yin Y et al (2024) Potential applications of ROS-responsive silk fibroin materials in smart drug delivery systems. ACS Appl Polym Mater 6(6):3413–3421. https://doi.org/10.1021/acsapm.4c00037
Lisuzzo L, Cavallaro G, Milioto S, Lazzara G (2019) Layered composite based on halloysite and natural polymers: a carrier for the pH controlled release of drugs. New J Chem 43(27):10887–10893. https://doi.org/10.1039/C9NJ02565K

Liu T, Dan W, Dan N, Liu X, Liu X, Peng X (2017) A novel grapheme oxide-modified collagen-chitosan bio-film for controlled growth factor release in wound healing applications. Mater Sci Eng C 77:202–211. https://doi.org/10.1016/j.msec.2017.03.256

Liu Y, Song R, Zhang X, Zhang D (2020) Enhanced antimicrobial activity and pH-responsive sustained release of chitosan/poly (vinyl alcohol)/graphene oxide nanofibrous membrane loading with allicin. Int J Biol Macromol 161:1405–1413. https://doi.org/10.1016/j.ijbiomac.2020.08.051

Lu B, Luo D, Zhao A, Wang H, Zhao Y, Maitz MF et al (2019) pH responsive chitosan and hyaluronic acid layer by layer film for drug delivery applications. Prog Org Coat 135:240–247. https://doi.org/10.1016/j.porgcoat.2019.06.012

Lu B, Liu T, Wang H, Wu C, Chen H, Liu Z et al (2022) Ionic liquid transdermal delivery system: progress, prospects, and challenges. J Mol Liq 351:118643. https://doi.org/10.1016/j.molliq.2022.118643

Ma J, Wang Y, Lu R (2022) Mechanism and application of chitosan and its derivatives in promoting permeation in transdermal drug delivery systems: a review. Pharmaceuticals 15(4):459. https://doi.org/10.3390/ph15040459

Mahmudi M, Ardhani R, Pidhatika B, Suyanta S, Swasono YA, Rudianto RP et al (2024) Development of a local drug delivery system for promoting the regeneration of infective bone defects: composite films with controlled properties. Polym Bull. https://doi.org/10.1007/s00289-024-05243-8

Malviya R, Tyagi A, Fuloria S, Subramaniyan V, Sathasivam K, Sundram S et al (2021) Fabrication and characterization of chitosan—tamarind seed polysaccharide composite film for transdermal delivery of protein/peptide. Polymers 13(9):1531. https://doi.org/10.3390/polym13091531

Manna S, Gupta P, Nandi G, Jana S (2023) Recent update on alginate based promising transdermal drug delivery systems. J Biomater Sci Polym Ed 34(16):2291–2318. https://doi.org/10.1080/09205063.2023.2230847

Mohanta V, Madras G, Patil S (2014) Layer-by-layer assembled thin films and microcapsules of nanocrystalline cellulose for hydrophobic drug delivery. ACS Appl Mater Interfaces 6(22):20093–20101. https://doi.org/10.1021/am505681e

Mousavi ST, Harper GR, Municoy S, Ashton MD, Townsend D, Alsharif GHK et al (2020) Electroactive silk fibroin films for electrochemically enhanced delivery of drugs. Macromol Mater Eng 305(6):2000130. https://doi.org/10.1002/mame.202000130

Nasseri R, Deutschman CP, Han L, Pope MA, Tam KC (2020) Cellulose nanocrystals in smart and stimuli-responsive materials: a review. Mater Today Adv 5:100055. https://doi.org/10.1016/j.mtadv.2020.100055

Pandey V, Haider T, Jain P, Gupta PN, Soni V (2020) Silk as a leading-edge biological macromolecule for improved drug delivery. J Drug Deliv Sci Technol 55:101294. https://doi.org/10.1016/j.jddst.2019.101294

Pavaloiu R-D, Stoica-Guzun A, Stroescu M, Jinga SI, Dobre T (2014) Composite films of poly(vinyl alcohol)–chitosan–bacterial cellulose for drug controlled release. Int J Biol Macromol 68:117–124. https://doi.org/10.1016/j.ijbiomac.2014.04.040

Puiggalí-Jou A, Cazorla E, Ruano G, Babeli I, Ginebra M-P, García-Torres J et al (2020) Electroresponsive alginate-based hydrogels for controlled release of hydrophobic drugs. ACS Biomater Sci Eng 6(11):6228–6240. https://doi.org/10.1021/acsbiomaterials.0c01400

Quadrado RFN, Burgo TAL, Fussinger TD, Rossi GG, de Campos MMA, Iglesias BA et al (2023) Design of a light-responsive porphyrin/polysaccharide hybrid film for smart release of hydrocortisone. Colloids Surf A Physicochem Eng Asp 674:131877. https://doi.org/10.1016/j.colsurfa.2023.131877

Quintão WSC, Silva-Carvalho AE, Hilgert LA, Gratieri T, Cunha-Filho M, Saldanha-Araújo F et al (2022) Anti-inflammatory effect evaluation of naringenin and its incorporation into a chitosan-based film for transdermal delivery. Int J Pharm 627:122231. https://doi.org/10.1016/j.ijpharm.2022.122231

Rodrigues LC, Ribeiro AP, Silva SS, Reis RL (2024) Chitosan/virgin coconut oil-based emulsions doped with photosensitive curcumin loaded capsules: a functional carrier to topical treatment. Polymers 16(5):641. https://doi.org/10.3390/polym16050641
Sabbagh F, Kim BS (2022) Recent advances in polymeric transdermal drug delivery systems. J Control Release 341:132–146. https://doi.org/10.1016/j.jconrel.2021.11.025
Saïdi L, Vilela C, Oliveira H, Silvestre AJD, Freire CSR (2017) Poly(N-methacryloyl glycine)/nanocellulose composites as pH-sensitive systems for controlled release of diclofenac. Carbohydr Polym 169:357–365. https://doi.org/10.1016/j.carbpol.2017.04.030
Saraiva MM, Campelo MS, Câmara Neto JF, Lima ABN, Silva GDA, Dias ATDFF et al (2023) Alginate/polyvinyl alcohol films for wound healing: advantages and challenges. J Biomed Mater Res B Appl Biomater 111(1):220–233. https://doi.org/10.1002/jbm.b.35146
Sarwar MS, Huang Q, Ghaffar A, Abid MA, Zafar MS, Khurshid Z et al (2020) A smart drug delivery system based on biodegradable chitosan/poly(allylamine hydrochloride) blend films. Pharmaceuticals 12(2):131. https://doi.org/10.3390/pharmaceutics12020131
Schafer N, Balwierz R, Biernat P, Ochędzan-Siodłak W, Lipok J (2023) Natural ingredients of transdermal drug delivery systems as permeation enhancers of active substances through the stratum corneum. Mol Pharm 20(7):3278–3297. https://doi.org/10.1021/acs.molpharmaceut.3c00126
Sheikhi A, Hayashi J, Eichenbaum J, Gutin M, Kuntjoro N, Khorsandi D et al (2019) Recent advances in nanoengineering cellulose for cargo delivery. J Control Release 294:53–76. https://doi.org/10.1016/j.jconrel.2018.11.024
Shiohara A, Prieto-Simon B, Voelcker NH (2021) Porous polymeric membranes: fabrication techniques and biomedical applications. J Mater Chem B 9(9):2129–2154. https://doi.org/10.1039/D0TB01727B
Sidat Z, Marimuthu T, Kumar P, du Toit LC, Kondiah PPD, Choonara YE et al (2019) Ionic liquids as potential and synergistic permeation enhancers for transdermal drug delivery. Pharmaceuticals 11(2):96. https://doi.org/10.3390/pharmaceutics11020096
Silva SS, Fernandes EM, Pina S, Silva-Correia J, Vieira S, Oliveira JM et al (2017) 2.11 Polymers of biological origin☆. In: Ducheyne P (ed) Comprehensive biomaterials II. Elsevier, Oxford, pp 228–252. https://doi.org/10.1016/B978-0-12-803581-8.10134-1
Silva SS, Rodrigues LC, Fernandes EM, Reis RL (2020) Chapter 6 – Biopolymer membranes in tissue engineering. In: de Moraes MA, da Silva CF, Vieira RS (eds) Biopolymer membranes and films. Elsevier, pp 141–163. https://doi.org/10.1016/B978-0-12-818134-8.00006-7
Silva SS, Reis RL, Kundu SC (2024) Chapter 2 – Trends in silk biomaterials. In: Kundu SC, Reis RL (eds) Silk-based biomaterials for tissue engineering, regenerative and precision medicine, 2nd edn. Woodhead Publishing, pp 9–39. https://doi.org/10.1016/B978-0-323-96017-5.00007-8
Soe MT, Pongjanyakul T, Limpongsa E, Jaipakdee N (2020) Modified glutinous rice starch-chitosan composite films for buccal delivery of hydrophilic drug. Carbohydr Polym 245:116556. https://doi.org/10.1016/j.carbpol.2020.116556
Sousa CFV, Monteiro LPG, Rodrigues JMM, Borges J, Mano JF (2023) Marine-origin polysaccharides-based free-standing multilayered membranes as sustainable nanoreservoirs for controlled drug delivery. J Mater Chem B 11(28):6671–6684. https://doi.org/10.1039/d3tb00796k
Toledano-Osorio M, Vallecillo C, Vallecillo-Rivas M, Manzano-Moreno FJ, Osorio R (2022) Antibiotic-loaded polymeric barrier membranes for guided bone/tissue regeneration: a mini-review. Polymers (Basel) 14(4). https://doi.org/10.3390/polym14040840
Touitou E, Natsheh H, Zailer J (2023) Film forming systems for delivery of active molecules into and across the skin. Pharmaceutics 15(2). https://doi.org/10.3390/pharmaceutics15020397
Trombino S, Curcio F, Poerio T (2021) Chitosan membranes filled with cyclosporine A as possible devices for local administration of drugs in the treatment of breast cancer. Molecules 26(7):1889. https://doi.org/10.3390/molecules26071889
Van Gheluwe L, Chourpa I, Gaigne C, Munnier E (2021) Polymer-based smart drug delivery systems for skin application and demonstration of stimuli-responsiveness. Polymers 13(8):1285. https://doi.org/10.3390/polym13081285

Wang F, Zhang Q, Li X, Huang K, Shao W, Yao D et al (2019) Redox-responsive blend hydrogel films based on carboxymethyl cellulose/chitosan microspheres as dual delivery carrier. Int J Biol Macromol 134:413–421. https://doi.org/10.1016/j.ijbiomac.2019.05.049
Xue Y, Zhu Z, Zhang X, Chen J, Yang X, Gao X et al (2021) Accelerated bone regeneration by MOF modified multifunctional membranes through enhancement of osteogenic and angiogenic performance. Adv Healthc Mater 10(6):2001369. https://doi.org/10.1002/adhm.202001369
Yan G, Chen B, Zeng X, Sun Y, Tang X, Lin L (2020) Recent advances on sustainable cellulosic materials for pharmaceutical carrier applications. Carbohydr Polym 244:116492. https://doi.org/10.1016/j.carbpol.2020.116492
Zhang Y, Liu C, Wang J, Ren S, Song Y, Quan P et al (2023) Ionic liquids in transdermal drug delivery system: current applications and future perspectives. Chin Chem Lett 34(3):107631. https://doi.org/10.1016/j.cclet.2022.06.054
Zou L-B, Gong J-Y, Ju X-J, Liu Z, Wang W, Xie R et al (2022) Smart membranes for biomedical applications. Chin J Chem Eng 49:34–45. https://doi.org/10.1016/j.cjche.2022.06.006

# Photo-Reactive Crosslinking Using Visible Light: Versatile Hydrogel Fabrication for Emerging Fields

5

Fulya Özdemir, Süveydanas Çakici, Sedat Odabaş, and Bora Garipcan

**Abstract**

Photo-crosslinkable hydrogels have a wide application potential, especially in the fields of bioengineering and tissue engineering. Crosslinking of these hydrogels occurs under ultraviolet (UV) or visible light. When UV light is used, photoinitiators in polymer structures are activated and initiate crosslinking reactions by forming free radicals. This mechanism provides high structural integrity and mechanical strength in the formation of hydrogel networks. On the other hand, visible light offers advantages such as lower energy consumption and reduced negative effects on biological systems. In particular, visible light photo-crosslinking methods developed in recent years show higher biocompatibility to cells and tissues. As a mechanism, the activation of photosensitive

F. Özdemir
Faculty of Science, Department of Chemistry, Biomaterials and Tissue Engineering Laboratory (bteLAB), Ankara University, Ankara, Turkey

Interdisciplinary Research Unit for Advanced Materials (INTRAM), Ankara University, Ankara, Turkey

Ufuk University Vocational School of Health Services, Department of Medical Services and Techniques, Ankara, Turkey

S. Çakici · S. Odabaş (✉)
Faculty of Science, Department of Chemistry, Biomaterials and Tissue Engineering Laboratory (bteLAB), Ankara University, Ankara, Turkey

Interdisciplinary Research Unit for Advanced Materials (INTRAM), Ankara University, Ankara, Turkey
e-mail: odabas@ankara.edu.tr

B. Garipcan (✉)
Biomimetics and Bioinspired Biomaterials Research Laboratory, Institute of Biomedical Engineering, Boğaziçi University, Istanbul, Turkey
e-mail: bora.garipcan@bogazici.edu.tr



J. M. Oliveira et al. (eds.), *Sustainable Scaffolds-based Strategies in Tissue Engineering and Regenerative Medicine*, Biomaterials, Bioengineering and Sustainability 8, https://doi.org/10.1007/978-3-031-96274-5_5

photoinitiators and polymerization processes are similarly based on free radical production. These hydrogels are used in bioengineering applications; such as in tissue scaffolds, 3D bioprinting, and regenerative medicine. Adaptable structures of photo-crosslinkable materials are critical in terms of preserving the cellular structure and increasing cell viability. This chapter sheds light on current and future applications of photo-crosslinking processes using UV and visible light in bioengineering, highlighting the potential of these methods in tissue engineering and regenerative medicine.

**Keywords**

Photo-crosslink · Biomaterials · Visible light · Hydrogel · Photo-reactive crosslinking

## 5.1 Introduction

The process of creating chemical links or bridges between two distinct polymer chains or molecules is known as crosslinking. Crosslinking within polymer chains can take two different forms: physical and chemical. The kinetics of gelation and the characteristics of the resultant hydrogels are significantly influenced by the crosslinking of the polymer. In physically crosslinked hydrogels, weak intermolecular interactions, such as hydrogen bonding and Van der Waals forces, are effective. On the other hand, covalent linkages are created in the hydrogel structures using the chemical crosslinking technique. The chemical crosslinking technique known as photo-reactive crosslinking (photopolymerization) has garnered a lot of interest lately because of its clear benefits.

Hydrogels that are covalently crosslinked can be produced in a simple, effective, and clean way via photo-reactive crosslinking. While UV radiation has been used for photo-reactive crosslinking for a long time, there are drawbacks, especially in tissue engineering applications because of the possibility of cellular damage. On the other hand, employing visible light increases cell compatibility and expands the possible uses. Moreover, deeper penetration of visible light promotes the creation of smoother hydrogel structures. Because of this, crosslinking hydrogels using visible light has several uses in fields like drug delivery systems, tissue engineering, and three-dimensional (3D) cell encapsulation (Zheng et al. 2021).

When exposed to UV or visible light, a photoinitiator produces free radicals, which starts the process. Following their reaction with the double bonds of monomers, these radical species cause a propagating polymerization process that eventually results in the development of crosslinked polymer networks (Hu et al. 2012).

Photo-reactive crosslinking is a versatile and efficient technique for the rapid synthesis of hydrogels. This method provides precise control over various network properties, such as topographic properties of the hydrogel, crosslinking density, and matrix stiffness, by adjusting factors such as light intensity, duration of curing, and illuminated area. Traditionally, photo-reactive crosslinking is based on a chain polymerization mechanism. However, this approach presents several significant challenges, including limitations in crosslinking kinetics, oxygen inhibition, and difficulties in controlling heterogeneity within the hydrogel network. These factors may have a negative impact on the hydrogels' mechanical strength and swelling behaviour.

One of the main shortcomings of this approach is oxygen inhibition. The generated peroxy radicals are involved in the polymerization process. This decreases the polymerization's efficiency and rate, which commonly leads to the formation of undesirable by-products with a high viscosity. Free radical photo-reactive crosslinking (FRP), cationic photo-reactive crosslinking (CP), and the hybrid initiation system (FRCP) are the three primary categories into which photo-reactive crosslinking can be broadly classified, with FRP being the method most frequently used.

In the photopolymerization of hydrogels, three distinct systems can be utilized:

(a) When hydrophilic reactive monomers and bifunctional crosslinkers are used together.
(b) The application of either reactive monomers or non-reactive multifunctional crosslinkers.
(c) The utilization of multifunctional polymer chains.

These technologies offer versatility in customizing the hydrogels' chemical and physical characteristics for particular uses (Samadian et al. 2020).

The breakdown of the photoinitiator by exposure to light of an appropriate wavelength, such as UV, visible, or near-infrared, is the first step in the process of photopolymerization (Fig. 5.1). This is an attempt to incite radicals. Polymers in this system are functionalized with groups capable of undergoing photoinitiator free radical polymerization. After this reaction, a liquid-to-solid phase change takes place, producing injectable polymers for *in vivo* uses.

Visible light photoinitiators, in contrast to UV photoinitiators, have the following advantages: These photoinitiators absorb less energy than UV photoinitiators when exposed to visible light, resulting in minimal heat generation and thus a reduced risk

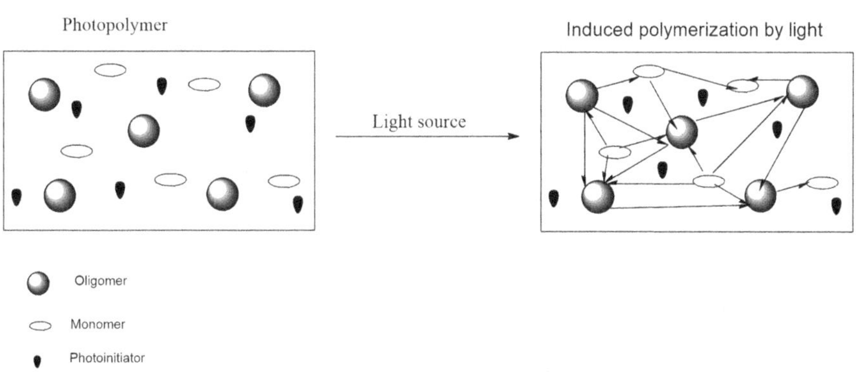


**Fig. 5.1** Stepwise process of photopolymerization, where light exposure triggers a chemical reaction converting monomers into a crosslinked polymer network

of cellular damage. Furthermore, visible light has a longer wavelength than UV radiation, allowing for deeper tissue penetration and more effective treatment. This increased transmission depth enhances the performance of photopolymerization in biological systems. Furthermore, the photo-crosslinking technology based on visible light has a higher bioactivity, which means that it is more compatible with living cells and can be used in tissue engineering and regenerative medicine (Hong et al. 2020).

### 5.1.1 Photoinitiators and Light Sources

The idea behind photo-reactive crosslinking is that when a photoinitiator is exposed to UV or visible light, it generates free radicals. To polymerize, these radicals engage with the monomers' double bonds. As the radicals propagate, they disperse and form crosslinked polymer networks, causing the structural organization of the hydrogel (Hu et al. 2012).

One of the two key components in a photo-polymerization system is the photoinitiator, which directly engages in chemical reactions upon radiation exposure. Various types of photoinitiators have been designed, containing one or more components depending on the specific light source used. Photoinitiation generally occurs through two distinct mechanisms. In the first, known as Type I photoinitiation, the photoinitiator undergoes cleavage upon light absorption, producing free radicals. In the second mechanism, classified as Type II photoinitiation, non-degradable molecules generate free radicals by abstracting hydrogen from hydrogen donor molecules, often with the help of co-initiators.

Type I photoinitiators, such as phenylbis(2,4,6-trimethylbenzoyl) phosphine oxide (BAPO, also known as Omnicure 819) and 2,4,6-trimethylbenzoyldiphenyl phosphine oxide (TPO), are widely used in commercial applications. Irgacure 819 is one of the most commonly used photoinitiators in 3D printing because of its adaptability and efficiency in starting polymerization under a variety of light sources, especially with both UV and 405 nm LED light (Bao 2022).

The first photoinitiators for visible-light photopolymerization were developed in the 1970s (Klee et al. 2015). These include the camphorquinone (CQ)/amine system, a two-component Type II photoinitiator system. In this system, camfurquinone acts as a color sensitizer (Max. absorption wavelength 468 nm). The amine acts as an electron and proton donor. These components initiate free radical polymerization upon exposure to visible light, making them widely used in applications such as dental resins and other light-curable materials (Haslinger et al. 2022). CQ-based photoinitiators have been utilized in dental applications, However, these two-component systems often present certain drawbacks. One notable disadvantage is their generally lower reaction rate, which can be attributed to the presence of the amine synergist. Additionally, these systems tend to exhibit higher toxicity levels, raising concerns regarding their safety in clinical applications. These limitations highlight the need for further development and optimization of photoinitiator systems to improve their performance and reduce associated health risks (Müller et al.

2022). In addition to CQ, several other compounds such as naphthalamide, benzoquinone, and titanocene-based materials have been employed as Type II photoinitiators. These alternative photoinitiators offer distinct advantages and properties that can enhance the efficiency and effectiveness of visible-light photo-polymerization. Their use expands the range of options available for optimizing photoinitiation processes in various applications, including dental materials, coatings, and other light-curable formulations (Kitano et al. 2013; Lougnot and Fouassier 1988; Zhang 2015).

While most Type I photoinitiators are predominantly sensitive to UV light, some compounds, such as amino-acetophenones (e.g., Irgacure 369) and diphenylphosphine oxides (e.g., 2,4,6-trimethylbenzoyl phosphine oxide), can be effectively activated by stimulation at 405 nm. This characteristic allows these photoinitiators to be used in applications that need visible-light sources, which increases their usefulness in a range of polymerization processes, including 3D printing and other light-curable systems. When they are activated at this wavelength, they become more adaptable and compatible with contemporary light sources (Hammoud et al. 2022).

Oxime esters, which are categorized as Type I photoinitiators, show strong photoreactivity when exposed to radiation because they cleave the N-O bond. Iminyl and acyloxy radicals are created by this cleavage, and they are essential for starting the polymerization process. The ability of oxime esters to rapidly and efficiently generate these reactive species makes them valuable in a variety of photopolymerization applications, increasing the overall efficiency and performance of the system (Pang 2020; Wang et al. 2021a). Although commercial photoinitiators such as O-benzoyl-α-oxooxime (OXE01) are effective in various applications, they have limitations due to their maximum absorption wavelength being less than 340 nm. This characteristic means that they are virtually transparent to wavelengths above 400 nm. Consequently, in the presence of OXE01, undesirable transformations of double bonds have been observed when exposed to light at 405 nm. These challenges highlight the need for the development of photoinitiators with broader absorption profiles to improve efficiency and minimize side reactions in photopolymerization processes (Fast et al. 2017; Liu et al. 2021; Qiu et al. 2020).

When functional groups of polymer chains are exposed to light, photo-crosslinking takes place regardless of the presence of photoinitiators in the system. By enabling the creation of crosslinked networks, the resulting polymers' mechanical qualities and stability are improved. An essential mechanism in hydrogels, coatings, and other light-curable systems is photo-crosslinking (Moon et al. 2023). The spectrum used for photopolymerization includes the ultraviolet (UV) region of 200–400 nm, the visible light region of 400–800 nm, and the near-infrared range of 800–1000 nm. Many different types of light sources can be employed in these systems. When choosing a light source, a variety of aspects must be taken into account, including the lamp's spectral characteristics, whether it is monochromatic or polychromatic, how it emits light (continuous, pulsed, or in a series of pulses), and its intensity. These elements are crucial in deciding how successful and efficient the photopolymerization process is (Pierau et al. 2022).

A light source is characterized by an emission spectrum that represents photon distribution as a function of wavelength, i.e., $(\lambda) = f(\lambda)$.

Although photopolymerization and crosslinking techniques began with UV light, visible light crosslinking techniques are more popular today (Fouassier et al. 2003). Visible light is a common method used to initiate chemical reactions. It is also economic and sustainable. Unlike thermal systems, light used as a chemical energy source allows the reaction to occur at lower temperatures (Ferraro et al. 2023).

Xenon lamps, mercury arc lamps, doped lamps, laser sources, pulsed light sources, UV lights, microwave lamps, excimer lamps, and home lamps are the most common types of photoinitiator light sources.

Medium-pressure mercury lamps release a strong photon flux in the UV/near-visible spectrum (280–445 nm) when exposed to an electric field; this corresponds to the absorption of most photoinitiator systems. A wide range of particular wavelengths (254, 313, 366, 405, 435, 546, and 579 nm) are produced by the tight transition between certain excited energy levels brought about by electrical currents in the mercury vapor. The wavelength that low-pressure mercury lamps emit is 254 nm (Fouassier and Lalevée 2012).

However, the visible light range (polychromatic light: 400–800 nm) has been added to the irradiation spectrum due to the usage of Xenon, doped mercury, and regular household halogen lamps. This extension is mainly due to the creation of new photoinitiators, which have improved photo-polymerization systems' light sensitivity and made it possible for them to react to longer wavelengths more successfully (Xiao et al. 2015).

Furthermore, xenon, iron, gallium, and indium halides can be added to mercury lamps to improve their emission range and extend it into the near UV and visible regions. Additionally, they can reach wavelengths as low as 420 nm. In contrast, xenon lamps provide broader coverage of the visible light or near UV spectrum, emitting light up to the 300–800 nm range (Fouassier and Lalevée 2012).

Household light bulbs exhibit high luminous output in the visible spectrum and are characterized by affordability, safety, and environmental friendliness. These attributes confer several advantages to domestic lighting solutions. Additionally, these bulbs minimize infrared (IR) and UV radiation, necessitate lower energy consumption, and provide a broad colour gamut. They are typically resistant to frequency interference and demonstrate relatively low beam divergence (Fouassier and Lalevée 2012; Lalevée et al. 2010).

However, bulbs often produce too much heat, which can lead to the formation of harmful photoproducts and/or overheating of samples, which affects the polymerization process and reduces its controllability. Various filter types can be used to prevent such situations, but this causes energy loss and the need for additional devices in the experimental setup (Anyaogu et al. 2007; Crivello 2015).

Recently developed LED technology has allowed for the employment of monochromatic light at specified wavelengths (405, 455, 470, 505, 530, 590, 660 nm) with different laser sources to extend the excitation range of photoinitiator systems from UV to near-infrared frequencies without sacrificing energy (Fouassier and Lalevée 2012; Fouassier et al. 2003).

Attractive features of laser sources include narrow emission bandwidth, monochromatic light, wide wavelength selection, high energy concentration on a small surface, instant curing, high power output, and less oxygen inhibition. In addition, advantages include simple design, easy adjustment possibilities, high focal power, the ability to be used as a light pen, and high spectral selectivity diode, femtosecond, pulsed, and continuous wave lasers are examples of several types of laser sources (Anyaogu et al. 2007; Fouassier and Lalevée 2012).

### 5.1.2 Crosslinking Mechanisms

Photo-crosslinking can be achieved through various mechanisms. The tyrosine molecule, an amino acid characterized by its phenolic ring and delocalized π electrons, can participate in the formation of hydrogels. This is accomplished through energy transfer processes that facilitate the creation of dityrosine bonds. These bonds form via electron transfer reactions that involve interactions with protons, allowing for the establishment of crosslinked networks within the hydrogel structure. Tyrosine's special characteristic makes it more useful for creating hydrogels and biocompatible polymers for a variety of purposes (Lee et al. 2019; Partlow et al. 2016). Hydrogels have an elastic structure because of dityrosine linkages. These connections can form not just between tyrosine-containing natural proteins or peptides, but also between chemically modified carbohydrates and tyrosine. The resultant materials' mechanical qualities are improved by their adaptability, which makes them appropriate for a range of uses in tissue scaffolding, drug delivery systems, and biomedical engineering. The capacity to establish dityrosine connections between several biomolecules highlights the possibility of developing robust and multifunctional hydrogel networks.

Dityrosine bonds can be formed using either UV or visible light. Upon UV excitation, tyrosine radicals are first generated, initiating a continuous polymerization process. These radicals undergo proton loss and radical isomerization, leading to the formation of diradicals. The diradicals then undergo tautomerization, a process that ultimately results in the formation of dityrosine bonds. This mechanism is essential for creating robust crosslinked networks in hydrogels and other biomaterials, contributing to their mechanical strength and flexibility (Fancy and Kodadek 1999; Min et al. 2016).

In the presence of visible light, the combination of ruthenium (Ru) and sodium persulfate (SPS) is utilized as a photocatalytic system (Fig. 5.2). In this system, ruthenium facilitates the generation of tyrosine radicals when exposed to visible light. These radicals then participate in the formation of dityrosine bonds through tautomerization and proton loss, leading to crosslinking within the hydrogel matrix. By using this method, photocrosslinking can be done effectively, improving the final materials' stability and mechanical qualities (Fancy and Kodadek 1999).

In one study, tyramine was added to alginate and carboxymethyl cellulose (CMC) to promote the creation of dityrosine bonds via Ru/SPS-mediated visible light photoinitiation, which led to the production of hydrogel. This study used an *in situ*

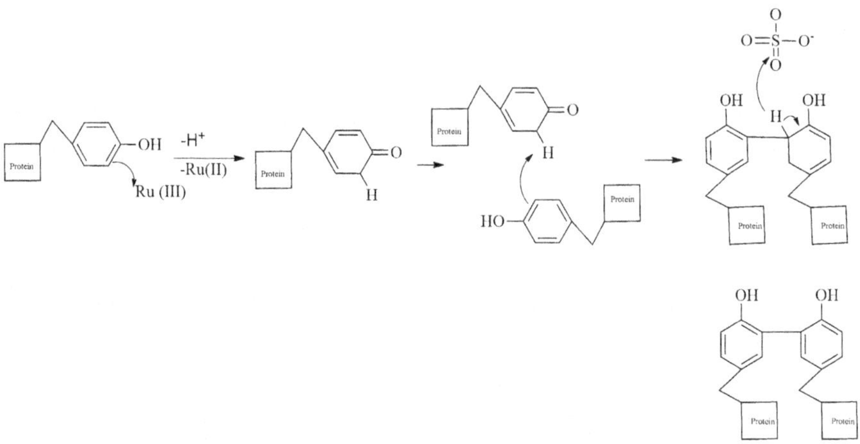


**Fig. 5.2** Mechanism of dityrosine with Ru/SPS

crosslinking method, which allowed the hydrogels to be printed three-dimensionally. The resultant CMC-Tyr-Gel and Alg-Tyr-Gel hydrogels demonstrated their potential for osteogenesis and chondrogenesis after 3 weeks of culture by successfully inducing osteogenic and chondrogenic differentiation. These findings highlight the potential applications of modified natural polymers in tissue engineering and regenerative medicine (Lim et al. 2020).

Methacrylate and its derivatives are commonly used in esterification reactions in light-activated crosslinking processes (Mu et al. 2020; Reis et al. 2009). Under UV light, methacryloyl side chains are excited, leading to the formation of radicals that create cross-links between each other. Methacrylate groups undergo esterification reactions by binding to free amine (–$NH_2$), carboxyl (–COOH), and hydroxyl (–OH) groups present in structures such as proteins and polysaccharides (Yue et al. 2017). These bonds are stable and exhibit high durability as they are insensitive to pH and temperature variations.

In UV light-activated photo-crosslinking, photoinitiators such as Irgacure 2959 and Lithium Acetophenone (LAP) are employed. When exposed to UV radiation, Irgacure 2959 is known to produce free radicals at a wavelength of 365 nm, whereas LAP begins to activate at 405 nm. These initiators generate free radicals, which facilitate the cross-linking of polymer chains and the creation of covalent bonds. Because of this property, they are indispensable in a number of applications, such as the creation of hydrogels and other light-curable materials (Basara et al. 2019).

Photoinitiators employed in visible light induced crosslinking include Ru/SPS and Eosin Y, which facilitate bond formation at wavelengths of 420–450 nm and 450–550 nm, respectively. In a study focused on promoting cartilage regeneration, bioinks suitable for 3D bioprinting were developed. A visible light-sensitive GelMA-Tyr bioink was designed, incorporating bifunctional groups, with Ru/SPS serving as the photoinitiator for both functional groups. The results demonstrated successful *in vitro* cartilage formation from embedded chondroprogenitor cells.

Moreover, the study found that in contrast to the UV light (365–400 nm range) commonly used in biofabrication, which can cause native chondrocytes in cartilage defects to die, in situ crosslinking using visible light did not harm the surrounding tissue. This emphasizes how visible light crosslinking may be used in tissue engineering applications as a safer substitute (Lim et al. 2020).

Norbornene is classified as a cyclic alkene, characterized by its unique cyclic structure and the presence of π electrons in the double bond. These features contribute to the inherent instability of the molecule, endowing it with a significant energy potential that allows it to respond to external stimuli. As a result, norbornene is recognized as a molecule that readily participates in chemical reactions, making it valuable in various applications, including polymer synthesis and materials science. Its reactivity is leveraged in the development of crosslinked networks and advanced polymeric materials (Devaraj et al. 2008).

The norbornene ring can be covalently linked to natural polymers through amide formation reactions (Fig. 5.3). Thiol-norbornene crosslinking is achieved via UV light exposure. When light activates a substance, radicals are produced that take protons out of thiol groups and form thiyl radicals. Norbornene radicals are produced when these thiyl radicals react with the norbornene ring's double bond. A thioether bond is created when the thiol's hydrogen is removed by the norbornene radical. Remarkably, Eosin Y acts as the photoinitiator in the visible light system, and the thiol-norbornene crosslinking reaction proceeds via the same mechanism under visible light as it does under UV light. Thiol-norbornene hydrogels have shown great versatility and potential in regenerative medicine through their use in a variety of applications, such as cancer cell biology, musculoskeletal tissue regeneration, and heart valve and vascular tissue engineering (Lin et al. 2015).

Triethanolamine possesses an electron-donating chemical structure. It forms hydrogels by crosslinking with natural polymers such as cellulose, gelatin, and chitosan under visible light (Haria and König 2014). In reactions photocatalyzed by Eosin Y under visible light, Triethanolamine (TEA) is used as an electron-donating group (Fig. 5.4). During light activation, Eosin Y generates an anionic radical, while Triethanolamine forms a cationic radical, which covalently bonds with the polymers' vinyl groups, thereby creating crosslinked hydrogels.

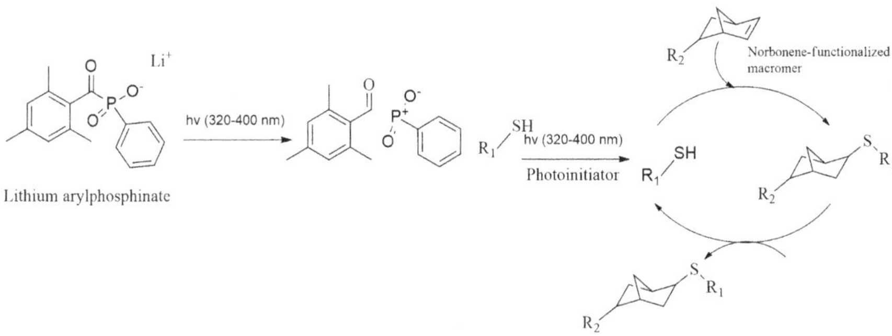


**Fig. 5.3** Norbornene ring, amide formation mechanism

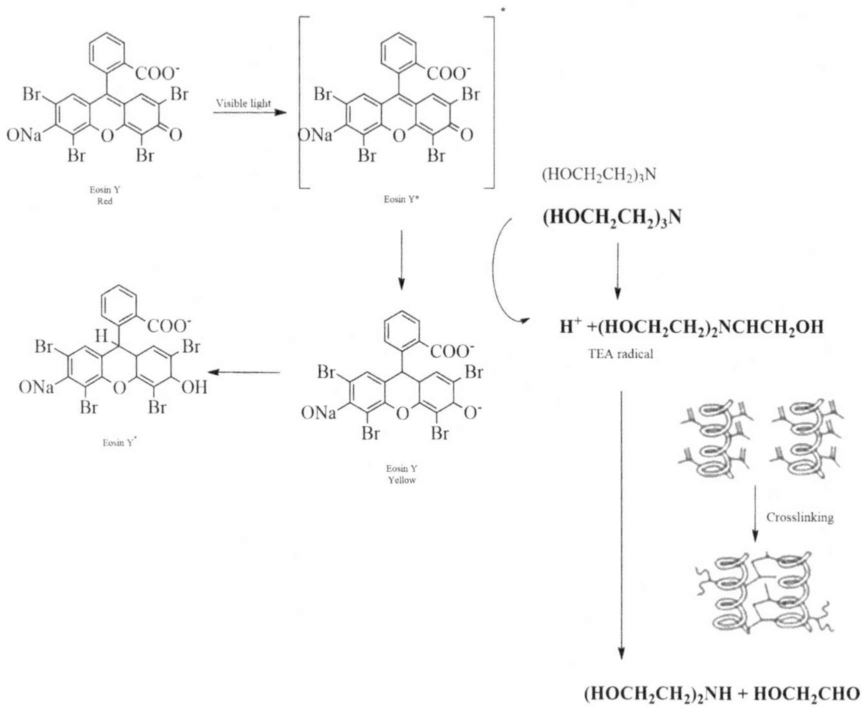


**Fig. 5.4** Mechanism of Eosin-Y/TEA

### 5.1.3 Visible Light Photo-Crosslinkable Hydrogels

Monomers or polymer chains are crosslinked to form hydrophilic and three-dimensional polymer networks, forming hydrogels. Hydrogels contain hydrogen bonds and thus exhibit electrostatic interactions, including host-guest complexes and covalent combinations of these interactions. This unique structure enables cells to survive within the hydrogel matrix. The similarity of hydrogels to the extracellular matrix (ECM) contributes to their seamless integration with the surrounding tissue. They can be applied to target sites to deliver drugs, growth factors, and cells via injection or surgical procedures, mediating the controlled release of materials. Furthermore, hydrogels provide high compatibility with the tissues in the applied area, minimizing the risk of inflammatory responses in surgical applications. This characteristic enhances their potential utility in various biomedical fields, including tissue engineering and regenerative medicine (Wang et al. 2021b).

Photo-crosslinked hydrogels are polymers that have two or more reactive groups and are water-soluble (Fig. 5.5). Photo-crosslinked hydrogels can be made from components that are naturally occurring or artificially manufactured. Alginate, hyaluronic acid, chitosan, collagen, silk fibroin, and gelatin are the most commonly used photo-crosslinked materials including many others (Table 5.1). Synthetic materials

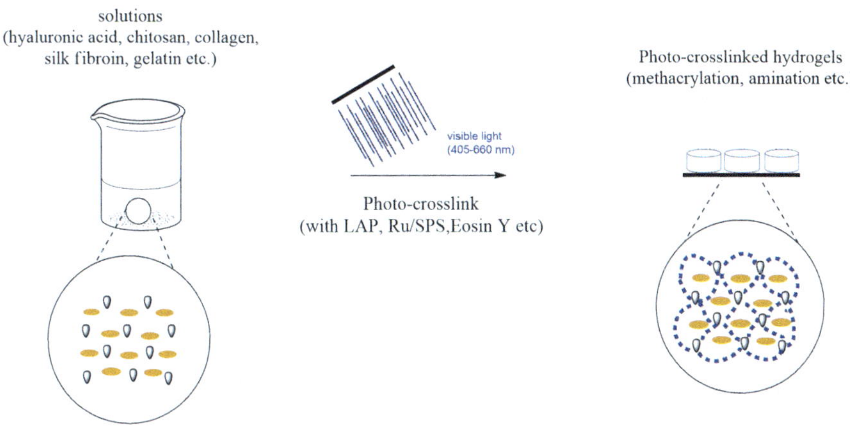


**Fig. 5.5** Schematic representation of hydrogels formed by visible light crosslinking of photo-crosslinkable biopolymers

include pluronic, polyacrylic acid (PAA), polyvinyl alcohol (PVA), polyethylene glycol (PEG), and polypeptides (Kimura et al. 2021; Wang et al. 2021b).

The most widely utilized hydrogel today is alginate. Alginate is a linear polysaccharide with water-soluble properties obtained from bacteria or brown algae. The gelation mechanisms are different. Alginate tends to gel spontaneously by creating intermolecular hydrogen bonds at pH levels lower than 3. Alginate can also react with the carboxylic acid groups of sugars and two or more trivalent cations, such as $Ca^{2+}$, $Mg^{2+}$, $Sr^{2+}$, $Ba^{2+}$, or $Al^{3+}$, to produce a hydrogel. The alginate chain is composed of subunits such as mannuronic acid (unit M) and guluronic acid (unit G). It is also configured in blocks, with G and M units swapping places. The greater the number of G blocks and molecular weight present in alginate hydrogels, the higher their crosslinking density. The hardness and viscosity of the gelled alginate solution are determined by the polymer concentration and molecular weight distribution. Divalent ions are lost during the slow breakdown of alginate hydrogels. Following disintegration, it loses its mechanical qualities (Li et al. 2020).

To increase stability and stop alginate hydrogels from disintegrating, methacrylate groups can be added to carboxyl or hydroxyl groups in the resulting methacrylate-alginate to create a photo-crosslinked product. Compared to $Ca^{2+}$ crosslinked alginate, this is more stable. Methacrylic anhydride, 2-aminoethyl methacrylate, and glycidyl methacrylate are a few of the many methacrylate compounds that fall under this category. This leads to the generation of three distinct methacrylate alginates. The first is alginate-glycidyl methacrylate; the second is alginate-2-aminoethyl methacrylate; and the third is alginate-methacrylic anhydride. Methylacrylate alginate can be photo-crosslinked using photoinitiators at 514 nm by argon-ion laser or 245 nm by UV light (Wang et al. 2021b). Łańcucka et al. created a novel hybrid material using hydrogels based on photo-crosslinked gelatin and alginate that were made from silica particles and showed promise for bone tissue regeneration. Methacrylate and methacrylamide were used to functionalize photo-crosslinked

**Table 5.1** Photo-crosslinked hydrogels derived from chemically modified natural biomaterials

| Biomaterial | Functional group | Commercial reagent | Chemical structure | Modified biomaterial |
|---|---|---|---|---|
| Alginate 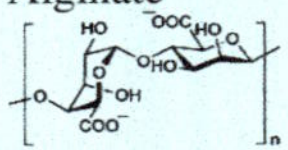 | Methacrylate | Glycidyl methacrylate (GMA) | 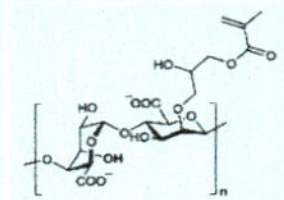 | Alginate-glycidyl methacrylate (AAGMA) (Wang et al. 2015) |
| Hyaluronic acid 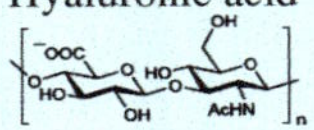 | Methacrylate | Methacrylic anhydride (MA) | 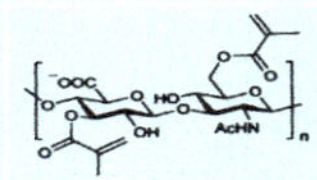 | Methacrylate hyaluronic Acid (HAMA) (Liu et al. 2019b) |
| | Tyramine | Tyramine | 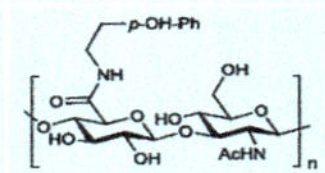 | HA-Tyr (Loebel et al. 2017; Grenier et al. 2015) |
| Chitosan 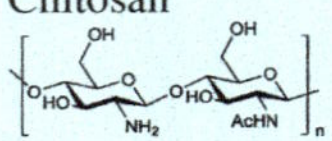 | Methacrylate | Methyl acryloyl glycine (MAG) | 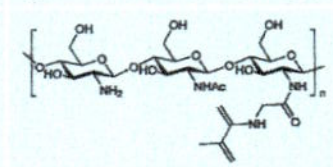 | CS-MAG (Qi et al. 2013) |
| | Maleate | Maleic anhydride | 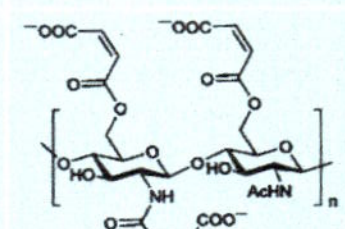 | Maleic CS (Zhou et al. 2018) |
| Collagen 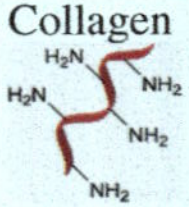 | Methacrylate | Methacrylic anhydride (MA) | 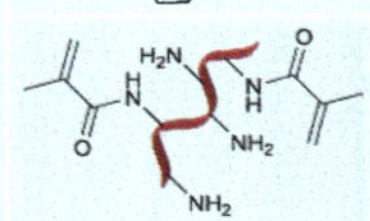 | Collagen methacrylamide (CMA) (Yang et al. 2018; Drzewiecki et al. 2017) |
| Silk fibroin 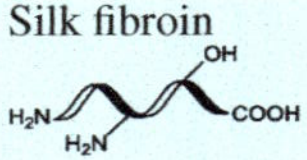 | Methacrylate | 2-Isocyanatoethyl methacrylate (IEMA) | 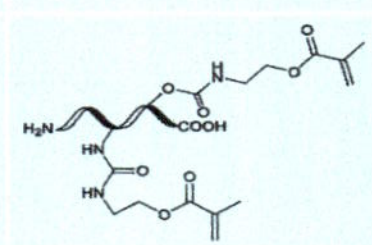 | Silk fibroin methacrylate (Kim et al. 2018) |

| | | | | |
|---|---|---|---|---|
| Gelatin 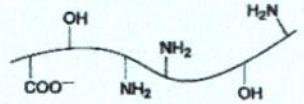 | Methacrylate | Glycidyl methacrylate | 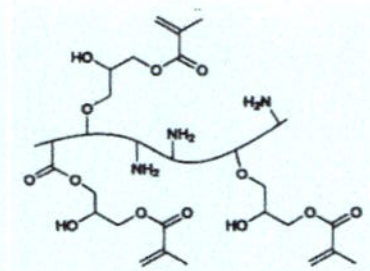 | Glycidyl methacrylate modified gelatin (Gel-GM) (Huang et al. 2020) |
| | Thiol | Cysteamine | 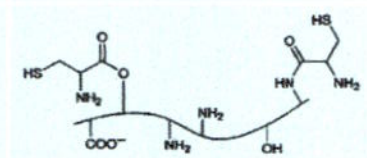 | Thiolated-gelatin (GE-SH) (Li et al. 2018) |
| | Furan | Furan containing chromophore MFVF | 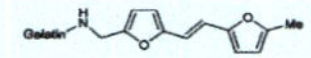 | Gel-MFVF (García-Astrain et al. 2015) |
| | Catechol | Dopamine | 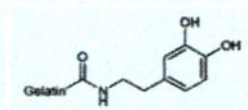 | Catechol-modified gelatin (PCG) (Liu et al. 2019a) |
| | Phenolation | Phloretic acid | 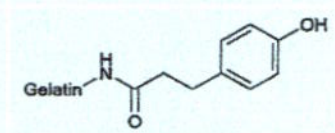 | Phenol-modified gelatin (PPG) (Liu et al. 2019a) |
| | Norbornene | Carbic anhydride | 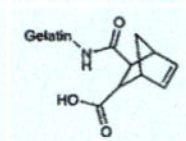 | GelNB (Greene and Lin 2015) |
| | Furfurylamine | Furfuryl glycidyl Ether | 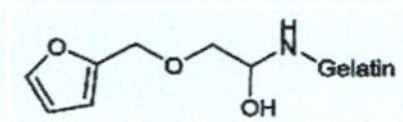 | Furfuryl-gelatin (AnilKumar et al. 2019; García-Astrain et al. 2014) |

alginate and gelatin, respectively. Three types of polymeric solutions are formulated with the two-dimensional sub-micron particles produced by the STÖBER method: photo-crosslinked alginate, gelatin, and gelatin/alginate. Afterward, the produced materials' mechanical characteristics, gel fraction, and swelling ratio were assessed and contrasted with those of pure hydrogel matrices devoid of silica particles (Lewandowska-Łańcucka et al. 2017).

In order to create 3D-printed bioinks, Lee et al. chemically joined tyrosine-like residues to alginate chains to create a non-toxic visible light-based photo-crosslinkable alginate conjugate. Tyramine and alginate were conjugated in the study, and the mixture was crosslinked at 452 nm using visible blue light. According to the study's findings, the photo-crosslinkable tyramine-conjugated alginate hydrogel outperformed the control group in terms of imprinting resolution and cell survivability. With improved cell migration and proliferation, the developed multi-cell-loaded photo-crosslinkable alginate-based microgel bioink is a biocompatible material that presents a workable way to create implantable, functional artificial organ structures for 3D bioprinting-based tissue engineering (Lee et al. 2023).

Another often used hydrogel is hyaluronic acid, a linear, non-sulphated, water-soluble glycosaminoglycan present in nearly all vertebrate tissues. In biological systems, lysosomal digestion and endocytosis take place subsequent to the hydrolysis of hyaluronic acid by free radicals or hyaluronidase. In biological processes such as angiogenesis, cellular migration, differentiation, and proliferation, hyaluronic acid plays a highly active role. It also contributes to the construction of the extracellular matrix, inflammation, tissue repair, wound healing, and metastatic metabolism. Clinicians utilize hyaluronic acids and their derivatives in medicinal products, and the significance of these compounds in the creation of functional goods is becoming increasingly apparent (Li et al. 2020).

Donelly et al. used RF as a photoinitiator to generate Tyr-modified and visible light photo-crosslinked HA hydrogels for articular cartilage regeneration. The Tyr amine group in the system was functionalized by the HA carboxyl group. Additionally, riboflavin 5′-phosphate (FMN) was used as a photoinitiator to cause crosslinking in visible light. FMN can either receive or shed two hydrogen atoms to decrease or oxidize recycling materials. The redox reactions of FMN can lead to crosslinking by generating radicals. Consequently, covalent linkages are created when exposed to visible light, resulting in the formation of photo-crosslinked HA-Tyr hydrogels. First, the behaviour of the gel was examined under various conditions of photo-crosslinking with varying degrees of HA-Tyr conjugate. In addition, the mechanical, chemical, and biological properties were assessed at different durations of irradiation to confirm the possible use of cross-linked HA-Tyr hydrogels as soft tissue fillers (Donnelly et al. 2017).

The HA-Tyr conjugate, which forms an amide bond between the amino group of Tyr and the carboxylic group of HA, is made by carbodiimide chemistry. One gram of HA was dissolved in 100 mL of deionized (di) water, and then two milligrams each of NHS and EDC were added to create a 1% by weight HA solution. The carboxyl group was activated by immersing HA in a cold bath for 5 minutes. TYR (0.25–2.5 mM) was dissolved in the activated HA solution and then allowed to sit at room temperature for the entire night. For three days, dialysis was carried out using a cellulose membrane (cutoff, 12,000–13,000) to eliminate unreacted NHS and EDC (Hong et al. 2020).

L. Fenn et al. looked into the methacrylate modification of hyaluronic acid. Different concentrations of HA-MA solutions [2%, 3%, and 4% (WT and VOL)] are ready for crosslinking under visible green light. 1-vinyl-2-pyrrolidinone (VP) was employed in the study as a photo-crosslinker and as a catalyst for triethanolamine (TEOA). For the next experiments, the solution with the maximum absorbance at 530 nm is chosen as the photostimulation system (Fenn and Florcani 2016).

In their study on wound closure, Ghaffari et al. created a printable hydrogel based on hyaluronic acid (HA) to transfer the ACE-inhibitory peptide of VLSTSFPPW (VW-9) generated from yeast to the wound site. To create a phenol-functional HA (HA-Tyr), tyramine (Tyr) was first conjugated to the carboxyl groups of HA. Ethylenediamine was then used to aminate the carboxylic acid groups of HA-Tyr (HA-Tyr-$NH_2$). They 3D printed the resulting hydrogel and employed visible light to crosslink it. Following the trial, the conjugated hydrogel showed better cell viability than the control group. It was then printed with a different geometry to determine whether it could be used to make a variety of wound dressings (Ghaffari-Bohlouli et al. 2023).

The history of chitosan, which is derived from chitin, began in the nineteenth century with Rouget's discussion of chitosan in its deacetylated form in 1859. After cellulose, chitin is one of the most prevalent biomaterials. It is the second most common biomaterial generated by biosynthesis every year. Chitin plays a major role in the formation of the exoskeleton of insects, mollusks, and hard-shelled marine animals. Furthermore, in the cell walls of certain fungi, it is the primary fibrillar polymer. Chitosan shares a structure with glycosaminoglycans. When chitin is N-deacetylated, a linear polymer is formed. Chitosan, composed of glucosamine and N-acetyl glucosamine units joined by β(1-4) glycosidic bonds, is a linear polysaccharide. The amount of glucosamine determines the degree of deacetylation (DD). The molecular weight of chitosan can vary from 300 to 1000 kD, while its DD can range from 30 to 95 Dec. Preparation and sourcing techniques determine the cause of these variations. It is only soluble in weak acids. This is due to the protonation of the free amino groups on glucosamine. The amino group and the primary and secondary hydroxyl groups located at positions C(2), C(3), and C(6), respectively, are the functional groups of chitosan. These groups contribute to the synthesis of different scaffolds for use in tissue engineering applications and to modifications for graft copolymerization similar to chitosan. Various chitosan derivatives with different properties have been created to increase the solubility and processability of chitosan. The chemical structure of chitosan is amenable to both covalent and ionic modification. Therefore, it allows for comprehensive tuning of biological and mechanical properties. Chitosan is approved by the U.S. Food and Drug Administration as non-toxic after oral administration in humans and has been widely researched for a variety of biomedical applications. Hydrogels can be formed by various ionic or chemical crosslinking methods (Donnelly et al. 2017; Fenn and Floreani 2016; Kim et al. 2008; Lewandowska-Łańcucka et al. 2017; Li et al. 2020).

Hyun H. and his colleagues developed a crosslinked glycol chitosan hydrogel in visible light to develop a controlled-release anticancer drug system against the side effects of chemotherapy (Hyun et al. 2018).

Additionally, visible light crosslinked hydrogel systems were demonstrated using three blue light initiators and methacrylate glycol chitosan (MeGC). These three initiators are riboflavin (RF), fluorescein (FR), and camphor quinone (CQ). In the study, a minimum irradiation time of 120 s was used to form MeGC gels that could encapsulate cells with CQ or FR. For CQ and FR, respectively, these conditions resulted in a 25% and 5% decrease in encapsulated chondrocyte viability. While the mechanical strength of the gels induced by CQ or FR increased to 600 s at the end of long-term irradiation. The compressive modulus was greatly improved by keeping the irradiation time at 300 s without compromising the cell viability of the RF-initiated MeGC gels. At longer irradiation times, the observed swelling and degradation rate was less (Hu et al. 2012).

The most common sources of collagen are marine and animal species. Applications for this protein in biomaterials are common. It is extensively present in tissues including blood vessels, ligaments, skin, bone, and cornea. Moreover, it comprises one-third of all the proteins in the human body (Li et al. 2020; Nair and Laurencin 2007). This protein's fibrous structure is favourable in terms of mechanical properties. It is made up of three parallel left-handed polypeptide chains arranged in a right-handed rotating triple helix (Li et al. 2020). Collagen can be processed into a variety of forms, including powder, sheet, tube, foam, sponge, and hydrogel. It is soluble in acidic aqueous solutions. The most prevalent extracellular matrix (ECM) component is crucial for preserving the biological and structural integrity of the cell. Collagen, as a biomaterial, can therefore provide a beneficial ECM-mimetic microenvironment for the functionalization of different cell processes, including adhesion, migration, propagation, proliferation, and differentiation (Gómez-Guillén et al. 2011).

Collagen is a 300,000-molecular-weight polymer that resembles a rod. Its length is around 300 nm. The human body has 27 distinct kinds of collagen that have been identified thus far. Type I collagen is the most prevalent and has been the subject of the most research. It is the most prevalent protein in mammals, which explains why. Three polypeptide subunits, each containing an average of 1050 amino acids, make up type I collagen. These subunits have comparable amino acid contents. About 33% glycine, 25% proline, and 25% hydroxyproline make up its composition. Type I collagen is the most prevalent kind of collagen in connective tissue, even though 27 other forms of collagen have been discovered so far (Gómez-Guillén et al. 2011; Nair and Laurencin 2007).

For application in meniscus tissue engineering, riboflavin-based photo-crosslinking has produced collagen hydrogel in recent years. In order to enhance the mechanical characteristics of a collagen-based scaffold and postpone its rate of deterioration, photoinitiation with riboflavin (RF) and photo-crosslinking with ultraviolet (UV) light were employed. After photo-crosslinking collagen with RF, it was found that the hydrogel's mechanical qualities increased and that the enzyme-induced deterioration of collagen scaffolds was delayed (Heo et al. 2016).

The goal of Fernandes et al.'s work was to seal corneal incisions and provide a sticky layer between the crosslinked gel and surrounding collagen. Using riboflavin (PhotoCol-RB) as a photoinitiator and ultraviolet-A (UV-A) light as an *in-situ* matrix, a collagen gel was created. The study revealed that PhotoCol-RB gels facilitate corneal epithelialization and have higher transparency than physically crosslinked collagen. These gels can be prepared in advance and put to the corneal wound for convenient handling at a later time, or they can be created immediately at the wound site to fit the geometry of a defect (Fernandes-Cunha et al. 2023).

Silk fibroin is a fibrous protein derived from Bombyx mori cocoons. As a natural biopolymer, it has been extensively studied for use in tissue engineering and regenerative medicine (Raia et al. 2017). Silk is non-immunogenic and is FDA-approved for certain medical devices, including sutures, and as a support structure during reconstructive surgery (Partlow et al. 2014). Owing to silk's adaptable nature, and prior research showing its dependability, effectiveness, and inherent pliability in tissue engineering, silk protein has emerged as a promising option for elastomeric biomaterials. Because of its mechanical strength, biocompatibility, and relatively gradual disintegration over several months, it can produce natural hydrogels. Silk has multiple applications as a biomaterial, including film, membrane, gel, sponge, powder, and scaffold. Burn wound dressings, meshes, vascular prostheses, enzyme immobilization matrices, and structural implants are among the applications (Cao and Wang 2009).

In the presence of visible light, Applegate et al. transformed a liquid silk solution containing flavin-mononucleotide (FMN), a water-soluble form of riboflavin, into a highly elastic hydrogel. It may be possible to alter the cornea's optical thickness with high spatial resolution to improve visual acuity in people with myopia, hyperopia, and corneal ectatic disorders without the risks associated with laser vision correction, as the study showed that silk hydrogels can adhere to the epithelium-free cornea in pig eyes. In the investigation, 450 nm of light was employed. The effect of accelerated riboflavin radicalizing tyrosine residues in the silk protein backbone and the creation of dityrosine complexes that bind silk molecules together resulted in transparent, elastic silk fibroin hydrogels at the end of the study. It was determined that people with ocular problems could benefit from using these hydrogels as lenses. They also showed how contemporary laser vision correction surgery techniques can be used to precisely sculpt the silk-gel layer on the cornea (Applegate et al. 2016).

Gelatin, one of the most well-known hydrogels, is a heterogeneous combination of water-soluble proteins formed by partially hydrolyzing collagen. It bears many similarities to collagen and is frequently utilized in tissue engineering applications as a biocompatible and biodegradable material. Gelatin has also been used as a scaffold material to lessen the hazards of immunogenicity and pathogen transmission associated with collagen because of its low cost, good biocompatibility, and biodegradability; yet, it can still replicate biological ECM (Mishra et al. 2019; Shalumon et al. 2019). Gelatin is a biomaterial widely used in tissue engineering today.

Wang et al. synthesized a photo-crosslinkable material from two distinct natural biomaterials, hyaluronic acid (HA) and gelatin, to create a bioink suitable for 3D printing tissue engineering scaffolds. The work involved the methacrylation of gelatin, followed by the combination of HA with GelMA. Hydrazone was used to crosslink HA, whereas lithium acyl phosphinate (LAP) was used to photo-crosslink JelMA. A biocompatibility test was conducted *in vitro* using mesenchymal stem cells from rats. Upon completion of the study, the novel hydrogel exhibited remarkable mechanical strength, self-healing capabilities, and biocompatibility. Tissue scaffolds with consistent filaments and dimensions can be created using the bio-ink that was obtained, making it appropriate for use in bioprinting, and *in vitro* studies demonstrated the biomaterial's biocompatibility (Wang et al. 2022).

The conjugate of gelatin and norbornene (GelNB) has been reported in recent investigations as a photo-crosslinkable bioink for 3D bioprinting. The photoinitiator employed in the gelatin-norborne hydrogel crosslinking process was lithium acyl phosphinate (LAP) at 405 nm of visible light. High cytometry and viability were demonstrated by 3D-printed GelNB hydrogels in the study when human umbilical vein endothelial cells (HUVECs) were encapsulated. With luminal structures, encapsulated HUVECs created a network of interconnected microvascular structures. Greatly enhanced hydrogel vasculogenesis and endothelialization. Because of this, it has been shown that the created bio-ink technology applies to a variety of biofabrication applications (Duong and Lin 2023).

## 5.2 Conclusion

Visible light crosslinking presents several advantages, including enhanced cellular compatibility, broader applicability, and increased tissue penetration depth compared to ultraviolet-based crosslinking methods. The use of photoinitiators in visible light enables the efficient, clean, and facile fabrication of covalently crosslinked hydrogels.

A wide range of hydrogels, including alginate, hyaluronic acid, chitosan, collagen, silk fibroin, and gelatin, are employed in photo-reactive crosslinking. These hydrogels can undergo chemical modifications to introduce photo-reactive groups, thereby enhancing their stability and crosslinking efficacy. The ability to precisely control the hydrogel network properties, such as polymerization, crosslink density, and matrix stiffness, offers significant versatility and potential for tailoring hydrogel synthesis.

Photoinitiators are critical to initiate the crosslinking reaction upon exposure to light. Various types, including Type I and Type II photoinitiators, play a pivotal role in photo-crosslinking systems. The selection of appropriate photoinitiators, based on the desired light wavelength and the specific requirements of the hydrogel system, is essential to the process. An essential aspect of this method is the selection of a suitable light source for the photoinitiator. Various visible light sources, such as xenon lamps, mercury arc lamps, laser sources, and even household lamps, offer sustainable and cost-effective solutions for triggering crosslinking reactions, making this approach highly attractive for practical applications.

In summary, this chapter highlights the potential of photoreactive crosslinking in hydrogel synthesis. The ability to precisely control the crosslinking process, fine-tune hydrogel properties, and improve biocompatibility offers promising opportunities in various fields such as tissue engineering, 3D cell encapsulation, and drug delivery. Continued advances in this field are expected to pave the way for the development of innovative hydrogel-based biomaterials.

**Funding, Acknowledgments, and Disclosures** *Funding*: This study was not supported by any organization or institution.

*Competing Interest*: The authors declare they have no competing interests.

*Non-Financial Competing Interests*: There are no financial competing interests (political, personal, religious, ideological, academic, intellectual, commercial, or any other) to declare in relation to this manuscript.

## References

AnilKumar S, Allen SC, Tasnim N, Akter T, Park S, Kumar A, Chattopadhyay M, Ito Y, Suggs LJ, Joddar B (2019) The applicability of furfuryl-gelatin as a novel bioink for tissue engineering applications. J Biomed Mater Res Part B Appl Biomater 107:314–323

Anyaogu KC, Ermoshkin AA, Neckers DC, Mejiritski A, Grinevich O, Fedorov AV (2007) Performance of the light emitting diodes versus conventional light sources in the UV light cured formulations. J Appl Polym Sci 105:803–808

Applegate MB, Partlow BP, Coburn J, Marelli B, Pirie C, Pineda R, Kaplan DL, Omenetto FG (2016) Photocrosslinking of silk fibroin using riboflavin for ocular prostheses. Adv Mater 28:2417–2420

Bao Y (2022) Recent trends in advanced photoinitiators for vat photopolymerization 3D printing. Macromol Rapid Commun 43:2200202

Basara G, Yue X, Zorlutuna P (2019) Dual crosslinked gelatin methacryloyl hydrogels for photolithography and 3D printing. Gels 5(3):34

Cao Y, Wang B (2009) Biodegradation of silk biomaterials. Int J Mol Sci 10:1514–1524

Crivello J (2015) Sensitization of cationic photopolymerizations. In: Dyes and chromophores in polymer science. Wiley, pp 45–79

Devaraj NK, Weissleder R, Hilderbrand SA (2008) Tetrazine-based cycloadditions: application to pretargeted live cell imaging. Bioconjug Chem 19:2297–2299. https://doi.org/10.1021/bc8004446

Donnelly PE, Chen T, Finch A, Brial C, Maher SA, Torzilli PA (2017) Photocrosslinked tyramine-substituted hyaluronate hydrogels with tunable mechanical properties improve immediate tissue-hydrogel interfacial strength in articular cartilage. J Biomater Sci Polym Ed 28:582–600

Drzewiecki KE, Malavade JN, Ahmed I, Lowe CJ, Shreiber DI (2017) A thermoreversible, photocrosslinkable collagen bio-ink for free-form fabrication of scaffolds for regenerative medicine. Technology 5:185–195

Duong VT, Lin C (2023) Digital light processing 3D bioprinting of gelatin-norbornene hydrogel for enhanced vascularization. Macromol Biosci:2300213

Fancy DA, Kodadek T (1999) Chemistry for the analysis of protein-protein interactions: rapid and efficient cross-linking triggered by long wavelength light. Proc Natl Acad Sci USA 96:6020–6024. https://doi.org/10.1073/pnas.96.11.6020

Fast DE, Lauer A, Menzel JP, Kelterer A-M, Gescheidt G, Barner-Kowollik C (2017) Wavelength-dependent photochemistry of oxime ester photoinitiators. Macromolecules 50:1815–1823

Fenn SL, Floreani R (2016) Visible light crosslinking of methacrylated hyaluronan hydrogels for injectable tissue repair. J Biomed Mater Res Part B Appl Biomater 104:1229–1236

Fernandes-Cunha GM, Brunel LG, Arboleda A, Manche A, Seo YA, Logan C, Chen F, Heilshorn SC, Myung D (2023) Collagen gels crosslinked by photoactivation of riboflavin for the repair and regeneration of corneal defects. ACS Appl Bio Mater 6:1787–1797

Ferraro V, Adam CR, Vranic A, Bräse S (2023) Recent advances of transition metal complexes for photopolymerization and 3D printing under visible light. Adv Funct Mater:2302157

Fouassier J-P, Lalevée J (2012) Photoinitiators for polymer synthesis: scope, reactivity, and efficiency. Wiley

Fouassier JP, Allonas X, Burget D (2003) Photopolymerization reactions under visible lights: principle, mechanisms and examples of applications. Prog Org Coat 47:16–36

García-Astrain C, Gandini A, Peña C, Algar I, Eceiza A, Corcuera M, Gabilondo N (2014) Diels–Alder "click" chemistry for the cross-linking of furfuryl-gelatin-polyetheramine hydrogels. RSC Adv 4:35578–35587

García-Astrain C, Peña-Rodriguez C, Retegi A, Eceiza A, Corcuera MA, Gabilondo N (2015) Green chemistry for the cross-linking of photo-sensitive furan modified gelatin. Mater Lett 160:142–145

Ghaffari-Bohlouli P, Simińska-Stanny J, Jafari H, Mirzaei M, Nie L, Delporte C, Shavandi A (2023) Printable hyaluronic acid hydrogel functionalized with yeast-derived peptide for skin wound healing. Int J Biol Macromol 232:123348

Gómez-Guillén MC, Giménez B, López-Caballero ME, Montero MP (2011) Functional and bioactive properties of collagen and gelatin from alternative sources: a review. Food Hydrocoll 25:1813–1827

Greene T, Lin C-C (2015) Modular cross-linking of gelatin-based thiol–norbornene hydrogels for in vitro 3D culture of hepatocellular carcinoma cells. ACS Biomater Sci Eng 1:1314–1323

Grenier S, Donnelly PE, Gittens J, Torzilli PA (2015) Resurfacing damaged articular cartilage to restore compressive properties. J Biomech 48:122–129

Hammoud F, Pavlou A, Petropoulos A, Graff B, Siskos MG, Hijazi A, Morlet-Savary F, Dumur F, Lalevée J (2022) Naphthoquinone-based imidazolyl esters as blue-light-sensitive type I photoinitiators. Polym Chem 13:4817–4831

Haria DP, König B (2014) Synthetic applications of eosin Y in photoredox catalysis. Chem Commun 50:6688–6699. https://doi.org/10.1039/c4cc00751d

Haslinger C, Leutgeb LP, Haas M, Baudis S, Liska R (2022) Synthesis and photochemical investigation of tetraacylgermanes. ChemPhotoChem 6:e202200108

Heo J, Koh RH, Shim W, Kim HD, Yim H-G, Hwang NS (2016) Riboflavin-induced photocrosslinking of collagen hydrogel and its application in meniscus tissue engineering. Drug Deliv Transl Res 6:148–158

Hong BM, Kim HC, Jeong JE, Park SA, Park WH (2020) Visible-light-induced hyaluronate hydrogel for soft tissue fillers. Int J Biol Macromol 165:2834–2844. https://doi.org/10.1016/j.ijbiomac.2020.10.155

Hu J, Hou Y, Park H, Choi B, Hou S, Chung A, Lee M (2012) Visible light crosslinkable chitosan hydrogels for tissue engineering. Acta Biomater 8:1730–1738. https://doi.org/10.1016/j.actbio.2012.01.029

Huang X, Zhang Y, Li F, Zhang J, Dai Y, Shi M, Zhao Y (2020) Highly efficient alginate-based macromolecular photoinitiator for crosslinking and toughening gelatin hydrogels. J Polym Sci 58:1439–1449

Hyun H, Park MH, Lim W, Kim SY, Jo D, Jung JS, Jo G, Um S, Lee D-W, Yang DH (2018) Injectable visible light-cured glycol chitosan hydrogels with controlled release of anticancer drugs for local cancer therapy in vivo: a feasible study. Artif Cells Nanomed Biotechnol 46:874–882

Kim I-Y, Seo S-J, Moon H-S, Yoo M-K, Park I-Y, Kim B-C, Cho C-S (2008) Chitosan and its derivatives for tissue engineering applications. Biotechnol Adv 26:1–21

Kim HH, Kim JW, Choi J, Park YH, Ki CS (2018) Characterization of silk hydrogel formed with hydrolyzed silk fibroin-methacrylate via photopolymerization. Polymer (Guildf) 153:232–240

Kimura A, Yoshida F, Ueno M, Taguchi M (2021) Application of radiation crosslinking technique to development of gelatin scaffold for tissue engineering. Radiat Phys Chem 180:109287. https://doi.org/10.1016/j.radphyschem.2020.109287

Kitano H, Ramachandran K, Bowden NB, Scranton AB (2013) Unexpected visible-light-induced free radical photopolymerization at low light intensity and high viscosity using a titanocene photoinitiator. J Appl Polym Sci 128:611–618

Klee JE, Maier M, Fik CP (2015) Dyes and chromophores in polymer science. Wiley

Lalevée J, Blanchard N, Tehfe M-A, Morlet-Savary F, Fouassier JP (2010) Green bulb light source induced epoxy cationic polymerization under air using tris (2, 2′-bipyridine) ruthenium (II) and silyl radicals. Macromolecules 24:10191–10195

Lee J, Ju M, Cho OH, Kim Y, Nam KT (2019) Tyrosine-rich peptides as a platform for assembly and material synthesis. Adv Sci 6. https://doi.org/10.1002/advs.201801255

Lee S, Choi G, Yang YJ, Joo KI, Cha HJ (2023) Visible light-crosslinkable tyramine-conjugated alginate-based microgel bioink for multiple cell-laden 3D artificial organ. Carbohydr Polym 313:120895

Lewandowska-Łańcucka J, Mystek K, Mignon A, Van Vlierberghe S, Łatkiewicz A, Nowakowska M (2017) Alginate-and gelatin-based bioactive photocross-linkable hybrid materials for bone tissue engineering. Carbohydr Polym 157:1714–1722

Li L, Lu C, Wang L, Chen M, White J, Hao X, McLean KM, Chen H, Hughes TC (2018) Gelatin-based photocurable hydrogels for corneal wound repair. ACS Appl Mater Interfaces 10:13283–13292

Li J, Wu C, Chu PK, Gelinsky M (2020) 3D printing of hydrogels: rational design strategies and emerging biomedical applications. Mater Sci Eng R Reps 140:100543

Lim KS, Abinzano F, Bernal PN, Albillos Sanchez A, Atienza-Roca P, Otto IA, Peiffer QC, Matsusaki M, Woodfield TBF, Malda J, Levato R (2020) One-step photoactivation of a dual-functionalized bioink as cell carrier and cartilage-binding glue for chondral regeneration. Adv Healthc Mater 9:1–13. https://doi.org/10.1002/adhm.201901792

Lin CC, Ki CS, Shih H (2015) Thiol-norbornene photoclick hydrogels for tissue engineering applications. J Appl Polym Sci 132:1–11. https://doi.org/10.1002/app.41563

Liu Y, Ng SC, Yu J, Tsai W-B (2019a) Modification and crosslinking of gelatin-based biomaterials as tissue adhesives. Colloids Surf B Biointerfaces 174:316–323

Liu J, Xiao Y, Wang X, Huang L, Chen Y, Bao C (2019b) Glucose-sensitive delivery of metronidazole by using a photo-crosslinked chitosan hydrogel film to inhibit Porphyromonas gingivalis proliferation. Int J Biol Macromol 122:19–28

Liu S, Graff B, Xiao P, Dumur F, Lalevée J (2021) Nitro-carbazole based oxime esters as dual photo/thermal initiators for 3D printing and composite preparation. Macromol Rapid Commun 42:2100207

Loebel C, Szczesny SE, Cosgrove BD, Alini M, Zenobi-Wong M, Mauck RL, Eglin D (2017) Cross-linking chemistry of tyramine-modified hyaluronan hydrogels alters mesenchymal stem cell early attachment and behavior. Biomacromolecules 18:855–864. https://doi.org/10.1021/acs.biomac.6b01740

Lougnot DJ, Fouassier JP (1988) Comparative reactivity of water soluble photoinitiators as viewed in terms of excited states processes. J Polym Sci Part A Polym Chem 26:1021–1033

Min K, Yun G, Jang Y, Kim K-R, Ko YH, Jang H, Lee Y, Kim K, Kim D (2016) Covalent self-assembly and one-step photocrosslinking of tyrosine-rich oligopeptides to form diverse nanostructures. Angew Chem 128:7039–7042. https://doi.org/10.1002/ange.201601675

Mishra R, Varshney R, Das N, Sircar D, Roy P (2019) Synthesis and characterization of gelatin-PVP polymer composite scaffold for potential application in bone tissue engineering. Eur Polym J 119:155–168. https://doi.org/10.1016/j.eurpolymj.2019.07.007

Moon SH, Hwang HJ, Jeon HR, Park SJ, Bae IS, Yang YJ (2023) Photocrosslinkable natural polymers in tissue engineering. Front Bioeng Biotechnol 11:1127757. https://doi.org/10.3389/fbioe.2023.1127757

Mu X, Sahoo JK, Cebe P, Kaplan DL (2020) Photo-crosslinked silk fibroin for 3d printing. Polymers (Basel) 12:1–18. https://doi.org/10.3390/polym12122936

Müller SM, Schlögl S, Wiesner T, Haas M, Griesser T (2022) Recent advances in Type I photoinitiators for visible light induced photopolymerization. ChemPhotoChem 6:e202200091

Nair LS, Laurencin CT (2007) Biodegradable polymers as biomaterials. Prog Polym Sci 32:762–798

Pang Y (2020) NIR-sensitized activated photoreaction between cyanines and oxime esters: free-radical photopolymerization. Angew Chem 132:11537–11544

Partlow BP, Hanna CW, Rnjak-Kovacina J, Moreau JE, Applegate MB, Burke KA, Marelli B, Mitropoulos AN, Omenetto FG, Kaplan DL (2014) Highly tunable elastomeric silk biomaterials. Adv Funct Mater 24:4615–4624

Partlow BP, Applegate MB, Omenetto FG, Kaplan DL (2016) Dityrosine cross-linking in designing biomaterials. ACS Biomater Sci Eng 2:2108–2121. https://doi.org/10.1021/acsbiomaterials.6b00454

Pierau L, Elian C, Akimoto J, Ito Y, Caillol S, Versace D-L (2022) Bio-sourced monomers and cationic photopolymerization–the green combination towards eco-friendly and non-toxic materials. Prog Polym Sci 127:101517

Qi Z, Xu J, Wang Z, Nie J, Ma G (2013) Preparation and properties of photo-crosslinkable hydrogel based on photopolymerizable chitosan derivative. Int J Biol Macromol 53:144–149

Qiu W, Li M, Yang Y, Li Z, Dietliker K (2020) Cleavable coumarin-based oxime esters with terminal heterocyclic moieties: photobleachable initiators for deep photocuring under visible LED light irradiation. Polym Chem 11:1356–1363

Raia NR, Partlow BP, McGill M, Kimmerling EP, Ghezzi CE, Kaplan DL (2017) Enzymatically crosslinked silk-hyaluronic acid hydrogels. Biomaterials 131:58–67. https://doi.org/10.1016/j.biomaterials.2017.03.046

Reis AV, Fajardo AR, Schuquel ITA, Guilherme MR, Vidotti GJ, Rubira AF, Muniz EC (2009) Reaction of glycidyl methacrylate at the hydroxyl and carboxylic groups of poly(vinyl alcohol) and poly(acrylic acid): is this reaction mechanism still unclear? J Org Chem 74:3750–3757. https://doi.org/10.1021/jo900033c

Samadian H, Maleki H, Allahyari Z, Jaymand M (2020) Natural polymers-based light-induced hydrogels: promising biomaterials for biomedical applications. Coord Chem Rev 420:213432. https://doi.org/10.1016/j.ccr.2020.213432

Shalumon KT, Liao HT, Kuo CY, Wong CB, Li CJ, Mini PA, Chen JP (2019) Rational design of gelatin/nanohydroxyapatite cryogel scaffolds for bone regeneration by introducing chemical and physical cues to enhance osteogenesis of bone marrow mesenchymal stem cells. Mater Sci Eng C 104:109855. https://doi.org/10.1016/j.msec.2019.109855

Wang X, Hao T, Qu J, Wang C, Chen H (2015) Synthesis of thermal polymerizable alginate-GMA hydrogel for cell encapsulation. J Nanomater 2015:970619

Wang W, Jin M, Pan H, Wan D (2021a) Remote effect of substituents on the properties of phenyl thienyl thioether-based oxime esters as LED-sensitive photoinitiators. Dyes Pigments 192:109435

Wang Y, Zhang S, Wang J (2021b) Photo-crosslinkable hydrogel and its biological applications. Chin Chem Lett 32:1603–1614. https://doi.org/10.1016/j.cclet.2020.11.073

Wang Y, Chen Y, Zheng J, Liu L, Zhang Q (2022) Three-dimensional printing self-healing dynamic/photocrosslinking gelatin-hyaluronic acid double-network hydrogel for tissue engineering. ACS Omega 7:12076–12088

Xiao P, Zhang J, Dumur F, Tehfe MA, Morlet-Savary F, Graff B, Gigmes D, Fouassier JP, Lalevée J (2015) Visible light sensitive photoinitiating systems: recent progress in cationic and radical photopolymerization reactions under soft conditions. Prog Polym Sci 41:32–66

Yang K, Sun J, Guo Z, Yang J, Wei D, Tan Y, Guo L, Luo H, Fan H, Zhang X (2018) Methacrylamide-modified collagen hydrogel with improved anti-actin-mediated matrix contraction behavior. J Mater Chem B 6:7543–7555

Yue K, Li X, Schrobback K, Sheikhi A, Annabi N, Leijten J, Zhang W, Zhang YS, Hutmacher DW, Klein TJ, Khademhosseini A (2017) Structural analysis of photocrosslinkable methacryloyl-modified protein derivatives. Biomaterials 139:163–171. https://doi.org/10.1016/j.biomaterials.2017.04.050

Zhang J (2015) Structure design of naphthalimide derivatives: toward versatile photoinitiators for near-UV/visible LEDs, 3D printing, and water-soluble photoinitiating systems. Macromolecules 48:2054–2063

Zheng Z, Eglin D, Alini M, Richards GR, Qin L, Lai Y (2021) Visible light-induced 3D bioprinting technologies and corresponding bioink materials for tissue engineering: a review. Engineering 7:966–978. https://doi.org/10.1016/j.eng.2020.05.021

Zhou Y, Zhao S, Zhang C, Liang K, Li J, Yang H, Gu S, Bai Z, Ye D, Xu W (2018) Photopolymerized maleilated chitosan/thiol-terminated poly (vinyl alcohol) hydrogels as potential tissue engineering scaffolds. Carbohydr Polym 184:383–389

# Part II

# Natural Materials and Eco-Wastes Used as Sacrificial Templates

# Sustainable Materials Derived from Lignin for Tissue Engineering and Regenerative Medicine

6

Liam O'Brien and Maurice N. Collins

**Abstract**

This chapter discusses sustainable materials derived from lignin for tissue engineering. Lignin is a complex, sustainable biopolymer that is abundantly available, making it an attractive alternative to traditional synthetic materials. Currently, lignin is used for low-value manufacturing, such as energy production by incineration. Lignin has emerged to provide unique mechanical and chemical properties for tissue engineering applications. Its antioxidant, antimicrobial, biodegradability, biocompatibility, and mechanical strength are of great interest and are discussed here. Lignin has excellent potential for biomedical engineering. However, further research and studies on appropriate animal models are needed to verify lignin-derived biomaterials as a safe and viable option for tissue engineering applications, particularly its long-term stability. Developing lignin-derived materials for tissue engineering can address some of the challenges associated with traditional biomaterials, including cost, availability, and sustainability.

**Keywords**

Lignin · Biomaterials · Sustainable · Tissue engineering · Lignin properties

L. O'Brien · M. N. Collins (✉)
School of Engineering, University of Limerick, Limerick, Ireland
e-mail: maurice.collins@ul.ie



J. M. Oliveira et al. (eds.), *Sustainable Scaffolds-based Strategies in Tissue Engineering and Regenerative Medicine*, Biomaterials, Bioengineering and Sustainability 8, https://doi.org/10.1007/978-3-031-96274-5_6

## 6.1 Introduction

Recently, there has been an increased emphasis on finding sustainable materials for engineering applications as the planet moves toward a greener and eco-friendlier environment (Serafin et al. 2025). Tissue engineering and regenerative medicine are involved in the regeneration of the function of biological structures lost due to congenital abnormalities, injury, and disease through using cells, tissues, and organs to repair, restore and replace defective tissue (Hellman 2008). Multiple studies have proven that lignin-based materials have a vast range of applications in tissue engineering and beyond. This chapter will discuss applications of lignin-based materials for bone tissue regeneration, scaffold fabrication, drug delivery and gene delivery, wound healing and skin tissue engineering, neural tissue engineering, and cartilage tissue engineering, amongst many other applications.

Lignin is an abundant phenolic biopolymer, interlaced between cellulose and hemicellulose in the cell wall of vascular plants (Liu et al. 2022a). This amorphous, heterogenous, low molecular weight material is the second most abundant natural polymer after cellulose. Lignin elicits excellent antibacterial behaviour and antioxidant properties (Collins et al. 2019). Lignin has been primarily regarded as a waste by-product from the paper and pulp industry however this is changing given the drive towards sustainable materials. The separated polysaccharides can then be hydrolyzed into valuable commodities such as paper, thickening and emulsifying agents for food and fuel. However, 98% of the lignin separated from cellulose is sent for incineration to produce steam for energy (Oriez et al. 2020; Rosenboom et al. 2022). Only 2% of lignin waste is recycled and used to create a viable product. These statistics have led engineers and scientists to experiment with lignin to determine its properties that can be valorized for other applications. With biomedical applications been promising as we discuss in the upcoming sections.

## 6.2 Biological Properties of Lignin

Lignin has been rigorously evaluated to determine possible applications in tissue engineering. Multiple studies have revealed that lignin has excellent antimicrobial, biocompatibility, biodegradability, and antioxidant properties.

Antimicrobial activity can be defined as inhibiting or killing the growth of pathogens in a body (Mackintosh 2001). It is well-documented that plants have antimicrobial extracts which inhibit the proliferation of bacteria, fungi, viruses, and protozoa (Vaou et al. 2021). The antimicrobial activity associated with lignin depends heavily on its extraction method, as it leads to variations in the chemical composition and structure of lignin. Lignin is mainly obtained from a kraft process but can also be acquired from organosolv, lignosulphonate, and soda processes (Ndaba et al. 2020). Kraft extraction comprises wood chips placed in a pressure vessel with elevated temperatures and a solution of sodium hydroxide and sodium sulphide, causing the separation of lignin from cellulose (Cheremisinoff and Rosenfeld 2010). Yang and colleagues have shown that lignin extracted using the kraft method possesses antimicrobial activity against bacteria such as Escherichia coli, Staphylococcus aureus, and Staphylococcus epidermidis by the use of lignin nanoparticles in an alginate hydrogel (Yang et al. 2018). Another paper used lignin nanoparticles in a polyvinyl/alcohol/chitosan hydrogel to show that lignin extracted by a kraft process can inhibit Salmonella enterica, Bacillus cereus, and Pseudomonas aeruginosa, which can be crucial for wound healing (Spasojević et al. 2016). The selection of which method to use depends on what properties of lignin you seek; if an increased antimicrobial activity is desired, the kraft or organosolv method is best suited (Lobo et al. 2021).

The phenolic structure of lignin varies in different forms of lignin-derived biomaterials, contributing to its antimicrobial behaviour. The phenolic system includes hydroxyl groups and aromatic rings (Ayad and Akkal 2019). The higher the phenolic hydroxyl content in lignin, the higher the resistance to bacterial adhesion. However, these hydroxyl groups can damage and kill the bacteria's cell wall leading to cell necrosis. They achieve this by creating hydrogen bonds with the bacteria cell wall. Other chemical groups in lignin that allow for the formation of hydrogen bonds are aldehyde and ketone groups (Chen et al. 2023). Also, the lower the molecular weight, the increased inhibition of Gram-negative and Gram-positive bacteria growth. This was shown using bamboo kraft lignin by organosolv fractionation (Yun et al. 2021). The molecular weight property is controlled by polymerization, which is the chemical binding of monomers to form chain-like structures called polymers. The longer these chains allow for higher molecular weights, short chains have low molecular weights.

Antioxidants are chemical scavengers of active free radicals and therefore promote oxidization (Koltover and Skipa 2023). If oxidization occurs pre or post-transplantation or implantation, it can induce oxidative stress. According to Pizzino and colleagues, oxidative stress arises due to an imbalance between the production and accumulation of reactive oxygen species in cells and tissues and the ability of a biological system to detoxify these reactive products (Pizzino et al. 2017). This stress can cause immediate and long-term graft dysfunction and potencies such as graft rejection, delayed graft function, primary nonfunction of the transplants, inflammation, and cancers. (Shi and Xue 2016). Many studies on lignin have shown many characteristics, such as scavenging free radicals and chelate metal ions that allow the biopolymer to be classified as an antioxidant. The primary sources of the antioxidant property in lignin arise from the phenolic hydroxyl groups such as syringyl and p-hydroxyphenyl, methoxy, aliphatic hydroxyl, and a double bond between the outermost carbon in the side chain (Lourençon et al. 2021). The phenolic hydroxyl group present in lignin can catch centred radicals along with hydroperoxyl radicals derived from the reaction of carbon-centred radicals with oxygen (Xiao et al. 2021). These properties can stabilize the free radicals delocalizing by unpaired electrons. Some of the non-phenolic compounds contribute to lignin's antioxidant ability. Flavonoids have been proven to share hydrogen atoms with free radicals to trap them and hinder radical chain reactions (Zheng et al. 2022). Decreased polydispersity, heterogeneity, and low molecular weight have been shown to increase lignin's antioxidant properties (Zahed et al. 2018).

Lignin-based nanoparticles and hydrogels have been found to scavenge free radicals and reduce oxidative stress in various cell types. Liang and colleagues investigated poly(lactic acid) (PLA) lignin nanofibers as an antioxidant biomaterial for cartilage regeneration. They demonstrated that PLA-lignin nanofibers could protect bone-marrow-derived mesenchymal stem cells from oxidative stress and promote chondrogenic differentiation with a series of *in vitro* tests. This indicates that PLA-lignin could be used for antioxidant tissue scaffolds (Liang et al. 2022). Furthermore, it was displayed by Bertolo and colleagues stated that lignin-derived nanoparticles from sugarcane bagasse had an antioxidant potential against DPPH radicals and reduced oxidative stress in skin fibroblasts as a result of the decreased level of reactive oxygen species (Bertolo et al. 2019). Saratable and colleagues have also explored the antioxidant behaviour of sodium lignosulfonate silver nanoparticles. They also have found that this lignin-derived biomaterial has immense free radical scavenging against DPPH (Saratale et al. 2022).

*In vivo*, studies were also conducted to evaluate the antioxidant properties of lignin. Xu and colleagues extracted lignin from coconut husks and manufactured a hydrogel produced from polyethylene glycol (PEG), polypropylene glycol (PPG), and polydimethylsiloxane (PDMS) and incorporated the lignin. This gel was placed into the burn wounds of mice, and its antioxidant properties were evaluated. It showed vigorous antioxidant activity as it reduced active oxygen levels, which protected cells from apoptosis due to oxidative stress (Xu et al. 2021). Similarly, Deng and colleagues conducted *in vivo* tests on bisphenol AF-induced zebrafish evaluating antioxidant properties by monitoring the activity of catalase, superoxide dismutase, and malondialdehyde in the model. For this study, they used lignin nanoparticles. They found that lignin returned increased catalase and superoxide dismutase levels to normal, which reduced the intracellular reactive oxygen species. Lignin also reduces abnormal malondialdehyde, which causes cell membrane damage due to oxidation. It was concluded that lignin possessed a scavenging ability for reactive oxygen species and free radicals (Deng et al. 2023).

Biocompatibility is one of the most crucial properties to be addressed to ensure safe and effective materials for tissue engineering applications. Biocompatibility can be defined as the ability of a material to perform as required with an appropriate response from the host (Jurak et al. 2021). Biocompatible biopolymers such as polysaccharides and proteins are highly researched and have shown significant advancements in tissue engineering applications. But lignin's biocompatibility still needs much more investigation before being deemed biocompatible. For a material to be considered biocompatible, it must undergo tests such as cytotoxicity and immunogenicity. These tests alone can't determine the biocompatibility of lignin precisely but will give a good representation of a material's biocompatible nature. Cytotoxicity refers to the degree to which a material can cause reversible cellular injury or progress to cell apoptosis or necrosis (Azqueta et al. 2022). This test is the most researched area of lignin's capacity to be biocompatible. The cytotoxicity of lignin varies due to different strains of lignin from different biomasses. Most studies on lignin's cytotoxic behaviours revealed lignin to be non-toxic, but these findings don't cover every type of lignin (Ullah et al. 2022). It has been documented that increased phenolic groups due to different extraction methods can cause increased antioxidant behaviour and cytotoxic activity (Zheng et al. 2021). Immunogenicity is the ability of a material to provoke an unwanted immune response (Ratanji et al. 2014). Several *in vitro* and *in vivo* studies have investigated this property of lignin and lignin-based biomaterials for tissue engineering applications. These studies found the lignin they assessed to be non-immunogenic as cytokine production and cell proliferation were minimal. Still, all agreed that high concentrations of phenolic groups might cause immunogenic responses (Chavooshi et al. 2023; Belgodere et al. 2019; Moohan et al. 2019). Ten and colleagues investigated using lignin nanotubes as a method for gene delivery into human cells. They concluded that the immunogenicity is influenced by the source of lignin and its extraction procedure (Ten et al. 2014).

Lignin's biocompatibility is still a new and rapidly developing area of research. However, there are significant indications that lignin possesses higher biocompatibility, lower toxicity, and low immunogenic responses than its synthetic peers. Therefore, it has exciting in regenerative medicine.

Biodegradability is a crucial factor that must be considered when placing any biomaterial into the human body, particularly for scaffold formation and drug delivery mechanisms. Biodegradability looks at the material's ability to be absorbed and disintegrated safely by bodily tissue without needing surgical removal. The biodegradability of lignin and lignin-derived materials can be modified to make them more soluble in water and reduce the complexity of their structure. Lignin can be chemically treated by processes such as sulfonation and hydrolysis, which increase solubility. Another method is to combine the lignin with other synthetic biodegradable polymers, such as polylactic-co-glycolic acid (PLGA) and polycaprolactone (PCL) (Makadia and Siegel 2011; Nevoralová et al. 2020). Several studies have evaluated the biodegradability of lignin and lignin-derived biomaterials for possible tissue engineering applications. Amini and colleagues investigated the biodegradability of electrospun PCL fibers with incorporated lignin nanoparticles *in vitro* by MTT assay. It was concluded that increased lignin in the PCL fibers leads to increased degradation, water intake, and cell viability (Amini et al. 2020). Similarly, Silva-Barrosa and colleagues assessed the degradation of tricalcium phosphate, sodium alginate, and lignin scaffold with variations of percentage lignin. They characterized the degradation by evaluating the absence or presence of lysozyme on the scaffolds incubated in the DMEM-F12 medium. It was found that there was a 12% weight loss of the scaffold in the presence of lysozyme after 21 days and only 8% without the presence of lysozyme. They determined that this slow degradation was due to the lack of enzymes specific to lignin (Silva-Barroso et al. 2023). Chandna and colleagues investigated the biodegradability of lignin-derived silver nanocomposite hydrogel by HRTEM, BET, XRD, and other analytical methods. They concluded that these lignin-derived hydrogels possessed excellent biodegradability qualities and were suitable for wound dressing and encapsulating materials (Chandna et al. 2022).

## 6.3 Bio Mechanical Properties of Lignin

Lignin mechanical properties heavily depend on the source of lignin and the isolation process. Lignin has been added to many scaffolds and hydrogels to increase the composite's Young's modulus, yield stress, flexural strength, and tensile strength. For example, Sameni and colleagues investigated the mechanical properties of different soda lignin/high-density polyethylene (HDPE) blends. They found that adding lignin to HDPE increased tensile, flexural strength, and Young's modulus compared to neat HDPE. They also found that lignin reduced thermal properties, such as degradation and HDPE melting temperatures (Sameni et al. 2018). Nan and colleagues reviewed many studies and found that lignin increases the mechanical strength of many composites. Adding lignin to a chitosan-PVA composite increased mechanical strength by 46.87 MPa. They also supplied another study that showed enhanced mechanical properties of cross-linked lignin-agarose-silk fibroin due to lignin's presence in the nanocomposite. This biomaterial was useful for tissue scaffold applications (Nan et al. 2022). Chen and colleagues prepared polyacrylamide/lignin nanoparticle hydrogel to assess the mechanical influence of lignin on the hydrogel. They found that lignin increased fracture strength by many orders of magnitude compared to neat polyacrylamide hydrogel. It also displayed excellent elastic behaviour as well as high tensile and compressive strength (Chen et al. 2019).

## 6.4 Tissue Engineering and Regenerative Medicine Applications

The possible use of lignin for tissue engineering applications has been widely researched in recent years—the main contributing reasons are its sustainability, biocompatible properties, vast availability, and reduced environmental impact (Winters et al. 2025). Scientists and engineers have evaluated lignin-based biomaterials for bone tissue regeneration, gene therapy, scaffold formation, wound therapy, neural tissue regeneration, and drug delivery.

### 6.4.1 Bone Tissue

Bone tissue engineering and regeneration is an emerging field of tissue engineering. Strategies are being developed to heal diseased and defective bone tissue using a combination of cells, growth factors, and biomaterials (Alonzo et al. 2021). The function of bone is to give structure to the body and protect vital organs such as the brain, heart, and lungs. Bone is an extraordinarily complex structure consisting of mineralized collagen fibers with four main cell types, osteoblasts, bone lining cells, osteocytes, and osteoclasts (Florencio-Silva et al. 2015). Bone defects such as osteoporosis, breaks, fractures, and cancers are becoming an evident problem amongst older adults as the life expectancy of humans is ever-increasing, leading to novel and minor procedures required to fix these ailments. Bone tissue engineering aims to solve this by adding new functional tissue to these areas and applying appropriate growth factors to regenerate the damaged area without causing any adverse immune response (Akter and Ibanez 2016). The use of lignin in tissue engineering has been widely researched, and many possible applications have been found. Lignin can be added to scaffolds, bioceramics, hydrogels, and composites to increase the material's biocompatibility, biodegradability, and anti-microbial properties.

Several studies investigate the addition of lignin to biomaterials for bone tissue-specific scaffolds. A bone tissue scaffold is a matrix that allows and entices the attachment and proliferation of osteoinductive cells on its surface, stimulating bone regrowth (Ghassemi et al. 2018). Mohammadalipour and colleagues added lignin to polyhydroxy butyrate (PHB) to create an electrospun PHB/Lignin scaffold. It was confirmed that lignin fixed the brittle and hydrophobic nature of the PHB. This allowed for the proliferation and adhesion of MG63 cells. It was also noted that lignin improved the mechanical properties of the scaffold by increasing the modulus, tensile strength, and elongation of the scaffold due to the hydrogen bond network. In addition, Lignin aided with reducing the diameter of the electrospun fibers, increasing the scaffold's biodegradability and permeability. It was concluded that the PHB/lignin scaffold outputs could provide excellent potential in bone tissue engineering applications (Mohammadalipour et al. 2022). Another study examined lignin nanoparticles synthesized from alkali lignin for bone tissue applications. These nanoparticles were incorporated with PCL to create nanofibers. Haider and colleagues discovered that adding lignin increased the nanofiber's diameters and reduced mechanical properties such as crystallinity and porosity. Furthermore, the lignin nanoparticles improved surface wetting and swelling properties and increased their anti-microbial activity against gram-positive bacteria. Lignin nanoparticles/PCL nanofibers displayed exciting proliferation and attachment of pre- osteoblastic cell line (MC3T3-E1) and improved osteogenic differentiation compared to normal PCL nanofibers. It was concluded that this material could be a promising potential biomaterial for bone tissue scaffolds (Haider et al. 2023). Other studies found that adding lignin to a synthetic polymer can increase and promote bone cell proliferation and adhesion to scaffolds and increase mechanical properties helping the scaffold (Luzi et al. 2020; Kai et al. 2016).

Hydrogels can also play a significant role in bone tissue engineering. These hydrogels can transport specific cells to precise locations, promote cell adhesion, and guarantee cell viability, vascularization, proliferation, and differentiation by mimicking the extracellular matrix of bone. The advantage of hydrogels for bone tissue engineering is their injectability. This means the hydrogel can be injected minimally invasively into the affected area, even for the most complex geometries of the affected area (Liu et al. 2022b).

Multiple studies have examined the viability of hydrogels in bone tissue engineering. Chen and colleagues studied the performance of a polyacrylamide/lignin nanoparticle nanocomposite hydrogel. They found that the gel possesses high compressive and tensile strength and high recoverability. In addition, the lignin endowed strong hydrogen bonds, allowing the gel to distribute applied loads evenly. This allows the gel to withstand forces endured during injection into the patient. It was concluded that this gel has potential applications in the bone tissue engineering field (Chen et al. 2019). Abdullah and colleagues investigated a crosslinked lignin and gelatine cryogel. They found that the gel had excellent antibacterial and antioxidant behaviour and could withstand high compressive stresses endured during the injection phase with over 90% recovery to its original shape. The cryogel also exhibited excellent cytocompatibility behaviour. It was concluded that this hydrogel shows great promise for potential bone tissue engineering applications (Abdullah et al. 2021). Another study examined an aerogel that consisted of lignin crosslinked with alginate. Quraishi and colleagues created the aerogel by using pressurized carbon dioxide for gelation. It was found that lignin reduces alginate's hydrophilicity to create a better microenvironment for the cells to proliferate and differentiate. They also reported that lignin helped reduce scaffold degradation rate, enabling it to be similar to new bone tissue regeneration. They also concluded that this hydrogel is a potential candidate for bone tissue engineering applications.

### 6.4.2 Skin Tissue

The skin is the largest organ in the body and consists of three main layers, the epidermis, dermis, and subcutaneous layer (Ho et al. 2017). Its principal function is to provide a barrier covering the entire body and provide a defense against harmful diseases and chemicals. The most common wound injuries to the skin result from burns, accidents, or scarring after invasive surgical procedures. Burns can be categorized into first, second, and third-degree burns. First-degree burns only cause slight damage to the epidermis layer of the skin. Second-degree burns cause blistering on the epidermis and also affect the dermis layer. Third-degree burns involve all three layers of the skin and can be extremely dangerous. The potential use of tissue engineering methods to heal wounds is being extensively researched. Lignin can play a massive role in preventing infection and microbial activity in these wounds. Several studies have investigated lignin's actual ability in wound healing and therapy.

Li and colleagues created a novel lignin polyurethane/silver nanoparticle composite foam for wound healing. They found that the lignin functioned as a reducing and capping agent to reduce silver ions, resulting in silver nanoparticles. They found this foam to have enhanced thermal, mechanical, and biodegradable properties. Adding lignin contributed to the outstanding antimicrobial behaviour, assessed *in vitro* and *in vivo* on rats' wounds. It concluded that this foam could potentially be the future of the biomass wound dressing field (Li et al. 2022). Lignin has also been combined with polyvinyl alcohol and chitosan to create a novel composite hydrogel for wound dressing applications. The introduction of lignin to the hydrogel vastly improved mechanical strength, protein absorption capacity, and wound environmental regulation ability (Zhang et al. 2019). Du and colleagues explored lignin nanoparticles with adjustable sizes for accelerated wound healing. They concluded that lignin nanoparticles demonstrated outstanding antioxidant absorption capacity of protein and bacterial activity. After *in vivo* studies on mice, they found that adding lignin nanoparticles accelerated the healing process. After 15 days, the wound was closed entirely, and granulation tissue formation was advanced (Du et al. 2022). This study further verifies the use of lignin in wound healing applications. While Decante et al. developed a stretchable gelatin lignin/gealtin hydrogel with potential for printing for a range of potential tissue engineering applications (Decante et al. 2024).

One of the most significant factors in wound healing is the ability to keep the wound sterile from pathogens. Dominguez-Robles and colleagues designed and 3D printed a PCL/lignin wound dressing loaded with curcumin and D-Panthenol. Once again, lignin's antimicrobial and antioxidant properties are justification for choice. The material was shown to be 98% resistant to *Staphylococcus aureus* due to the addition of lignin. *In vivo* studies on Wistar rats confirmed the biomaterials application to the wound dressing. It was concluded that PCL/lignin material enhanced epidermal differentiation and showed excellent results in the testing parameters, epithelisation, inflammatory reaction, and proliferation rate of fibroblasts (Domínguez-Robles et al. 2023). Perez-Rafael and colleagues created a hyaluronic acid and silver-lignin nanoparticle hydrogel for chronic wound healing. They found that this hydrogel inhibited the major enzymes myeloperoxidase and matrix metalloproteinases responsible for wound chronicity in patient. The hydrogel also exhibited strong resistance to *Staphylococcus aureus* and *Pseudomonas aeruginosa*. The hydrogel was assessed *in vivo* on diabetic mouse models, and it revealed no inflammation due to toxicity. They concluded that hyaluronic acid and silver-lignin nanoparticle hydrogel are excellent possibilities for wound healing (Pérez-Rafael et al. 2021).

### 6.4.3 Drug Delivery/Gene Therapy Delivery in Regenerative Medicine

Drug delivery can be defined as a process of administrating a pharmaceutical compound to achieve a medicinal effect for humans and animals alike. There are many common routes of entry to the body. Enteral routes include oral, sublingual, buccal, and rectal administration. Parenteral routes consist of intravenous injections, intramuscular medication, and subcutaneous injections. Other forms are transnasal, transdermal, and inhalation. Tissue engineering techniques have been explored for drug delivery. These techniques consist of physical or chemical absorption of the drug onto the tissue scaffold or drug encapsulation directly on the scaffold or by creating a drug delivery system on the scaffold. The diffusion or degradation of the tissue scaffold controls the drug release. Drug delivery in tissue engineering can accelerate the regeneration of new cells (Rambhia and Ma 2015). Lignin may provide biodegradability properties to drug delivery systems, allowing for quicker breakdown of scaffolds or encapsulations and more rapid drug release. Several studies investigate lignin application to drug delivery systems.

Alqahtani and colleagues investigated lignin nanoparticles for oral drug delivery of curcumin. The nanoparticles were sourced from organosolv lignin and prepared using a modified phases separation method. They found the lignin nanoparticles to increase the permeability of curcumin by fivefold in comparison to ordinary curcumin. *In vivo* studies on rats were conducted for pharmacokinetic analysis and found bioavailability of curcumin increased tenfold. They concluded that lignin nanoparticles are an effective method of oral drug delivery, in particular for low solubility and low bioavailable drugs (Alqahtani et al. 2019). Ali and Mehanna investigated lignin nanoparticles for anticancer drug delivery. They found that lignin possessed excellent absorption capacity, biodegradability, and non- toxicity while also applying medicinal effects such as anti-cancer, antimicrobial, and antioxidant. They also found that lignin nanoparticles' drug-delivering ability is affected by the lignin extraction method, purification, and quantitation method (Ali and Mehanna 2022). Garg and colleagues further verified lignin-based materials for drug delivery, particularly for anticancer therapy. They concluded that lignin's excellent physical and chemical properties make the biomaterial an ideal candidate for any drug delivery application (Garg et al. 2022).

Gels are also an effective method of drug delivery. Hydrogels can offer drug encapsulation, injectability properties, and minimally invasive pain relief. Chiani and colleagues fabricated a lignin/gelatin hydrogel for carrying Ribavirin. They found that lignin caused variation in pore structure, viscoelastic behavior, and swelling ability. Large pores resulted from increased lignin content in the hydrogel. Lignin also enhanced the Ribavirin release rate showing exciting potential for hydrogel drug delivery (Chiani et al. 2023). Gujjala and colleagues also investigated hydrogel's application in drug delivery systems. They reviewed a lignin-based hydrogel consisting of lignin, GANTREZ, and PEG/glycerol and found it to have the very functional capacity for the drug curcumin for drug- delivering (Gujjala et al. 2022). Zhao and colleagues prepared lignin-based nanoparticles loaded with Doxorubicin Hydrochloride as shown in Figure 6.1. They discovered that the surface of the lignin-hollowed nanoparticles could respond to the magnetic field and folic acid. The folic-magnetic functionalized lignin hollow nanoparticles had increased cellular uptake in the presence of HeLa cells. This research proved lignin nanoparticle variations as effective nano-delivery mechanisms and an inhibitor for the proliferation of cancer cells (Carrion et al. 2021; Zhou et al. 2019).

Gene therapy is a technique where the genetic material of cells is manipulated and modified to treat and prevent disease. This technique applies to tissue engineering, as cells can create new tissues to replace damaged cells. Gene therapy aims to either replace mutated genes, fix mutated genes, or make the immune system more aware of diseased cells in the body. Several studies have been conducted to explore lignin-derived biomaterials as gene carriers for gene therapy.

Liu and colleagues investigated the use of lignin for nonviral gene carriers. Lignin was converted into a macroinitiator using 2-bromoisobutyryl. Atom transfer radical polymerization was utilized for culturing poly[2-(dimethylamino)ethyl methacrylate] (PDMAEMA). This material's *in vitro* gene transfection capability was assessed by luciferase assay in HeLa, COS7, and MDA-MB-231 cell lines. They found that the grafting degree and cationic chain length significantly affected the performance of the gene transfection efficiency. Adjusting these factors

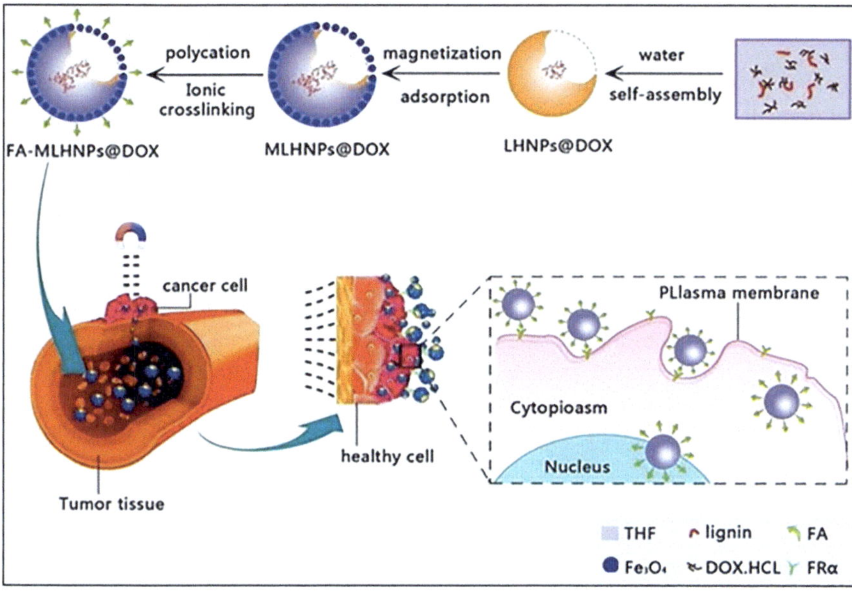


**Fig. 6.1** Schematically representation of lignin-based hollow nanoparticles for delivering Doxorubicin Hydrochloride for cancer cell therapy Reused from Ref. (Carrion et al. 2021) (CC BY 4.0)

precisely for different cell lines can lead to lignin/PDMAEMA being an attractive choice for gene delivery applications (Liu et al. 2022c). Furthermore, Ullah and colleagues examined lignin nanoparticles for gene delivery for diabetic diseases. They concluded that lignin particles exhibit excellent antimicrobial, biocompatibility, and antioxidant behaviour, making them an attractive gene delivery material for diabetic conditions (Ullah et al. 2022). Figueiredo and colleagues examined the potential of lignin-based nanoparticles for gene delivery and enhanced antiproliferation effect on cancer cells. They developed three nanoparticles for this study, pure lignin nanoparticles, iron(III)- complexed lignin nanoparticles, and $Fe_3O_4$-infused nanoparticles. The pure lignin nanoparticles exhibited excellent drug loading capacity for poorly insoluble drugs such as sorafenib and benzazulene. The pure lignin nanoparticles loaded with benzazulene displayed an enhanced antiproliferation effect on cancer cells compared to pure benzazulene. This material showed maximum inhibitory concentration after 24-hours incubation (Figueiredo et al. 2017). Finally, Sathasivam and colleagues observed using lignin-derived nanoparticles for cancer therapy. They found that eucalyptus kraft lignin showed inhibitory responses in mouse models for melanoma and hepatoma. This type of lignin possesses secondary metabolites, such as phenolic constituents, suppressing tumour growth. They also looked at how peptide-conjugated NL combined with a drug carrier could guide itself toward the specific tumour cells (Sathasivam et al. 2023).

### 6.4.4 Neural Tissue

Neural tissue engineering involves designing and fabricating scaffolds, hydrogels, and films that aid the regeneration of damaged and diseased neural tissue. The nervous system in the body consists of two parts, the central nervous system and the peripheral nervous system. The central nervous system consists of the brain and the spinal cord, while the peripheral nervous system consists of all the nerves which are branched off the spinal cord. A nerve consists of a group of specialized cells called neurons. A neuron comprises three main parts axon, cell body, and dendrites. Neural tissue injuries are one of the most debilitating injuries on the planet. These injuries range from spinal cord injuries and brain injuries resulting from traumatic accidents to neurodegenerative diseases such as Parkinson's, Motor Neuron, and Alzheimer's. Lignin-based biomaterials can play an effective role in treating these diseases, according to recent studies, due to their properties and cost-effectiveness (Winters et al. 2024).

Wang and colleagues examined polycaprolactone/lignin bioscaffolds for application in the nerve tissue regeneration field. The material was synthesized by using solvent-free ring opening polymerization. A nanofibrous scaffold was constructed to promote the growth of neurons and Schwann cells. They found the lignin/PCL to show higher cell viability than neat PCL after oxidative stress was induced on the scaffold. Schwann cells and dorsal root ganglion neurons were found to proliferate on the scaffold with enhanced myelin basic protein expressions of the Schwann cells and neurite growth of the dorsal root ganglion neurons (Wang et al. 2018). Furthermore, Saudi and colleagues studied the effect of adding different concentrations of lignin (1, 3 and 5%wt) to polyvinyl alcohol/polyglycerol sebacate for nerve tissue engineering applications. They used MTT assay and scanning electron microscopy to evaluate PC12 cell proliferation and attachment. The cell culture had enhanced adhesion and proliferation due to the lignin presence. The mRNA expression level for Gfap, β-Tub, and Map2 showed the positive effect of lignin on neural cell differentiation. It was concluded that polyvinyl alcohol/polyglycerol sebacate/5% lignin exhibited excellent potential as a material for neural tissue engineering (Saudi et al. 2019). Dhania and colleagues investigated modifications of polyhydroxyalkanoate as a viable material for neural tissue engineering. They discussed the addition of lignin to the polymer and the enhanced mechanical properties such as elongation, biodegradability, and thermal stability. They looked at the applications of polyhydroxyalkanoate as a nerve conduit for peripheral nerve and spinal cord treatments. They found lignin-induced polyhydroxyalkanoate to possess flexibility, non-toxic, and enhanced chemical and mechanical properties. It was also shown that lignin-induced polyhydroxyalkanoate supported axonal regeneration and nerve survival in both peripheral nerves and the spinal cord (Dhania et al. 2022). They have proved its potential use as a sustainable composite for neural tissue engineering. Lastly, Liu and colleagues reviewed the application of lignin-based composites for neural regeneration. They discussed the advances in materials testing, particularly in animal models. Polycaprolactone/lignin was electrospun to create a conduit to replace sciatic nerve in rats. The findings found no significant difference between neat polycaprolactone and lignin/polycaprolactone over nerve function, conductive nerve speed, and expression of cell differentiation markers. They also presented another study that found that successful integration of lignin into polyvinyl alcohol and polyglycerol sebacate gave rise to increase nerve growth and cell differentiation (Liu et al. 2022a; Amini et al. 2020; Saudi et al. 2019).

### 6.4.5 Cartilage Tissue

Cartilage can be described as connective tissue comprised of chondrocytes, collagen fibers, and proteoglycans, giving cartilage strength and elastic behaviour. Three types of cartilage are found in the body: hyaline, fibrocartilage, and elastic cartilage. Hyaline is the most commonly found cartilage, as it appears at all joints in the body. Its function is to provide frictionless movement between bones. Fibrocartilage is found in areas needing strengthening and stability, such as the meniscus in the knee. Elastic cartilage is located in the ear and nose. Common injuries associated with Cartilage are an avascular structure making it extremely difficult to repair and regenerate new tissue on its own. Currently, the most popular method to repair damaged cartilage is surgical repair. For this reason, tissue engineering has become an attractive alternative to surgical repair. There currently is cartilage repair being completed using autografts and allografts. This entails pieces of cartilage being harvested from less-loading areas and grafted onto the damaged area to restore full function. Lignin-derived materials can have potential solutions to cartilage damage and injury.

Liang and colleagues investigated the appropriateness of polylactic acid/lignin nanofibers for cartilage regeneration. The study was conducted on the knee joints of Sprague-Dawley rats. They compared the control, Polylactic acid, against using polylactic acid/lignin after two, four-, and six weeks post- transplantation. The control had very little new tissue growth. On the contrary, the polylactic acid/lignin showed extensive new tissue development, well compacted, and integrated into the cartilage. This shows polylactic acid/lignin as a potential material for cartilage tissue engineering (Liang et al. 2022).

Bonifacio and colleagues developed a gellan gum/lignin hydrogel for cartilage regeneration. The hydrogel consisted of two blends of gellan gum with different concentrations of acyl, which embedded lignin (0.4%w/v) and crosslinked with magnesium ions. Rheology, static compression tests, Solid State Nuclear Magnetic Resonance, and assays were methods used to characterize the hydrogel on human mesenchymal stem cells. It was found that lignin enhanced the mesenchymal stem cell 3D constructs chondrogenesis and upregulation of collagen type II, aggrecan, and chondrogenic genes. Furthermore, the hydrogel exhibited antibacterial behaviour against Staphylococcus aureus and Staphylococcus epidermis. They concluded that the developed gellan gum/lignin hydrogel displays excellent potential for cartilage regeneration applications (Bonifacio et al. 2022).

Liang and colleagues investigated the potential of poly(ε-caprolactone)/lignin nanofibers for osteoarthritis treatment. The nanofibers consisted of poly (ε-caprolactone), and polycaprolactone grafted lignin. It was determined that lignin of the PCL-lignin fibers demonstrated excellent antioxidant, anti-inflammatory, and mechanical integrity through *in vivo* studies on osteoarthritis in rabbits. In addition, the nanofibers inhibited reactive oxygen species and stimulated chondrocyte regeneration providing potential treatment for cartilage-degenerative diseases such as osteoarthritis (Liang et al. 2020).

## 6.5 Conclusion

In conclusion, sustainable materials derived from lignin have emerged as a potential yet promising alternative to traditional synthetic materials for tissue engineering and regenerative medicine applications. Lignin has several advantages over synthetic materials, including its enhanced antimicrobial, antioxidant, biodegradability, biocompatibility, and tunable mechanical properties. Furthermore, its sustainability, abundant availability, and cost efficiency make it a highly attractive material for fabricating composites. In addition, Lignin materials are easily processable, allowing for the easy incorporation of bioactive molecules such as drugs and growth factors for accelerated tissue regeneration and healing. Several studies have verified lignin-derived materials' potential application in bone tissue engineering, drug delivery and gene delivery, skin tissue engineering, tissue scaffold formation, neural tissue engineering, and cartilage repair engineering. All results gathered from these studies have proven that lignin-derived biomaterials can promote cell adhesion, proliferation, and differentiation. Moreover, lignin-based biomaterials exhibit excellent mechanical properties such as high Young's modulus and flexural strength, allowing for the imitation of natural native tissue in the body. However, with such exciting results, there are drawbacks. Currently, no lignin-derived material for tissue engineering applications is readily available on the market. These materials need optimization for processing techniques, which allow for exact properties, such as biodegradation and morphology, of these materials to be replicated. Furthermore, regulating bodies such as the FDA and EMA must draft and approve standardized procedures and protocols before use in a clinical setting. Ultimately, sustainable materials derived from lignin provide great optimism for the future. With further research and development, these biomaterials can provide beneficial and efficient solutions, allowing for improved patient welfare and a more sustainable healthcare system.

## References

Abdullah T, Colombani T, Alade T, Bencherif SA, Memić A (2021) Injectable lignin-co-gelatin Cryogels with antioxidant and antibacterial properties for biomedical applications. Biomacromolecules 22(10):4110–4121

Akter F, Ibanez J (2016) Chapter 8 – bone and cartilage tissue engineering. In: Akter F (ed) Tissue engineering made easy. Academic, pp 77–97

Ali DA, Mehanna MM (2022) Role of lignin-based nanoparticles in anticancer drug delivery and bioimaging: an up-to-date review. Int J Biol Macromol 221:934–953

Alonzo M, Primo FA, Kumar SA, Mudloff JA, Dominguez E, Fregoso G et al (2021) Bone tissue engineering techniques, advances and scaffolds for treatment of bone defects. Curr Opin Biomed Eng 17:100248

Alqahtani MS, Alqahtani A, Al-Thabit A, Roni M, Syed R (2019) Novel lignin nanoparticles for oral drug delivery. J Mater Chem B 7(28):4461–4473

Amini S, Saudi A, Amirpour N, Jahromi M, Najafabadi SS, Kazemi M et al (2020) Application of electrospun polycaprolactone fibers embedding lignin nanoparticle for peripheral nerve regeneration: in vitro and in vivo study. Int J Biol Macromol 159:154–173

Ayad R, Akkal S (2019) Chapter 12 – Phytochemistry and biological activities of algerian Centaurea and related genera. In: Atta-ur R (ed) Studies in natural products chemistry. Elsevier, 63, pp 357–414

Azqueta A, Stopper H, Zegura B, Dusinska M, Møller P (2022) Do cytotoxicity and cell death cause false positive results in the in vitro comet assay? Mutat Res Genet Toxicol Environ Mutagen 881:503520

Belgodere JA, Zamin SA, Kalinoski RM, Astete CE, Penrod JC, Hamel KM et al (2019) Modulating mechanical properties of Collagen–Lignin composites. ACS Appl Bio Mater 2(8):3562–3572

Bertolo MRV, Brenelli de Paiva LB, Nascimento VM, Gandin CA, Neto MO, Driemeier CE et al (2019) Lignins from sugarcane bagasse: renewable source of nanoparticles as Pickering emulsions stabilizers for bioactive compounds encapsulation. Ind Crop Prod 140:111591

Bonifacio MA, Cometa S, Cochis A, Scalzone A, Gentile P, Scalia AC et al (2022) A bioprintable gellan gum/lignin hydrogel: a smart and sustainable route for cartilage regeneration. Int J Biol Macromol 216:336–346

Carrion CC, Nasrollahzadeh M, Sajjadi M, Jaleh B, Soufi GJ, Iravani S (2021) Lignin, lipid, protein, hyaluronic acid, starch, cellulose, gum, pectin, alginate and chitosan-based nanomaterials for cancer nanotherapy: challenges and opportunities. Int J Biol Macromol 178:193–228

Chandna S, Paul S, Kaur R, Gogde K, Bhaumik J (2022) Photodynamic lignin hydrogels: a versatile self-healing platform for sustained release of photosensitizer Nanoconjugates. ACS Appl Polym Mater 4(12):8962–8976

Chavooshi R, Ranjkesh MR, Hashemi B, Roshangar L (2023) Cellulose and lignin-derived scaffold and their biological application in tissue engineering, drug delivery, and wound healing: a review. Cell J (Yakhteh) 25(3):158–164

Chen Y, Zheng K, Niu L, Zhang Y, Liu Y, Wang C et al (2019) Highly mechanical properties nanocomposite hydrogels with biorenewable lignin nanoparticles. Int J Biol Macromol 128:414–420

Chen M, Li Y, Liu H, Zhang D, Shi QS, Zhong XQ et al (2023) High value valorization of lignin as environmental benign antimicrobial. Mater Today Bio 18:100520

Cheremisinoff NP, Rosenfeld PE (2010) Chapter 6 – sources of air emissions from pulp and paper mills. In: Cheremisinoff NP, Rosenfeld PE (eds) Handbook of pollution prevention and cleaner production. William Andrew Publishing, Oxford, pp 179–259

Chiani E, Beaucamp A, Hamzeh Y, Azadfallah M, Thanusha AV, Collins MN (2023) Synthesis and characterization of gelatin/lignin hydrogels as quick release drug carriers for ribavirin. Int J Biol Macromol 224:1196–1205

Collins MN, Nechifor M, Tanasă F, Zănoagă M, McLoughlin A, Stróżyk MA et al (2019) Valorization of lignin in polymer and composite systems for advanced engineering applications – a review. Int J Biol Macromol 131:828–849

Decante G, Cengiz IF,Costa J,Collins MN,Reis RL,Silva-Correia J, Oliveira JM (2024) Sustainable highly stretchable and tough gelatin-alkali lignin hydrogels for scaffolding and 3D printing applications Materials Today Communications 39:108875. https://doi.org/10.1016/j.mtcomm.2024.108875

Deng J, Gu J, Zhao X, Yan B, Wang L, Ji G et al (2023) Improving the protective ability of lignin against vascular and neurological development in BPAF-induced zebrafish by high-pressure homogenization technology. Int J Biol Macromol 231:123356

Dhania S, Bernela M, Rani R, Parsad M, Grewal S, Kumari S et al (2022) Scaffolds the backbone of tissue engineering: advancements in use of polyhydroxyalkanoates (PHA). Int J Biol Macromol 208:243–259

Domínguez-Robles J, Cuartas-Gómez E, Dynes S, Utomo E, Anjani QK, Detamornrat U et al (2023) Poly(caprolactone)/lignin-based 3D-printed dressings loaded with a novel combination of bioactive agents for wound-healing applications. Sustain Mater Technol 35:e00581

Du B, Li W, Bai Y, Pan Z, Wang Q, Wang X et al (2022) Fabrication of uniform lignin nanoparticles with tunable size for potential wound healing application. Int J Biol Macromol 214:170–180

Figueiredo P, Lintinen K, Kiriazis A, Hynninen V, Liu Z, Bauleth-Ramos T et al (2017) In vitro evaluation of biodegradable lignin-based nanoparticles for drug delivery and enhanced antiproliferation effect in cancer cells. Biomaterials 121:97–108

Florencio-Silva R, Sasso GR, Sasso-Cerri E, Simões MJ, Cerri PS (2015) Biology of bone tissue: structure, function, and factors that influence bone cells. Biomed Res Int 2015:421746

Garg J, Nee Chiu M, Krishnan S, Kumar Tripathi L, Pandit S, Farasati Far B et al (2022) Applications of lignin nanoparticles for cancer drug delivery: an update. Mater Lett 311:131573

Ghassemi T, Shahroodi A, Ebrahimzadeh MH, Mousavian A, Movaffagh J, Moradi A (2018) Current concepts in scaffolding for bone tissue engineering. Arch Bone Jt Surg 6(2):90–99

Gujjala LKS, Kim J, Won W (2022) Technical lignin to hydrogels: an eclectic review on suitability, synthesis, applications, challenges and future prospects. J Clean Prod 363:132585

Haider MK, Kharaghani D, Sun L, Ullah S, Sarwar MN, Ullah A et al (2023) Synthesized bioactive lignin nanoparticles/polycaprolactone nanofibers: a novel nanobiocomposite for bone tissue engineering. Biomater Adv 144:213203

Hellman KB (2008) Tissue engineering: translating science to product. In: Ashammakhi N, Reis RL, Chiellini F (eds) Topics in tissue engineering, pp 1–28

Ho J, Walsh C, Yue D, Dardik A, Cheema U (2017) Current advancements and strategies in tissue engineering for wound healing: a comprehensive review. Adv Wound Care (New Rochelle). 6(6):191–209

Jurak M, Wiącek AE, Ładniak A, Przykaza K, Szafran K (2021) What affects the biocompatibility of polymers? Adv Colloid Interf Sci 294:102451

Kai D, Ren W, Tian L, Chee PL, Liu Y, Ramakrishna S et al (2016) Engineering poly(lactide)–lignin nanofibers with antioxidant activity for biomedical application. ACS Sustain Chem Eng 4(10):5268–5276

Koltover VK, Skipa TA (2023) Chapter 16 – antioxidant pharmacology. In: Koltover VK (ed) Antiaging pharmacology. Academic, pp 341–365

Li S, Zhang Y, Ma X, Qiu S, Chen J, Lu G et al (2022) Antimicrobial lignin-based polyurethane/ag composite foams for improving wound healing. Biomacromolecules 23(4):1622–1632

Liang R, Zhao J, Li B, Cai P, Loh XJ, Xu C et al (2020) Implantable and degradable antioxidant poly(ε-caprolactone)-lignin nanofiber membrane for effective osteoarthritis treatment. Biomaterials 230:119601

Liang R, Yang X, Yew PYM, Sugiarto S, Zhu Q, Zhao J et al (2022) PLA-lignin nanofibers as antioxidant biomaterials for cartilage regeneration and osteoarthritis treatment. J Nanobiotechnol 20(1):327

Liu Y, Wang X, Wu Q, Pei W, Teo MJ, Chen ZS et al (2022a) Application of lignin and lignin-based composites in different tissue engineering fields. Int J Biol Macromol 222:994–1006

Liu X, Sun S, Wang N, Kang R, Xie L, Liu X (2022b) Therapeutic application of hydrogels for bone- related diseases. Front Bioeng Biotechnol 10:998988

Liu X, Yin H, Song X, Zhang Z, Li J (2022c) Lignin-based nonviral gene carriers functionalized by poly[2-(Dimethylamino)ethyl methacrylate]: effect of grafting degree and cationic chain length on transfection efficiency. Biomolecules [Internet] 12(1):102

Lobo FCM, Franco AR, Fernandes EM, Reis RL (2021) An overview of the antimicrobial properties of Lignocellulosic materials. Molecules 26(6):1749

Lourençon TV, de Lima GG, Ribeiro CSP, Hansel FA, Maciel GM, da Silva K et al (2021) Antioxidant, antibacterial and antitumoural activities of Kraft lignin from hardwood fractionated by acid precipitation. Int J Biol Macromol 166:1535–1542

Luzi F, Tortorella I, Di Michele A, Dominici F, Argentati C, Morena F et al (2020) Novel nanocomposite PLA films with lignin/zinc oxide hybrids: design, characterization, interaction with mesenchymal stem cells. Nanomaterials [Internet]. 10(11):2176

Mackintosh JA (2001) The antimicrobial properties of melanocytes, melanosomes and melanin and the evolution of black skin. J Theor Biol 211(2):101–113

Makadia HK, Siegel SJ (2011) Poly lactic-co-glycolic acid (PLGA) as biodegradable controlled drug delivery carrier. Polymers (Basel) 3(3):1377–1397

Mohammadalipour M, Behzad T, Karbasi S, Mohammadalipour Z (2022) Optimization and characterization of polyhydroxybutyrate/lignin electro-spun scaffolds for tissue engineering applications. Int J Biol Macromol 218:317–334

Moohan J, Stewart SA, Espinosa E, Rosal A, Rodríguez A, Larrañeta E, Donnelly RF, Domínguez-Robles J (2019) Cellulose nanofibers and other biopolymers for biomedical applications. A review. Appl Sci 10(1):65

Nan N, Hu W, Wang J (2022) Lignin-based porous biomaterials for medical and pharmaceutical applications. Biomedicines 10(4):747

Ndaba B, Roopnarain A, Daramola MO, Adeleke R (2020) Influence of extraction methods on antimicrobial activities of lignin-based materials: a review. Sustain Chem Pharm 18:100342

Nevoralová M, Koutný M, Ujčić A, Starý Z, Šerá J, Vlková H et al (2020) Structure characterization and biodegradation rate of poly(ε-caprolactone)/starch blends. Front Mater 7:7

Oriez V, Peydecastaing J, Pontalier P-Y (2020) Lignocellulosic biomass mild alkaline fractionation and resulting extract purification processes: conditions, yields, and purities. Clean Technol [Internet]. 2(1):91–115

Pérez-Rafael S, Ivanova K, Stefanov I, Puiggalí J, del Valle LJ, Todorova K et al (2021) Nanoparticle-driven self-assembling injectable hydrogels provide a multi-factorial approach for chronic wound treatment. Acta Biomater 134:131–143

Pizzino G, Irrera N, Cucinotta M, Pallio G, Mannino F, Arcoraci V et al (2017) Oxidative stress: harms and benefits for human health. Oxidative Med Cell Longev 2017:8416763

Rambhia KJ, Ma PX (2015) Controlled drug release for tissue engineering. J Control Release 219:119–128

Ratanji KD, Derrick JP, Dearman RJ, Kimber I (2014) Immunogenicity of therapeutic proteins: influence of aggregation. J Immunotoxicol 11(2):99–109

Rosenboom J-G, Langer R, Traverso G (2022) Bioplastics for a circular economy. Na Rev Mater 7(2):117–137

Sameni J, Jaffer SA, Sain M (2018) Thermal and mechanical properties of soda lignin/HDPE blends. Compos A: Appl Sci Manuf 115:104–111

Saratale RG, Cho S-K, Saratale GD, Kadam AA, Ghodake GS, Magotra VK et al (2022) Lignin-mediated silver nanoparticle synthesis for photocatalytic degradation of reactive yellow 4G and in vitro assessment of antioxidant, diabetic, and antibacterial activities. Polym Int 14(3):648

Sathasivam T, Kai J, Sugiarto S, Yu Y, Soo DXY, Zhu Q et al (2023) Nano-strategies for lignin biomaterials toward cancer therapy. Adv Healthc Mater 12(19):e2300024

Saudi A, Amini S, Amirpour N, Kazemi M, Zargar Kharazi A, Salehi H et al (2019) Promoting neural cell proliferation and differentiation by incorporating lignin into electrospun poly(vinyl alcohol) and poly(glycerol sebacate) fibers. Mater Sci Eng C 104:110005

Serafin A, Casanova CR, Chandel AKS, Reis RL, Oliveira JM, Collins MN (2025) Conductive biological materials for in vitro models: properties and sustainability implications Abstract In vitro models https://doi.org/10.1007/s44164-025-00088-5

Shi S, Xue F (2016) Current antioxidant treatments in organ transplantation. Oxidative Med Cell Longev 2016:8678510

Silva-Barroso AS, Cabral CSD, Ferreira P, Moreira AF, Correia IJ (2023) Lignin-enriched tricalcium phosphate/sodium alginate 3D scaffolds for application in bone tissue regeneration. Int J Biol Macromol 239:124258

Spasojević D, Zmejkoski D, Glamočlija J, Nikolić M, Soković M, Milošević V et al (2016) Lignin model compound in alginate hydrogel: a strong antimicrobial agent with high potential in wound treatment. Int J Antimicrob Agents 48(6):732–735

Ten E, Ling C, Wang Y, Srivastava A, Dempere LA, Vermerris W (2014) Lignin nanotubes as vehicles for gene delivery into human cells. Biomacromolecules 15(1):327–338
Ullah I, Chen Z, Xie Y, Khan SS, Singh S, Yu C et al (2022) Recent advances in biological activities of lignin and emerging biomedical applications: a short review. Int J Biol Macromol 208:819–832
Vaou N, Stavropoulou E, Voidarou C, Tsigalou C, Bezirtzoglou E (2021) Towards advances in medicinal plant antimicrobial activity: a review study on challenges and future perspectives. Microorganisms 9(10):2041
Wang J, Tian L, Luo B, Ramakrishna S, Kai D, Loh XJ et al (2018) Engineering PCL/lignin nanofibers as an antioxidant scaffold for the growth of neuron and Schwann cell. Colloids Surf B: Biointerfaces 169:356–365
Winters C, Mario C,Collins MN (2025) Electrically conductive lignin reinforced PAA/HA scaffolds with enhanced biological activity via polyelectrolyte multilayer coatings for tissue engineering applications Scientific Reports 15(1). https://doi.org/10.1038/s41598-025-99440-8
Winters C, Carsi M, Sanchis MJ, Culebras M, Collins MN (2024) Electrically Conductive Lignin Reinforced Polyacrylic Acid/Hyaluronic Acid Scaffolds With PEDOT:HANanoparticles ABSTRACT Polymers for Advanced Technologies 35(10). https://doi.org/10.1002/pat.v35.10; https://doi.org/10.1002/pat.6607
Xiao L, Liu W, Huang J, Lou H, Qiu X (2021) Study on the antioxidant activity of lignin and its application performance in SBS elastomer. Ind Eng Chem Res 60(1):790–797
Xu J, Xu JJ, Lin Q, Jiang L, Zhang D, Li Z et al (2021) Lignin-incorporated Nanogel serving as an antioxidant biomaterial for wound healing. ACS Appl Bio Mater 4(1):3–13
Yang W, Fortunati E, Bertoglio F, Owczarek JS, Bruni G, Kozanecki M et al (2018) Polyvinyl alcohol/chitosan hydrogels with enhanced antioxidant and antibacterial properties induced by lignin nanoparticles. Carbohydr Polym 181:275–284
Yun J, Wei L, Li W, Gong D, Qin H, Feng X et al (2021) Isolating high antimicrobial ability lignin from bamboo Kraft lignin by Organosolv fractionation. Front Bioeng Biotechnol 9:683796
Zahed M, Muhammad Y, Muhammad J, Muhammad R, Abdul G, Irum J (2018) Lignin as natural antioxidant capacity. In: Matheus P (ed) Lignin. IntechOpen, Rijeka. p. Ch. 8
Zhang Y, Jiang M, Zhang Y, Cao Q, Wang X, Han Y et al (2019) Novel lignin–chitosan–PVA composite hydrogel for wound dressing. Mater Sci Eng C 104:110002
Zheng L, Lu G, Pei W, Yan W, Li Y, Zhang L et al (2021) Understanding the relationship between the structural properties of lignin and their biological activities. Int J Biol Macromol 190:291–300
Zheng Y-Z, Deng G, Zhang Y-C (2022) Multiple free radical scavenging reactions of flavonoids. Dyes Pigments 198:109877
Zhou Y, Han Y, Li G, Yang S, Xiong F, Chu F (2019) Preparation of targeted lignin–based hollow nanoparticles for the delivery of doxorubicin. Nanomaterials [Internet] 9(2):188

# Revolutionizing Regenerative Medicine: Designing Cutting-Edge Polysaccharide Scaffolds for Tissue Engineering Applications

7

Fang Chen, Ying Zheng, John Harmon, and Guoming Sun

**Abstract**

Tissue engineering and regenerative medicine (TERM) rely on three pivotal components: scaffolds, cells, and growth signaling molecules. Scaffolds play a paramount role by offering structural support for cell growth, facilitating the spatiotemporal release of biomolecules, and coordinating tissue growth and regeneration. Polysaccharides have emerged as versatile scaffolds in tissue regeneration due to their excellent biocompatibility and inherent structural characteristics. Comprising components found in the natural extracellular matrix, such as glycosaminoglycans, proteoglycans, and glycoproteins, polysaccharide compounds provide vital biological cues that foster cell growth and promote tissue regeneration. Beyond their remarkable biological properties, polysaccharides boast abundant sources, diverse functionalities, and easy tunability. Recent strides in molecular understanding and advancements in material fabrication technologies have enabled the creation of intricate 3D structures with desirable properties. The latest advancements in the extensively researched polysaccharide-based scaffolds for TERM, with a particular emphasis on those constructing

F. Chen
First Department of Bone Injury, Luzhou Municipal Hospital of Traditional Chinese Medicine, Luzhou, Sichuan, China

Y. Zheng
Department of Stomatology, Peking Union Medical College Hospital, Peking Union Medical College, Chinese Academy of Medical Sciences, Beijing, China

J. Harmon
Department of Surgery and the Hendrix Burn Lab, Johns Hopkins University School of Medicine, Baltimore, MD, USA

G. Sun (✉)
Sunogel Biotechnologies Inc., Lutherville Timonium, MD, USA
e-mail: gsun@sunogel.com



J. M. Oliveira et al. (eds.), *Sustainable Scaffolds-based Strategies in Tissue Engineering and Regenerative Medicine*, Biomaterials, Bioengineering and Sustainability 8, https://doi.org/10.1007/978-3-031-96274-5_7

advanced architectures tailored for tissue regeneration, are discussed herein. Polysaccharides are thus experiencing growing exploration and utilization in the realm of TERM.

**Keywords**

Tissue engineering · Regenerative medicine · Scaffolds · Polysaccharides · Extracellular matrix · Cell growth

## 7.1 Introduction

Lower species possess robust regenerative capacities, while higher species experience a decline in regenerative abilities during biological evolution, leaving humans with limited abilities to restore damaged tissues or organs once injured or traumatized. Timely tissue repair and regeneration is crucial to maintain the health and normal physiological function of an organism. When organs experience trauma, their innate healing and repair processes are initiated. Nevertheless, this endogenous healing is often insufficient to fully restore the damaged tissues. In many situations, tissue regeneration can be facilitated by medical intervention. Classical tissue engineering (TE) aims to meticulously recreate the intricacies of living tissues *in vitro* and transplant them into patients to facilitate the repair of damaged tissues or organs. Typically, TE involves the hierarchical design and construction of biologically active scaffolds with living cells and regulated biomolecules. During the past few decades, TE has demonstrated remarkable success in the restoration and repair of diverse tissues and organs, ranging from blood vessels to bones and skin (Sun et al. 2018). Despite these achievements, the repaired organs via classical TE approaches still fall short of the ideal when compared to the original counterparts. TE treatments frequently result in incomplete structures, such as the formation of fibrosis. Moreover, the newly formed tissues often exhibit inadequate functions or properties. The wound healing skin after deep cutaneous injuries, for instance, can only attain a maximum of 80% of the strength exhibited by the original tissue (Sun et al. 2011a). With advancements in biology and a profound understanding of the interaction between organs and engineered scaffolds, regenerative medicine, which integrates TE with other approaches such as immunomodulation (Han et al. 2020; Wu et al. 2021a), i.e., tissue engineering and regenerative medicine (TERM), aims to regenerate organs or tissues that closely mimic both the structure and function of the original ones, emerges as a beacon of hope (Mason and Dunnill 2008).

Biomaterial scaffolds play a crucial role by offering structural support for cell growth and facilitating the controlled release of biomolecules, thereby coordinating tissue growth and regeneration. Tissue-engineered scaffolds have the capability to replicate the structure and function of the extracellular matrix (ECM), creating a biomimetic growth environment for cells to adhere to specific patterns. Simultaneously, these scaffolds offer support to maintain the shape and size of original tissues and organs, preventing deformation or collapse due to external forces or gravity. They serve as attachment sites for cells, minimizing cell inactivation, guiding specialized cell functions, and regulating cell-cell interactions (O'Brien 2011). Therefore, tissue-engineered scaffolds play a crucial role in guiding the regeneration of damaged tissues and organs. As time progresses, the scaffold undergoes gradual degradation, eventually replaced by newly generated tissue, achieving the goal of repair and regeneration. A plethora of biological materials, including both natural and synthetic ones, have been transformed into ECM-like scaffolds to establish a conducive microenvironment for tissue and organ repair and regeneration (Hussey et al. 2018).

Polysaccharides, owing to their unique structure and outstanding biocompatibility (Sun and Mao 2012; Su et al. 2022), have been extensively explored as scaffolds for tissue regeneration (Sun et al. 2011a; Sun 2017; Jin et al. 2021; Wu et al. 2021b). Polysaccharides are nontoxic natural polymers that can degrade into small molecules and are absorbed by the body with low foreign body reaction. In addition to their biological properties, polysaccharides have rich sources and diverse functional groups (e.g., $-COOH$, $-NH_2$ or $-OH$), making them easily tailorable. This capability enhances the overall physical and biological properties of polysaccharide scaffolds. Polysaccharide scaffolds have found extensive applications in diverse areas of tissue regeneration, encompassing skin regeneration, cartilage repair, nerve regeneration, and angiogenesis (Sun et al. 2011a; Hanjaya-Putra et al. 2011; Moreira Teixeira et al. 2012; Xuan et al. 2023). Polysaccharides form a diverse family with remarkable renewability. However, even within a single polysaccharide, variations in molecular weight or composition can result in distinct outcomes in tissue repair or regeneration (Longinotti 2014), and screening and selecting of the appropriate polysaccharide thereby become essential. The design of polysaccharide scaffolds further necessitates meticulous consideration of essential factors such as biocompatibility, mechanical properties, degradability, and the scaffold ability to support cell growth. The exploration and utilization of polysaccharides open new avenues for enhancing the effectiveness of regenerative therapies and bringing us closer to the realization of truly perfect organ regeneration.

## 7.2 Properties of Polysaccharides

Polysaccharides represent the most naturally occurring, highly sustainable, and eco-friendly macromolecules that are widely investigated for numerous biomedical applications (Gericke et al. 2024). Formed by linking numerous monosaccharide molecules through glycosidic bonds, these complex carbohydrate substances (Table 7.1) exhibit large and intricate molecular structures (Mohammed et al. 2021). Polysaccharides, such as hyaluronic acid, dextran, alginate, chitosan, and pullulan, constitute a diverse family. They exhibit minimal immune rejection or toxicity, allowing for interaction with human tissues and cells without causing harm. These polysaccharides play essential roles as cell components in both connective and hard tissues within the human body (Ma et al. 2024). Modulating their structure through

**Table 7.1** The chemical structure and bioactivities of the most studied natural polysaccharides

| Type | Chemical structure | Source | Extraction techniques | Bioactivities | References |
|---|---|---|---|---|---|
| Hyaluronic acid | 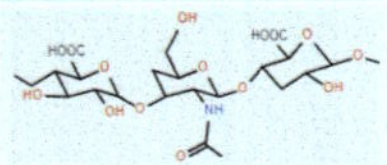 | Cartilage or eyeballs of animals | Acid-base method, enzymatic hydrolysis method | Moisturizing, lubricating, promoting cell proliferation and migration, anti-inflammatory, immunomodulating | Graciela et al. (2023); Luo et al. (2023) |
| Chitosan | 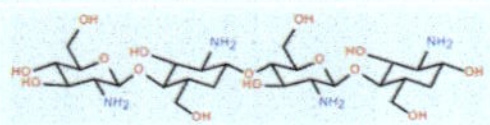 | Shrimp or crab shells | Acid-base method, enzymatic hydrolysis method | Natural cationic polysaccharides, antioxidant activity, antibacterial activity, immunomodulatory effect, antitumor activity, adsorption | Hisham et al. (2024); Wang et al. (2020) |
| Heparin | 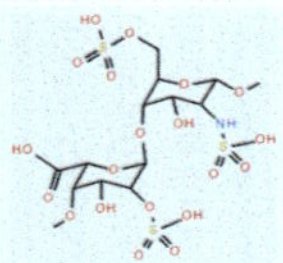 | Lung mucosa or small intestinal mucosa of animals | Acid-base method, enzymatic hydrolysis method | Anticoagulant activity, regulation of blood lipids, inhibition of smooth muscle growth, regulation of vascular growth, antiviral, anti-inflammatory activity | Alizadeh et al. (2019) |
| Cellulose | 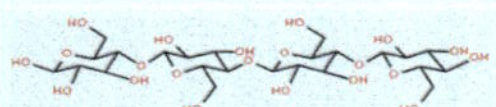 | The cell wall of a plant | Acid-base method, enzymatic hydrolysis method, mechanical peeling method, solvent method | Biocompatible, regulates cell activity, good mechanical properties, high fatigue resistance, easy to modify | Gabriel et al. (2020); Hickey and Pelling (2019); Magalhães et al. (2023); Alven and Aderibigbe (2020) |
| Chondroitin sulfate | 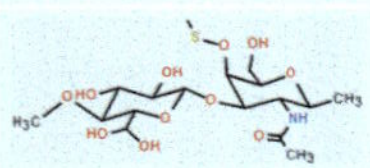 | The cartilage of animals | Acid-base method, enzymatic hydrolysis method | Anti-inflammatory, antioxidant, promote chondrocyte proliferation and differentiation, lipid-lowering, anticoagulant, immunomodulatory | Sharma et al. (2022); Shen et al. (2023) |
| Starch | 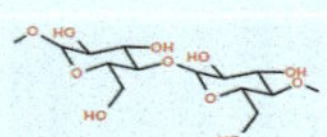 | Beans, potatoes | Washing method | Biocompatible, biodegradable, plastic, regulation of cellular sexual behavior, pro-angiogenesis | Palanisamy et al. (2022); El Halal et al. (2019) |

| | | | | | |
|---|---|---|---|---|---|
| Cyclodextrin | 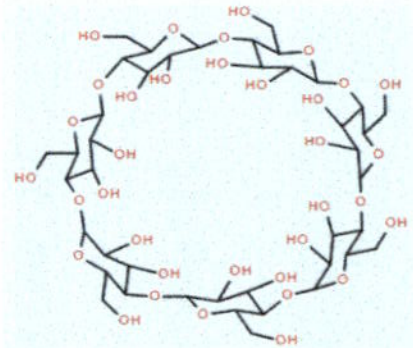 | Presence of cyclodextrin glucosyl transferase by bacillus | Enzymatic hydrolysis method | Increase drug solubility, improve drug stability, regulate drug release rate, and regulate drug release rate | Zhao et al. (2024); Wang et al. (2023a) |
| Alginic acid | 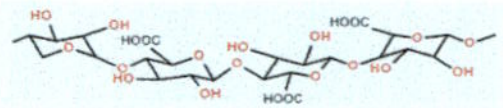 | Algae | Washing method, acid-base method, enzymatic hydrolysis method | Antioxidant activity, anti-inflammatory effect, hemostatic function, biocompatibility, biodegradability | Wang et al. (2023a); Bojorges et al. (2023); Otero et al. (2021) |
| Dextran | 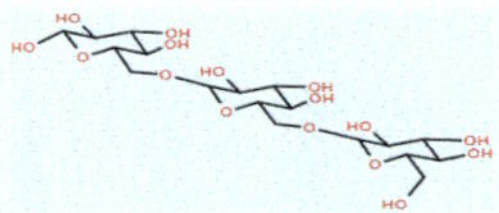 | Mesenteric staphylococcus | Fermentation | Immunomodulatory effects, anti-inflammatory, antitumor effects, antioxidant and anti-aging effects, pro-angiogenic effects, pro-cell adhesion and proliferation | Lin et al. (2024); Yang and Huang (2021) |

chemical modifications or forming complexes with other biomolecules can further influence the biological activity of polysaccharides. Surface properties of polysaccharides, for instance, can be altered to regulate cell adhesion, proliferation, and differentiation (Chen et al. 2024a). This adaptability enables polysaccharides to closely mimic and facilitate natural biological processes in tissue engineering and regenerative medicine.

### 7.2.1 Sources of Polysaccharides Utilized in TERM

Polysaccharides are abundantly present in nature (Fig. 7.1). On the basis of their sources, polysaccharides are typically grouped as of animal, plant, or microbial origins (Mohammed et al. 2021). Though existed naturally, the commercially available polysaccharides may be derived differently. Some polysaccharides (e.g., chitosan) are mainly obtained from natural resources, while some (e.g., dextran) are

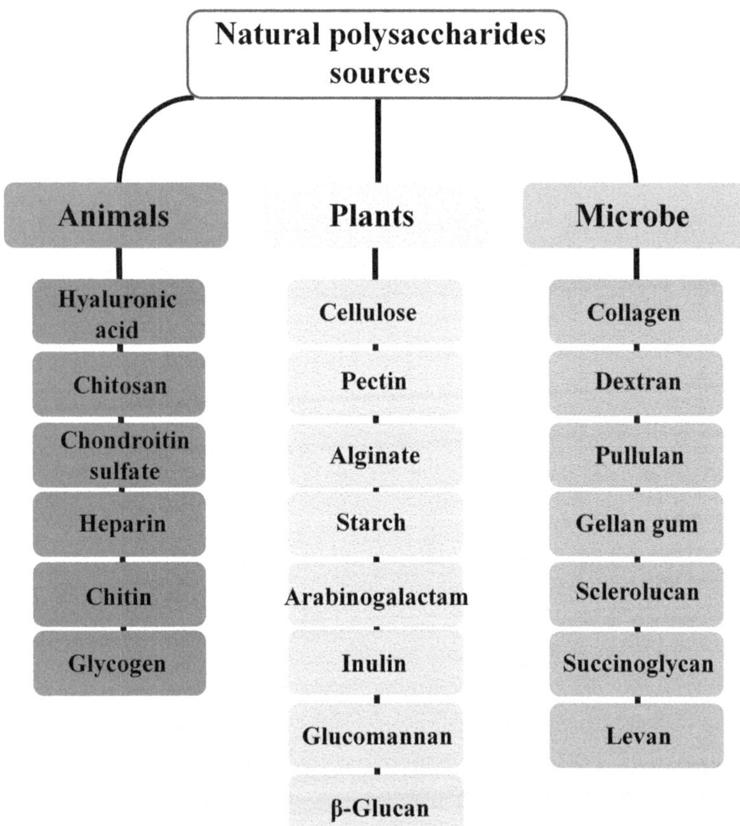


**Fig. 7.1** Classification of polysaccharides from natural sources

mainly are mainly produced by bacterial fermentation (Hisham et al. 2024; Murphy et al. 2023). The resource and their production method of polysaccharides can lead to variations in their bioactivity. The molecular weight of polysaccharides not only affects their biocompatibility, but also influences their bioactivity (Longinotti 2014; Mukherjee et al. 2022; Wang et al. 2023b). Therefore, the selection of the appropriate polysaccharide becomes crucial, particularly in the context of specific applications.

#### 7.2.1.1 Animal-Derived Polysaccharides

Animal-derived polysaccharides constitute a diverse group of complex compounds extracted from animals, possessing intricate structures and varied physiological functions. They are generally able to regulate the immune system, enhance antioxidant capacity, and facilitate cell signaling, hence promoting tissue repair and regeneration (Tudu and Samanta 2023). These polysaccharides are widely distributed in animal tissues and fluids, including skin, bones, cartilage, eyeballs, blood, and milk (Yang et al. 2024). Notable polysaccharides of animal origin include heparin, hyaluronic acid (HA), chitosan, and chondroitin sulfate.

Hyaluronic acid, a glycan naturally occurring in human and animal tissues, is commonly sourced from animal cartilage and eyeball tissue. Hyaluronic acid exhibits excellent water retention and mechanical support properties. It regulates cell interactions and signaling pathways, and also modulates inflammatory responses, thereby largely investigated for tissue regeneration (Hussain et al. 2017). Derived mainly from crab and shrimp shells, chitosan has remarkable biocompatibility, antimicrobial properties, and gelation ability, and has found applications in almost all TERM (Younes and Rinaudo 2015). Recent trends include microbial fermentation technology for chitosan production, minimizing dependence on marine resources and contributing to environmental protection (Liu et al. 2023a). Chondroitin sulfate, a sulfated glycosaminoglycan, is widely distributed in animal extracellular matrices. Mainly extracted from animal cartilage, it stimulates chondrocytes to synthesize components crucial for cartilage repair and regeneration (Wang et al. 2007). Heparin, a mucopolysaccharide sulfate, is derived from bovine lungs or porcine small intestines. Possessing various biological activities, including anticoagulant, anti-inflammatory, antioxidant, and antifibrotic properties, heparin plays a vital role in tissue regeneration. It promotes vascular endothelial cell proliferation and migration, accelerates wound healing, inhibits inflammatory responses, and reduces tissue fibrosis, making it a valuable treatment for conditions such as burns, wounds, and chronic ulcers (Pruett et al. 2021). It also shows promise in addressing fibrosis-related chronic diseases like cirrhosis, renal fibrosis, and pulmonary fibrosis (Saito et al. 2022).

#### 7.2.1.2 Plant-Derived Polysaccharides

Polysaccharides of plant origin refer to those extracted from various plants, often possessing a wide range of biological activities and nutritional functions. Typically found in plant cell walls, interstitial spaces, and other tissue structures, these plant-derived polysaccharides exhibit good biocompatibility, making them compatible with human tissues without causing significant immune rejection (Guru et al. 2023; Wang et al. 2023c). This characteristic makes them promising for applications in human tissue regeneration, particularly as biomaterials in tissue engineering and regenerative medicine. Plant-derived polysaccharides may contain bioactive components, such as unique polysaccharide structures and polyphenolic compounds, influencing the proliferation, differentiation, and repair of human cells and aiding the tissue regeneration process (Jafari et al. 2022).

Polysaccharides derived from plants comprise a variety of compounds such as cellulose, pectin, alginate, and starch (Guru et al. 2023). Cellulose, commonly found in plant cell walls, offers mechanical strength and stability to plant cells, making it essential for maintaining their form and resistance to external stresses (Hickey and Pelling 2019). Pectin, a natural polymer found in plant cell walls, has water-soluble properties and is utilized in drug delivery systems and wound dressings due to its gelling and emulsification stabilization abilities (Li et al. 2021). Alginate is a complex polymer extracted from brown seaweed, known for its renewability and sustainability. It exhibits excellent biocompatibility, eliciting no immune reactions, and is both degradable and environmentally friendly. With notable water solubility, alginate readily dissolves in water to form a solution, facilitating easy processing and application. Starch is a natural component found in various plants, such as fruits, vegetables, and grains, and serves as energy storage. Starch is a complex polysaccharide, and modified starch is widely used in biomedical and pharmaceutical fields (Garcia et al. 2020).

#### 7.2.1.3 Microbially Derived Polysaccharides

Microbially derived polysaccharides encompass those produced by microorganisms of monosaccharides and their derivatives. They are mainly categorized as bacterial polysaccharides, yeast polysaccharides, and fungal polysaccharides. These polysaccharides are typically obtained through such methods as fermentation and genetic engineering. Fermentation, for instance, a common method for industrial-scale production, involves microorganisms such as lactic acid bacteria producing polysaccharide enzymes, or yeast generating compounds like β-glucan through fermentation. This approach offers advantages like a short production cycle, low costs, and readily available raw materials. Advancements in genetic engineering have led to the creation of genetically modified microbial strains capable of more efficiently producing specific types of polysaccharides. This involves modifying the genetic information of microorganisms to induce the production of desired polysaccharides during fermentation (Mahmoud et al. 2021; Chaisuwan et al. 2020). Microorganism-derived polysaccharides often exhibit diverse biological activities, including immunomodulation, antioxidant properties, anti-tumor effects, hypoglycemic and blood lipid regulation (Zhang et al. 2024). An extensively investigated microbially derived polysaccharide is dextran. Dextran is naturally present in trace amounts in various foods, such as refined crystalline sugar and honey, and also found in dental plaque. Industrially used dextran is produced through the bacterial fermentation of sucrose, particularly by *Leuconostoc mesenteroides* (Prechtl et al. 2018; Vettori et al. 2012). Dextran exhibits various biological activities, and has been largely used in various clinical conditions such as thrombosis, radiological imaging, wound healing, and eye lubricant, making it valuable in biomedical engineering for tissue repair and regeneration (Luanda and Badalamoole 2023).

### 7.2.2 Structural Features of Polysaccharides in TERM

Polysaccharides are formed by the linkage of many monosaccharides through glycosidic bonds, either in the α- or β-configuration. The specific glycosidic bond connections and positions vary among different types of polysaccharides, influencing their overall structure and properties. In addition to the main backbone, polysaccharides often exhibit a branched structure, with branching sites typically linked by α-1,6- or α-1,3-glycosidic bonds. These branches provide additional sites for interactions with other molecules. Moreover, polysaccharide chains contain distinct functional groups such as carboxyl, hydroxyl and amine groups, offering extensive modification possibilities for additional potentials (Xu et al. 2019; Zhao et al. 2023). Examining their fundamental structures enables us to effectively identify suitable polysaccharides for various TERM applications.

The chemical structure of HA comprises two monosaccharide units, N-acetylglucosamine and glucuronic acid, connected by β-1,4-glycosidic bonds. These repeating units create a helical contour structure in the HA molecule, forming a highly polymerized linear chain structure with thousands of repeating units. HA possesses a molecular weight up to 20,000 kDa, significantly impacting its biological properties (Frasheri et al. 2023; Lee et al. 2021). HA is rich in carboxyl and hydroxyl groups, conferring hydrophilicity, viscosity, and the ability to adsorb and retain water molecules. Moreover, the hydroxyl groups allow for convenient and purposeful modifications (Khetan et al. 2013).

Chitosan, derived from the N-deacetylation of chitin, possesses a molecular weight of up to 2000 kDa and a degree of deacetylation ranging from 50% to 100%. These characteristics greatly impact the physical, chemical and biological properties of chitosan (Yuan et al. 2011). The protonation of amino groups of chitosan gives rise to its excellent antimicrobial properties, which play a crucial role in biological processes like interaction with molecules on the cell surface and regulation of cell signaling (Verlee et al. 2017). Chitosan is thereby one of the most investigated biomaterials used in TERM applications.

Alginic acid, composed of α-glucuronic acid (GlcA) and β-glucuronic acid (M-GlcA) units, adopts a helical-shaped structure, demonstrating elasticity and flexibility. The negatively charged carboxyl groups in alginic acid confer ionization in aqueous solution, ensuring water solubility, absorption, and the ability to form gel structures (Lee and Mooney 2012). The monosaccharide unit of alginic acid contains carboxyl groups (−COOH), imparting a negative charge to the molecule and causing it to ionize readily in aqueous solutions. This ionization property contributes to alginic acid's strong water solubility and water absorption, enabling it to absorb a significant amount of water and form a gelatinous substance. Additionally, the carboxyl groups in the alginic acid molecule can form complexes with metal ions (e.g., calcium), and the carboxyl groups allows additional chemical modification, resulting in hydrogel structures with tunable stability and viscosity. Consequently, alginic acid has found numerous TERM applications (Elosegui-Artola et al. 2023).

Dextran is a hydrophilic homopolysaccharide primarily composed of linear α-1,6-linked D-glucopyranose residues, with a low percentage of α-1,2, α-1,3, and α-1,4-linked side chains. Unlike the aforementioned polysaccharides, dextran is not a charged polymer; however, it has a great amount of hydroxyl groups, enabling multiple chemical modifications to introduce various functional groups for various TERM applications (Sun and Mao 2012). Additionally, the distinctive planar helical structure of β-(1,3)-dextran enables it to bind to cell membrane receptors, thereby activating macrophages to produce various cytokines, epidermal growth factor, tumor necrosis factor, and angiogenesis factors (Chan et al. 2009).

Heparin, consisting of repeating gluconic acid and N-acetylglucosamine units, possesses sulfate groups that impact its spatial structure and biological activity. These polysaccharide chains can vary in length and branching degrees, resulting in diverse heparin structures. The location, number, and arrangement of these sulfate groups significantly influence the spatial structure of the heparin molecule, determining its biological activity and interactions with other molecules. Additionally, the sulfate groups confer a negative charge to the heparin molecule, contributing to its anticoagulant and other biologically active properties (Banik et al. 2021).

Cellulose is a linear polymer comprising β-D-glucose molecules linked by β-1,4-glycosidic bonds. Cellulose molecules form a highly ordered crystalline structure via hydrogen bonds, so it is insoluble in aqueous solution. The presence of hydroxyl groups on the cellulose increases its hydrophilicity and repels cell attachments, while the modification of cellulose through hydroxyl groups not only enhances its biocompatibility and cell affinity, but also enable enormous TERM applications (Hickey and Pelling 2019).

Starch exists in two forms, namely amylose and amylopectin. Amylose is composed of α-D-glucose units linked by α-1,4-glucosidic bonds, while amylopectin contains branches with α-1,6-glucosidic bonds within the amylose molecule. The presence of these α-glucose groups imparts biologically significant properties to starch, including degradability and the development of semi-crystalline structures (Singh et al. 2016).

Cyclodextrin molecules are composed of several glucose units linked by β-1,4-glucosidosidic bonds, forming a cyclic structure. This ring-like configuration provides cyclodextrins with a hollow central channel capable of enveloping and creating complexes with diverse substances. Due to its hollow nature, cyclodextrin can form inclusion complexes by encapsulating various molecules, showcasing promising applications in drug delivery. This property enhances the solubility, stability, and bioavailability of drugs, and also facilitates drug protection and controlled release (Liu et al. 2021a; Jansook et al. 2018).

Polysaccharides exhibit a diverse array of structures, with their monosaccharide composition, molecular weight, functional groups, and the overall molecular structure collectively influences their properties and functions. These structural traits play a pivotal role in leveraging polysaccharides for various TERM applications. The rich structural and functional characteristics of polysaccharides offer extensive possibilities and resources for advancing regenerative medicine, bearing profound significance in the realms of tissue repair, regenerative therapies, and personalized treatment.

### 7.2.3 Modification of Polysaccharide to Enhance Applicability in TERM

In the realm of TERM, a 4D tissue-engineered scaffold that features not only intricate structure and desired bioactivity but also finely tuned biodegradation over time is indispensable in effectively restoring tissues and organs (Francis and DeForest 2023). Natural polysaccharides exhibit great biological properties. Nevertheless, their applications are frequently constrained without appropriate modifications. Therefore, engineering polysaccharides becomes crucial and highly desirable. Chemical modifications of polysaccharides offer a viable approach, not only to enhance their bioactivity but also to improve ease of processing and expand their applications. Moreover, modifications may even enable polysaccharides to boost their role in cell signaling to enhance tissue repairs (Chen et al. 2020a). Polysaccharides have been engineered as biomaterials to support cell attachment and growth, promoting cell migration, proliferation, and differentiation, thereby contributing to tissue regeneration and repair (Wu et al. 2021b; Zhu et al. 2019). Studies have demonstrated that the rheological properties, gelation behavior, and cross-linking ability of polysaccharides can be regulated through appropriate functionalization modifications, enabling precise control of scaffold materials for constructing more intricate and functional tissue structures (Elosegui-Artola et al. 2023; Sun et al. 2010, 2011b). Currently, chemical modification methods, such as methylation, phosphorylation, carboxymethylation, sulfation, and acetylation, are commonly employed to alter the chemical structure and conformation of polysaccharides (Fig. 7.2).

#### 7.2.3.1 Methylation

Methylation modification of polysaccharides involves introducing methyl groups onto polysaccharide molecules, and this process can be accomplished through various methods (Temple et al. 2022; Sims et al. 2018), with chemical methylation being the more common approach. In the reaction, hydrogen in the hydroxyl groups will be substituted by methyl groups, and degree of substitution can reach up to 100%. Methylation alters the physical and chemical properties of polysaccharides (Gericke et al. 2024), enhancing solubility, stability, and biocompatibility, thereby broadening their applications.

#### 7.2.3.2 Phosphorylation

Phosphorylation modification of polysaccharides involves introducing a phosphate group into the polysaccharide molecule, and this can be achieved via different methods (Xia et al. 2021; Zhou and Huang 2021). Phosphorylation modification of

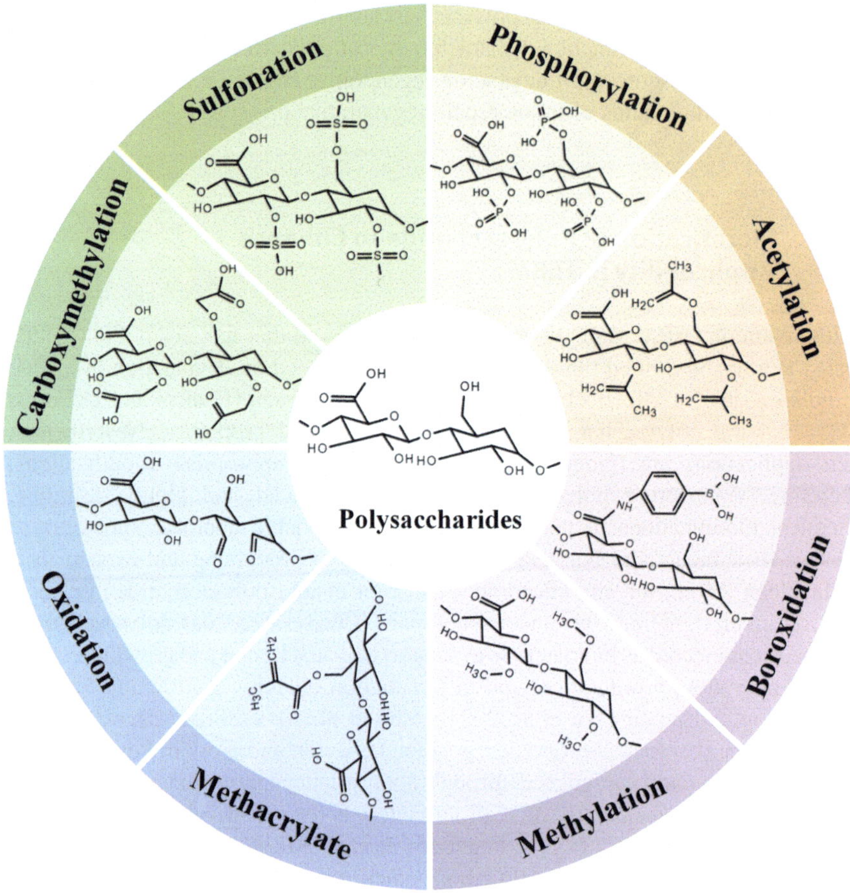


**Fig. 7.2** Methods of polysaccharide modification to expand their applicability in TERM

polysaccharides brings about various beneficial effects, such as antioxidant, antimicrobial, and anti-tumoral properties (Laffargue et al. 2023). In general, phosphorylation enhances their hydrophilicity, solubility, stability, and binding ability to biomolecules, which broadens the scope of polysaccharide applications in various fields. Moreover, phosphorylation modification enables polysaccharides to modulate intracellular signaling pathways. Within the cell, phosphorylation improves the interaction of polysaccharides with other proteins or nucleic acids, influencing the activation or inhibition of cell signaling pathways (Zhang et al. 2022). This, in turn, regulates physiological functions like cell metabolism, proliferation, and differentiation. Additionally, phosphorylation modification can impact the immunogenicity and immunomodulatory function of polysaccharides, rendering them potentially promising in the field of TERM.

#### 7.2.3.3 Carboxymethylation

Carboxymethylation of polysaccharides involves introducing carboxymethyl groups onto the hydroxyl groups in the polysaccharide molecule through a chemical modification process. This modification may increase the hydrophilicity and nucleophilicity of polysaccharides (Xie et al. 2021). Carboxymethylated polysaccharides typically exhibit high thermal stability, contributing to greater stability during the processing of medical biomaterials and helping to maintain the structure and properties of these materials. Moreover, carboxymethylation improves the biocompatibility of polysaccharides and has also been employed to improve various bioactivities, which make carboxymethylated polysaccharides great candidates in pharmaceutical and biomedical fields (Chakka and Zhou 2020).

#### 7.2.3.4 Sulfation

Polysaccharide sulfation is a chemical sulfonation, in which sulfate groups are incorporated into the hydroxyl groups of polysaccharides (Zhao et al. 2023). Sulfated polysaccharides generally exhibit improved biocompatibility, reducing the likelihood of immune rejection in humans. Additionally, sulfation can enhance the biological activity of polysaccharides, providing properties such as anticoagulation, anti-inflammatory effects, and promotion of cell adhesion. Given these enhanced biological properties, sulfated polysaccharides hold great potentials for many TERM applications (Dinoro et al. 2019; Arlov et al. 2021). Meanwhile, these bioactivities also make them suitable as carriers or auxiliary materials for anticoagulants and anti-inflammatory drugs.

#### 7.2.3.5 Acetylation

Polysaccharide acetylation involves the chemical modification process of introducing acetyl groups into the polysaccharide molecule. This typically includes reacting acetic anhydrides or other acetyl-containing compounds with polysaccharides, thereby introducing acetyl groups onto the hydroxyl groups of the polysaccharides. The degree of acetylation can be controlled, influencing both the physical and biological properties of the polysaccharide (Wang et al. 2024). Acetylation has the potential to alter the solubility, crystallinity, and stability of polysaccharides. Acetylated polysaccharides generally exhibit improved biocompatibility as well as immunomodulatory and antioxidant bioactivity (Li et al. 2023a), making them promising for various applications in the biomedical field.

#### 7.2.3.6 Other Modifications

In addition to the aforementioned modifications, incorporating various functional groups into polysaccharides can undoubtedly impart unique properties. For instance, methacrylic polysaccharide modification involves the chemical reaction of polysaccharides with methacrylic anhydride to introduce methacrylic groups into the polysaccharide molecule. This modification not only enhances the water solubility of polysaccharides but also imparts photochemical crosslinking functionality. When exposed to ultraviolet light in the presence of photo initiators, methacrylated polysaccharides can rapidly crosslink into hydrogel, which has found many applications in tissue repair and regeneration (Schnellmann et al. 2022; Rape et al. 2015). The oxidation of polysaccharides can convert hydroxyl groups into aldehyde groups in the presence of oxidants (Muhammad et al. 2020). This oxidation process enables the formation of reversible Schiff base bonds or hydrazone bonds between the aldehyde groups in oxidized polysaccharides and certain polymers rich in amino or hydrazide groups. Moreover, multiple reaction also allow integration of different functional groups into polysaccharides (Sun et al. 2010). This capability allows polysaccharides to contribute to the development of responsive tissue engineering scaffolds sensitive to pH, temperature, or other external stimuli (Gholamali 2021).

## 7.3 Polysaccharide-Based Scaffolds

### 7.3.1 Characteristics of Polysaccharide-Based Scaffolds

Polysaccharide scaffolds constitute a prevalent category of tissue-engineered scaffolds, primarily composed of natural or synthetic polysaccharide materials. Polysaccharides exhibit excellent biocompatibility, degradability, and unique biological activity. These attributes render polysaccharide-based scaffolds highly valuable for a broad spectrum of applications and research endeavors in the fields of tissue engineering and regenerative medicine (Jin et al. 2021; Luo et al. 2021).

#### 7.3.1.1 Cytotoxicity

Biocompatibility, the fundamental ability of tissue-engineered scaffolds to interact harmoniously with living tissues, is a critical characteristic that determines the success of tissue repair and regeneration. Typically structured as 3D micro- or nanoscale networks, polysaccharide-based scaffolds feature broad, rough, and charged surface structures that significantly influence protein adsorption and cell adhesion, and these scaffolds closely resemble the structure of biological ECMs in the human body. This unique architecture promotes cell attachment, proliferation, and differentiation, crucial for supporting the intricate processes of tissue regeneration and repair. Owing to their chemical composition, favorable surface properties, and ECM-like 3D structures, polysaccharide-based scaffolds thereby exhibit excellent cytocompatibility (Su et al. 2022) and histocompatibility (Hu et al. 2021). As a result, polysaccharide-based scaffolds can fulfill their intended functions within the organism without triggering noticeable immune reactions, fostering compatibility with human tissues.

#### 7.3.1.2 Biodegradation

The essence of polysaccharide-based scaffolds lies in natural polysaccharides, which usually undergo hydrolysis and enzyme-mediated degradation within living organisms. Throughout the degradation process of polysaccharide scaffolds, the metabolites generated include sugars, amino acids, or other natural compounds. In general, these metabolites exhibit biocompatibility, causing no allergic or toxic reactions. They do not interfere with the normal physiological functions of surrounding tissues and can be safely absorbed or excreted by the body (Chen et al. 2023). The degradation rate of polysaccharide scaffolds can be tailored (e.g., cross-linking density) to align with the pace of tissue repair and regeneration. Consequently, these scaffolds gradually break down and get absorbed in the body while actively promoting the generation of new tissue. Certain polysaccharide scaffolds demonstrate sensitivity to specific physiological conditions or environments, such as pH and temperature (Chowhan and Giri 2020). Leveraging these properties allows for precise modulation of the degradation rate and relevant property of the scaffold, enhancing fine control over the tissue repair process.

#### 7.3.1.3 Biological Activity

The presence of functional groups of the polysaccharide-based scaffolds, such as carboxyl and amine groups, enables them to interact with other biomolecules within the organism through mechanisms like hydrogen bonds and ionic bonds, showcasing a wide range of biological activities. These scaffolds can engage with receptors, enzymes, and other biomacromolecules within living organisms. In addition, polysaccharide scaffolds consist of various glycan units interconnected in specific patterns, forming complex branched or linear structures. The diversity of glycan units and their arrangement confers a distinctive spatial conformation and charge properties upon the polysaccharide scaffold, influencing its interactions with biomolecules, such as proteins and cell membrane receptors. As such, polysaccharide-based scaffolds boast diverse bioactivities, including immunomodulation, antioxidant, antitumor, anti-inflammatory, and metabolic regulation on a broad scale (Duceac and Coseri 2022; Yu et al. 2018). Frequently, polysaccharide scaffolds are fabricated as carriers for the controlled release of bioactive substances, including growth factors, drugs, and genes, to regulate cell proliferation, differentiation, and the generation of new tissues (Wu et al. 2021b).

#### 7.3.1.4 Ease of Manipulation

The performance of polysaccharide scaffolds can be precisely engineered and meticulously tuned through various strategies in the design and preparation process to cater to diverse tissue regeneration and repair scenarios. Key methods encompass chemical and physical modification, and the controlled release of various biomolecules. Chemical regulation of polysaccharide scaffolds primarily involves material screening and selection, the incorporation of functional groups, and control of chemical cross-linking. The choice of chemical composition impact biocompatibility, biodegradation, and controlled release capacity of scaffolds. Modulating cross-linking density alters mechanical properties and degradation rate (Sun et al. 2011b), enabling adaptation to different tissue environments. Introducing functional groups imparts desired biological activity and controlled release of the scaffolds. The physical structure of polysaccharide scaffolds, encompassing parameters like pore structure, pore size, and surface morphology, can be tailored via various fabrication methods. Altering pore structure and size influences cell migration, biomolecule transportation, and nutrient exchange efficiency. Surface morphology adjustments impact the interaction and attachment between the scaffold and surrounding tissues.

As tissue-engineered scaffolds, polysaccharide scaffolds also facilitate controlled release. Polysaccharide scaffolds are also capable of carrying bioactive factors like growth factors and cytokines, influencing tissue regeneration and repair processes by regulating the release behavior of these factors.

### 7.3.2 The Role and Design Principles of Polysaccharide-Based Scaffolds

A solid understanding of the comprehensive role of polysaccharides in tissue engineering and regenerative medicine is crucial for designing appropriate scaffolds. Both the physical and chemical properties of porous scaffolds influence the rate of cell behavior within them. Other than serving as intricate 3D porous structures, polysaccharide scaffolds offer crucial support to cells and provide directional channels that guide cell migration along specific pathways. Renowned for their biocompatibility and biodegradability, polysaccharide scaffolds bind to migrating receptors on cell surfaces, fostering cell adhesion. These binding triggers specific signaling pathways, regulating cell differentiation and functional expression. For instance, growth factors embedded in scaffolds can bind to cell surface receptors, initiating associated signaling pathways that encourage differentiation and functional expression in specific cell types. As carriers for bioactive molecules like chemokines, growth factors, and extracellular matrix components, polysaccharide scaffolds enable controlled release, activating specific signaling pathways in cells, which significantly contributes to the control and coordination of these cells and growth factors that may determine the outcome of tissue regeneration. Furthermore, the physical properties of polysaccharide scaffolds, including stiffness and elasticity, significantly influence cell behavior and function. Appropriate mechanical stimulation can modulate cell differentiation and functional expression by altering cell morphology and mechanical tension (Chen et al. 2024b). In essence, understanding these cellular mechanisms sheds light on the multifaceted role of polysaccharide scaffolds in orchestrating tissue regeneration. With all these improved insights, a pro-regenerative polysaccharide niche can be effectively created, enabling cells to self-assemble into optimal structures, i.e., tissue regeneration.

### 7.3.3 Fabrication of Polysaccharide Scaffolds to Revolutionize TERM

Polysaccharide scaffolds exhibit diverse structural manifestations, each offering unique advantages and applicability in TERM (Nie et al. 2019). The selection and design of these scaffolds can be tailored to specific application requirements. Simultaneously, customization of structure and performance through material combinations and precise process controls allows the scaffolds to meet the varied needs of different tissue engineering domains (Datta et al. 2020). When faced with irregularly shaped and sizable wounds, such as cuts and lacerations, traditional immobilization materials may struggle to fully conform to the wound surface or even cause additional damage to surrounding tissues. Polysaccharide scaffolds, being malleable and moldable, present potential advantages in adapting to irregular trauma and promoting wound repair (Taylor and in het Panhuis 2016; Bertsch et al. 2022). Several methods can be employed to shape polysaccharide scaffolds.

The thermoforming method involves softening and flowing the polysaccharide scaffold through the application of high temperature and pressure, followed by pressing it into the desired shape. This technique is often employed in the preparation of various thin-film polysaccharide scaffolds. In the mold molding method, melted polysaccharide material or a precursor solution for the polysaccharide scaffold is injected into a pre-designed mold and cured under suitable conditions to achieve the desired shape and size (Chuaynukul et al. 2017; Truckenmüller et al. 2011). The electrospinning method creates polysaccharide scaffolds of different shapes on various spinning receiving devices through electrospinning, commonly used for tubular and flake polysaccharide hydrogels (Lee et al. 2009). With the advancements in digital technology, 3D printing technology can produce high-precision, personalized, and customized polysaccharide scaffolds (Shokrani et al. 2023). This involves stacking and solidifying polysaccharide materials layer by layer to create intricate three-dimensional structures using digital design data. Each fabrication method and its resulting scaffold have their own advantages and provide a distinct microenvironment suitable for specific applications (Fig. 7.3).

#### 7.3.3.1 Electrospinning Scaffolds

Electrospinning stands out as a transformative method wherein a polymer solution or molten polymer is propelled onto a collector under a high electric field, forming a fibrous structure (Min et al. 2019). The process begins with the preparation of polysaccharides in an appropriate solvent, with the addition of a crosslinker or other auxiliary agents to fine-tune fiber properties. Under the influence of a high-voltage electric field, the solution is expelled from the spinning nozzle, rapidly solidifying into fibers in the air. Subsequently, further curing can be performed to enhance the mechanical properties and stability of the fibers. The non-woven fibers form a three-dimensional scaffold with a multi-layered structure, mimicking the hierarchical complexity of natural tissues. The scaffold features open pore structures, creating space for cell growth and allowing the diffusion of nutrients, oxygen, and waste products. Control over pore size and distribution is achieved by adjusting spinning conditions and scaffold design (Su et al. 2024). Typically, electrospun fibers exhibit incredibly fine diameter, ranging from a few nanometers to tens of microns. Their surfaces are rich in micro-nano structures, varying from folded to granular or nanoscale textures. This surface morphology substantially increases the specific surface area of fibrous materials, enhancing the likelihood of cell attachment and growth while facilitating the adsorption and release of biomolecules (Jayasinghe 2022). Electrospinning scaffolds frequently serve as extracellular matrix scaffolds, providing vital support and guidance to damaged tissues, making electrospinning a pivotal technique in advancing the field of tissue engineering (Xu et al. 2024).

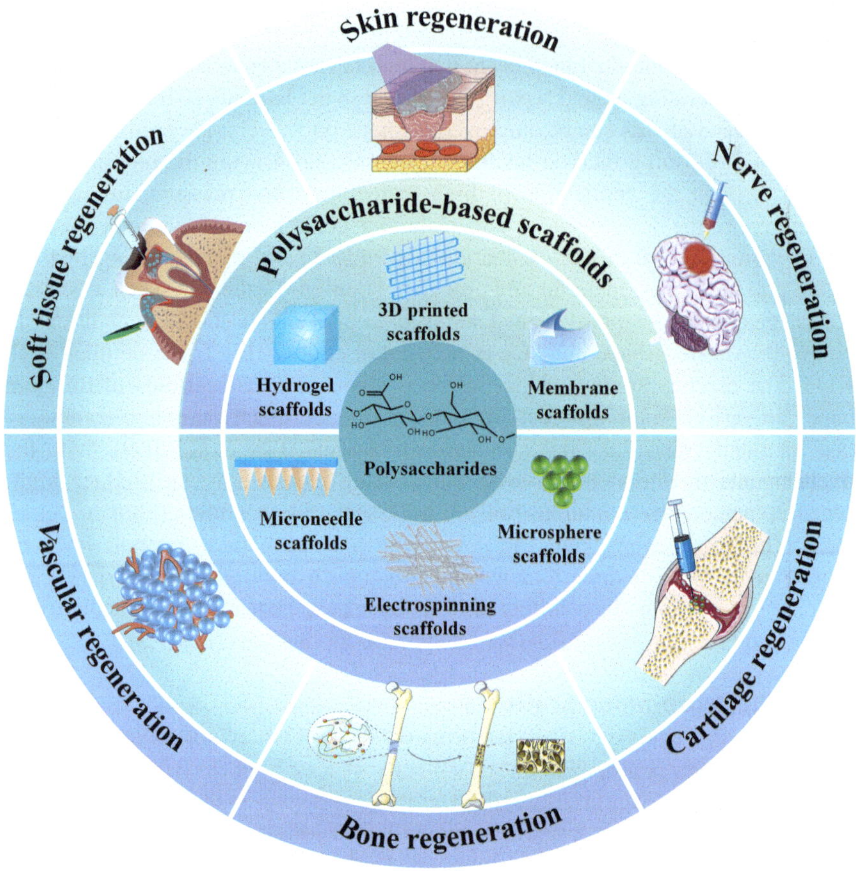


**Fig. 7.3** Polysaccharide-based TERM applications. Attributed to their excellent biocompatibility and inherent structural characteristics, polysaccharides have found been explored for all TERM applications

### 7.3.3.2 Microsphere Scaffolds

Polysaccharide microsphere scaffolds are three-dimensional structures composed of tiny spherical particles made from polysaccharide materials. These scaffolds boast a high specific surface area, offering extensive contact points for interaction with cells and bioactive substances. This characteristic amplifies cell attachment and growth while efficiently facilitating drug adsorption and release. The utilization of three-dimensional porous microspheres as scaffolds in tissue engineering has been extensively documented (Cai et al. 2013). However, these spherical scaffolds have limited tunability in terms of their porous structure. In recent years, microporous annealed particles (MAP) has evolved an promising approach to fabricate porous scaffolds (Qazi et al. 2022). The microspheres create numerous voids and pores, forming a porous structure whose size and distribution can be fine-tuned by adjusting their size, shape, and packing (Darling et al. 2020; Muir et al. 2023). This adaptability enables the design and fabrication of microsphere scaffolds tailored to precise application requirements, allowing controlled manipulation of cell behavior and drug release. Furthermore, microsphere scaffolds can undergo specific functionalization through surface modification or the incorporation of functional substances. For instance, the surface of microspheres can be adorned with bioactive molecules like cell adhesion peptides or growth factors (Muir et al. 2022), amplifying cell attachment and growth. Additionally, the embedding of nanoparticles or nanomaterials into microspheres imparts specific physical, chemical, or biological properties to the scaffolds.

#### 7.3.3.3 Membrane Scaffolds

Membrane scaffolds, characterized by their flat, thin, and flexible structure, offer easy attachment to tissue or organ surfaces. Various methods are employed to craft membrane scaffolds (Sun et al. 2007; Iijima et al. 2017; Deng et al. 2010), which can be manipulated through such parameters as material composition, film thickness, and pore structure. This flexibility enables the design and fabrication of membrane scaffolds tailored into sheets or elongated strips of various shapes and sizes to meet diverse application requirements. Leveraging natural polysaccharides in membrane scaffold preparation ensures biocompatibility and degradability, allowing gradual *in vivo* degradation while providing crucial support for emerging tissues. Furthermore, the biological activity of membrane scaffolds can be augmented through surface modification or the incorporation of bioactive substances (Caridade et al. 2015). These enhancements promote cell attachment, proliferation, and differentiation, expediting the tissue repair process. In essence, membrane scaffolds emerge as versatile platforms with the potential to revolutionize tissue engineering applications.

#### 7.3.3.4 Hydrogel Scaffolds

Hydrogel are 3D hydrophilic polymer scaffolds that can effectively retain a substantial amount of water. With their ECM-mimicking structures and high-water content, polysaccharide hydrogel scaffolds create an environment akin to biological tissues, supporting cell attachment, growth, and spread, thereby leading to diverse applications in TERM (Slaughter et al. 2009; Mantha et al. 2019). In general, hydrogels can be constructed by adjusting parameters such as precursor modification, concentration and crosslinking degree to mimic the *in vivo* microenvironment, facilitating cell growth and the formation of new tissues. Other than improving their biocompatibility and immunomodulating capacity (Wu et al. 2021a; Sun 2017), fine-tuning material properties such as stiffness and elastic modulus, polysaccharide scaffolds deliver appropriate mechanical stimulation to foster cell proliferation, differentiation, and functional expression (Forget et al. 2013). Meanwhile, polysaccharide hydrogels can be modified to introduce or encapsulate specific bioactive molecules and/or growth factors, which regulate cell signaling pathways and functional expression, interacting with cells to promote adhesion, proliferation, differentiation, and the orchestrated formation of new tissues (Chen et al. 2020a; Sun et al. 2011b). These tuning capacity enables us to manipulate polysaccharide hydrogel to be the versatile scaffolds. In essence, hydrogel scaffolds offer a nurturing milieu for tissue engineering endeavors.

#### 7.3.3.5 Microneedle Scaffolds for Targeted Tissue Repair

Microneedle scaffolds are composed of minuscule needle-shaped structures within a supporting matrix that are characterized by non-invasive and painless administration, have found various applications (Lyu et al. 2023; Yang et al. 2022a). Many materials are utilized to form the microneedle structures, while polysaccharides needle structures typically feature micropores or microchannels at their tips showing potentials for tissue repair and regeneration (Lyu et al. 2023; An et al. 2024; Zhang et al. 2023). These scaffolds are able to promote angiogenesis, enhancing cell proliferation, and guiding cellular differentiation, which collectively boost the regenerative environment, thereby facilitating the repair and regeneration of damaged tissues. A notable capability of microneedle scaffolds is their ability to penetrate the skin or tissues, enabling the direct delivery of bioactive agents (e.g., growth factors) to the targeted site. This targeted and controlled release greatly improve and regulate the rate and quality of tissue regeneration. Despite their potential, challenges persist in the design, preparation, and application of microneedles. These challenges include ensuring the stability of microneedles, enhancing transdermal penetration capabilities, and fine-tuning the regulation of drug release rates. Addressing these challenges will further elevate the efficacy and versatility of microneedle scaffolds in advancing targeted tissue repair.

#### 7.3.3.6 3D Printed Polysaccharide Scaffolds

The utilization of advanced 3D printing technology in constructing tissue scaffolds signifies a breakthrough in engineering scaffolds with precisely designed structures and functions. The versatility of 3D printing allows for the customization of scaffold attributes such as shape, internal structure, and porosity (Sadeghianmaryan et al. 2024). Computer-aided design (CAD) tailors these features to fit the specific injured tissue site, offering precise support and guidance (Jamee et al. 2021). Moreover, these scaffolds can be engineered to facilitate controlled release of bioactive substances, such as cells and growth factors, i.e., 3D bioprinting. The regulated release enables precise control over the speed and quality of tissue regeneration, thereby enhancing the overall tissue repair and regeneration process (Zhang and Song 2018; Vijayavenkataraman et al. 2018). Polysaccharide 3D printed scaffolds leverage the qualities of natural polysaccharides, known for their remarkable biocompatibility and biodegradability, with the injured tissue anatomy, which ensures seamless integration with injured sites, thereby fostering the growth and targeted repair of new tissues (Shokrani et al. 2023). As a result, 3D printed polysaccharide scaffolds are assuming an increasingly pivotal role in advancing the frontiers of regenerative medicine.

## 7.4 Advancements in Polysaccharide Scaffolds for Tissue Repair and Regeneration

### 7.4.1 Advancements in Blood Vessel Engineering

Blood vessels serve as crucial conduits, supplying nutrients and oxygen to the tissues and organs. In instances of tissue repair and regeneration, the formation of new blood vessels becomes essential in supporting cell growth and differentiation, thereby determining the outcome of tissue regeneration (Sun et al. 2011b). Accordingly, many vascularization strategies have been proposed and investigated for TERM (Lovett et al. 2009). Polysaccharide-based scaffolds, particularly hydrogel scaffolds (Sun and Mao 2012), have been extensively studied for their potential to promote angiogenesis and vascularization (Wu et al. 2021b). Vascularization is primarily achieved either through scaffold engineering, cellular engineering, and the controlled the release of growth factors (Song et al. 2018). With an optimized scaffold structure, we demonstrated that the encapsulation of angiogenic growth factors (GFs) can induce rapid and functional neovascularization within the dextran hydrogel scaffold (Sun et al. 2011b). Even in the absence of growth factors, we proved that our the dextran hydrogel promoted robust vascularization and the regeneration of full skin with skin appendages (Sun et al. 2011a). While Feng et al. employed tyramine to modify konjac glucomannan (KGM) and heparin (Hep), resulting in the creation of an injectable hydrogel scaffold (KGM-TA/Hep-TA) (Feng et al. 2017). This innovative hydrogel scaffold not only sequesters exogenous growth factors (GFs) but also activates macrophages, prompting the secretion of endogenous GFs, thus inducing angiogenesis. Subcutaneous encapsulation of the hydrogel scaffolds in mice revealed abundant and mature angiogenesis within 14 days.

Some polysaccharides, such as low molecular weight HA and lentinan, have inherent angiogenic potentials and are also utilized in vascular engineering (Longinotti 2014; Jia et al. 2024). Silva et al. synthesized a spongy hydrogel based on HA with a semi-interpenetrating network (GG-HA) (Silva et al. 2016). GG-HA hydrogel scaffolds demonstrated excellent biocompatibility and facilitated the release of HA through enzyme-mediated hydrolysis, promoting human umbilical vein endothelial cell (HUVEC) proliferation *in vitro*. In an ischemic hind-limb mouse model, implanted GG-HA sponge-like hydrogel scaffolds effectively promoted angiogenesis in a substance-dependent manner, leading to neovascularization and improved blood perfusion in the ischemic hind limbs. A bio-multifunctional composite that leverages the unique properties of alginate, chitosan and fucoidan suppressed inflammation and enhanced vascularization by upregulating CD31-specific protein expression, thereby promoting the healing of full-thickness wounds (Hao et al. 2020). Recently, Li et al. developed a functionalized chitosan hydrogel (CS/TA@PAs-Eu) that features good mechanical stability and ionic conductivity (Li et al. 2024). The chitosan hydrogel network offers a biocompatible platform with strong mechanical and electrical properties. Additionally, the synergistic effects of TA, PAs, and $Eu^{3+}$ enhance antioxidant activity by improving mitochondrial membrane potential, permeability, and function. This effectively promotes myocardial vascularization in infarcted areas and facilitates myocardial reperfusion, presenting a promising alternative for adaptive myocardial infarction repair. Collectively, polysaccharides, whether through inherent angiogenic effects or engineered angiogenic properties, can promote vascularization and facilitate the repair or regeneration of tissues and organs.

### 7.4.2 Soft Tissue Regeneration

Soft tissues are frequently subjected to various injuries, including physical traumas and pathogen-related damage, necessitating prompt and effective treatment. Soft tissues injuries often result in the formation of fibrotic scars during the repair process, thereby compromising the functionality of the affected tissues and organs or even death (Huang and Ogawa 2020). The regeneration of deep skin wounds poses a significant challenge, for instance, often resulting in unsightly scars and impaired functionality. Extensive research has been conducted on polysaccharide-based scaffolds for constructing advanced architectures as medical interventions for various tissues and organs (Wu et al. 2021b). As aforementioned, angiogenesis and neovascularization play crucial roles in repairing and regenerating soft tissues. Our research has shown that our dextran-based hydrogel system, which encourages endothelial cell infiltration, significantly enhances dermal wound healing in a third-degree burn model. This system not only substantially reduces scar formation but also surpasses the performance of existing commercial products. Remarkably, the dextran hydrogel devoid of either cell- or growth factor-delivery was sufficient to induce various levels of dermal regeneration up to anatomic restoration of full-thickness skin with substantial matrix synthesis and appendage formation for both deep wounds and pre-existing scarred skins (Sun et al. 2011a, b). These findings highlight the potential of dextran-based hydrogels as effective therapeutic interventions for soft tissue regeneration.

Vocal folds (VF), the soft and delicate tissues within the larynx, govern the production of sound but are susceptible to damages from a variety of environmental factors (Li et al. 2016). Injuries to the vocal folds often result in scarring, leading to deterioration in function. Thereby, regenerative therapy is highly desired. The development of pro-regenerative biomaterials advanced the regeneration of VF (Coburn et al. 2022). Zou et al. developed a dual-network hydrogel scaffold using carboxymethyl chitosan and sodium alginate (Zou et al. 2023), which promoted fibroblast activity, collagen and elastin secretion, and ECM formation, showcasing great biocompatibility. With mechanical properties akin to vocal folds, this scaffold enhanced vibration and vocalization. Injecting this hydrogel into damaged vocal folds in rabbits led to superior vocal fold reconstruction over time.

According to the World Health Organization, nearly half of the world's population is affected by oral diseases. Oral situations like such as pulpitis or periodontitis pose a significant global health challenge (Wolf et al. 2021). Dental pulp is a highly vascularized connective tissue located beneath the dentin layer, providing essential nutrients crucial for tooth health (Yu and Abbott 2007). Root canal therapy is a conventional approach to treat infections in the tooth pulp, but it commonly results in the fragility and fracturing of the tooth due to loss of vitality (Sugiaman et al. 2023). Regenerative engineering using biomaterial scaffolds has shown promising potential in re revitalizing dental pulp (Yang et al. 2022b; Athirasala et al. 2017). Pulp regeneration is primarily accomplished through the utilization of cells, growth factors and exosome. Polysaccharide-based hydrogels not only promote angiogenesis or vasculature to facilitate pulp regeneration (Wang et al. 2023d), but they could also contribute to the restoration of dentin and enamel (Alsalhi 2024). Some clinical trials have demonstrated the possibility of pulp regeneration (Nakashima et al. 2017), but it still remains a challenge. Gums, vital for tooth stability (Monteiro and Yelick 2017), require specialized tissue engineering materials for restoration. Feng et al. crafted a novel chitosan-based 3D lamellar scaffold mimicking gingival structure (Feng et al. 2021), which exhibited high water absorption and superior mechanical properties. The *in vitro* studies demonstrated its ability to facilitate bone marrow mesenchymal stem cell adhesion, migration, proliferation, and revascularization, while the *in vivo* studies further proved that the scaffold accelerated vascular network formation and accelerated gingival regeneration through pro-angiogenic factor secretion.

### 7.4.3 Innovative Approaches for Nerve Tissue Regeneration

Nerves are specialized soft tissues tasked with receiving, transmitting, and processing information from both internal and external sources. They play a crucial role in regulating behavior, emotions, and physiological functions throughout the body. Nerve tissue damage is a serious clinical condition that can significantly impact the quality of life for patients. Peripheral nerves have limited regenerative capacity, but surgical treatments are commonly required to treat larger nerve damage (Schmidt and Leach 2003). Traditional surgical suture techniques, although commonly used, often result in incomplete restoration (Pfister et al. 2011). In response to these challenges, Xue et al. introduced a groundbreaking nerve adhesive, the dual-network nerve adhesive (DNNA), designed to replace traditional sutures (Xue et al. 2023). DNNA combines natural polyphenol-modified HA (HA-TU-CAT) with acellular peripheral nerve matrix (DPN). This innovative adhesive offers strong adhesion through bonding polyphenol groups with tissue amino and carboxyl groups. DNNA effectively anastomoses damaged nerves, promoting axonal junction, remyelination, and reducing intraneural inflammation and fibrosis, ultimately facilitating optimal nerve regeneration. Both *in vitro* and *in vivo* data demonstrated that this adhesive hydrogel robustly facilitated nerve recovery.

Scar-induced adhesions between nerve tissue and surrounding areas can impede nerve recovery post-repair. Chen et al. devised a nanofiber membrane (BSP/WPU NFM) composed of *Bletilla striata* polysaccharide (BSP) and waterborne polyurethane (WPU) (Chen et al. 2020b). This membrane that exhibited good biocompatibility and tunable biodegradation was able to achieve anti-adhesion effects for over

6 weeks. The neurite outgrowth tests demonstrated that the nanofiber membrane effectively accelerated neurite growth. In rabbit sciatic nerve repair models, BSP/WPU NFM encapsulation significantly promoted nerve regeneration, resulting in elevated compound muscle action potential, muscle contractility, and reduced perineural adhesion. Recognizing the limited self-repair ability of neural tissue, Xuan et al. explored external stimuli to enhance repair capabilities. They developed a self-healing conductive hydrogel scaffold (HASPy) by modifying pyrrole-1-propionic acid (Py-COOH) with HA (Xuan et al. 2023). This injectable, biodegradable, and biocompatible hydrogel scaffold enhanced Schwann cell survival, proliferation, and differentiation through various signaling pathways. Application of HASPy in rat sciatic nerve repair experiments demonstrated its effectiveness in promoting nerve regeneration. Taken together, these advancements on the basis of advanced polysaccharide scaffolds offer promising avenues for addressing challenges in nerve tissue repair, emphasizing improved outcomes and patient well-being.

### 7.4.4 Cartilage Repair and Regeneration

Articular cartilage plays a crucial role in the smooth motion of joints, and its impairment can lead to debilitating conditions such as osteoarthritis (OA) (Buckwalter et al. 2005). This condition not only cause pain and restrict mobility but also poses a significant challenge for medical treatments. Advances in tissue engineering, particularly the use of polysaccharide scaffolds have shown promising results in the repair and regeneration of damaged cartilage. These hydrogels mimic the natural extracellular matrix, providing a conducive environment for chondrocyte activity, which is essential for cartilage repair. Chen et al. developed an innovative HA methacrylate hydrogel scaffold (AHAMA) to treat cartilage defects (Chen et al. 2021). They demonstrated that AHAMA scaffolds supported cell adhesion, migration, and proliferation *in vitro*, ultimately leading to the restoration of cartilage structure and function in a rat model. While biomaterial scaffolds alone have limited capacity to fully restore structure and function, they are often enhanced through the combination of factors such as cells (Liu et al. 2021b). Feng et al. synthesized highly biocompatible microgels (Gel-HA) from vinyl sulfonated hyaluronic acid (HA-VS) and vulcanized gelatin (Gel-SH) (Feng et al. 2019). Gel-HA could self-assemble into microporous scaffolds that facilitated cell viability, proliferation, and chondrogenic differentiation of bone mesenchymal stem cells (BMSCs). Moreover, Gel-HA could self-assemble into cartilage-like scaffolds through cell-cell interconnection. *In vivo* experiments further demonstrated that the Gel-HA microgels loaded with BMSCs could effectively suppress vascularization and hypertrophy, thereby promoting cartilage repair. The integration of such biomaterials into therapeutic strategies offers hope for patients suffering from joint disorders, potentially reducing the need for invasive procedures and improving overall joint health.

Additionally, cartilage damage often extends deep into the associated subchondral bone, resulting in osteochondral defects that lead to severe pain, movement disorders, and degenerative joint disease. While microfracture procedures and autologous chondrocyte implantation have been introduced for repairing articular surface cartilage, they are not ideal for addressing subchondral bone repair due to the distinct biological lineages and properties of cartilage and subchondral bone tissue. To address this challenge, Li et al. utilized 3D printing technology to create an exosome-laden bilayer scaffold that mimics the chondrogenic and osteogenic microenvironment found in natural osteochondral tissue (Li et al. 2023b). They prepared a bioink by blending hyaluronic acid-graft-dopamine (HA-DA), oxidized hyaluronic acid (OHA), and methacrylated gelatin (GelMA), and then incorporated decellularized extracellular matrix (dECM) derived from porcine cartilage (DCM) and bone (DBM) into the hydrogel bioinks to form microenvironment-specific biomimetic hydrogel scaffolds. The 3D-printed bilayer scaffold, with continuous exosome release, effectively facilitated MSC cell bioactivity, as well as chondrogenic and osteogenic differentiation *in vitro*. Additionally, they further showed that the gel accelerated the simultaneous regeneration of cartilage and subchondral bone tissues in a rat preclinical model. This innovative biomimetic approach demonstrates the design of structured 3D biomimetic scaffolds for targeted regeneration of complex tissues.

### 7.4.5 Bone Tissue Regeneration

Bone tissue, akin to skin and liver, exhibits remarkable regenerative capabilities. However, addressing large defects or managing chronic conditions such as infections, tumors, and osteoporosis presents significant challenges. Bone tissue engineering scaffolds emerge as a reliable solution, addressing large bone defects and facilitating the regeneration process. Regenerating bone defects is a multifaceted process involving the modulation of the immune microenvironment, activation of vascular growth, and subsequent osteogenic differentiation. Wu et al. developed a multifunctional hydrogel scaffold by incorporating black phosphorus-based nanocomposites (BPPD) into a gelatin methacrylic acid (GelMA)/sodium alginate methacrylate (Alg-MA) hybrid hydrogel (Wu et al. 2024). This intelligent hydrogel scaffold exhibited outstanding immunomodulatory properties, effectively promoting the differentiation of macrophages into the M2 type, which created a conducive immune microenvironment for the early stages of bone regeneration. Additionally, the hydrogel scaffold demonstrated the ability to stimulate the growth of vascular tissue and recruit endogenous stem cells, thereby accelerating the bone repair process. An *in vivo* study in a rat model confirmed accelerated osteogenesis and bone tissue remodeling of critical size bone defects, highlighting the scaffold's efficacy in bone regeneration.

The periosteum plays essential roles in maintaining the bone functions and repairing bone defects. Mimicking the remarkable osteogenic and bone regenerative capabilities of the periosteum has emerged as a novel approach for repairing and regenerating bone defects (Dai et al. 2022). Liu et al. synthesized a composite hydrogel scaffold using a polydopamine-coated PCL porous scaffold (BM-g-DPCL) as the hard structure and dopamine-modified HA (HADA) and gelatin as the soft structure (Liu et al. 2023b). This biomimetic structure, mirroring the soft-hard combination of natural bone, accelerated mineral deposition, provided a favorable microenvironment for cell adhesion, growth, and osteogenic differentiation, and facilitated rapid vascularization and bone regeneration.

For unreparable intra-articular bone tissue, prosthetic implantation is a common clinical solution. However, periprosthetic intra-articular infection (PJI) poses a significant challenge (Lucenti et al. 2024). To address this challenge, Ouyang et al. developed a nanowire-like composite scaffold (P@C) from chitosan and Puerarin that exhibits excellent antibacterial properties (Ouyang et al. 2023). This simple yet effective scaffold closely mimicked the biological function of antimicrobial peptides (AMPs). The *in vitro* antimicrobial studies demonstrated that the scaffold effectively countered bacterial infections. Puerarin and chitosan collaboratively disrupted bacterial membrane structures, showcasing both antibacterial efficacy and superior bone regeneration capacity in *E. coli*-infected rats. This innovative approach presents a promising avenue for addressing complications related to prosthetic implantation.

## 7.5 Challenges and Future Perspectives in TERM

The intrinsic capability of the human body to repair and regenerate damaged tissues forms the foundation of natural healing processes. However, the extent of regenerative potential varies among different tissues and organs, influenced by factors such as stem cells, immune response, and the gender and age of human. When faced with severe injuries, external interventions become essential to promote tissue regeneration. TERM emerges as a promising avenue to overcome these challenges. Central to TERM are biological scaffolds, which offer a three-dimensional microenvironment mimicking native tissues, thereby providing necessary physical and chemical cues for cell bioactivities. A vast array of studies, ranging from basic research to clinical trials, have revealed various factors crucial in designing more advanced scaffolds (Sun et al. 2018; Place et al. 2009; Qin et al. 2022). Despite the potential benefits, the production and utilization of biological scaffolds may pose environmental and social challenges, conflicting with sustainable development goals. To address this, considerations of environmental friendliness and sustainability guide material selection. Polysaccharides, with their unique biocompatibility, biodegradability, and versatility for modification, stand out as preferred materials for biological scaffold preparation. However, translating polysaccharide scaffolds into clinical applications presents challenges.

Polysaccharides, even when they are the same type, can exhibit variable properties due to differences in extraction methods, processing techniques, and the purity of natural polysaccharides. These variations can affect scaffold performance and their pro-regenerative capacity. Furthermore, the incorporation of toxic molecules or additives in polysaccharide modification raises concerns about their impact on the body. Addressing these challenges demands rational design, precise control, multi-step preparation, and careful consideration of environmental implications. Understanding the exact mechanisms underlying tissue regeneration facilitated by polysaccharide scaffolds remains an ongoing exploration, involving factors like cytokines, signaling pathways, and hormones. Furthermore, in the context of societal progress, the emphasis on sustainable development and environmental protection necessitates efforts to reduce manufacturing costs, simplify production processes, and minimize environmental pollution.

## 7.6 Conclusion: Shaping the Future of Tissue Regeneration with Polysaccharide Scaffolds

The quest for effective tissue regeneration has led to significant advancements in the field of regenerative medicine, with a particular focus on the pivotal role of biological scaffolds. Among these, polysaccharide scaffolds have emerged as a beacon of promise because of their unique biocompatibility, biodegradability, and modifiability. These 3D scaffolds offer a guiding framework for cellular activities. The adaptability of polysaccharide scaffolds to diverse tissue types, defect sizes, and treatment goals makes them indispensable in the pursuit of effective tissue regeneration. While challenges remain, which encompass the variability in polysaccharide extraction and processing, concerns about the use of potentially toxic additives, and the imperative to understand the intricate cellular mechanisms driving tissue regeneration. The shift towards sustainable practices adds another layer of complexity, urging researchers to balance innovation with eco-friendly considerations. As science and sustainability converge, further studies addressing these challenges will pave the new way for the development of more advanced scaffolds for tissue engineering and regenerative medicine.

## References

Alizadeh A, Dyck SM, Karimi-Abdolrezaee S (2019) Traumatic spinal cord injury: an overview of pathophysiology, models and acute injury mechanisms. Front Neurol 10:282

Alsalhi A (2024) Applications of selected polysaccharides and proteins in dentistry: a review. Int J Biol Macromol 260:129215

Alven S, Aderibigbe BA (2020) Chitosan and cellulose-based hydrogels for wound management. Int J Mol Sci 21:9656

An H, Gu Z, Huang Z, Huo T, Xu Y, Dong Y et al (2024) Novel microneedle platforms for the treatment of wounds by drug delivery: a review. Colloids Surf B: Biointerfaces 233:113636

Arlov Ø, Rütsche D, Asadi Korayem M, Öztürk E, Zenobi-Wong M (2021) Engineered sulfated polysaccharides for biomedical applications. Adv Funct Mater 31:2010732

Athirasala A, Lins F, Tahayeri A, Hinds M, Smith AJ, Sedgley C et al (2017) A novel strategy to engineer pre-vascularized full-length dental pulp-like tissue constructs. Sci Rep 7:3323

Banik N, Yang SB, Kang TB, Lim JH, Park J (2021) Heparin and its derivatives: challenges and advances in therapeutic biomolecules. Int J Mol Sci 22:10524

Bertsch P, Diba M, Mooney DJ, Leeuwenburgh SCG (2022) Self-healing injectable hydrogels for tissue regeneration. Chem Rev 123:834–873

Bojorges H, López-Rubio A, Martínez-Abad A, Fabra MJ (2023) Overview of alginate extraction processes: impact on alginate molecular structure and techno-functional properties. Trends Food Sci Technol 140:104142
Buckwalter JA, Mankin HJ, Grodzinsky AJ (2005) Articular cartilage and osteoarthritis. Instr Course Lect 54:465–480
Cai Y, Chen Y, Hong X, Liu Z, Yuan W (2013) Porous microsphere and its applications. Int J Nanomedicine 8:1111–1120
Caridade SG, Monge C, Almodóvar J, Guillot R, Lavaud J, Josserand V et al (2015) Myoconductive and osteoinductive free-standing polysaccharide membranes. Acta Biomater 15:139–149
Chaisuwan W, Jantanasakulwong K, Wangtueai S, Phimolsiripol Y, Chaiyaso T, Techapun C et al (2020) Microbial exopolysaccharides for immune enhancement: fermentation, modifications and bioactivities. Food Biosci 35:100564
Chakka VP, Zhou T (2020) Carboxymethylation of polysaccharides: synthesis and bioactivities. Int J Biol Macromol 165:2425–2431
Chan GC, Chan WK, Sze DM (2009) The effects of beta-glucan on human immune and cancer cells. J Hematol Oncol 2:25
Chen G, He L, Zhang P, Zhang J, Mei X, Wang D et al (2020a) Encapsulation of green tea polyphenol nanospheres in PVA/alginate hydrogel for promoting wound healing of diabetic rats by regulating PI3K/AKT pathway. Mater Sci Eng C Mater Biol Appl 110:110686
Chen S-H, Chou P-Y, Chen Z-Y, Chuang DC-C, Hsieh S-T, Lin F-H (2020b) An electrospun nerve wrap comprising Bletilla striata polysaccharide with dual function for nerve regeneration and scar prevention. Carbohydr Polym 250:116981
Chen J, Yang J, Wang L, Zhang X, Heng BC, Wang D-A et al (2021) Modified hyaluronic acid hydrogels with chemical groups that facilitate adhesion to host tissues enhance cartilage regeneration. Bioact Mater 6:1689–1698
Chen SK, Wang X, Guo YQ, Song XX, Yin JY, Nie SP (2023) Exploring the partial degradation of polysaccharides: structure, mechanism, bioactivities, and perspectives. Compr Rev Food Sci Food Saf 22:4831–4870
Chen X, Shen M, Yu Q, Chen Y, Xie J (2024a) Recent advance in chemistry modified methods of natural polysaccharides and their applications. Trends Food Sci Technol 144:104317
Chen F, Wu P, Zhang H, Sun G (2024b) Signaling pathways triggering therapeutic hydrogels in promoting chronic wound healing. Macromol Biosci 24:e2300217
Chowhan A, Giri TK (2020) Polysaccharide as renewable responsive biopolymer for in situ gel in the delivery of drug through ocular route. Int J Biol Macromol 150:559–572
Chuaynukul K, Nagarajan M, Prodpran T, Benjakul S, Songtipya P, Songtipya L (2017) Comparative characterization of bovine and fish gelatin films fabricated by compression molding and solution casting methods. J Polym Environ 26:1239–1252
Coburn PT, Li X, Li J, Kishimoto Y, Li-Jessen NYK (2022) Progress in vocal fold regenerative biomaterials: an immunological perspective. Adv Nanobiomed Res 2:2100119
Dai K, Deng S, Yu Y, Zhu F, Wang J, Liu C (2022) Construction of developmentally inspired periosteum-like tissue for bone regeneration. Bone Res 10:1
Darling NJ, Xi W, Sideris E, Anderson AR, Pong C, Carmichael ST et al (2020) Click by click microporous annealed particle (MAP) scaffolds. Adv Healthc Mater 9:e1901391
Datta LP, Manchineella S, Govindaraju T (2020) Biomolecules-derived biomaterials. Biomaterials 230:119633
Deng H, Zhou X, Wang X, Zhang C, Ding B, Zhang Q et al (2010) Layer-by-layer structured polysaccharides film-coated cellulose nanofibrous mats for cell culture. Carbohydr Polym 80:474–479
Dinoro J, Maher M, Talebian S, Jafarkhani M, Mehrali M, Orive G et al (2019) Sulfated polysaccharide-based scaffolds for orthopaedic tissue engineering. Biomaterials 214:119214
Duceac IA, Coseri S (2022) Biopolymers and their derivatives: key components of advanced biomedical technologies. Biotechnol Adv 61:108056
El Halal SLM, Kringel DH, Zavareze ER, Dias ARG (2019) Methods for extracting cereal starches from different sources: a review. Starch – Stärke 71:1900128

Elosegui-Artola A, Gupta A, Najibi AJ, Seo BR, Garry R, Tringides CM et al (2023) Matrix viscoelasticity controls spatiotemporal tissue organization. Nat Mater 22:117–127

Feng Y, Li Q, Wu D, Niu Y, Yang C, Dong L et al (2017) A macrophage-activating, injectable hydrogel to sequester endogenous growth factors for in situ angiogenesis. Biomaterials 134:128–142

Feng Q, Li Q, Wen H, Chen J, Liang M, Huang H et al (2019) Injection and self-assembly of bioinspired stem cell-laden gelatin/hyaluronic acid hybrid microgels promote cartilage repair in vivo. Adv Funct Mater 29:1906690

Feng Y, Gao H-L, Wu D, Weng Y-T, Wang Z-Y, Yu S-H et al (2021) Biomimetic lamellar chitosan scaffold for soft gingival tissue regeneration. Adv Funct Mater 31:2105348

Forget A, Christensen J, Lüdeke S, Kohler E, Tobias S, Matloubi M et al (2013) Polysaccharide hydrogels with tunable stiffness and provasculogenic properties via α-helix to β-sheet switch in secondary structure. Proc Natl Acad Sci 110:12887–12892

Francis RM, DeForest CA (2023) 4D biochemical photocustomization of hydrogel scaffolds for biomimetic tissue engineering. Acc Mater Res 4:704–715

Frasheri I, Tsakiridou ND, Hickel R, Folwaczny M (2023) The molecular weight of hyaluronic acid influences metabolic activity and osteogenic differentiation of periodontal ligament cells. Clin Oral Investig 27:5905–5911

Gabriel T, Belete A, Syrowatka F, Neubert RHH, Gebre-Mariam T (2020) Extraction and characterization of celluloses from various plant byproducts. Int J Biol Macromol 158:1248–1258

Garcia MAVT, Garcia CF, Faraco AAG (2020) Pharmaceutical and biomedical applications of native and modified starch: a review. Starch – Stärke 72:1900270

Gericke M, Amaral AJR, Budtova T, De Wever P, Groth T, Heinze T et al (2024) The European Polysaccharide Network of Excellence (EPNOE) research roadmap 2040: advanced strategies for exploiting the vast potential of polysaccharides as renewable bioresources. Carbohydr Polym 326:121633

Gholamali I (2021) Stimuli-responsive polysaccharide hydrogels for biomedical applications: a review. Regen Eng Transl Med 7:91–114

Graciela C-Q, José Juan E-C, Gieraldin C-L, Xóchitl Alejandra P-M, Gabriel A-Á (2023) Hyaluronic acid: extraction methods, sources and applications. Polymers 15:3473

Guru PR, Kar RK, Nayak AK, Mohapatra S (2023) A comprehensive review on pharmaceutical uses of plant-derived biopolysaccharides. Int J Biol Macromol 233:123454

Han F, Wang J, Ding L, Hu Y, Li W, Yuan Z et al (2020) Tissue engineering and regenerative medicine: achievements, future, and sustainability in Asia. Front Bioeng Biotechnol 8:83

Hanjaya-Putra D, Bose V, Shen YI, Yee J, Khetan S, Fox-Talbot K et al (2011) Controlled activation of morphogenesis to generate a functional human microvasculature in a synthetic matrix. Blood 118:804–815

Hao Y, Zhao W, Zhang L, Zeng X, Sun Z, Zhang D et al (2020) Bio-multifunctional alginate/chitosan/fucoidan sponges with enhanced angiogenesis and hair follicle regeneration for promoting full-thickness wound healing. Mater Des 193:108863

Hickey RJ, Pelling AE (2019) Cellulose biomaterials for tissue engineering. Front Bioeng Biotechnol 7:45

Hisham F, Maziati Akmal MH, Ahmad F, Ahmad K, Samat N (2024) Biopolymer chitosan: potential sources, extraction methods, and emerging applications. Ain Shams Eng J 15:102424

Hu X, Wang Y, Zhang L, Xu M (2021) Simple ultrasonic-assisted approach to prepare polysaccharide-based aerogel for cell research and histocompatibility study. Int J Biol Macromol 188:411–420

Huang C, Ogawa R (2020) The vascular involvement in soft tissue fibrosis-lessons learned from pathological scarring. Int J Mol Sci 21:2542

Hussain Z, Thu HE, Katas H, Bukhari SNA (2017) Hyaluronic acid-based biomaterials: a versatile and smart approach to tissue regeneration and treating traumatic, surgical, and chronic wounds. Polym Rev 57:594–630

Hussey GS, Dziki JL, Badylak SF (2018) Extracellular matrix-based materials for regenerative medicine. Nat Rev Mater 3:159–173

Iijima K, Tsuji Y, Kuriki I, Kakimoto A, Nikaido Y, Ninomiya R et al (2017) Control of cell adhesion and proliferation utilizing polysaccharide composite film scaffolds. Colloids Surf B: Biointerfaces 160:228–237
Jafari H, Ghaffari-Bohlouli P, Niknezhad SV, Abedi A, Izadifar Z, Mohammadinejad R et al (2022) Tannic acid: a versatile polyphenol for design of biomedical hydrogels. J Mater Chem B 10:5873–5912
Jamee R, Araf Y, Naser IB, Promon SK (2021) The promising rise of bioprinting in revolutionalizing medical science: advances and possibilities. Regen Ther 18:133–145
Jansook P, Ogawa N, Loftsson T (2018) Cyclodextrins: structure, physicochemical properties and pharmaceutical applications. Int J Pharm 535:272–284
Jayasinghe SN (2022) Unspooling the history of cell electrospinning. Matter 5:4–7
Jia Z, Chen L, Gu D, Li X, Wen T, Li W (2024) Lentinan-loaded GelMA hydrogel accelerates diabetic wound healing through enhanced angiogenesis and immune microenvironment modulation. Int J Biol Macromol 264:130716
Jin M, Shi J, Zhu W, Yao H, Wang DA (2021) Polysaccharide-based biomaterials in tissue engineering: a review. Tissue Eng Part B Rev 27:604–626
Khetan S, Guvendiren M, Legant WR, Cohen DM, Chen CS, Burdick JA (2013) Degradation-mediated cellular traction directs stem cell fate in covalently crosslinked three-dimensional hydrogels. Nat Mater 12:458–465
Laffargue T, Moulis C, Remaud-Siméon M (2023) Phosphorylated polysaccharides: applications, natural abundance, and new-to-nature structures generated by chemical and enzymatic functionalisation. Biotechnol Adv 65:108140
Lee KY, Mooney DJ (2012) Alginate: properties and biomedical applications. Prog Polym Sci 37:106–126
Lee KY, Jeong L, Kang YO, Lee SJ, Park WH (2009) Electrospinning of polysaccharides for regenerative medicine. Adv Drug Deliv Rev 61:1020–1032
Lee BM, Park SJ, Noh I, Kim C-H (2021) The effects of the molecular weights of hyaluronic acid on the immune responses. Biomater Res 25:27
Li L, Stiadle JM, Lau HK, Zerdoum AB, Jia X, Thibeault SL et al (2016) Tissue engineering-based therapeutic strategies for vocal fold repair and regeneration. Biomaterials 108:91–110
Li D-Q, Li J, Dong H-L, Li X, Zhang J-Q, Ramaswamy S et al (2021) Pectin in biomedical and drug delivery applications: a review. Int J Biol Macromol 185:49–65
Li H, Wang Y, Zhao P, Guo L, Huang L, Li X et al (2023a) Naturally and chemically acetylated polysaccharides: structural characteristics, synthesis, activities, and applications in the delivery system: a review. Carbohydr Polym 313:120746
Li Q, Yu H, Zhao F, Cao C, Wu T, Fan Y et al (2023b) 3D printing of microenvironment-specific bioinspired and exosome-reinforced hydrogel scaffolds for efficient cartilage and subchondral bone regeneration. Adv Sci (Weinh) 10:e2303650
Li Z, Li Q, Cao W, Zhan J, He Y, Xing X et al (2024) A strongly robust chitosan-based programmed control functional hydrogel improved mitochondrial function and pro-vascularization for adaptive repair of myocardial infarction. Adv Funct Mater 34:2312631
Lin H, Han R, Wu W (2024) Glucans and applications in drug delivery. Carbohydr Polym 332:121904
Liu Z, Ye L, Xi J, Wang J, Feng Z-G (2021a) Cyclodextrin polymers: structure, synthesis, and use as drug carriers. Prog Polym Sci 118:101408
Liu Y, Wang M, Luo Y, Liang Q, Yu Y, Chen F et al (2021b) Enhancing stem cell therapy for cartilage repair in osteoarthritis – a hydrogel focused approach. Gels 7:263
Liu K, Guo L, Chen X, Liu L, Gao C (2023a) Microbial synthesis of glycosaminoglycans and their oligosaccharides. Trends Microbiol 31:369–383
Liu Q, Chen M, Gu P, Tong L, Wang P, Zhu J et al (2023b) Covalently grafted biomimetic matrix reconstructs the regenerative microenvironment of the porous gradient polycaprolactone scaffold to accelerate bone remodeling. Small 19:2206960
Longinotti C (2014) The use of hyaluronic acid based dressings to treat burns: a review. Burns Trauma 2:162–168

Lovett M, Lee K, Edwards A, Kaplan DL (2009) Vascularization strategies for tissue engineering. Tissue Eng Part B Rev 15:353–370

Luanda A, Badalamoole V (2023) Past, present and future of biomedical applications of dextran-based hydrogels: a review. Int J Biol Macromol 228:794–807

Lucenti L, Testa G, Caldaci A, Sammartino F, Cicio C, Ilardo M et al (2024) Preoperative risk factors for periprosthetic joint infection: a narrative review of the literature. Healthcare (Basel) 12:666

Luo M, Zhang X, Wu J, Zhao J (2021) Modifications of polysaccharide-based biomaterials under structure-property relationship for biomedical applications. Carbohydr Polym 266:118097

Luo Z, Wang Y, Li J, Wang J, Yu Y, Zhao Y (2023) Tailoring hyaluronic acid hydrogels for biomedical applications. Adv Funct Mater 33:2306554

Lyu S, Dong Z, Xu X, Bei HP, Yuen HY, James Cheung CW et al (2023) Going below and beyond the surface: microneedle structure, materials, drugs, fabrication, and applications for wound healing and tissue regeneration. Bioact Mater 27:303–326

Ma Y, Morozova SM, Kumacheva E (2024) From nature-sourced polysaccharide particles to advanced functional materials. Adv Mater 36:e2312707

Magalhães S, Fernandes C, Pedrosa JFS, Alves L, Medronho B, Ferreira PJT et al (2023) Eco-friendly methods for extraction and modification of cellulose: an overview. Polymers 15:3138

Mahmoud YAG, El-Naggar ME, Abdel-Megeed A, El-Newehy M (2021) Recent advancements in microbial polysaccharides: synthesis and applications. Polymers 13:4136

Mantha S, Pillai S, Khayambashi P, Upadhyay A, Zhang Y, Tao O et al (2019) Smart hydrogels in tissue engineering and regenerative medicine. Materials 12:3323

Mason C, Dunnill P (2008) A brief definition of regenerative medicine. Regen Med 3:1–5

Min L, Pan H, Chen S, Wang C, Wang N, Zhang J et al (2019) Recent progress in bio-inspired electrospun materials. Compos Commun 11:12–20

Mohammed ASA, Naveed M, Jost N (2021) Polysaccharides; classification, chemical properties, and future perspective applications in fields of pharmacology and biological medicine (a review of current applications and upcoming potentialities). J Polym Environ 29:2359–2371

Monteiro N, Yelick PC (2017) Advances and perspectives in tooth tissue engineering. J Tissue Eng Regen Med 11:2443–2461

Moreira Teixeira LS, Bijl S, Pully VV, Otto C, Jin R, Feijen J et al (2012) Self-attaching and cell-attracting in-situ forming dextran-tyramine conjugates hydrogels for arthroscopic cartilage repair. Biomaterials 33:3164–3174

Muhammad M, Willems C, Rodríguez-Fernández J, Gallego-Ferrer G, Groth T (2020) Synthesis and characterization of oxidized polysaccharides for in situ forming hydrogels. Biomolecules 10:1185

Muir VG, Qazi TH, Weintraub S, Torres Maldonado BO, Arratia PE, Burdick JA (2022) Sticking together: injectable granular hydrogels with increased functionality via dynamic covalent inter-particle crosslinking. Small 18:e2201115

Muir VG, Weintraub S, Dhand AP, Fallahi H, Han L, Burdick JA (2023) Influence of microgel and interstitial matrix compositions on granular hydrogel composite properties. Adv Sci 10:2206117

Mukherjee S, Jana S, Khawas S, Kicuntod J, Marschall M, Ray B et al (2022) Synthesis, molecular features and biological activities of modified plant polysaccharides. Carbohydr Polym 289:119299

Murphy EJ, Rezoagli E, Collins C, Saha SK, Major I, Murray P (2023) Sustainable production and pharmaceutical applications of β-glucan from microbial sources. Microbiol Res 274:127424

Nakashima M, Iohara K, Murakami M, Nakamura H, Sato Y, Ariji Y et al (2017) Pulp regeneration by transplantation of dental pulp stem cells in pulpitis: a pilot clinical study. Stem Cell Res Ther 8:61

Nie J, Pei B, Wang Z, Hu Q (2019) Construction of ordered structure in polysaccharide hydrogel: a review. Carbohydr Polym 205:225–235

O'Brien FJ (2011) Biomaterials & scaffolds for tissue engineering. Mater Today 14:88–95

Otero P, Carpena M, Garcia-Oliveira P, Echave J, Soria-Lopez A, Garcia-Perez P et al (2021) Seaweed polysaccharides: emerging extraction technologies, chemical modifications and bioactive properties. Crit Rev Food Sci Nutr 63:1901–1929

Ouyang L, Chen B, Liu X, Wang D, Li Y, Liao Y et al (2023) Puerarin@Chitosan composite for infected bone repair through mimicking the bio-functions of antimicrobial peptides. Bioact Mater 21:520–530

Palanisamy CP, Cui B, Zhang H, Gunasekaran VP, Ariyo AL, Jayaraman S et al (2022) A critical review on starch-based electrospun nanofibrous scaffolds for wound healing application. Int J Biol Macromol 222:1852–1860

Pfister BJ, Gordon T, Loverde JR, Kochar AS, Mackinnon SE, Cullen DK (2011) Biomedical engineering strategies for peripheral nerve repair: surgical applications, state of the art, and future challenges. Crit Rev Biomed Eng 39:81–124

Place ES, Evans ND, Stevens MM (2009) Complexity in biomaterials for tissue engineering. Nat Mater 8:457–470

Prechtl RM, Janssen D, Behr J, Ludwig C, Kuster B, Vogel RF et al (2018) Sucrose-induced proteomic response and carbohydrate utilization of Lactobacillus sakei TMW 1.411 during dextran formation. Front Microbiol 9:2796

Pruett LJ, Jenkins CH, Singh NS, Catallo KJ, Griffin DR (2021) Heparin microislands in microporous annealed particle scaffolds for accelerated diabetic wound healing. Adv Funct Mater 31:2104337

Qazi TH, Wu J, Muir VG, Weintraub S, Gullbrand SE, Lee D et al (2022) Anisotropic rod-shaped particles influence injectable granular hydrogel properties and cell invasion. Adv Mater 34:2109194

Qin J, Chen F, Wu P, Sun G (2022) Recent advances in bioengineered scaffolds for cutaneous wound healing. Front Bioeng Biotechnol 10:841583

Rape AD, Zibinsky M, Murthy N, Kumar S (2015) A synthetic hydrogel for the high-throughput study of cell–ECM interactions. Nat Commun 6:8129

Sadeghianmaryan A, Ahmadian N, Wheatley S, Alizadeh Sardroud H, Nasrollah SAS, Naseri E et al (2024) Advancements in 3D-printable polysaccharides, proteins, and synthetic polymers for wound dressing and skin scaffolding – a review. Int J Biol Macromol 266:131207

Saito T, Kotani T, Suzuka T, Matsuda S, Takeuchi T, Sato T (2022) Adipose-derived stem/stromal cells with heparin-enhanced anti-inflammatory and antifibrotic effects mitigate induced pulmonary fibrosis in mice. Biochem Biophys Res Commun 629:135–141

Schmidt CE, Leach JB (2003) Neural tissue engineering: strategies for repair and regeneration. Annu Rev Biomed Eng 5:293–347

Schnellmann R, Ntekoumes D, Choudhury MI, Sun S, Wei Z, Gerecht S (2022) Stiffening matrix induces age-mediated microvascular phenotype through increased cell contractility and destabilization of adherens junctions. Adv Sci (Weinh) 9:e2201483

Sharma R, Kuche K, Thakor P, Bhavana V, Srivastava S, Mehra NK et al (2022) Chondroitin sulfate: emerging biomaterial for biopharmaceutical purpose and tissue engineering. Carbohydr Polym 286:119305

Shen Q, Guo Y, Wang K, Zhang C, Ma Y (2023) A review of chondroitin sulfate's preparation, properties, functions, and applications. Molecules 28:7093

Shokrani H, Shokrani A, Seidi F, Mashayekhi M, Kar S, Nedeljkovic D et al (2023) Polysaccharide-based biomaterials in a journey from 3D to 4D printing. Bioeng Transl Med 8:e10503

Silva LP, Pirraco RP, Santos TC, Novoa-Carballal R, Cerqueira MT, Reis RL et al (2016) Neovascularization induced by the hyaluronic acid-based spongy-like hydrogels degradation products. ACS Appl Mater Interfaces 8:33464–33474

Sims IM, Carnachan SM, Bell TJ, Hinkley SFR (2018) Methylation analysis of polysaccharides: technical advice. Carbohydr Polym 188:1–7

Singh RS, Kaur N, Rana V, Kennedy JF (2016) Recent insights on applications of pullulan in tissue engineering. Carbohydr Polym 153:455–462

Slaughter BV, Khurshid SS, Fisher OZ, Khademhosseini A, Peppas NA (2009) Hydrogels in regenerative medicine. Adv Mater 21:3307–3329

Song HHG, Rumma RT, Ozaki CK, Edelman ER, Chen CS (2018) Vascular tissue engineering: progress, challenges, and clinical promise. Cell Stem Cell 22:340–354

Su C, Chen Y, Tian S, Lu C, Lv Q (2022) Research progress on emerging polysaccharide materials applied in tissue engineering. Polymers (Basel) 14:3268

Su W, Chang Z, Yuyu E, Feng Y, Yao X, Wang M et al (2024) Electrospinning and electrospun polysaccharide-based nanofiber membranes: a review. Int J Biol Macromol 263:130335

Sugiaman VK, Jeffrey, Naliani S, Pranata N, Djuanda R, Saputri RI (2023) Polymeric scaffolds used in dental pulp regeneration by tissue engineering approach. Polymers (Basel) 15:1082

Sun G (2017) Pro-regenerative hydrogel restores scarless skin during cutaneous wound healing. Adv Healthc Mater 6:1700659

Sun G, Mao JJ (2012) Engineering dextran-based scaffolds for drug delivery and tissue repair. Nanomedicine 7:1771–1784

Sun G, Zhang XZ, Chu CC (2007) Formulation and characterization of chitosan-based hydrogel films having both temperature and pH sensitivity. J Mater Sci Mater Med 18:1563–1577

Sun G, Shen Y-I, Ho CC, Kusuma S, Gerecht S (2010) Functional groups affect physical and biological properties of dextran-based hydrogels. J Biomed Mater Res A 93A:1080–1090

Sun G, Zhang X, Shen Y-I, Sebastian R, Dickinson LE, Fox-Talbot K et al (2011a) Dextran hydrogel scaffolds enhance angiogenic responses and promote complete skin regeneration during burn wound healing. Proc Natl Acad Sci 108:20976–20981

Sun G, Shen Y-I, Kusuma S, Fox-Talbot K, Steenbergen CJ, Gerecht S (2011b) Functional neovascularization of biodegradable dextran hydrogels with multiple angiogenic growth factors. Biomaterials 32:95–106

Sun G, Shen Y-I, Harmon JW (2018) Engineering pro-regenerative hydrogels for scarless wound healing. Adv Healthc Mater 7:1800016

Taylor DL, in het Panhuis M (2016) Self-healing hydrogels. Adv Mater 28:9060–9093

Temple H, Phyo P, Yang W, Lyczakowski JJ, Echevarría-Poza A, Yakunin I et al (2022) Golgi-localized putative S-adenosyl methionine transporters required for plant cell wall polysaccharide methylation. Nat Plants 8:656–669

Truckenmüller R, Giselbrecht S, Rivron N, Gottwald E, Saile V, van den Berg A et al (2011) Thermoforming of film-based biomedical microdevices. Adv Mater 23:1311–1329

Tudu M, Samanta A (2023) Natural polysaccharides: chemical properties and application in pharmaceutical formulations. Eur Polym J 184:111801

Verlee A, Mincke S, Stevens CV (2017) Recent developments in antibacterial and antifungal chitosan and its derivatives. Carbohydr Polym 164:268–283

Vettori MHPB, Franchetti SMM, Contiero J (2012) Structural characterization of a new dextran with a low degree of branching produced by Leuconostoc mesenteroides FT045B dextransucrase. Carbohydr Polym 88:1440–1444

Vijayavenkataraman S, Yan W-C, Lu WF, Wang C-H, Fuh JYH (2018) 3D bioprinting of tissues and organs for regenerative medicine. Adv Drug Deliv Rev 132:296–332

Wang D-A, Varghese S, Sharma B, Strehin I, Fermanian S, Gorham J et al (2007) Multifunctional chondroitin sulphate for cartilage tissue–biomaterial integration. Nat Mater 6:385–392

Wang W, Xue C, Mao X (2020) Chitosan: structural modification, biological activity and application. Int J Biol Macromol 164:4532–4546

Wang S, Wei Y, Wang Y, Cheng Y (2023a) Cyclodextrin regulated natural polysaccharide hydrogels for biomedical applications – a review. Carbohydr Polym 313:120760

Wang Z, Zhou X, Shu Z, Zheng Y, Hu X, Zhang P et al (2023b) Regulation strategy, bioactivity, and physical property of plant and microbial polysaccharides based on molecular weight. Int J Biol Macromol 244:125360

Wang H, Li H, Hou Y, Zhang P, Tan M (2023c) Plant polysaccharides: sources, structures, and antidiabetic effects. Curr Opin Food Sci 51:101013

Wang S, Xing X, Peng W, Huang C, Du Y, Yang H et al (2023d) Fabrication of an exosome-loaded thermosensitive chitin-based hydrogel for dental pulp regeneration. J Mater Chem B 11:1580–1590

Wang X, Wang Z, Shen M, Yi C, Yu Q, Chen X et al (2024) Acetylated polysaccharides: synthesis, physicochemical properties, bioactivities, and food applications. Crit Rev Food Sci Nutr 64:4849–4864
Wolf TG, Cagetti MG, Fisher JM, Seeberger GK, Campus G (2021) Non-communicable diseases and oral health: an overview. Front Oral Health 2:725460
Wu P, Liang Y, Sun G (2021a) Engineering immune-responsive biomaterials for skin regeneration. Biomater Transl 2:61–71
Wu P, Xi X, Li R, Sun G (2021b) Engineering polysaccharides for tissue repair and regeneration. Macromol Biosci 21:e2100141
Wu M, Liu H, Li D, Zhu Y, Wu P, Chen Z et al (2024) Smart-responsive multifunctional therapeutic system for improved regenerative microenvironment and accelerated bone regeneration via mild photothermal therapy. Adv Sci (Weinh) 11:e2304641
Xia S, Zhai Y, Wang X, Fan Q, Dong X, Chen M et al (2021) Phosphorylation of polysaccharides: a review on the synthesis and bioactivities. Int J Biol Macromol 184:946–954
Xie L, Shen M, Wang Z, Xie J (2021) Structure, function and food applications of carboxymethylated polysaccharides: a comprehensive review. Trends Food Sci Technol 118:539–557
Xu Y, Wu YJ, Sun PL, Zhang FM, Linhardt RJ, Zhang AQ (2019) Chemically modified polysaccharides: synthesis, characterization, structure activity relationships of action. Int J Biol Macromol 132:970–977
Xu Y, Saiding Q, Zhou X, Wang J, Cui W, Chen X (2024) Electrospun fiber-based immune engineering in regenerative medicine. Smart Med 3:e20230034
Xuan H, Wu S, Jin Y, Wei S, Xiong F, Xue Y et al (2023) A bioinspired self-healing conductive hydrogel promoting peripheral nerve regeneration. Adv Sci (Weinh) 10:e2302519
Xue W, Shi W, Kuss M, Kong Y, Alimi OA, Wang H et al (2023) A dual-network nerve adhesive with enhanced adhesion strength promotes transected peripheral nerve repair. Adv Funct Mater 33:2209971
Yang W, Huang G (2021) Extraction methods and activities of natural glucans. Trends Food Sci Technol 112:50–57
Yang C, Sheng T, Hou W, Zhang J, Cheng L, Wang H et al (2022a) Glucose-responsive microneedle patch for closed-loop dual-hormone delivery in mice and pigs. Sci Adv 8:eadd3197
Yang Z, Liu W, Liu H, Li R, Chang L, Kan S et al (2022b) The applications of polysaccharides in dentistry. Front Bioeng Biotechnol 10:970041
Yang B, Yang C, Liu R, Sui W, Zhu Q, Jin Y et al (2024) The relationship between preparation and biological activities of animal-derived polysaccharides: a comprehensive review. Foods 13:173
Younes I, Rinaudo M (2015) Chitin and chitosan preparation from marine sources. Structure, properties and applications. Mar Drugs 13:1133–1174
Yu C, Abbott P (2007) An overview of the dental pulp: its functions and responses to injury. Aust Dent J 52:S4–S6
Yu Y, Shen M, Song Q, Xie J (2018) Biological activities and pharmaceutical applications of polysaccharide from natural resources: a review. Carbohydr Polym 183:91–101
Yuan Y, Chesnutt BM, Haggard WO, Bumgardner JD (2011) Deacetylation of chitosan: material characterization and in vitro evaluation via albumin adsorption and pre-osteoblastic cell cultures. Materials (Basel) 4:1399–1416
Zhang B, Song J (2018) 3D-printed biomaterials for guided tissue regeneration. Small Methods 2:1700306
Zhang X, Liu T, Wang X, Zhou L, Qi J, An S (2022) Structural characterization, antioxidant activity and anti-inflammatory of the phosphorylated polysaccharide from Pholiota nameko. Front Nutr 9:976552
Zhang Y, Xu Y, Kong H, Zhang J, Chan HF, Wang J et al (2023) Microneedle system for tissue engineering and regenerative medicine. Exploration 3:20210170
Zhang H, Li Y, Fu Y, Jiao H, Wang X, Wang Q et al (2024) A structure-functionality insight into the bioactivity of microbial polysaccharides toward biomedical applications: a review. Carbohydr Polym 335:122078

Zhao T, Yang M, Ma L, Liu X, Ding Q, Chai G et al (2023) Structural modification and biological activity of polysaccharides. Molecules 28:5416

Zhao Y, Zheng Z, Yu C-Y, Wei H (2024) Engineered cyclodextrin-based supramolecular hydrogels for biomedical applications. J Mater Chem B 12:39–63

Zhou S, Huang G (2021) Preparation, structure and activity of polysaccharide phosphate esters. Biomed Pharmacother 144:112332

Zhu T, Mao J, Cheng Y, Liu H, Lv L, Ge M et al (2019) Recent progress of polysaccharide-based hydrogel interfaces for wound healing and tissue engineering. Adv Mater Interfaces 6:1900761

Zou C-Y, Hu J-J, Lu D, Li Q-J, Jiang Y-L, Wang R et al (2023) A self-fused hydrogel for the treatment of glottic insufficiency through outstanding durability, extracellular matrix-inducing bioactivity and function preservation. Bioact Mater 24:54–68

# Weaving a Green and Healthy Future: Silk-Based Biomaterials for Sustainability

8

Naresh Kasoju, K. C. Sumitha, K. S. Siju, A. M. Varsha, Piyush Mohapatra, and K. Sreelakshmi

**Abstract**

Sourced from silkworms, silk-based biomaterials offer a continuous and replenishable supply, promoting resource sustainability. Unlike the energy-intensive processing associated with synthetic materials, the production of silk generally involves lower environmental impact, encompassing reduced water usage and chemical inputs. Moreover, one of its key attributes lies in biodegradability, addressing concerns associated with the persistent environmental footprint of non-biodegradable materials. Silk's versatility enhances its significance as a sustainable biomaterial. Its applications span a spectrum of industries, including medicine, textiles, and beyond. In the medical field, silk's biocompatibility and ability to support tissue regeneration make it a valuable component in various implants, paving the way for advancements in regenerative medicine. Over time, its extraordinary features attracted the attention of many and its applications in drug delivery, biosensing and other allied biomedical fields are expanding. A comprehensive overview of how silk addresses sustainability issues in biomaterials and healthcare towards achieving sustainable development goals is presented herein.

**Keywords**

Sustainable development goals · Healthcare · Biomaterials · Silk fibroin · Biomedical applications

---

N. Kasoju (✉) · K. C. Sumitha · K. S. Siju · A. M. Varsha · P. Mohapatra · K. Sreelakshmi
Division of Tissue Culture, Department of Applied Biology, Biomedical Technology Wing, Sree Chitra Tirunal Institute for Medical Sciences and Technology, Thiruvananthapuram, Kerala, India
e-mail: naresh.kasoju@sctimst.ac.in



J. M. Oliveira et al. (eds.), *Sustainable Scaffolds-based Strategies in Tissue Engineering and Regenerative Medicine*, Biomaterials, Bioengineering and Sustainability 8, https://doi.org/10.1007/978-3-031-96274-5_8

## 8.1 Introduction

Sustainable Development Goals (SDGs) serve as a global blueprint for a more inclusive, equitable, and sustainable future. Within this framework, sustainable biomaterials play a pivotal role in achieving environmental and social objectives. By prioritizing materials derived from renewable sources with minimal ecological impact, sustainable biomaterials contribute directly to SDGs related to responsible consumption, climate action, and life below water and on land. Integrating these biomaterials into various industries fosters innovation reduces environmental harm, and aligns with the overarching mission of the SDGs, ultimately promoting a harmonious coexistence between human activities and the planet (Poland and Dooris 2010). The significance of sustainable biomaterials in the contemporary era is paramount, as society endeavors to shift towards environmentally conscious practices. Among these biomaterials, silk stands out as a unique and eco-friendly option with profound implications across various industries. Derived from the natural protein fibers produced by silkworms during the sericulture process, silk possesses inherent qualities that align seamlessly with sustainability objectives. As global attention shifts towards responsible and sustainable practices, the utilization of silk as a biomaterial represents a conscientious choice (Kasoju and Bora 2012a). Its eco-friendly properties, combined with adaptability and performance, position silk as a catalyst for positive environmental change across diverse sectors. Embracing silk as an eco-friendly biomaterial is not merely an option; it signifies a strategic commitment to a greener and more sustainable future.

Herein, we provide an overview of biomaterials, including definition, classes, and a details about polymeric biomaterials with special attention to natural polymeric biomaterials. Then, we discuss silk and their unique properties that makes it an excellent biomaterial. An outline to the international scenario and a special mention to the Indian scenario on the silk-based biomaterials research is also given. The applications of silk based biomaterials are also discussed. Subsequently, we explore how silk addresses the sustainability issues in biomaterials with brief notes on SDGs, sustainability in biomedical engineering field, natural polymers as sustainable biomaterials. We also compare silk-based biomedical materials with traditional ones in the context of SDGs. Moreover, the current challenges in the adoption of silk based biomaterials, and the emerging research and innovations to overcome those challenges, are respectively discussed.

## 8.2 Overview of Biomaterials

Biomaterials constitute a cornerstone in the intersection of science, medicine, and engineering, offering transformative solutions to address complex challenges in healthcare. These materials, designed to interact with biological systems, play a pivotal role in the development of medical devices, implants, drug delivery systems, and tissue engineering constructs. As we traverse the diverse landscape of biomaterials, it becomes evident that these materials are integral to the fabric of modern healthcare. Biomaterials encompass a diverse array of materials classified based on their composition, origin, and intended applications. The primary classes include metals, ceramics, polymers, and composites, each tailored to meet specific needs in the realm of medical science. (a) *Metallic biomaterials*, such as titanium and stainless steel, feature prominently in orthopaedics and dental applications owing to their robust mechanical properties, corrosion resistance, and biocompatibility (Williams 2005). Implants made from these materials provide structural support and withstand biomechanical forces in load-bearing scenarios. (b) *Ceramics biomaterials*, exemplified by hydroxyapatite and tricalcium phosphate, mimic the mineral composition of natural bone. Commonly used in bone grafts and dental implants, ceramics exhibit bioactivity, promoting integration with surrounding tissues and facilitating bone regeneration (Hench and Polak 2002). (c) *Polymeric biomaterials*, characterized by their long-chain molecules, offer versatility and tunable properties. Synthetic polymers like polyethylene and polyvinyl chloride find applications in medical devices, while natural polymers contribute to the burgeoning field of tissue engineering and drug delivery (Langer and Tirrell 2004). (d) *Composite biomaterials* merge different material classes to optimize properties for specific applications. Combining ceramics with polymers, for instance, enhances mechanical strength, making composites suitable for load-bearing implants (Sandler et al. 1999).

Polymeric biomaterials, a significant subset of biomaterials, boast a diverse range of applications due to their tunable mechanical properties, biocompatibility, and controlled release capabilities. They are further categorized into synthetic and natural polymers. (a) *Synthetic Polymers:* Engineered for specific applications, synthetic polymers play a pivotal role in modern biomaterial design. Polyethylene glycol (PEG) and poly(lactic-co-glycolic acid) (PLGA) are exemplary synthetic polymers used in drug delivery due to their controlled release kinetics (Torchilin 2011). (b) *Natural Polymers:* Derived from biological sources, natural polymers offer inherent biocompatibility and biodegradability. Collagen, chitosan, silk fibroin, and hyaluronic acid are examples of natural polymers contributing to the development of sustainable and biomimetic biomaterials (Place et al. 2009). Particularly, the natural polymers, sourced from biological entities, stand out for their biomimetic properties, making them invaluable in diverse biomedical applications. Collagen, the most abundant protein in the human body, serves as a structural scaffold in tissues. Its biocompatibility and ability to mimic the extracellular matrix make collagen a preferred choice in tissue engineering and wound healing (Lee et al. 2001). Derived from chitin, chitosan possesses antimicrobial properties and is biocompatible. Its applications range from wound dressings to drug delivery systems (Jayakumar et al. 2010). Extracted from silkworm silk, silk fibroin combines remarkable mechanical strength with biocompatibility. This makes it ideal for wound healing, drug delivery, and tissue engineering (Kundu et al. 2013). Naturally present in the body, hyaluronic acid is known for its hydrating properties. It is employed in dermal fillers, ophthalmic solutions, and drug delivery systems (Burdick and Prestwich 2011). Natural polymers, inspired by the elegance of nature, represent a paradigm shift towards sustainable and biocompatible biomaterials.

## 8.3 Silk Based Biomaterials

Silks, originating from insects like silkworms (*Bombyx mori*) and spiders, represent nature's ingenuity in material design. The primary structural component of silk fibers is fibroin, a protein that imparts exceptional strength and flexibility (Kasoju and Bora 2012a). Silk, a natural protein fiber produced by certain arthropods, has captivated human civilization for centuries with its luxurious texture and versatility. Beyond its traditional use in textiles, silk has emerged as a remarkable biomaterial, finding applications in a diverse range of medical and technological fields (Hakimi et al. 2007; Vepari and Kaplan 2007). Silkworm silk, in particular, is renowned for its fine texture and has been cultivated for centuries to produce textiles. Spider silk, on the other hand, is known for its extraordinary tensile strength, surpassing many synthetic materials. The molecular structure and mechanical properties of silk make it an intriguing candidate for biomaterial applications (Sun et al. 2021).

Silk possesses a remarkable set of properties that render it distinct and highly suitable for various biomedical applications (Fig. 8.1). (1) *Biocompatibility:* Silk is inherently biocompatible, meaning it elicits minimal immune response when introduced into living organisms. This property is crucial for biomaterials intended for medical use, ensuring compatibility with biological systems (DeBari et al. 2021). (2) *Mechanical strength:* Silk fibers exhibit impressive mechanical strength, making them comparable to or even surpassing certain synthetic materials. The combination of tensile strength and flexibility makes silk an excellent choice for applications requiring durability and resilience (Shao and Vollrath 2002). (3) *Biodegradability:* Silk is biodegradable, meaning it can be naturally broken down by biological processes over time. This feature is advantageous for medical implants or devices, as they can be designed to degrade gradually, minimizing long-term foreign body reactions (Cao and Wang 2009). (4) *Versatility:* Silk biomaterials are highly versatile and can be engineered into various forms, including films, scaffolds, hydrogels, and

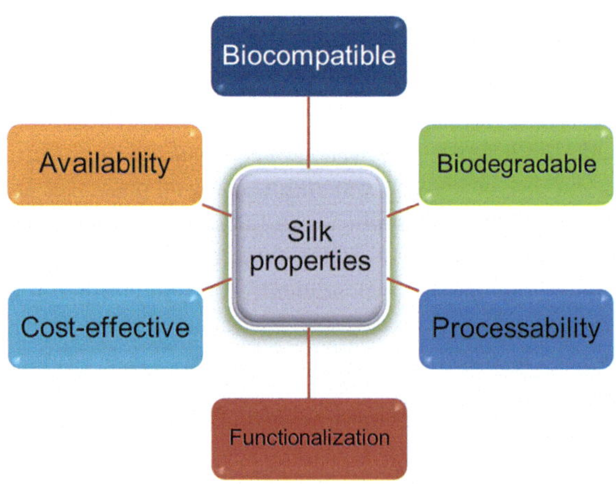


**Fig. 8.1** Schematic representation of unique properties of silk biomaterials that make it attractive for numerous biomedical applications

fibers. This versatility allows for tailored designs to suit specific biomedical applications, ranging from tissue engineering to drug delivery (Qi et al. 2017). (5) *Other special features:* Silk can promote biomineralization, facilitating the deposition of minerals like calcium phosphate. This property is crucial for applications in bone tissue engineering, where the integration of biomaterials with the natural bone structure is essential (Neubauer et al. 2021).

***International Scenario*** The international research landscape on silk-based biomaterials is marked by extensive exploration, with scientists and researchers investigating diverse applications and pushing the boundaries of silk's potential. Key areas of focus include (a) *Tissue engineering:* Silk scaffolds have garnered significant attention for their use in tissue engineering (Kundu et al. 2013). Researchers are exploring silk's ability to support cell growth and tissue regeneration, making it a promising material for constructing artificial tissues and organs. (b) *Drug delivery systems:* The controlled release properties of silk make it an ideal candidate for drug delivery systems (Tsioris et al. 2012). Researchers are developing silk-based carriers for targeted and sustained drug delivery, with potential applications in cancer therapy, among others. (c) *Biosensors and biodevices:* Silk's compatibility with biological systems has led to its use in biosensors and biodevices (Wang et al. 2021). These applications leverage silk's unique properties for sensing and monitoring biological signals. Collaborations between research institutions and industry players are driving innovations in silk biomaterials across the globe.

***Indian Scenario*** India has a rich tradition of sericulture, particularly in states like Karnataka, West Bengal, and Andhra Pradesh. This legacy provides a strong foundation for research into silk-based biomaterials, as the country has a well-established sericulture industry. In India, research on silk-based biomaterials is gaining momentum, driven by the country's rich heritage in silk production and the growing recognition of silk's potential in biomedical applications. Key areas of focus in the Indian research landscape include: (a) *Silk processing and engineering:* Indian researchers are developing innovative methods for silk processing and engineering to tailor silk biomaterials for specific biomedical applications (Biswal et al. 2022). (b) *Traditional medicine integration:* India's traditional medicine systems, such as Ayurveda and Siddha, are being integrated with modern biomedical research to explore the therapeutic potential of silk-based biomaterials in regenerative medicine (Kasoju and Bora 2012c). (c) *Clinical translation:* Indian researchers are actively involved in translational research to bridge the gap between benchtop discoveries and clinical applications of silk-based biomaterials (Mukherjee et al. 2022). Government initiatives in India have supported silk-based biomaterials research through funding and infrastructure development. These initiatives aim to foster innovation, create a conducive research environment, and propel India to the forefront of biomaterials research.

## 8.4 Applications of Silk-Based Biomaterials

### 8.4.1 Silk in Tissue Engineering

Tissue engineering has emerged as a ground-breaking field that combines principles of biology, engineering, and materials science to develop functional substitutes for damaged or diseased tissues. The success of tissue engineering relies heavily on the choice of biomaterials that can provide a suitable microenvironment for cell growth, support tissue regeneration, and eventually integrate with the host organism. Silk-based biomaterials, particularly those derived from the silkworm *B. mori*, have gained substantial attention due to their unique combination of mechanical strength, biocompatibility, and tunable degradation characteristics (Beena et al. 2022). They have been processed into a variety of forms and have been explored in numerous tissue engineering applications (Fig. 8.2). Here, we provide an overview of silk-based biomaterials in tissue engineering, focusing on their use in bone, neural, and cardiovascular tissue regeneration.

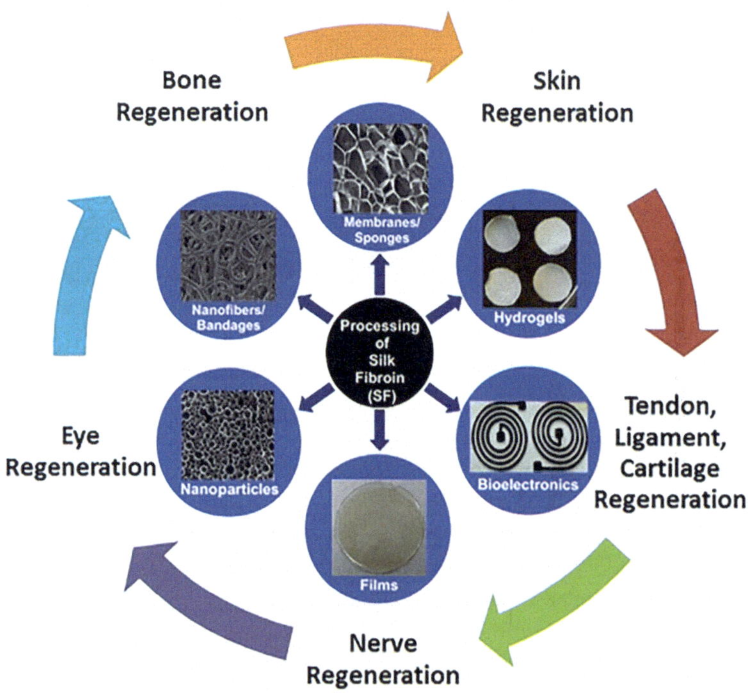


**Fig. 8.2** Graphical representation indicating numerous forms of silk based biomaterials and their applications in various tissue engineering applications. (Reproduced from (Jao et al. 2016). © 2016 by the authors. Licensee MDPI, Basel, Switzerland)

***Bone Tissue Engineering*** Bone tissue engineering aims to develop scaffolds that can mimic the structural and functional properties of natural bone tissue, promoting effective regeneration. Silk fibroin, the protein extracted from silkworm silk, has proven to be an exceptional candidate for bone tissue engineering applications. Studies have demonstrated the ability of silk fibroin scaffolds to support cell adhesion, proliferation, and osteogenic differentiation (Kim et al. 2005). The tunable mechanical properties of silk-based scaffolds allow for the creation of matrices that match the stiffness of native bone tissue, a critical factor in promoting appropriate cellular responses (Karageorgiou and Kaplan 2005). One notable study by Wang et al. explored the use of silk fibroin scaffolds in cartilage tissue engineering. The researchers successfully cultured human articular chondrocytes on silk scaffolds, demonstrating their potential in creating functional cartilage tissues (Wang et al. 2006). The controlled degradation of silk over time aligns with the gradual regeneration process of bone tissues, making silk-based biomaterials suitable for long-term applications.

***Neural Tissue Engineering*** Silk-based biomaterials have shown considerable promise in neural tissue engineering, where the challenge lies in creating scaffolds that support nerve cell growth and guide axonal regeneration. Silk fibroin, owing to its biocompatibility and mechanical properties, has been utilized to develop nerve conduits for the repair of peripheral nerve injuries (Das et al. 2015). The porous structure of silk scaffolds facilitates nutrient diffusion and cell migration, crucial factors for neural tissue regeneration. A study by Yonesi et al. investigated the use of silk fibroin scaffolds for neural tissue engineering applications. The researchers demonstrated that silk-based scaffolds supported the attachment and proliferation of neural cells and promoted neurite outgrowth, suggesting their potential to repair damaged neural tissues (Yonesi et al. 2021). The ability to fine-tune the properties of silk allow for the incorporation of bioactive molecules, further enhancing their neuro-regenerative capabilities (Uebersax et al. 2007).

***Cardiovascular Tissue Engineering*** Cardiovascular tissue engineering seeks to develop biomaterials that can replicate the complex structure and function of native cardiovascular tissues, including blood vessels and heart valves. Silk-based biomaterials have been explored for creating vascular grafts, patches, and heart valve scaffolds due to their tunable mechanical properties and biocompatibility (Kasoju and Bora 2012a). In an interesting proof-of-concept study by Tufan et al., induced pluripotent stem cells were differentiated into cardiomyocytes on silk-based porous scaffolds capable of harvesting energy from simulated cardiac motion. These scaffolds, engineered to replicate the mechanical and physical properties of cardiac tissue, also function as triboelectric nanogenerator electrodes, offering a dual purpose as both regenerative patches and energy-harvesting devices (Tufan et al. 2023). Additionally, silk scaffolds have been utilized to create vascular grafts with controlled degradation rates, aligning with the dynamic needs of vascular tissue regeneration (Gupta et al. 2020).

***Wound Healing and Skin Tissue Engineering*** Silk-based biomaterials have also found applications in wound healing and skin tissue engineering (Mukherjee et al. 2022). Silk fibroin scaffolds have been investigated for their ability to support the adhesion and proliferation of skin cells, promoting the regeneration of functional skin tissues (Mazurek et al. 2022). The biodegradability of silk ensures that the scaffold gradually breaks down as new skin tissue forms, avoiding the need for scaffold removal procedures. Several studies have demonstrated the potential of silk fibroin scaffolds in promoting wound healing. The researchers observed enhanced epithelialization and collagen deposition in wounds treated with silk-based scaffolds, emphasizing their role in supporting the natural healing process (Patil et al. 2020). The modifiable properties of silk allow for the incorporation of antimicrobial agents, making them suitable for applications in wound dressings (Si et al. 2023).

While silk-based biomaterials showcase immense potential in tissue engineering, certain challenges need to be addressed. One significant challenge lies in the careful selection and modification of silk materials to create scaffolds with specific qualities appropriate for the intended tissue type. The optimization of degradation rates, surface properties, and mechanical strength requires a thorough understanding of the intricate relationship between structure and function in silk-based biomaterials (Rockwood et al. 2011; Dan et al. 2023). Future directions in silk-based tissue engineering involve advancements in processing techniques, such as electrospinning and 3D printing, to create scaffolds with enhanced structural complexity and biomimicry (Kasoju and Bora 2012b; Mohamed Ameer et al. 2020; Mukherjee et al. 2022). The incorporation of bioactive molecules, growth factors, and cells into silk scaffolds aims to further enhance their regenerative capabilities (Wang et al. 2006). Collaborative efforts between material scientists, bioengineers, and clinicians are crucial to translating these advancements from the laboratory to clinical applications.

### 8.4.2 Silk-Based Drug Delivery Systems

Drug delivery systems play a pivotal role in modern medicine, aiming to enhance the efficacy, safety, and targeted delivery of therapeutic agents. In recent years, silk-based biomaterials have emerged as promising candidates for drug delivery applications, capitalizing on their unique properties, biocompatibility, and controlled release capabilities. Silk fibroin can be processed into various forms, including films, nanoparticles, microparticles, and hydrogels, providing versatility in designing drug delivery platforms (Rockwood et al. 2011). The controlled and sustained release of therapeutic agents from silk-based carriers has been a focal point in research, aiming to optimize drug pharmacokinetics and minimize side effects. Silk fibroin's biodegradability aligns with the principles of sustainable drug delivery, as the material naturally breaks down over time, eliminating the need for removal procedures. The controlled release properties of silk-based carriers have been demonstrated in various studies. For instance, silk fibroin nanoparticles have been utilized to encapsulate and release drugs in a sustained manner, offering prolonged therapeutic effects and minimizing fluctuations in drug concentrations.

Silk-based drug delivery systems have shown promise in enhancing the bioavailability of poorly water-soluble drugs. The amphiphilic nature of silk fibroin allows for the encapsulation of hydrophobic drugs, improving their solubility and absorption. Additionally, silk-based carriers can be functionalized with targeting ligands, enabling site-specific delivery of drugs to enhance therapeutic efficacy while minimizing systemic side effects (Wang et al. 2010; Kasoju and Bora 2012c). For instance, silk hydrogels have emerged as a versatile platform for localized drug delivery, particularly in wound healing and tissue regeneration applications (You et al. 2025). These hydrogels can encapsulate drugs and growth factors, facilitating their controlled release at the site of application. For instance, silk fibroin hydrogels loaded with antibiotics have demonstrated efficacy in preventing infections and promoting wound healing (Guo et al. 2023).

In cancer therapy, silk-based nanoparticles have shown significant potential for targeted drug delivery. The modular nature of silk fibroin allows for the incorporation of chemotherapeutic agents within nanoparticles. These silk-based nanocarriers can be engineered to respond to specific stimuli, such as pH or enzymes present in the tumor microenvironment, enabling site-specific drug release (Kundu et al. 2013). The ability of silk nanoparticles to encapsulate a variety of drugs, including small molecules and biomolecules, contributes to their versatility in cancer therapy (Fig. 8.3). Silk-based drug delivery systems have also been explored for managing

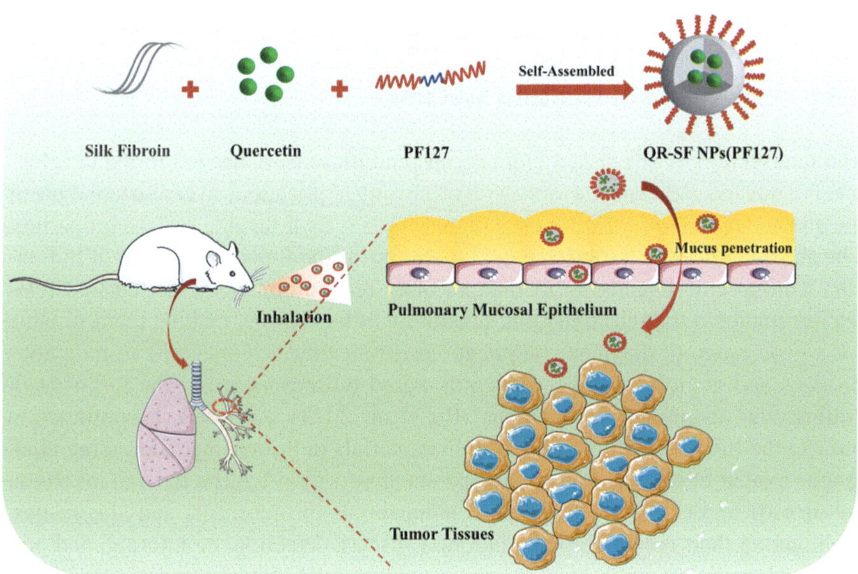


**Fig. 8.3** A schematic representation showing effective use of PF127-modified, quercetin-loaded silk-based nanoparticles for lung cancer therapy. (Reproduced from (Tang et al. 2023). © 2023 Shenyang Pharmaceutical University. Published by Elsevier B.V.)

inflammatory diseases. Silk fibroin carriers can encapsulate anti-inflammatory agents, allowing for the controlled release of drugs to modulate the inflammatory response. This application has potential implications for diseases such as rheumatoid arthritis and inflammatory bowel disease, where localized drug delivery can enhance therapeutic outcomes (Pham et al. 2022; Diez-Echave et al. 2021). Further, silk-based biomaterials have been investigated for antimicrobial drug delivery applications. Silk fibroin nanoparticles loaded with antimicrobial agents have demonstrated efficacy in inhibiting bacterial growth, presenting potential applications in treating bacterial infections. The sustained release properties of silk carriers contribute to prolonged antimicrobial activity (Ghalei and Handa 2022).

Recent advancements in silk-based drug delivery platforms include the integration of nanotechnology and the development of smart drug delivery systems. Silk nanoparticles with tunable sizes and surface properties have been engineered to enhance drug-loading capacities and improve cellular uptake (Wang et al. 2010). Additionally, stimuli-responsive silk-based carriers, such as those responsive to temperature or enzymatic activity, have been designed to achieve on-demand drug release, further optimizing therapeutic outcomes (Lin et al. 2023). While the potential of silk-based biomaterials in drug delivery is evident, their successful translation from laboratory research to clinical applications poses certain challenges. Ensuring regulatory approval, scalability of production, and addressing potential immunogenicity concerns are critical aspects that need to be addressed for widespread adoption in clinical settings (Hong et al. 2020). Collaborative efforts between material scientists, pharmacologists, and clinicians are essential to overcoming these challenges.

### 8.4.3 Silk-Based Biosensing Systems

The intersection of silk-based biomaterials and biosensors has led to the development of advanced diagnostic tools with promising applications in healthcare. One of the distinctive features of silk-based biomaterials is their tunable surface properties. The ability to modify the surface of silk allows for the functionalization of biosensors with specific ligands, antibodies, or other recognition elements. This customization enhances the selectivity and sensitivity of biosensors, making them suitable for a wide range of diagnostic applications. From cancer biomarkers to infectious diseases and neurological disorders, silk-based biosensors are at the forefront of cutting-edge diagnostic applications (Ru et al. 2023). As research continues to unravel the full potential of silk-based biomaterials in biosensing, addressing challenges related to clinical translation and standardization will be paramount. Here, we provide an overview of silk-based biomaterials in biosensors and diagnostics, highlighting their potential to revolutionize disease detection, monitoring, and personalized medicine.

Silk-based biosensors have shown promise in the early detection of cancer biomarkers. By functionalizing silk surfaces with antibodies specific to certain cancer-related proteins, these biosensors can selectively capture and quantify biomarker concentrations. This approach offers a non-invasive and rapid means of diagnosing cancer at its earliest stages. On the other hand, silk-based biosensors have been employed in the detection of infectious diseases. By incorporating silk fibroin as a matrix for immobilizing antigens or nucleic acid probes, these biosensors can selectively recognize and detect pathogens such as bacteria or viruses. The versatility of silk allows for the development of point-of-care diagnostics for rapid and accurate identification of infectious agents (Márquez et al. 2024). Further, Silk-based biosensors have been explored for the detection of biomarkers associated with neurological disorders. By integrating silk as a stable and biocompatible platform, researchers can create biosensors that selectively capture and quantify specific biomolecules related to conditions like Alzheimer's disease or Parkinson's disease. This opens avenues for early diagnosis and monitoring of neurological disorders (Ru et al. 2023).

Silk biomaterials have been explored in a variety of forms for the design and development of biosensors. For instance, silk based hydrogel loaded with enzymes such as glucose oxidase and horseradish peroxidase was used along with graphene filed-effect transistors to detect glucose in a physiological buffer solution (Fig. 8.4) (Wang et al. 2021). Silk-based nanoparticles have been utilized in immuno-sensors for the detection of antibodies or antigens. The unique properties of silk nanoparticles, including stability and ease of functionalization, make them suitable for capturing and detecting specific immune responses. This application has implications for autoimmune disease diagnostics and monitoring (Moraes et al. 2019). On the other hand, silk nanoparticles have been employed in biosensors designed for monitoring drug delivery. By encapsulating drugs within silk nanoparticles and incorporating biosensing elements, these systems can provide real-time feedback on drug release kinetics. This technology holds promise for optimizing therapeutic

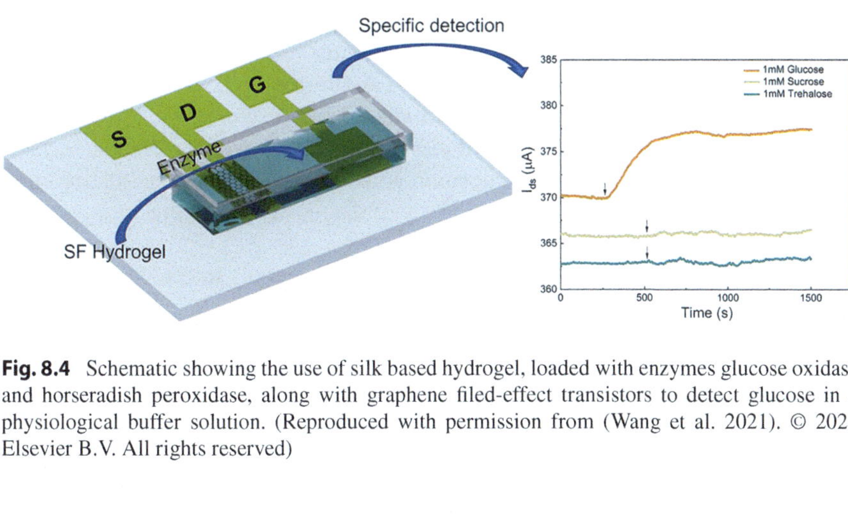


**Fig. 8.4** Schematic showing the use of silk based hydrogel, loaded with enzymes glucose oxidase and horseradish peroxidase, along with graphene filed-effect transistors to detect glucose in a physiological buffer solution. (Reproduced with permission from (Wang et al. 2021). © 2021 Elsevier B.V. All rights reserved)

interventions and ensuring precise drug administration (Kim et al. 2013). Further, the stable and biocompatible nature of silk allows for the immobilization of DNA probes onto nanoparticles, enabling the selective detection of complementary sequences; his capability is crucial in applications such as genetic testing and pathogen detection (Benvidi et al. 2018).

Silk's versatility allows for the creation of biosensors capable of detecting multiple analytes simultaneously, offering a comprehensive diagnostic approach. While the integration of silk-based biomaterials in biosensors and diagnostics holds immense promise, certain challenges need to be addressed for widespread adoption. Standardizing production methods, optimizing surface functionalization strategies, and ensuring regulatory approval are crucial steps toward translating these technologies from the laboratory to clinical settings. Future directions in silk-based biosensors involve advancements in nanotechnology, the incorporation of artificial intelligence for data analysis, and the development of multiplexed sensing platforms.

### 8.4.4 Other Biomedical Applications of Silk Biomaterials

Silk-based biomaterials, derived from the natural protein fibroin produced by silkworms, have demonstrated remarkable versatility beyond their well-known applications in tissue engineering and drug delivery. The exploration of silk in bio-imaging, medical textiles, flexible electronics, wearable devices, and bioresorbable medical implants showcases the material's adaptability and potential to revolutionize diverse fields in healthcare (Bandyopadhyay et al. 2019). As research continues to unveil new properties and functionalities of silk, interdisciplinary collaborations between material scientists, engineers, and medical professionals will drive the development of innovative and patient-centric solutions.

***Bio-imaging*** Silk-based biomaterials have found application in the field of bio-imaging, offering unique advantages for both diagnostic and therapeutic purposes. Silk nanoparticles, due to their biocompatibility and ability to encapsulate imaging agents, have been explored as contrast agents for various imaging modalities, including magnetic resonance imaging (MRI) and fluorescence imaging (Xu et al. 2018). The controlled release properties of silk enable sustained imaging contrast, enhancing the diagnostic capabilities of these agents. Moreover, the tunable surface properties of silk allow for functionalization with targeting ligands, facilitating site-specific imaging and improving the accuracy of diagnostic procedures.

***Medical Textiles*** The integration of silk-based biomaterials into medical textiles has ushered in a new era of smart and functional fabrics for healthcare applications (Murugesh Babu et al. 2020). Silk fibers can be woven into textiles with inherent properties such as breathability, moisture-wicking, and antimicrobial effects. These textiles find applications in wound dressings, compression garments, and even wearable sensing devices. Silk's natural properties make it comfortable for the wearer, and ongoing research explores the incorporation of additional functionalities, such as drug release capabilities, directly into the fabric. This convergence of textiles and biomaterials holds potential for patient-specific and personalized healthcare solutions (Omollo Oduor et al. 2021).

***Flexible Electronics*** Silk-based biomaterials have gained attention in the field of flexible electronics, contributing to the development of biocompatible and flexible electronic devices. Silk's intrinsic flexibility and biocompatibility make it an ideal substrate for electronic components that need to conform to the contours of the human body. Silk fibroin can be processed into thin films or membranes that serve as flexible and biocompatible platforms for electronic circuits and sensors (Zhu et al. 2016). These silk-based electronic devices hold promise for applications in wearable health monitoring, smart textiles, and implantable electronics.

***Wearable Devices*** The integration of silk-based biomaterials into wearable devices has opened up new possibilities in the realm of health monitoring and personalized medicine. Silk's compatibility with the human body, combined with its mechanical strength, enables the creation of comfortable and durable wearable devices. These devices can incorporate sensors for monitoring vital signs, drug delivery systems, or even biometric authentication features. Silk-based wearables offer the advantage of being lightweight, flexible, and skin-friendly, making them suitable for continuous and unobtrusive health monitoring (Fig. 8.5).

***Bioresorbable Implants*** Silk-based biomaterials are increasingly being explored for the development of bioresorbable medical implants, providing temporary support or functionality before gradually dissolving in the body. Silk's controllable degradation kinetics and excellent biocompatibility make it suitable for applications such as orthopedic screws, plates, and scaffolds (Perrone et al. 2014). These implants can provide mechanical support during the initial stages of tissue healing and regeneration and then naturally degrade as the tissue regains its strength. This approach minimizes the need for additional surgical interventions for implant removal.

## 8.5 Exploration of how Silk Addresses Sustainability Issues in Biomaterials

***SDGs—with a Focus on Healthcare*** A complex and diverse concept, "sustainability" embodies the necessity of meeting current needs while preserving the interests of future generations (Church et al. 2022). The attainment of sustainability requires the alleviation of problems related to overconsumption, underconsumption, and insufficient consumption abilities, with a focus on the development of visually appealing and visually stimulating sustainable futures (Grant 2010). The SDGs, a set of seventeen goals, were established to achieve a better and more sustainable future for all by 2030 (Collier 2020). This project is a crucial component of a larger

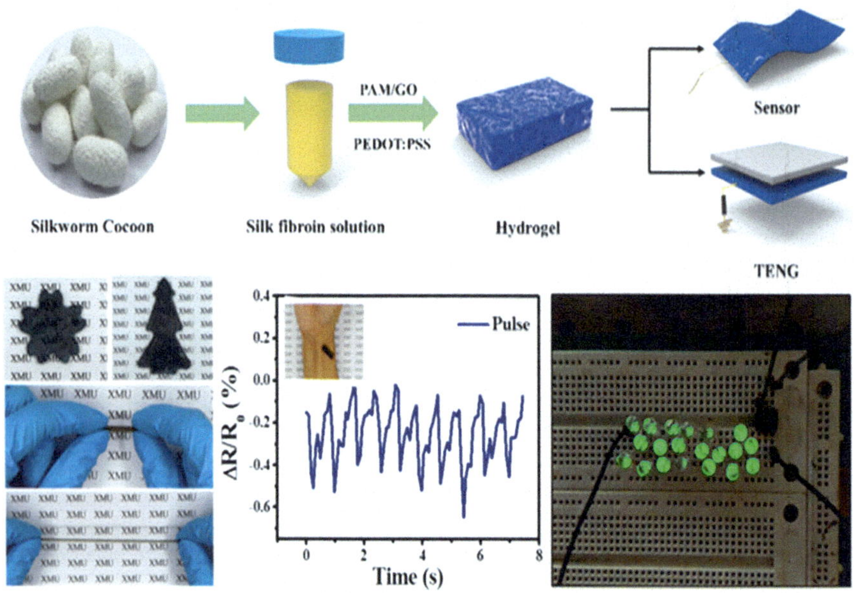


**Fig. 8.5** Schematic showing silk fibroin based soft conductive hydrogel comprising of polyacrylamide, graphene oxide and poly(3,4-ethylenedioxythiophene): poly(styrenesulfonate) and their use as strain/pressure sensors in health and exercise monitors, soft robots, and power sources. (Reproduced with permission from (He et al. 2020). Copyright © 2020 American Chemical Society)

strategy meant to tackle major worldwide problems related to sustainable development. The idea of a "shared blueprint for peace and prosperity for people and the planet" has been expressed from a variety of angles (Comfort 2023). The healthcare industry plays a critical role in achieving many SDGs, and incorporating sustainability principles is essential to guaranteeing long-term effectiveness. In particular, objectives like Good Health and Well-being (SDG 3), Responsible Consumption and Production (SDG 12), Climate Action (SDG 13), and Partnerships for the Goals (SDG 17) bear relevance to the healthcare domain.

***Biomedical Engineering and the Scope for Sustainability*** The diverse field of biomedical engineering applies engineering concepts to solve the problems in the healthcare sector, focusing on the creation and use of prosthetics, implants, and medical equipment. Apart from its concrete benefits, biomedical engineering is essential to the progress of health research and development, particularly in the areas of illness diagnosis, treatment, and prevention. The field's dedication to novel approaches is demonstrated by approaches like tissue engineering, affordable point-of-care devices and many others. For the future of biomedical engineers and educators in this sector, achieving the 2030 Agenda—with a particular emphasis on the SDGs about "Good Health and well-being" and "Quality Education"—is of utmost importance. Adopting this agenda emphasizes how crucial it is for the biomedical device industry to have a holistic viewpoint, which calls for research that takes social, economic, and environmental factors into account (Lalor et al. 2019). This strategy encourages cooperation between patient advocacy groups, academic institutions, and businesses to advance sustainability in biomedical research (Salvetti et al. 2018). A range of sustainable biomaterials has been explored for use in the biomedical field, including those derived from renewable resources (Jana et al. 2022). The pursuit of sustainability in biomedical applications has prompted the investigation of a variety of biomaterials, such as those made from waste products, renewable resources, and materials particularly engineered to coexist peacefully with biological systems (Biswal et al. 2020). Sustainable biomaterials would be useful in a variety of biomedical contexts which will eventually enhance patient outcomes, demonstrating the potential for both cost-effectiveness and environmental preservation (Pesode et al. 2023).

***Natural Polymers as Sustainable Biomaterials*** Natural polymers, sourced from diverse origins like plants, animals, and microbes, constitute a structurally intricate class of materials (Matei et al. 2022). Abundant in sugars, plant materials, and bacterial biomass, these polymers, characterized by hydrophilic properties due to an abundance of hydroxyl groups, hold immense potential across various applications (Rayner et al. 2016). Polysaccharides, proteins, and polyesters represent the three primary groups of natural polymers, including notable examples like collagen, silk fibroin, cellulose, and chitosan (Aravamudhan et al. 2014). These polymers, recognized for their sustainability, are garnering significant scientific attention (Silva et al. 2021). Incorporating natural fibers into polymer matrices enhances the sustainability of resulting materials, aligning with global efforts towards environmentally friendly practices (Fuentes Molina et al. 2021). In the biomedical realm, natural polymers play a crucial role in scaffold creation for tissue regeneration and healing. Their flexibility and biodegradability make them particularly appealing, allowing the design of adaptable scaffolds to meet evolving healing needs (Doppalapudi et al. 2015). However, meticulous material selection and modification remain challenges, especially in tailoring scaffolds for specific wound types (Ter Horst et al. 2019). This scientific exploration of natural polymers underscores their adaptability, ushering in an era of advanced material creation with enhanced performance and reduced environmental impact. As researchers delve into the intricate relationship between structure and function in natural polymers, new opportunities for innovation emerge, spanning biomedical engineering and sustainable materials. The ongoing scientific inquiry promises a future marked by materials that not only meet high-performance standards but also adhere to global sustainability objectives.

***Silk as a Sustainable Biomaterial*** Silk stands as an exemplary sustainable biomaterial, woven with ecological virtues and versatile applications. Derived from the silkworm's natural protein secretion during the sericulture process, silk embodies biodegradability, addressing the imperative for responsible material life cycles (Cao and Wang 2009). Its renewability further elevates its sustainability profile, as silkworms continuously produce this biomaterial, aligning with circular and regenerative principles. The sericulture process is notably eco-friendly, requiring less energy and water compared to the production of some synthetic biomaterials, contributing to a reduced environmental footprint. Silk's multifaceted nature extends to diverse applications, from textiles to medicine, showcasing its adaptability and reducing the need for disparate materials (Kasoju and Bora 2012a). In the textile industry, silk fabrics are celebrated for their eco-friendly appeal, meeting the burgeoning demand for sustainable fashion. Biocompatibility renders silk a preferred choice in medicine, fostering sustainable healthcare practices. Additionally, silk's role in tissue regeneration and regenerative medicine highlights its significance in sustainable approaches to health and well-being. Beyond its material attributes, sericulture often plays a role in local economies, contributing to social sustainability and community development. As awareness of environmental concerns grows, silk's unique properties make it an increasingly attractive choice for various industries seeking more sustainable alternatives. In essence, silk emerges as a beacon of sustainability, intertwining biological ingenuity with environmental responsibility, offering a harmonious solution in the quest for eco-conscious biomaterials (Kasoju and Sunilkumar 2024).

Comparing silk-based biomedical materials with traditional biomedical materials in the context of SDGs involves assessing their environmental impact, social implications, and contribution to sustainable practices (Fig. 8.6). As presented in

**Fig. 8.6** Silk based biomaterials, related biomedical products and technologies, in many ways, contribute to several SDGs, and therefore, offer a harmonious solution in the quest for the sustainable development

**Table 8.1** Comparing silk-based biomedical materials with traditional biomedical materials in the context of sustainable development goals

| SDG | Relevant aspect | Silk biomaterials | Traditional biomaterials |
|---|---|---|---|
| SDG 1—no poverty | Community empowerment | Sericulture can empower local communities through income generation. | The impact on poverty reduction may vary. |
| SDG 3—good health and Well-being | Health impact | Biocompatible, reducing risks of adverse health effects in medical applications. | May include materials with potential health concerns or allergic reactions. |
| SDG 5—gender equality | Gender equality | Sericulture often involves female workers, contributing to gender inclusivity. | Employment opportunities may not always be gender-neutral. |
| SDG 6—clean water and sanitation | Water conservation | Sericulture is often considered more water-efficient than certain traditional biomaterials. | Production processes may involve substantial water usage. |
| SDG 9—industry, innovation, and infrastructure | Circular economy | Supports circular economies through its renewability and biodegradability. | May involve linear production processes with less emphasis on circularity. |
| SDG 12—responsible consumption and production | Biodegradability | Naturally biodegradable, aligning with goals of reducing environmental impact. | This may include non-biodegradable materials, contributing to waste and pollution. |
| SDG 13—climate action | Reduced environmental footprint | Lower energy consumption and environmental impact in production. | Some may have higher carbon footprints, contributing to climate change. |
| SDG 15—life on land | Renewable sourcing | Sourced from silkworms in a renewable and sustainable manner. | Depending on the source, may involve processing from finite resources. |
| SDG 15—life on land | Biodiversity conservation | Sericulture can coexist with biodiversity, especially in eco-friendly farming practices. | Synthesis and associated processes may impact biodiversity. |
| SDG 17—partnerships for the goals | Innovation and technology transfer | Encourages innovation in biomaterials and technology transfer. | Transitioning to more sustainable alternatives requires collaborative efforts and partnerships. |

Table 8.1, considering silk-based biomedical materials within the framework of SDGs reveals its potential to contribute significantly to a more sustainable and equitable future, especially when compared to certain traditional biomedical materials.

## 8.6 Current Challenges in the Widespread Adoption of Silk-Based Biomaterials

The widespread adoption of silk-based biomaterials faces several challenges across technological, economic, and regulatory domains. Standardizing processing methods remains a critical hurdle, as the lack of uniform techniques limits broader acceptance and utilization across industries. Establishing standardized protocols will enhance consistency and facilitate integration into various applications. Scaling up production presents another challenge, requiring innovations in sericulture and raw material sourcing. Ensuring a steady supply of silkworms and mulberry plants, while optimizing production processes, will be key to meeting large-scale demand. Additionally, high production costs further hinder adoption, as silk extraction and processing involve labor-intensive agricultural practices. Developing cost-effective methods and achieving economies of scale will make silk-based biomaterials more competitive. Overcoming competition with well-established materials also requires demonstrating clear advantages, such as improved sustainability, performance, and cost-effectiveness. Industries accustomed to synthetic or traditional biomaterials need compelling evidence to transition to silk-based alternatives.

Further, the mechanical strength remains a limiting factor for high-stress applications, despite silk's impressive tensile properties. Researchers must develop strategies to enhance strength without compromising other desirable characteristics, making silk more suitable for load-bearing and industrial applications. In medical fields, ensuring optimal biocompatibility is crucial. While silk is generally well-tolerated by the human body, specific medical applications demand precise compatibility and safety standards, necessitating further research and clinical validation. Besides, controlling degradation rates adds another layer of complexity. Silk's biodegradability is advantageous, but tailoring its degradation profile for different applications requires a fine balance between maintaining structural integrity and ensuring timely breakdown post-use. Public awareness also plays a pivotal role in adoption. Compared to synthetic materials, silk biomaterials remain less recognized. Educating consumers, industries, and healthcare professionals about their benefits will drive acceptance and market integration.

Ensuring consistent quality presents a challenge due to natural variations in silk production, influenced by factors such as diet and environmental conditions. Implementing stringent quality control measures and standardizing production will help meet industry expectations. Navigating regulatory hurdles further complicates commercialization, as approval processes vary across regions and applications. Streamlining regulations and working closely with policymakers will accelerate the market entry of silk-based biomaterials. Overall, addressing these challenges demands a coordinated effort from researchers, industry leaders, and policymakers. Advancing research, refining production techniques, and promoting awareness will unlock the full potential of silk-based biomaterials, paving the way for their widespread adoption across diverse sectors.

## 8.7 Emerging Research and Innovations to Overcome Challenges

Widespread adoption of silk-based biomaterials requires a strategic approach that integrates research advancements, technological innovations, industry collaboration, and regulatory support. Addressing key challenges begins with standardizing processing methods, where researchers and industry stakeholders must work together to establish uniform techniques. Developing clear guidelines and best practices will ensure consistency across various applications and seri-industries. Scaling production is equally crucial, necessitating innovations in sericulture to secure a stable and scalable raw material supply. Exploring sustainable and efficient farming methods will further support large-scale silk production while maintaining environmental responsibility. Besides, reducing production costs remains a priority, and investing in research and development can optimize sericulture and silk extraction processes to make production more cost-effective. Innovations in processing and manufacturing will play a significant role in achieving affordability without compromising quality. Competing with established materials requires thorough cost-benefit analyses and comparative studies to showcase the advantages of silk-based biomaterials over traditional or synthetic alternatives. Highlighting successful applications in industries where silk offers unique benefits will strengthen its position in the market.

Next, enhancing the mechanical strength of silk-based biomaterials is essential for their application in high-stress environments. Collaborations between material scientists and engineers should focus on developing methods to reinforce silk without compromising its other desirable properties. Research into composite materials and structural modifications can significantly enhance performance. In medical applications, extensive biomedical research is needed to fine-tune silk biomaterials for specific uses. Partnering with healthcare professionals will help ensure optimal biocompatibility and safety, paving the way for broader clinical adoption. Further, understanding and controlling degradation rates is another critical aspect of silk biomaterial development. Research in material science should aim to tailor degradation profiles for specific applications through precise material modifications. Public awareness also plays a vital role in market acceptance. Educational campaigns should target consumers, industries, and healthcare professionals to highlight the benefits and applications of silk-based biomaterials, emphasizing their eco-friendly and sustainable nature.

Next, ensuring product quality through standardization is necessary for industry-wide adoption. Implementing stringent quality control measures throughout production will guarantee consistency and reliability. Developing industry standards and certification systems will further enhance trust in silk-based biomaterials. Navigating regulatory hurdles is equally important, and engaging with regulatory bodies to advocate for streamlined approval processes will facilitate market entry. Providing strong evidence of safety and efficacy will support policy discussions and regulatory approvals, ultimately accelerating the adoption of these materials. Overall, by actively implementing these strategies, researchers, industry leaders, policymakers, and advocacy groups can drive the widespread adoption of silk-based biomaterials. A collaborative effort across these sectors will promote sustainability and innovation, ensuring silk-based materials play a significant role in advancing diverse industries.

## 8.8 Conclusions

Silk-based biomaterials have emerged as a promising and sustainable alternative with diverse applications across various industries. Derived from the natural protein fibers produced by silkworms during the sericulture process, silk possesses unique properties that make it an attractive biomaterial. Its inherent biocompatibility, biodegradability, and versatility position silk as a viable option in medical, textile, and other high-performance applications. In the context of sustainability, silk-based biomaterials address environmental concerns associated with conventional materials. Being a renewable resource, silk is sourced from silkworms through sericulture, a process that can be integrated with eco-friendly agricultural practices. The biodegradable nature of silk further contributes to sustainability, as it naturally breaks down over time, reducing long-term environmental impact. However, the widespread adoption of silk-based biomaterials faces challenges. The cost of production is a significant hurdle, requiring innovations in sericulture processes and silk extraction techniques to achieve cost-effective large-scale production. Ensuring consistent quality poses another challenge, given the natural variability in silk production influenced by factors such as the silkworm's diet and environmental conditions. Overcoming these challenges requires collaborative efforts from researchers, industry stakeholders, and policymakers. Looking ahead, the future potential of silk-based biomaterials is vast. Ongoing research aims to enhance the mechanical strength of silk for applications in high-stress environments, such as load-bearing structures. Biomedical research continues to explore tailored silk biomaterials for medical applications, including drug delivery systems and tissue engineering. Advancements in processing methods and the establishment of industry standards will contribute to scalability and standardization. Moreover, the unique properties of silk, including its lightweight nature, thermal regulation, and adaptability, make it an attractive candidate for emerging technologies and sustainable innovations. As public awareness grows, and with continued scientific exploration, silk-based biomaterials are poised to play a crucial role in fostering sustainability and meeting the demands of various industries, promising a greener and more sustainable future.

## References

Aravamudhan A, Ramos DM, Nada AA, Kumbar SG (2014) Natural polymers. In: Natural and synthetic biomedical polymers. Elsevier, pp 67–89

Bandyopadhyay A, Chowdhury SK, Dey S et al (2019) Silk: a promising biomaterial opening new vistas towards affordable healthcare solutions. J Indian Inst Sci 99:445–487. https://doi.org/10.1007/s41745-019-00114-y

Benvidi A, Abbasi Z, Dehghan Tezerjani M et al (2018) A Highly Selective DNA Sensor Based on Graphene Oxide-Silk Fibroin Composite and AuNPs as a Probe Oligonucleotide Immobilization Platform. Acta Chim Slov 65(2):278–288. https://doi.org/10.17344/acsi.2017.3667. PMID: 29993090.

Beena M, Ameer JM, Kasoju N (2022) Optically clear silk fibroin films with tunable properties for potential corneal tissue engineering applications: a process–property–function relationship study. ACS Omega 7:29634–29646. https://doi.org/10.1021/acsomega.2c01579

Biswal T, BadJena SK, Pradhan D (2020) Sustainable biomaterials and their applications: a short review. Mater Today Proc 30:274–282. https://doi.org/10.1016/j.matpr.2020.01.437

Biswal B, Dan AK, Sengupta A et al (2022) Extraction of silk fibroin with several Sericin removal processes and its importance in tissue engineering: a review. J Polym Environ 30:2222–2253. https://doi.org/10.1007/s10924-022-02381-w

Burdick JA, Prestwich GD (2011) Hyaluronic acid hydrogels for biomedical applications. Adv Mater 23:H41–H56. https://doi.org/10.1002/adma.201003963

Cao Y, Wang B (2009) Biodegradation of silk biomaterials. Int J Mol Sci 10:1514–1524. https://doi.org/10.3390/ijms10041514

Church JM, Tirrell A, Moomaw WR, Ragueneau O (2022) Sustainability. In: Routledge handbook of global environmental politics, 2nd edn. Routledge, London, pp 217–227

Collier P (2020) Embedding the sustainability development goals in *survey review*'s remit. Surv Rev 52:287–288. https://doi.org/10.1080/00396265.2020.1746020

Comfort K (2023) Introduction. In: A global humanities approach to the United Nations' sustainable development goals, 1st edn. Routledge, London, pp 1–22

Dan FR, Varsha SH et al (2023) A study on source dependent batch to batch variations in silk fibroin films for potential applications in corneal tissue engineering. MedComm – Biomater Appl 2:e48. https://doi.org/10.1002/mba2.48

Das S, Sharma M, Saharia D et al (2015) In vivo studies of silk based gold nano-composite conduits for functional peripheral nerve regeneration. Biomaterials 62:66–75. https://www.sciencedirect.com/science/article/pii/S0142961215004342

DeBari MK, King CI, Altgold TA, Abbott RD (2021) Silk fibroin as a green material. ACS Biomater Sci Eng 7:3530–3544. https://doi.org/10.1021/acsbiomaterials.1c00493

Diez-Echave P, Ruiz-Malagón AJ, Molina-Tijeras JA et al (2021) Silk fibroin nanoparticles enhance quercetin immunomodulatory properties in DSS-induced mouse colitis. Int J Pharm 606:120935. https://doi.org/10.1016/j.ijpharm.2021.120935. Epub 2021 Jul 24. PMID: 34310954

Doppalapudi S, Katiyar S, Domb AJ, Khan W (2015) Biodegradable natural polymers. In: Puoci F (ed) Advanced polymers in medicine. Springer International Publishing, Cham, pp 33–66

Fuentes Molina N, Fragozo Brito Y, Polo Benavides JM (2021) Recycling of residual polymers reinforced with natural fibers as a sustainable alternative: a review. Polymers 13:3612. https://doi.org/10.3390/polym13213612

Ghalei S, Handa H (2022) A Review on Antibacterial Silk Fibroin-based Biomaterials: Current State and Prospects. Mater Today Chem 23:100673. https://doi.org/10.1016/j.mtchem.2021.100673. Epub 2021 Dec 9. PMID: 34901586; PMCID: PMC8664245.

Grant LK (2010) Sustainability: from excess to aesthetics. Behav Soc Issues 19:7–47. https://doi.org/10.5210/bsi.v19i0.2789

Gupta P, Lorentz KL, Haskett DG et al (2020) Bioresorbable silk grafts for small diameter vascular tissue engineering applications: in vitro and in vivo functional analysis. Acta Biomater 105:146–158. https://doi.org/10.1016/j.actbio.2020.01.020

Guo Z, Yan L, Zhou B et al (2023) In situ photo-crosslinking silk fibroin based hydrogel accelerates diabetic wound healing through antibacterial and antioxidant. Int J Biol Macromol 242(Pt 3):125028. https://doi.org/10.1016/j.ijbiomac.2023.125028

Hakimi O, Knight DP, Vollrath F, Vadgama P (2007) Spider and mulberry silkworm silks as compatible biomaterials. Compos Part B Eng 38:324–337. https://doi.org/10.1016/j.compositesb.2006.06.012

He F, You X, Gong H et al (2020) Stretchable, biocompatible, and multifunctional silk fibroin-based hydrogels toward wearable strain/pressure sensors and triboelectric Nanogenerators. ACS Appl Mater Interfaces 12:6442–6450. https://doi.org/10.1021/acsami.9b19721

Hench LL, Polak JM (2002) Third-generation biomedical materials. Science 295:1014–1017. https://doi.org/10.1126/science.1067404

Hong S, Choi DW, Kim HN et al (2020) Protein-based nanoparticles as drug delivery systems. Pharmaceutics 12:604. https://doi.org/10.3390/pharmaceutics12070604

Jana S, Das P, Mukherjee J et al (2022) Waste-derived biomaterials as building blocks in the biomedical field. J Mater Chem B 10:489–505. https://doi.org/10.1039/D1TB02125G

Jao D, Mou X, Hu X (2016) Tissue regeneration: a silk road. J Funct Biomater 7:22. https://doi.org/10.3390/jfb7030022

Jayakumar R, Prabaharan M, Nair SV, Tamura H (2010) Novel chitin and chitosan nanofibers in biomedical applications. Biotechnol Adv 28:142–150. https://doi.org/10.1016/j.biotechadv.2009.11.001

Karageorgiou V, Kaplan D (2005) Porosity of 3D biomaterial scaffolds and osteogenesis. Biomaterials 26:5474–5491. https://doi.org/10.1016/j.biomaterials.2005.02.002

Kasoju N, Bora U (2012a) Silk fibroin in tissue engineering. Adv Healthc Mater 1:393–412. https://doi.org/10.1002/adhm.201200097

Kasoju N, Bora U (2012b) Silk fibroin based biomimetic artificial extracellular matrix for hepatic tissue engineering applications. Biomed Mater 7:045004. https://doi.org/10.1088/1748-6041/7/4/045004

Kasoju N, Bora U (2012c) Fabrication and characterization of curcumin-releasing silk fibroin scaffold. J Biomed Mater Res B Appl Biomater 100B:1854–1866. https://doi.org/10.1002/jbm.b.32753

Kasoju N, Sunilkumar A (2024) Convergence of tissue engineering and sustainable development goals. Biotechnol Sustain Mater 1:20. https://doi.org/10.1186/s44316-024-00021-y

Kim HJ, Kim U-J, Vunjak-Novakovic G et al (2005) Influence of macroporous protein scaffolds on bone tissue engineering from bone marrow stem cells. Biomaterials 26:4442–4452. https://doi.org/10.1016/j.biomaterials.2004.11.013

Kim K, Lam J, Lu S et al (2013) Osteochondral tissue regeneration using a bilayered composite hydrogel with modulating dual growth factor release kinetics in a rabbit model. J Control Release 168:166–178. https://doi.org/10.1016/j.jconrel.2013.03.013

Kundu B, Rajkhowa R, Kundu SC, Wang X (2013) Silk fibroin biomaterials for tissue regenerations. Adv Drug Deliv Rev 65:457–470. https://doi.org/10.1016/j.addr.2012.09.043

Lalor F, Fitzpatrick J, Sage C, Byrne E (2019) Sustainability in the biopharmaceutical industry: seeking a holistic perspective. Biotechnol Adv 37:698–707. https://doi.org/10.1016/j.biotechadv.2019.03.015

Langer R, Tirrell DA (2004) Designing materials for biology and medicine. Nature 428:487–492. https://doi.org/10.1038/nature02388

Lee CH, Singla A, Lee Y (2001) Biomedical applications of collagen. Int J Pharm 221:1–22. https://doi.org/10.1016/S0378-5173(01)00691-3

Li C, Vepari C, Jin H-J et al (2006) Electrospun silk-BMP-2 scaffolds for bone tissue engineering. Biomaterials 27:3115–3124. https://doi.org/10.1016/j.biomaterials.2006.01.022

Lin X, Cai L, Cao X, Zhao Y (2023) Stimuli-responsive silk fibroin for on-demand drug delivery. Smart Med 2(2):e20220019. https://doi.org/10.1002/SMMD.20220019. PMID: 39188280; PMCID: PMC11235688

Márquez A, Santiago S, Dos Santos MV et al (2024) Reusable Colorimetric Biosensors on Sustainable Silk-Based Platforms. ACS Appl Bio Mater 7(2):853–862. https://doi.org/10.1021/acsabm.3c00872. Epub 2024 Jan 25. PMID: 38270977; PMCID: PMC10880051

Matei E, Predescu AM, Râpă M et al (2022) Natural polymers and their nanocomposites used for environmental applications. Nano 12:1707. https://doi.org/10.3390/nano12101707

Mazurek Ł, Szudzik M, Rybka M, Konop M (2022) Silk fibroin biomaterials and their beneficial role in skin wound healing. Biomol Ther 12:1852. https://doi.org/10.3390/biom12121852

Mohamed Ameer J, Venkatesan RB, Damodaran V et al (2020) Fabrication of co-cultured tissue constructs using a dual cell seeding compatible cell culture insert with a clip-on scaffold for potential regenerative medicine and toxicological screening applications. J Sci Adv Mater Devices 5:207–217. https://doi.org/10.1016/j.jsamd.2020.04.004

Moraes ML, Lima LR, Vicentini-Oliveira JC et al (2019) Immunosensor for the Diagnostics of Autoimmune Hemolytic Anemia (AIHA) Based on Immobilization of a Monoclonal Antibody on a Layer of Silk Fibroin. J Nanosci Nanotechnol 19(7):3772–3776. https://doi.org/10.1166/jnn.2019.16305. PMID: 30764933

Mukherjee S, Krishnan A, Athira RK et al (2022) Silk fibroin and silk sericin in skin tissue engineering and wound healing: retrospect and prospects. In: Natural polymers in wound healing and repair. Elsevier, pp 301–331

Murugesh Babu K, Sahana N, Anitha DV, Kavya BS (2020) Silk fibroin coated antimicrobial textile medical products. J Text Inst 112(8):1199–1207. https://doi.org/10.1080/00405000.2020.1806634

Neubauer VJ, Döbl A, Scheibel T (2021) Silk-based materials for hard tissue engineering. Materials 14:674. https://doi.org/10.3390/ma14030674

Omollo Oduor E, Wanjiru Ciera L, Kamalha E (2021) Applications of Silk in Biomedical and Healthcare Textiles [Internet]. Textiles for Functional Applications. IntechOpen. Available from: https://doi.org/10.5772/intechopen.96644

Patil PP, Reagan MR, Bohara RA (2020) Silk fibroin and silk-based biomaterial derivatives for ideal wound dressings. Int J Biol Macromol 164:4613–4627. https://www.sciencedirect.com/science/article/pii/S0141813020341283

Perrone GS, Leisk GG, Lo TJ et al (2014) The use of silk-based devices for fracture fixation. Nat Commun 5:3385. https://doi.org/10.1038/ncomms4385. PMID: 24594992

Pesode P, Barve S, Wankhede SV, Ahmad A (2023) Sustainable materials and Technologies for Biomedical Applications. Adv Mater Sci Eng 2023:1–22. https://doi.org/10.1155/2023/6682892

Pham DT, Thao NTP, Thuy BTP et al (2022) Silk fibroin hydrogel containing Sesbania sesban L. extract for rheumatoid arthritis treatment. Drug Deliv 29(1):882–888. https://doi.org/10.1080/10717544.2022.2050848. PMID: 35277106; PMCID: PMC8920400

Place ES, Evans ND, Stevens MM (2009) Complexity in biomaterials for tissue engineering. Nat Mater 8:457–470. https://doi.org/10.1038/nmat2441

Poland B, Dooris M (2010) A green and healthy future: the settings approach to building health, equity and sustainability. Crit Public Health 20:281–298. https://doi.org/10.1080/09581596.2010.502931

Qi Y, Wang H, Wei K et al (2017) A review of structure construction of silk fibroin biomaterials from single structures to multi-level structures. Int J Mol Sci 18:237. https://doi.org/10.3390/ijms18030237

Rayner M, Östbring K, Purhagen J (2016) Application of natural polymers in food. In: Olatunji O (ed) Natural polymers. Springer International Publishing, Cham, pp 115–161

Rockwood DN, Preda RC, Yücel T et al (2011) Materials fabrication from Bombyx mori silk fibroin. Nat Protoc 6:1612–1631. https://doi.org/10.1038/nprot.2011.379

Ru M, Hai AM, Wang L et al (2023) Recent progress in silk-based biosensors. Int J Biol Macromol. 224:422–436. https://doi.org/10.1016/j.ijbiomac.2022.10.134. Epub 2022 Oct 18. PMID: 36270404
Salvetti M, Lubetzki C, Kapoor R et al (2018) Steps towards collective sustainability in biomedical research. Trends Mol Med 24:429–432. https://doi.org/10.1016/j.molmed.2018.03.001
Sandler J, Shaffer MSP, Prasse T et al (1999) Development of a dispersion process for carbon nanotubes in an epoxy matrix and the resulting electrical properties. Polymer 40:5967–5971. https://doi.org/10.1016/S0032-3861(99)00166-4
Shao Z, Vollrath F (2002) Surprising strength of silkworm silk. Nature 418:741–741. https://doi.org/10.1038/418741a
Si R, Chen W, Chen J et al (2023) Green chemistry fabrication of durable antimicrobial peptide-immobilized silk fibroin films for accelerated full-thickness wound healing. Mater Today Chem 29:101468. https://doi.org/10.1016/j.mtchem.2023.101468
Silva ACQ, Silvestre AJD, Vilela C, Freire CSR (2021) Natural polymers-based materials: a contribution to a greener future. Molecules 27:94. https://doi.org/10.3390/molecules27010094
Sun W, Gregory DA, Tomeh MA, Zhao X (2021) Silk fibroin as a functional biomaterial for tissue engineering. Int J Mol Sci 22:1499. https://doi.org/10.3390/ijms22031499
Tang Y, Zhang L, Sun R et al (2023) Pulmonary delivery of mucus-traversing PF127-modified silk fibroin nanoparticles loading with quercetin for lung cancer therapy. Asian J Pharm Sci 18:100833. https://doi.org/10.1016/j.ajps.2023.100833
Ter Horst B, Moiemen NS, Grover LM (2019) Natural polymers. In: Biomaterials for skin repair and regeneration. Elsevier, pp 151–192
Torchilin V (2011) Tumor delivery of macromolecular drugs based on the EPR effect. Adv Drug Deliv Rev 63:131–135. https://doi.org/10.1016/j.addr.2010.03.011
Tsioris K, Raja WK, Pritchard EM et al (2012) Fabrication of silk microneedles for controlled-release drug delivery. Adv Funct Mater 22:330–335. https://doi.org/10.1002/adfm.201102012
Tufan Y, Öztatlı H, Doganay D et al (2023) Multifunctional Silk Fibroin/Carbon Nanofiber Scaffolds for In Vitro Cardiomyogenic Differentiation of Induced Pluripotent Stem Cells and Energy Harvesting from Simulated Cardiac Motion. ACS Appl Mater Interfaces 15(36):42271–42283. https://pubs.acs.org/doi/10.1021/acsami.3c08601
Uebersax L, Mattotti M, Papaloïzos M et al (2007) Silk fibroin matrices for the controlled release of nerve growth factor (NGF). Biomaterials 28(30):4449–4460. https://www.sciencedirect.com/science/article/pii/S0142961207005297
Vepari C, Kaplan DL (2007) Silk as a biomaterial. Prog Polym Sci 32:991–1007. https://doi.org/10.1016/j.progpolymsci.2007.05.013
Wang Y, Blasioli D, Kim H et al (2006) Cartilage tissue engineering with silk scaffolds and human articular chondrocytes. Biomaterials 27:4434–4442. https://doi.org/10.1016/j.biomaterials.2006.03.050
Wang X, Yucel T, Lu Q et al (2010) Silk nanospheres and microspheres from silk/pva blend films for drug delivery. Biomaterials 31:1025–1035. https://doi.org/10.1016/j.biomaterials.2009.11.002
Wang Z, Yu H, Zhao Z (2021) Silk fibroin hydrogel encapsulated graphene filed-effect transistors as enzyme-based biosensors. Microchem J 169:106585. https://doi.org/10.1016/j.microc.2021.106585
Williams D (2005) In: Buddy D. Ratner, editors, Academic Press (2004) ISBN 0-12-582463-7 (864pp., $95/49.99). Biomaterials 26:5093–5093. https://doi.org/10.1016/j.biomaterials.2005.01.034
Xu HL, ZhuGe DL, Chen PP et al (2018) Silk fibroin nanoparticles dyeing indocyanine green for imaging-guided photo-thermal therapy of glioblastoma. Drug Deliv 25(1):364–375. https://doi.org/10.1080/10717544.2018.1428244. PMID: 29366360; PMCID: PMC6058481

Yonesi M, Garcia-Nieto M, Guinea GV et al (2021) Silk fibroin: an ancient material for repairing the injured nervous system. Pharmaceutics 13:429. https://doi.org/10.3390/pharmaceutics13030429

You C, Wang C, Ma Z, Yu Q, Liu S (2025) Review on application of silk fibroin hydrogels in the management of wound healing. Int J Biol Macromol 298:140082 https://doi.org/10.1016/j.ijbiomac.2025.140082

Zhu B, Wang H, Leow WR et al (2016) Silk fibroin for flexible electronic devices. Adv Mater 28:4250–4265. https://doi.org/10.1002/adma.201504276

# Marine Biopolymers from Fish and Seafood Processing By-Products

# 9

Jesus Valcarcel, Adrián Pedreira, and José Antonio Vázquez

**Abstract**

The supply chain of seafood products creates a large amount of waste, from fishing discards to by-products of industrial processing. This biomass contains however valuable materials with potential in tissue engineering and regenerative medicine. The present chapter focuses on the recovery of biopolymers from seafood waste with applications in this field from the point of view of sustainability in a broad sense. Such perspective allows in the first place the identification of raw materials suitable for the isolation of the biopolymers of interest, based on their abundance as by-products and discards, and leaving aside endangered species. Conversely, this selection narrows the number of biopolymers amenable to sustainable production, which concentrate on chondroitin sulfate, chitin, chitosan, collagen and gelatin. And secondly, it provides the scope to analyse how

J. Valcarcel (✉)
Grupo de Reciclado y Valorización de Materiales Residuales (REVAL), Instituto de Investigaciones Marinas (IIM-CSIC), Vigo, Galicia, Spain

Unidad de apoyo científico-técnico (UACT), Instituto de Investigaciones Marinas (IIM-CSIC), Vigo, Galicia, Spain
e-mail: jvalcarcel@iim.csic.es

A. Pedreira
Grupo de Reciclado y Valorización de Materiales Residuales (REVAL), Instituto de Investigaciones Marinas (IIM-CSIC), Vigo, Galicia, Spain

Grupo de Biosistemas e Ingeniería de Bioprocesos (Bio2eng), Instituto de Investigaciones Marinas (IIM-CSIC), Vigo, Galicia, Spain

J. A. Vázquez
Grupo de Reciclado y Valorización de Materiales Residuales (REVAL), Instituto de Investigaciones Marinas (IIM-CSIC), Vigo, Galicia, Spain



J. M. Oliveira et al. (eds.), *Sustainable Scaffolds-based Strategies in Tissue Engineering and Regenerative Medicine*, Biomaterials, Bioengineering and Sustainability 8, https://doi.org/10.1007/978-3-031-96274-5_9

green are the most widely used valorization processes to recover each biomaterial, proposing cleaner alternatives and outlining steps towards their integral valorization by finding uses for the effluents stemming from the isolation processes. This ultimately contributes to the viability of the valorization processes and places seafood production closer to the zero-waste goal.

**Keywords**

Seafood products · Biopolymer recovery · Sustainability · Tissue engineering · Valorization processes

## 9.1 Introduction

The marine environment represents a rich source of materials with a number of properties useful for their potential application in the biomedical field. Although the greatest efforts have been carried out in the discovery of new drugs, polysaccharides, proteins and mineral matter extracted from algae, animals and marine microorganisms have been characterized in recent decades and proposed as components of systems for the controlled release of drugs, and of tissue engineering scaffolds and constructs (Wan et al. 2021; Zhang et al. 2022a). However, the current overexploitation of the marine environment imposes restrictions on the viability of obtaining biomaterials from threatened species, or when their extraction involves a considerable environmental impact.

In this scenario, obtaining these materials from fish discards and waste generated in industrial food processing becomes especially relevant, as it implies a better use of the marine resources, with the consequent environmental and economic benefits. Fisheries and aquaculture production peaked at 214 million tonnes of aquatic animals in 2020 (FAO 2022). Typically, between 36% and 53% of this biomass comprise non-edible fractions (head, viscera, skeletons, bones, etc.) (FAO 2022), and in the case of wild capture, discards account for an extra 9.1 million tonnes (10% of the global catch estimates) (Gilman et al. 2020), leaving the figure of underutilized material in excess of 60 million tonnes. Traditionally, some fish by-products have ended in the production of fishmeal and oil, but extraction of higher value materials would aid in the economic viability of valorisation processes, thus enhancing the sustainability of the fishing industry.

With this vision, we have identified biomaterials amenable to sustainable extraction from seafood waste and at the same time with potential as high-value products in tissue engineering and regenerative medicine, due to their bioactivity, accessibility, abundance, and performance. It is worth mentioning here that although algae contain a wealth of useful compounds, the amount of by-products resultant of farming and processing is rather limited, and therefore we have not considered them herein.

## 9.2 Chondroitin Sulfate

Glycosaminoglycans (GAGs) designate a family of negatively charged polydisperse linear polysaccharides. Although also present in prokaryotes, GAGs ubiquity manifest mainly on animal tissue, both on the surface of cells and in the extracellular matrix, where they participate in fundamental biological processes, such as cellular signalling, tissue development, anticoagulation, wound repair, and disease progression (Casale and Crane 2019; Perez et al. 2023). Their common feature is the presence of an amino sugar (*N*-hexosamine, usually acetylated) linked to an acid sugar (uronic acid) or galactose, forming disaccharide units. The glyosidic linkage of disaccharides units forms polymers with varying degrees of complexity, not only due to diversity in the sequence of units, but also because sulfation in different positions of both rings commonly occurs. Hence, GAGs can be distinguished based on the building disaccharide blocks, from simple hyaluronic acid, through to keratan sulfate, chondroitin sulfate, and dermatan sulfate, up to complex heparin and heparin sulfate.

The extensive presence of GAGs in animal tissue has led to the discovery of an array of structures in both vertebrates and invertebrates, some widespread across species but others containing rare variations peculiar to a particular organism. In many instances such unique GAGs exist only in marine animals, reflecting the evolutionary differences with terrestrial counterparts (Yamada et al. 2011; Valcarcel et al. 2017). Singular GAG structures occur mainly in marine invertebrates, with associated bioactivities potentially useful in tissue regeneration and pharmacology, for example, neurite outgrowth due to unusual sulfation in chondroitin sulfate (Higashi et al. 2015); or anticoagulant activity of branched chondroitin sulfate, oversulfated dermatan sulfate, and heparin analogues (Ustyuzhanina et al. 2021; Lan et al. 2023). An important drawback that compromise the viability and sustainability of many of these biomolecules comes from the fact that they are sourced from wild animals (starfish, sand dollar), sometimes also marketed as food (clams, sea cucumbers), and when extracted from seafood by-products yields are low (heparin from shrimp heads, chondroitin sulfate from squid pen cartilage) (Valcarcel et al. 2017).

Considering these premises, sustainably sourced GAGs must come from abundant waste materials rich in the biomolecule of interest. From this point of view, probably chondroitin sulfate represents the most prevalent GAG from marine waste, as it is one of the main components in the cartilage of chondrichthyan fish. This class of animals features a skeleton made of cartilage instead of bones, and includes sharks, rays, and chimaeras, but only the first two are commercially fished. In both cases, industrial processing results in by-products (more than 50% of the whole fish) containing a large fraction of cartilage. For instance, cartilage comprises around 25% of the waste generated in blue shark processing (Vázquez et al. 2016a, 2020; López-Álvarez et al. 2020).

Furthermore, individuals below the minimum legal size, or belonging to non-commercial species, are considered discards, and have traditionally been returned to the sea. A number of countries have implemented legislation to tackle this wasteful practice, such as the landing obligation in the EU (European Comission 2020), which forces fishing vessels to land all catched fish under certain conditions. As a result, there is an important residual biomass susceptible to valorization to obtain a number of added-value compounds, ideally from a biorefinery perspective, including chondroitin sulfate (García-Santiago et al. 2020).

Chondroitin sulfate, as other members of the glycosaminoglycan family, participate in fundamental cellular processes, such as communication, differentiation and cell growth (Yamada and Sugahara 2008). These biological properties make chondroitin sulfate interesting in cartilage regeneration, either as a nutraceutical or incorporated into tissue engineering scaffolds, in nerve regeneration, and as an anti-inflammatory, antimetastatic, and anticoagulant agents (Valcarcel et al. 2017).

The bioactivity of chondroitin sulfate depends largely on its ability to interact with proteins, which is determined by its chemical properties and therefore by its structure (Benito-Arenas et al. 2018; Vallet et al. 2022). Such structure is based on a sequence of glucuronic acid (GlcA) and N-acetylgalactosamine (GalNAc) linked by alternating β-(1 → 4) and β-(1 → 3) glycosidic bonds. Both rings can be sulfated in different positions, giving rise to different disaccharide units arranged sequentially in polymer chains of variable length, which results in considerable heterogeneity in molecular mass, sulfation pattern and charge density (Fig. 9.1) (Valcarcel et al. 2017; Kjellén and Lindahl 2018).

Marine chondroitin sulfate differs from that extracted from terrestrial animals in both sulfation and molecular weight, being molecular weight generally higher in marine materials. With regard to sulfation, in terrestrial sources disaccharides predominate sulfated in position −4 of GalNAc (CS-A), while in marine animals the majority of units are sulfated in position −6 (CS-C), accompanied by a significant proportion of disulfated units (López-Álvarez et al. 2019). Table 9.1 displays molecular weight and sulfation pattern of CS extracted from marine and terrestrial species. Noteworthy, we have not included in the list marine endangered species, but only those abundant and either commercially fished or captured as by-catch.

Molecular weight as "number average molecular weight" (Mn) in kDa with associated polydispersity index in parentheses. Unsulfated: GlcA-GalNAc 0S; A-units: GlcA-GalNAc 4S; C-units: GlcA-GalNAc 6S; B-units: GlcA 2S-GalNAc 4S; D-units: GlcA 2S-GalNAc 6S; E-units: GlcA -GalNAc 4,6S.

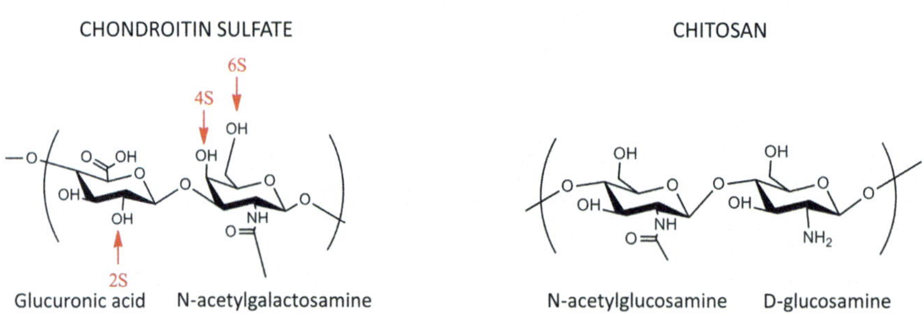


**Fig. 9.1** Chemical structures of chondroitin sulfate and chitosan

**Table 9.1** Characterization of chondroitin sulfate from cartilage from fish vs terrestrial species

| Source | | % of disaccharide units | | | | | | Ref |
|---|---|---|---|---|---|---|---|---|
| | Mn | 0 | A | C | D | E | B | |
| Blue shark (*Prionace glauca*) | 60 (1.2) | 16 | 10 | 64 | 9 | – | – | Novoa-Carballal et al. (2017) |
| Blackmouth catshark (*Galeus melastomus*) | 55 (1.3) | 4 | 24 | 55 | 15 | 1 | 1 | Vázquez et al. (2018a) |
| Small-spotted catshark (*Scyliorhinus canicula*) | 43 (1.5) | 23 | 32 | 32 | 16 | – | – | Novoa-Carballal et al. (2017) |
| Thornback skate (*Raja clavata*) | 44 (1.3) | 16 | 20 | 56 | 9 | – | – | Novoa-Carballal et al. (2017) |
| Rabbit fish (*Chimaera monstrosa*) | 51 (1.4) | 4 | 22 | 53 | 17 | 2 | 1 | Vázquez et al. (2019b) |
| Bovine | 10–15 (1.8–2.2) | 6 | 60 | 33 | – | – | – | Volpi (2007) |
| Porcine | 9–13 (1.4–1.7) | 6 | 80 | 14 | – | – | – | Volpi (2007) |
| Chicken | 9–13 (1.6–2.0) | 8 | 72 | 20 | – | – | – | Volpi (2007) |

As can be seen, there are differences in both molecular weigth and composition between the selected species, with the exception of Rabbit fish (*C. monstrosa*) and Blackmouth catshark (*G. melastomus*), which present very similar characteristics. Blue shark (*P. glauca*) and Small-spotted catshark (*S. canicula*) are notable, since the former shows the highest molecular weight and an exceptionally high percentage of CS-C, while the characteristics of the latter would be found halfway between marine and terrestrial chondroitin sulfate, with a similar percentage of CS-A and CS-C.

Another important aspect related with the sustainability of the materials concerns the process to isolate and purify the biopolymers, bearing also in mind its feasibility for upscaling. As is common in the isolation of natural macromolecules, protocols to produce natural polysaccharides include homogenization of the tissue, hydrolysis to break down the matrix, and release of the target molecule followed by purification. In the case of CS, initial separation of the cartilage helps in the following steps. With the idea of reducing the environmental impact, the CS process design may replace the initial chemical deproteinization conventionally used, with enzymatic hydrolysis. A number of enzymes have proven suitable for this purpose, such as papain, trypsin, or subtisilin (Panagos et al. 2014; Maccari et al. 2015; Lin et al. 2017). Ultrasound treatment applied to both chemical and enzymatic deproteinization have proven capable of increasing CS yields (Yang et al. 2021; Tsai et al. 2023), although scale-up of this technology is problematic (Al Khawli et al. 2019). Methods used for subsequent purification include alcoholic or cetylpyridinium chloride precipitation, anion exchange chromatography and dialysis (Zhao et al. 2013; Panagos et al. 2014; Maccari et al. 2015; Bougatef et al. 2020; Zhou et al. 2020; Wang et al. 2021).

Most published protocols, however, do not operate under optimal conditions, leading to excessive use of reagents and time, and incorporate time-consuming techniques such as dialysis for several days. From a green perspective, the efficient and cheaper isolation may be achieved by optimizing enzymatic proteolysis, followed by hydroalcoholic-saline precipitation, and final purification with ultrafiltration membranes, establishing in each case the optimal operating parameters (Murado et al. 2010; Vázquez et al. 2016a, 2018a, 2019b). This approach manages to reduce volume and concentration of reagents, making also possible to recover bioapatites for biomedical applications and protein hydrolysates for aquaculture feed, increasing the sustainability of the process (Fig. 9.1).

## 9.3 Chitin and Chitosan

Chitin is a linear polysaccharide formed by N-acetylglucosamine monomers linked by $\beta(1 \rightarrow 4)$ glycosidic bonds (Fig. 9.1). Its function is mainly structural, being a fundamental component of the skeletons and scales of some animals, and the cell walls of fungi. It is plentiful in nature, ranking as the second most abundant polymer after cellulose (Piekarska et al. 2023). Due to its very limited solubility, it is generally partially deacetylated to give rise to chitosan, a derivative soluble in acidic medium.

The applications of chitosan are very diverse, ranging from agriculture, food, or medicine, to cosmetics (Hamed et al. 2016). Within the biomedical field, research on chitosan and derivatives in biomaterials to reconstruct bone and articular cartilage defects stands out (LogithKumar et al. 2016; Li et al. 2018), as well as in nanocomposites for the controlled release of drugs (Ali and Ahmed 2018).

Commercial chitosan is generally obtained from chitin extracted from the shell of crustaceans or isolated after fermentation of fungi. In the chitin from these sources, the polymer chains are in an antiparallel fashion (α-chitin), however in the squid pen, these are arranged in parallel (β-chitin), giving rise to lower inter- and intra-molecular forces that result in greater water absorption capacity and a more homogeneous chitosan than that obtained from α-chitin (Domard 2011). Some studies have also found differences between the two in tissue engineering applications (Reys et al. 2013; Shavandi et al. 2015). It is however difficult to pinpoint the differences observed, because besides chain arrangement, others factor such as molecular weight or degree of acetylation affect chitosan properties.

Both crustacean and squid chitosan represent sustainable sources, since chitin is isolated from food processing by-products, i.e. crustacean shells and squid pens. On the one hand, the process to isolate squid chitin is more environmentally-friendly than the crustacean counterpart, because the latter requires additional acid demineralization and decolouration to remove mineral matter, mainly calcium carbonate, and carotenoid pigments in the shells, steps usually not necessary for squid pens (Cuong et al. 2016; Mohan et al. 2021) (Fig. 9.2).

Although decolouration complicates the process and requires organic solvents, the astaxanthin recovered, a carotenoid with antioxidant properties, can contribute

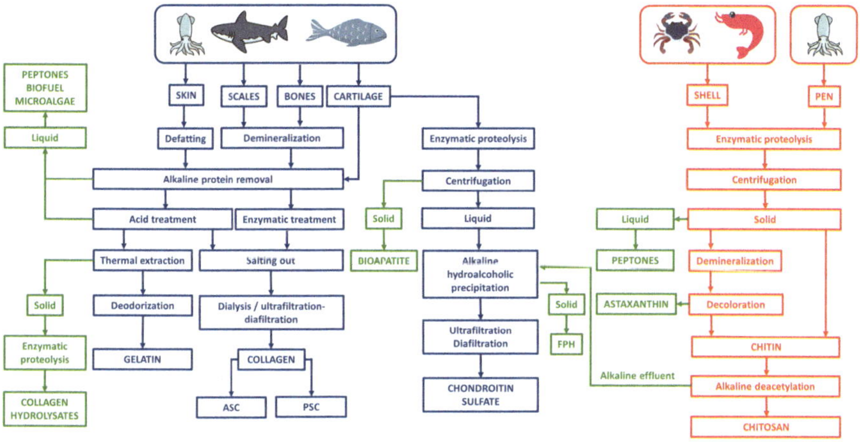


**Fig. 9.2** Valorization processes to obtain biopolymers

to the economic viability of the valorization of crustacean shells (Vázquez et al. 2019a). In both cases, chitin is linked to proteins, requiring therefore hydrolysis to release the polysaccharide. Traditionally, proteolysis has been carried out by alkaline hydrolysis, but enzymatic reactions have proven suitable to replace the chemical methods. Furthermore, the liquid effluents rich in protein can be used as nitrogen source (peptones) in microbial culture (Vázquez et al. 2016b), contributing to the sustainability of the process.

Emerging technologies for deproteinization, such as subcritical water extraction have shown good performance, reaching 96% deproteinization and yielding 82% of chitin from shrimp cephalotorax, although the process takes place at high temperature (260 °C) and pressure (Espíndola-Cortés et al. 2017). Among these technologies, deep eutectic solvents have probably received most of the attention, as demineralization and deproteinization can be achieved in one step (Zhu et al. 2017; Saravana et al. 2018), although depolimerization occurs to a larger extent than in conventional treatments (Hong et al. 2018). Choline chloride-malonic acid combinations are possibly the most suitable systems, showing good yields of chitin and moderate depolymerisation (Zhu et al. 2017; Hong et al. 2018; Saravana et al. 2018). The sustainability of these solvents is however questionable because of the toxicity and limited biodegradability shown by certain mixtures (McReynolds et al. 2022).

As a final step, a heterogeneous reaction of solid chitin with a concentrated alkaline solution at temperatures above 70 °C partially removes the acetyl groups from chitin to obtain chitosan. The process can be assisted by microwave radiation, reducing reaction time while increasing yield (Alishahi et al. 2011; Nouri et al. 2016). Deep eutectic solvents capable of dissolving chitin also facilitate the reaction (Vicente et al. 2020). Alkali concentration, temperature and reaction time influence the degree of deacetylation, but also the molecular weight due to side reactions of depolymerisation (Galed et al. 2005; Gîjiu et al. 2022). Table 9.2 lists molecular

**Table 9.2** Molecular weight and deactylation degree of chitosan from different sources. Molecular weight as "number average molecular weight" (Mn) in kDa. with associated polydispersity index in parentheses; DD (%): degree of deacetylation

| Source | Mn | DD (%) | Ref |
|---|---|---|---|
| Squid pen (*Illex argentinus*) | 143 (1.3) – 339 (1.3) | >93 | Vázquez et al. (2017a) |
| Squid pen (*Loligo opalescens*) | 139 (1.4) – 300 (1.7) | >94 | Vázquez et al. (2018b) |
| Squid | 88 (3.3) | 98 | Weinhold et al. (2009) |
| Prawn cephalotorax (*Penaeus vannamei*) | 58 (1.8) – 93 (1.3) | >92 | Vázquez et al. (2017b) |
| Prawn shells (*Penaeus vannamei*) | 2.4–9.9 | 94–98 | Mohammed et al. (2013) |
| Crab | 53 (3.2) – 288 (2.2) | 87–97 | Weinhold et al. (2009) |
| Crab | 10.6 (3.8) | 96 | Alexeev et al. (2000) |

weight and degree of deacetylation of chitosan isolated from squid pen and crustacean shells. In both cases, high degrees of deacetylation can be achieved, which favours the solubility of the polymers.

While enzymatic deproteinization of pens and shells to obtain chitin is feasible, realistic green alternatives to strong alkali treatment do not currently exist for the deacetylation of chitin to obtain chitosan. Some microorganisms (e.g., *Mucor rouxi*, *Rhizopus stolonifera*, *Aspergillus nidulans*) produce enzymes capable of attacking chitin, but so far yields and degrees of deacetylation achieved do not match the alkaline treatment for high molecular weight chitin (Win and Stevens 2001; Chatterjee et al. 2005; Ladchumananandasivam et al. 2012; Harmsen et al. 2019; Amer et al. 2022; Yang et al. 2022a; Rakshit et al. 2023). This is mainly because of chitin crystallinity impairing enzymatic reactions (Jaworska and Roberts 2016).

Sustainability of the deacetylation of chitin can be increased by using the resulting alkaline effluents as input for other processes following a biorefinery concept, where a number of valorization processes are integrated aiming at zero waste (Veríssimo et al. 2021). Within this framework, the deacetylation effluent can be salvaged to impart alkalinity to the selective precipitation of chondroitin sulfate from fish cartilage (Vázquez et al. 2018b), as depicted in Fig. 9.2.

## 9.4 Collagen and Gelatin

Collagen is a structural protein and fundamental component in the connective tissue of animals, being especially abundant in skin, bones, and ligaments. It is made up of a trifibrillar structure (triple helix) with a molecular weight of around 300 kDa composed mainly of glycine, proline, and hydroxyproline. This collagen motif is capable of self-assembling to form fibril networks, with mechanical and biological properties dependent on how aggregation takes place, which gives rise to more than 20 types of collagen (Brodsky and Persikov 2005; Ricard-Blum 2011). However, only a few types are commercially available, mainly type-I from the skin and tendons and type-II from the cartilage of bovine, porcine, and sometimes avian origins. These conventional sources have been put into questions due to sanitary reasons (bovine spongiform encephalopathy) or are not acceptable because of cultural reasons (religion, vegetarianism), presenting collagen of marine origin as an alternative (Coppola et al. 2020).

Collagen displays a rather limited solubility, restricted to acidic media, which may be increased and widened by partial thermal hydrolysis, rendering water-soluble gelatin. The properties of gelatin will vary depending on the source and the degree of breakdown of the initial collagen (Alipal et al. 2021). Collagen has been extracted from a range of marine organisms including vertebrates and invertebrates (Salvatore et al. 2020), while gelatin in most cases is produced from fish tissues (Gómez-Guillén et al. 2002; Milovanovic and Hayes 2018). Collagen from invertebrates such as sea urchins, sponges, sea cucumbers, and jellyfish have received a great deal of attention for biomedical applications (Lim et al. 2019; Senadheera et al. 2020; Fassini et al. 2021), but generally these organisms are not commercial species. From the point of view of sustainability, processing by-products and discards would be preferred as collagen and gelatin sources, which narrows the selection to certain fish and cephalopods. Some examples are listed in Table 9.3.

It must be noted that the collagen extracted from these fish by-products is for the most part type-I in skin, scales and bone, and type-II in cartilage, while types –III and –IV are present in sponges and jellyfish (Jafari et al. 2020). However, type-I is the most commonly found in animals and the most widely used in biomedicine (Rezvani Ghomi et al. 2021). Another important consideration is thermal stability, as the denaturation temperatures of collagen and related strength of the gels formed with gelatin are lower than those from mammalian sources, as displayed in Table 9.3, and may limit their use at body temperature. Such differences relate to the amino-acid composition, in particular to the proportion and position of the imino acids proline and hydroxyproline, which in turn influence the stability of the collagen triple helix (Brodsky and Persikov 2005). In this regard, cold-water fish species show lower proportion of imino acids than warm-water counterparts, resulting generally in reduced thermal stability (Harvey et al. 2021).

Besides species and tissue, the extraction process also affects the properties of the resulting collagen and gelatin. Both processes require an initial pre-treatment to remove non-collagenous protein, usually carried out with NaOH for a period of time varying from hours to days (Liu et al. 2014; Wang et al. 2018). For species with high fat content (e.g., salmon skin) the process must include a defatting step with alcohols, such as butyl alcohol or isopropanol, being also necessary demineralization for scales and bones, typically with hydrochloric acid or EDTA (Skierka et al. 2007; Liu et al. 2014; Mahboob 2015).

In the case of collagen, two main extraction methods exist to solubilize the material. The first method uses dilute acid to break inter- and intra- molecular bonds to release the collagen molecules, leading to what is known as acid-soluble collagen (ASC). The collagenous material undergoes similar acid treatments for gelatin production, but with different conditions such as shorter treatment times (Fig. 9.2).

**Table 9.3** Collagen and gelatin properties from sustainable marine sources. DT: denaturation temperature. Gel strength expressed as *g* Bloom

| Tissue | Organism | Collagen DT (°C) | Gelatin gel strength | Ref |
|---|---|---|---|---|
| Skin | Nile Tilapia (*Oreochromis niloticus*) | 32–34 | 128–382 | Zhang et al. (2016); Song et al. (2019); Liao et al. (2021) |
| | Barramundi (*Lates calcarifer*) | 33 | 200–369 | Sinthusamran et al. (2013, 2014) |
| | Turbot (*Scophthalmus maximus*) | 30 | 82–177 | Valcarcel et al. (2021a) |
| | Sturgeon (*Acipenser schrenckii*) | 32–33 | 105–141 | Wang et al. (2014); Tang et al. (2015) |
| | Cuttlefish (*Sepia lycidas*) | 27 | | Nagai et al. (2001) |
| | Cuttlefish (*Sepia officinalis*) | | 181–259 | Balti et al. (2011) |
| | Salmon (*Salmo salar*) | 30 | 45–98 | Alves et al. (2017); Vázquez et al. (2021) |
| | Spanish mackerel (*Scomberomorous niphonius*) | 15 | – | Li et al. (2013) |
| | Bigeye tuna (*Thunnus obesus*) | 32–34 | 30–260 | Azhara et al. (2018); Ahmed et al. (2019) |
| | Seabream (*Sparus aurata*) | 34 | 230 | Valcarcel et al. (2021b) |
| | Seabass (*Dicentrarchus labrax*) | – | 181 | Valcarcel et al. (2021b) |
| | Seabass (*Lates calcarifer*) | 33 | – | Sinthusamran et al. (2013) |
| | Rainbow trout (*Oncorhynchus mykiss*) | 19–31 | 95 | Lee et al. (2016); Valcarcel et al. (2021b) |
| | Cod (*Gadus morhua*) | 16 | 15–83 | Carvalho et al. (2018); Alves et al. (2022) |
| Scales | Horse mackerel (*Trachurus japonicus*) | 25 | – | Okazaki and Osako (2014) |
| | Grey mullet (*Mugil cephalis*) | 27 | – | Okazaki and Osako (2014) |
| | Sardine (*Sardinops melanostictus*) | 28 | – | Nagai et al. (2004) |
| | Seabass (*Lateolabrax japonicus*) | 38 | – | Nagai et al. (2004) |
| | Seabass (*Lates calcarifer*) | | 270 | Cao et al. (2017) |
| | Bigeye tuna (*Thunnus obesus*) | 32 | | Ahmed et al. (2019) |
| | Yellowfin tuna (*Thunnus albacares*) | – | 290 | Karayannakidis and Zotos (2015) |
| | Nile Tilapia (*Oreochromis niloticus*) | 32–33 | 280–370 | El-Rashidy et al. (2015); Chen et al. (2016a); Peng et al. (2022) |
| Bones | Nile Tilapia (*Oreochromis niloticus*) | 33 | 295 | Liu and Huang (2016); Tinrat and Sila-asna (2017) |

(continued)

**Table 9.3** (continued)

| Tissue | Organism | Collagen DT (°C) | Gelatin gel strength | Ref |
|---|---|---|---|---|
| | Bigeye tuna (Thunnus obesus) | 32 | – | Ahmed et al. 2019) |
| | Atlantic bluefin tuna (*Thunnus thynnus*) | – | 109 | Haddar et al. (2011) |
| | Spanish mackerel (*Scomberomorous niphonius*) | 17–18 | 196 | Li et al. (2013); Mardawati et al. (2018) |
| Cartilage | Sturgeon (*Acipenser schrenckii*) | 33–36 | – | Liang et al. (2014) |
| | Sturgeon (*Acipenser baerii*) | – | 206 | Zhang et al. (2022b) |
| | Blue shark (*Prionace glauca*) | 42 | – | Pan et al. (2023) |
| Skin | Bovine | 41 | 193–270 | Hafidz et al. (2011); Subhan et al. (2015) |
| Skin | Porcine | 37 | 326–415 | Hafidz et al. (2011); Subhan et al. (2015) |

Commonly this is performed with acetic acid, although citric, lactic and hydrochloric acids have also been used (Sadowska et al. 2003; Blanco et al. 2019; Shaik et al. 2021; Sukeri et al. 2021). The second method uses enzymes to remove non-helical protein and increase the solubility of collagen. Pepsin is by far the most common, leading to pepsin soluble collagen (PSC), although others such as papain have also been used (Liu et al. 2012; Zeng et al. 2012; González-González et al. 2022). Proteolysis takes place at low pH where pepsin is most active, and sometimes the enzyme is added after acidic extraction to improve efficiency (Veeruraj et al. 2013; Dong and Dai 2022). After the extraction, collagen is precipitated by the addition of NaCl, centrifuged, and the residue redissolved in acetic acid. Final purification by dialysis or membrane ultrafiltration/diafiltration renders collagen ready for freeze-drying (Tang et al. 2015; Chen et al. 2016b).

Gelatin extraction mainly differs from the collagen process in the partial thermal denaturation applied to the pre-treated tissue, although gelatin pre-treatments may include additional acid treatment and different conditions to collagen (temperature, time, solid:liquid ratios, etc.) (Alfaro et al. 2015; Usman et al. 2022). Thermal extraction is usually carried out in water at 40–80 °C for several hours (Arnesen and Gildberg 2007; Cheow et al. 2007; Nagarajan et al. 2012), breaking down the triple helix of collagen. Upon cooling, the released protein coils partially rearrange into a helical structure forming gels with characteristic thermal stabilities (Yang et al. 2022b) (gel strengths in Table 9.3). The extent of the thermal conversion is related to the temperature and time applied, with higher temperatures and longer times producing more degradation of the collagen structure, therefore favouring gelatin yield at the expense of mechanical properties (Nagarajan et al. 2012). Once extracted, the gelatin solution is filtered and may undergo deodorization if necessary, e.g. with activated charcoal or β-cyclodextrin (Zhang et al. 2017; Valcarcel et al. 2021a), prior to drying.

Advanced techniques are being developed to increase extraction efficiency of collagen and gelatin (Al Khawli et al. 2019; Usman et al. 2022), but so far their development is still limited. In bighead carp scales, thermal extraction aided by ultrasound managed to recover more gelatin than the conventional method, while maintaining its physical properties, provided short times are used (Tu et al. 2015; Huang et al. 2017).

In the case of collagen, ultrasonic treatment increased ASC yield in seabass skin while reducing acid concentration, but amplitude and time must be carefully controlled to prevent collagen degradation (Kim et al. 2013). Collagen yield was also increased by extraction in deep eutectic solvents (e.g. lactic acid—urea in the skin of cod; citric acid—xylitol in the skin of blue shark) with similar or even superior quality to ASC, but toxicity and limited biodegradability of the solvents may compromise its sustainability (Bisht et al. 2021; Batista et al. 2022). Another proposed alternative is supercritical fluid extraction, which manages to reduce extraction times and simplify pre-treatment steps. (Sousa et al. 2020; Phon et al. 2023). Scaling-up of this technique is however difficult because of the cost and complexity of the equipment (Al Khawli et al. 2019).

Considering the raw materials, collagen and gelatin extraction starts from the right point in terms of sustainability, since skin, scales, bones and cartilage represent abundant by-products from the fish processing industry. However, the sustainability of the processes displays certain limitations. For instance, the isolation of collagen requires low temperatures, usually 4 °C, to avoid its denaturation, and gelatin extraction needs at least 40 °C. Furthermore, solid to liquid ratios are sometimes high, especially in the collagen process, surpassing in many cases 1:10 ratios. This leads to a considerable consumption of energy and high volumes of effluents (Vázquez et al. 2019a). Also, collagen yields from scales and bones are lower than those from skin (Pal and Suresh 2016; Salvatore et al. 2020), requiring also additional steps (demineralization). Since scales, bones and skin all contain type-I collagen, the latter might be preferred as collagen and gelatin sources.

One way to turn the processes more sustainable is to use the resulting effluents as raw materials for other processes. With this view, protein-rich alkaline solutions from deproteinization of fish skins retain value as nitrogen sources for microbial cultures with biotechnological applications (Vázquez et al. 2023a, b), for the cultivation of microalgae (Blanco et al. 2020) or biofuel production (Tawfik et al. 2021). On the other hand, the controlled enzymatic hydrolysis, with endo and/or exoproteases, of both collagen and gelatin gives rise to smaller peptides –generally in the range of 500–10,000 Da– usually known as collagen hydrolysates, widely marketed nowadays in the nutraceutical and cosmetic industries (Hong et al. 2019; Xu et al. 2023). The remaining solids after gelatin extraction can be equally subjected to proteolysis, resulting in a protein hydrolysate retaining a considerable amount of collagen peptides, as shown by its proline and hydroxyproline proportions (Valcarcel et al. 2021a). Such additional routes to make the processes more sustainable are depicted in green in Fig. 9.2.

## 9.5 Conclusions

Seafood waste holds great potential to become a sustainable source of valuable biopolymers for tissue engineering and regenerative medicine. Source selection becomes a major consideration from the sustainability perspective, and should focus on abundant and readily available by-products and discards from the seafood industry, as well as on species not under threat. Considering this, we have identified key biopolymers — chondroitin sulfate, chitin/chitosan, collagen, and gelatin — amenable to sustainable extraction, which possess the necessary bioactivity, abundance, and performance for applications in tissue regeneration and related fields.

A critical aspect of sustainability lies in the valorization processes themselves. While conventional extraction methods often rely on harsh chemicals and energy-intensive procedures, greener alternatives exist. For chondroitin sulfate, enzymatic deproteinization coupled with optimized precipitation and ultrafiltration offers a more environmentally friendly approach, allowing for the recovery of valuable by-products like bioapatites and protein hydrolysates. For chitin and chitosan, while chitin isolation from crustacean shells and squid pens benefits from enzymatic hydrolysis, the deacetylation of chitin to chitosan still presents a challenge in terms of developing truly green alternatives to alkaline treatment. However, integrating the alkaline effluent into other processes, such as chondroitin sulfate purification, moves towards a biorefinery model. Similarly, while collagen and gelatin extraction from fish by-products like skin, scales, and bones offers a sustainable starting point, the processes themselves require optimization. Exploring enzymatic extraction, ultrasound-assisted techniques, and the utilization of effluents as nutrient sources or for collagen hydrolysate production are crucial steps toward enhancing the sustainability of collagen and gelatin recovery.

Ultimately, achieving a truly sustainable seafood industry requires a holistic approach. This includes not only sourcing biopolymers from waste but also developing and implementing greener extraction and purification technologies. The biorefinery concept, integrating multiple valorization processes to minimize waste and maximize resource utilization, is essential for moving closer to a zero-waste seafood production model. Further research and development in these areas will be crucial for realizing the full potential of seafood waste as a valuable resource for biomedical applications while minimizing environmental impact.

## References

Ahmed R, Haq M, Chun B-S (2019) Characterization of marine derived collagen extracted from the by-products of bigeye tuna (Thunnus obesus). Int J Biol Macromol 135:668–676. https://doi.org/10.1016/j.ijbiomac.2019.05.213

Al Khawli F, Pateiro M, Domínguez R et al (2019) Innovative green technologies of intensification for valorization of seafood and their by-products. Mar Drugs 17

Alexeev VL, Kelberg EA, Evmenenko GA, Bronnikov SV (2000) Improvement of the mechanical properties of chitosan films by the addition of poly (ethylene oxide). Polym Eng Sci 40:1211–1215

Alfaro AT, Balbinot E, Weber CI et al (2015) Fish gelatin: characteristics, functional properties, applications and future potentials. Food Eng Rev 7:33–44. https://doi.org/10.1007/s12393-014-9096-5

Ali A, Ahmed S (2018) A review on chitosan and its nanocomposites in drug delivery. Int J Biol Macromol 109:273–286

Alipal J, Mohd Pu'ad NAS, Lee TC et al (2021) A review of gelatin: properties, sources, process, applications, and commercialisation. Mater Today Proc 42:240–250. https://doi.org/10.1016/j.matpr.2020.12.922

Alishahi A, Mirvaghefi A, Tehrani MR et al (2011) Enhancement and characterization of chitosan extraction from the wastes of shrimp packaging plants. J Polym Environ 19:776–783. https://doi.org/10.1007/s10924-011-0321-5

Alves AL, Marques ALP, Martins E et al (2017) Cosmetic potential of marine fish skin collagen. Cosmetics 4:39

Alves AL, Fraguas FJ, Carvalho AC et al (2022) Characterization of codfish gelatin: a comparative study of fresh and salted skins and different extraction methods. Food Hydrocoll 124:107238. https://doi.org/10.1016/j.foodhyd.2021.107238

Amer OA, Ali SS, Azab M et al (2022) Exploring new marine bacterial species, Alcaligenes faecalis Alca F2018 valued for bioconversion of shrimp chitin to chitosan for concomitant biotechnological applications. Int J Biol Macromol 196:35–45. https://doi.org/10.1016/j.ijbiomac.2021.12.033

Arnesen JA, Gildberg A (2007) Extraction and characterisation of gelatine from Atlantic salmon (Salmo salar) skin. Bioresour Technol 98:53–57. https://doi.org/10.1016/j.biortech.2005.11.021

Azhara AM, Thusharia GGN, Jayamannea SC, Asokab MDC (2018) Production of gelatin from bigeye tuna (Thunnus obesus) skin and characterisation of gelatin properties: an alternative to minimise fish waste in tuna fish processing industries in Sri Lanka

Balti R, Jridi M, Sila A et al (2011) Extraction and functional properties of gelatin from the skin of cuttlefish (Sepia officinalis) using smooth hound crude acid protease-aided process. Food Hydrocoll 25:943–950. https://doi.org/10.1016/j.foodhyd.2010.09.005

Batista MP, Fernández N, Gaspar FB et al (2022) Extraction of biocompatible collagen from blue shark skins through the conventional extraction process intensification using natural deep eutectic solvents. Front Chem 10:937036

Benito-Arenas R, Doncel-Pérez E, Fernández-Gutiérrez M et al (2018) A holistic approach to unravelling chondroitin sulfation: correlations between surface charge, structure and binding to growth factors. Carbohydr Polym 202:211–218. https://doi.org/10.1016/j.carbpol.2018.08.120

Bisht M, Martins M, Dias ACRV et al (2021) Uncovering the potential of aqueous solutions of deep eutectic solvents on the extraction and purification of collagen type I from Atlantic codfish (Gadus morhua). Green Chem 23:8940–8948

Blanco M, Vázquez JA, Pérez-Martín RI, Sotelo CG (2019) Collagen extraction optimization from the skin of the small-spotted catshark (S. canicula) by response surface methodology. Mar Drugs 17:40

Blanco GC, Stablein MJ, Tommaso G (2020) Cultivation of Chlorella vulgaris in anaerobically digested gelatin industry wastewater. Water Supply 21:1953–1965. https://doi.org/10.2166/ws.2020.263

Bougatef H, Ghlissi Z, Kallel R et al (2020) Chondroitin/dermatan sulfate purified from corb (Sciaena umbra) skin and bone: in vivo assessment of anticoagulant activity. Int J Biol Macromol 164:131–139. https://doi.org/10.1016/j.ijbiomac.2020.07.096

Brodsky B, Persikov AV (2005) Molecular structure of the collagen triple helix. In: Fibrous proteins: coiled-coils, collagen and elastomers. Academic Press, London, pp 301–339

Cao TH, Nguyen TTO, Nguyen TMH et al (2017) Characteristics and physicochemical properties of gelatin extracted from scales of seabass (Lates calcarifer) and Grey mullet (Mugil cephalus) in Vietnam. J Aquat Food Prod Technol 26:1293–1302. https://doi.org/10.1080/10498850.2017.1390026

Carvalho AM, Marques AP, Silva TH, Reis RL (2018) Evaluation of the potential of collagen from codfish skin as a biomaterial for biomedical applications. Mar Drugs 16

Casale J, Crane JS (2019) Biochemistry, glycosaminoglycans
Chatterjee S, Adhya M, Guha AK, Chatterjee BP (2005) Chitosan from Mucor rouxii: production and physico-chemical characterization. Process Biochem 40:395–400. https://doi.org/10.1016/j.procbio.2004.01.025
Chen J, Li L, Yi R et al (2016a) Extraction and characterization of acid-soluble collagen from scales and skin of tilapia (Oreochromis niloticus). LWT-Food Sci Technol 66:453–459
Chen S, Chen H, Xie Q et al (2016b) Rapid isolation of high purity pepsin-soluble type I collagen from scales of red drum fish (Sciaenops ocellatus). Food Hydrocoll 52:468–477. https://doi.org/10.1016/j.foodhyd.2015.07.027
Cheow CS, Norizah MS, Kyaw ZY, Howell NK (2007) Preparation and characterisation of gelatins from the skins of sin croaker (Johnius dussumieri) and shortfin scad (Decapterus macrosoma). Food Chem 101:386–391. https://doi.org/10.1016/j.foodchem.2006.01.046
Coppola D, Oliviero M, Vitale GA et al (2020) Marine collagen from alternative and sustainable sources: extraction, processing and applications. Mar Drugs 18:214
Cuong HN, Minh NC, Van Hoa N, Trung TS (2016) Preparation and characterization of high purity β-chitin from squid pens (Loligo chenisis). Int J Biol Macromol 93(Part A):442–447. https://doi.org/10.1016/j.ijbiomac.2016.08.085
Domard A (2011) A perspective on 30 years research on chitin and chitosan. Carbohydr Polym 84:696–703. https://doi.org/10.1016/j.carbpol.2010.04.083
Dong Y, Dai Z (2022) Physicochemical, structural and antioxidant properties of collagens from the swim bladder of four fish species. Mar Drugs 20:550
El-Rashidy AA, Gad A, Abu-Hussein AE-HG et al (2015) Chemical and biological evaluation of Egyptian Nile Tilapia (Oreochromis niloticas) fish scale collagen. Int J Biol Macromol 79:618–626
Espíndola-Cortés A, Moreno-Tovar R, Bucio L et al (2017) Hydroxyapatite crystallization in shrimp cephalothorax wastes during subcritical water treatment for chitin extraction. Carbohydr Polym 172:332–341. https://doi.org/10.1016/j.carbpol.2017.05.055
European Comission (2020) Discarding and the landing obligation
FAO (2022) The state of world fisheries and aquaculture 2022. Towards Blue Transformation. FAO Rome. https://doi.org/10.4060/cc0461en
Fassini D, Wilkie IC, Pozzolini M et al (2021) Diverse and productive source of biopolymer inspiration: marine collagens. Biomacromolecules 22:1815–1834. https://doi.org/10.1021/acs.biomac.1c00013
Galed G, Miralles B, Paños I et al (2005) N-Deacetylation and depolymerization reactions of chitin/chitosan: influence of the source of chitin. Carbohydr Polym 62:316–320. https://doi.org/10.1016/j.carbpol.2005.03.019
García-Santiago X, Franco-Uría A, Antelo LT et al (2020) Eco-efficiency of a marine biorefinery for valorization of cartilaginous fish biomass. J Ind Ecol 25:789–801. https://doi.org/10.1111/jiec.13066
Gîjiu CL, Isopescu R, Dinculescu D et al (2022) Crabs marine waste—a valuable source of chitosan: tuning chitosan properties by chitin extraction optimization. Polymers (Basel) 14:4492
Gilman E, Perez Roda A, Huntington T et al (2020) Benchmarking global fisheries discards. Sci Rep 10:14017. https://doi.org/10.1038/s41598-020-71021-x
Gómez-Guillén MC, Turnay J, Fernández-Díaz MD et al (2002) Structural and physical properties of gelatin extracted from different marine species: a comparative study. Food Hydrocoll 16:25–34. https://doi.org/10.1016/S0268-005X(01)00035-2
González-González DC, Rodríguez-Félix DE, García-Sifuentes CO et al (2022) Collagen scaffold derived from tilapia (Oreochromis niloticus) skin: Obtention, structural and Physicochemical properties. J Aquat Food Prod Technol 31:374–387. https://doi.org/10.1080/10498850.2022.2048332
Haddar A, Bougatef A, Balti R et al (2011) Physicochemical and functional properties of gelatin from tuna (Thunnus thynnus) head bones. J Food Nutr Res 50:150–159
Hafidz R, Yaakob CM, Amin I, Noorfaizan A (2011) Chemical and functional properties of bovine and porcine skin gelatin. Int Food Res J 18:787–791

Hamed I, Özogul F, Regenstein JM (2016) Industrial applications of crustacean by-products (chitin, chitosan, and chitooligosaccharides): a review. Trends Food Sci Technol 48:40–50. https://doi.org/10.1016/j.tifs.2015.11.007

Harmsen RAG, Tuveng TR, Antonsen SG et al (2019) Can we make chitosan by enzymatic Deacetylation of chitin? Molecules 24:3862

Harvey VL, Wogelius RA, Manning PL, Buckley M (2021) Experimental taphonomy of fish bone from warm and cold water species: testing the effects of amino acid composition on collagen breakdown in modern fish bone using thermal maturation experiments. J Archaeol Sci 126:105318. https://doi.org/10.1016/j.jas.2020.105318

Higashi K, Okamoto Y, Mukuno A et al (2015) Functional chondroitin sulfate from Enteroctopus dofleini containing a 3-O-sulfo glucuronic acid residue. Carbohydr Polym 134:557–565. https://doi.org/10.1016/j.carbpol.2015.07.082

Hong S, Yuan Y, Yang Q et al (2018) Versatile acid base sustainable solvent for fast extraction of various molecular weight chitin from lobster shell. Carbohydr Polym 201:211–217. https://doi.org/10.1016/j.carbpol.2018.08.059

Hong H, Fan H, Chalamaiah M, Wu J (2019) Preparation of low-molecular-weight, collagen hydrolysates (peptides): current progress, challenges, and future perspectives. Food Chem 301:125222. https://doi.org/10.1016/j.foodchem.2019.125222

Huang T, Tu Z, Xinchen-Shangguan et al (2017) Rheological and structural properties of fish scales gelatin: effects of conventional and ultrasound-assisted extraction. Int J Food Prop 20:1210–1220. https://doi.org/10.1080/10942912.2017.1295388

Jafari H, Lista A, Siekapen MM et al (2020) Fish collagen: extraction, characterization, and applications for biomaterials engineering. Polymers (Basel) 12:2230

Jaworska M, Roberts GAF (2016) The influence of chitin structure on its enzymatic deacetylation. Chem Process Eng 37:261–267

Karayannakidis PD, Zotos A (2015) Physicochemical properties of yellowfin tuna (Thunnus albacares) skin gelatin and its modification by the addition of various Coenhancers. J Food Process Preserv 39:530–538. https://doi.org/10.1111/jfpp.12258

Kim HK, Kim YH, Park HJ, Lee NH (2013) Application of ultrasonic treatment to extraction of collagen from the skins of sea bass Lateolabrax japonicus. Fish Sci 79:849–856. https://doi.org/10.1007/s12562-013-0648-z

Kjellén L, Lindahl U (2018) Specificity of glycosaminoglycan–protein interactions. Curr Opin Struct Biol 50:101–108. https://doi.org/10.1016/j.sbi.2017.12.011

Ladchumananandasivam R, da Rocha BG, Belarmino DD, Galv AO (2012) The use of exoskeletons of shrimp (Litopenaeus vanammei) and crab (Ucides cordatus) for the extraction of chitosan and production of nanomembrane

Lan D, Zhang J, Shang X et al (2023) Branch distribution pattern and anticoagulant activity of a fucosylated chondroitin sulfate from Phyllophorella kohkutiensis. Carbohydr Polym 321:121304. https://doi.org/10.1016/j.carbpol.2023.121304

Lee JK, Kang SI, Kim YJ et al (2016) Comparison of collagen characteristics of sea-and freshwater-rainbow trout skin. Food Sci Biotechnol 25:131–136

Li Z-R, Wang B, Chi C et al (2013) Isolation and characterization of acid soluble collagens and pepsin soluble collagens from the skin and bone of Spanish mackerel (Scomberomorous niphonius). Food Hydrocoll 31:103–113. https://doi.org/10.1016/j.foodhyd.2012.10.001

Li H, Hu C, Yu H, Chen C (2018) Chitosan composite scaffolds for articular cartilage defect repair: a review. RSC Adv 8:3736–3749

Liang Q, Wang L, Sun W et al (2014) Isolation and characterization of collagen from the cartilage of Amur sturgeon (Acipenser schrenckii). Process Biochem 49:318–323. https://doi.org/10.1016/j.procbio.2013.12.003

Liao W, Zhu Y, Lu Y et al (2021) Effect of extraction variables on the physical and functional properties of tilapia gelatin. LWT 146:111514. https://doi.org/10.1016/j.lwt.2021.111514

Lim Y-S, Ok Y-J, Hwang S-Y et al (2019) Marine collagen as a promising biomaterial for biomedical applications. Mar Drugs 17:467

Lin N, Mo X, Yang Y, Zhang H (2017) Purification and sequence characterization of chondroitin sulfate and dermatan sulfate from fishes. Glycoconj J 34:241–253. https://doi.org/10.1007/s10719-016-9759-y

Liu H, Huang K (2016) Structural characteristics of extracted collagen from tilapia (Oreochromis mossambicus) bone: effects of ethylenediaminetetraacetic acid solution and hydrochloric acid treatment. Int J Food Prop 19:63–75. https://doi.org/10.1080/10942912.2014.951939

Liu D, Liang L, Regenstein JM, Zhou P (2012) Extraction and characterisation of pepsin-solubilised collagen from fins, scales, skins, bones and swim bladders of bighead carp (Hypophthalmichthys nobilis). Food Chem 133:1441–1448. https://doi.org/10.1016/j.foodchem.2012.02.032

Liu D, Zhou P, Li T, Regenstein JM (2014) Comparison of acid-soluble collagens from the skins and scales of four carp species. Food Hydrocoll 41:290–297

LogithKumar R, KeshavNarayan A, Dhivya S et al (2016) A review of chitosan and its derivatives in bone tissue engineering. Carbohydr Polym 151:172–188. https://doi.org/10.1016/j.carbpol.2016.05.049

López-Álvarez M, López-Senra E, Valcárcel J et al (2019) Quantitative evaluation of sulfation position prevalence in chondroitin sulphate by Raman spectroscopy. J Raman Spectrosc 50:656–664. https://doi.org/10.1002/jrs.5563

López-Álvarez M, González P, Serra J et al (2020) Chondroitin sulfate and hydroxyapatite from Prionace glauca shark jaw: physicochemical and structural characterization. Int J Biol Macromol 156:329–339. https://doi.org/10.1016/j.ijbiomac.2020.04.010

Maccari F, Galeotti F, Volpi N (2015) Isolation and structural characterization of chondroitin sulfate from bony fishes. Carbohydr Polym 129:143–147

Mahboob S (2015) Isolation and characterization of collagen from fish waste material-skin, scales and fins of Catla catla and Cirrhinus mrigala. J Food Sci Technol 52:4296–4305

Mardawati E, Sugandi H, Kayaputri IL et al (2018) Study and characterization of powder mackerel (Scomberomorus commerson) bone gelatin through hydrolysis of hydrochloric acid. AIP Conf Proc 1927:30002. https://doi.org/10.1063/1.5021195

McReynolds C, Adrien A, Castejon N, Fernandes SCM (2022) Green in the deep blue: deep eutectic solvents as versatile systems for the processing of marine biomass. Green Chem Lett Rev 15:383–404. https://doi.org/10.1080/17518253.2022.2065890

Milovanovic I, Hayes M (2018) Marine Gelatine from rest raw materials. Appl Sci 8:2407

Mohammed MH, Williams PA, Tverezovskaya O (2013) Extraction of chitin from prawn shells and conversion to low molecular mass chitosan. Food Hydrocoll 31:166–171. https://doi.org/10.1016/j.foodhyd.2012.10.021

Mohan K, Muralisankar T, Jayakumar R, Rajeevgandhi C (2021) A study on structural comparisons of α-chitin extracted from marine crustacean shell waste. Carbohydr Polym Technol Appl 2:100037. https://doi.org/10.1016/j.carpta.2021.100037

Murado MA, Fraguas J, Montemayor MI et al (2010) Preparation of highly purified chondroitin sulphate from skate (Raja clavata) cartilage by-products. Process optimization including a new procedure of alkaline hydroalcoholic hydrolysis. Biochem Eng J 49:126–132. https://doi.org/10.1016/j.bej.2009.12.006

Nagai T, Yamashita E, Taniguchi K et al (2001) Isolation and characterisation of collagen from the outer skin waste material of cuttlefish (Sepia lycidas). Food Chem 72:425–429

Nagai T, Izumi M, Ishii M (2004) Fish scale collagen. Preparation and partial characterization. Int J Food Sci Technol 39:239–244

Nagarajan M, Benjakul S, Prodpran T et al (2012) Characteristics and functional properties of gelatin from splendid squid (Loligo formosana) skin as affected by extraction temperatures. Food Hydrocoll 29:389–397. https://doi.org/10.1016/j.foodhyd.2012.04.001

Nouri M, Khodaiyan F, Razavi SH, Mousavi M (2016) Improvement of chitosan production from Persian Gulf shrimp waste by response surface methodology. Food Hydrocoll 59:50–58. https://doi.org/10.1016/j.foodhyd.2015.08.027

Novoa-Carballal R, Pérez-Martín R, Blanco M et al (2017) By-products of Scyliorhinus canicula, Prionace glauca and Raja clavata: a valuable source of predominantly 6S sulfated chondroitin sulfate. Carbohydr Polym 157:31–37. https://doi.org/10.1016/j.carbpol.2016.09.050

Okazaki E, Osako K (2014) Isolation and characterization of acid-soluble collagen from the scales of marine fishes from Japan and Vietnam. Food Chem 149:264–270

Pal GK, Suresh PV (2016) Sustainable valorisation of seafood by-products: recovery of collagen and development of collagen-based novel functional food ingredients. Innov Food Sci Emerg Technol 37:201–215. https://doi.org/10.1016/j.ifset.2016.03.015

Pan Z, Ge B, Wei M et al (2023) Isolation and biochemical properties of type II collagen from blue shark (Prionace glauca) cartilage. Mar Drugs 21:260

Panagos CG, Thomson D, Moss C et al (2014) Characterisation of hyaluronic acid and chondroitin/dermatan sulfate from the lumpsucker fish, C lumpus. Carbohydr Polym 106:25–33. https://doi.org/10.1016/j.carbpol.2014.01.090

Peng J, Zi Y, Xu J et al (2022) Effect of extraction methods on the properties of tilapia scale gelatins. Int J Biol Macromol 221:1150–1160. https://doi.org/10.1016/j.ijbiomac.2022.09.094

Perez S, Makshakova O, Angulo J et al (2023) Glycosaminoglycans: what remains to be deciphered? JACS Au 3:628–656. https://doi.org/10.1021/jacsau.2c00569

Phon S, Pradana AL, Thanasupsin SP (2023) Recovery of collagen/gelatin from fish waste with carbon dioxide as a green solvent: an optimization and characterization. Recycling 8:30

Piekarska K, Sikora M, Owczarek M et al (2023) Chitin and chitosan as polymers of the future—obtaining, modification, life cycle assessment and main directions of application. Polymers (Basel):15, 793

Rakshit S, Pal K, Mondal S et al (2023) Extraction of chitosan from biologically-derived chitin by bacterial chitin deacetylase: process optimization and product quality assessment. Int J Biol Macromol 244:125389. https://doi.org/10.1016/j.ijbiomac.2023.125389

Reys LL, Silva SS, Oliveira JM et al (2013) Revealing the potential of squid chitosan-based structures for biomedical applications. Biomed Mater 8:45002

Rezvani Ghomi E, Nourbakhsh N, Akbari Kenari M et al (2021) Collagen-based biomaterials for biomedical applications. J Biomed Mater Res Part B Appl Biomater 109:1986–1999. https://doi.org/10.1002/jbm.b.34881

Ricard-Blum S (2011) The collagen family. Cold Spring Harb Perspect Biol 3:a004978–a004978. https://doi.org/10.1101/cshperspect.a004978

Sadowska M, Kołodziejska I, Niecikowska C (2003) Isolation of collagen from the skins of Baltic cod (Gadus morhua). Food Chem 81:257–262. https://doi.org/10.1016/S0308-8146(02)00420-X

Salvatore L, Gallo N, Natali ML et al (2020) Marine collagen and its derivatives: versatile and sustainable bio-resources for healthcare. Mater Sci Eng C 113:110963. https://doi.org/10.1016/j.msec.2020.110963

Saravana PS, Ho TC, Chae S-J et al (2018) Deep eutectic solvent-based extraction and fabrication of chitin films from crustacean waste. Carbohydr Polym 195:622–630. https://doi.org/10.1016/j.carbpol.2018.05.018

Senadheera TRL, Dave D, Shahidi F (2020) Sea cucumber derived type I collagen: a comprehensive review. Mar Drugs 18:471

Shaik MI, Asrul Effendi NF, Sarbon NM (2021) Functional properties of sharpnose stingray (Dasyatis zugei) skin collagen by ultrasonication extraction as influenced by organic and inorganic acids. Biocatal Agric Biotechnol 35:102103. https://doi.org/10.1016/j.bcab.2021.102103

Shavandi A, Bekhit AE-DA, Ali MA, Sun Z (2015) Bio-mimetic composite scaffold from mussel shells, squid pen and crab chitosan for bone tissue engineering. Int J Biol Macromol 80:445–454. https://doi.org/10.1016/j.ijbiomac.2015.07.012

Sinthusamran S, Benjakul S, Kishimura H (2013) Comparative study on molecular characteristics of acid soluble collagens from skin and swim bladder of seabass (Lates calcarifer). Food Chem 138:2435–2441. https://doi.org/10.1016/j.foodchem.2012.11.136

Sinthusamran S, Benjakul S, Kishimura H (2014) Characteristics and gel properties of gelatin from skin of seabass (*Lates calcarifer*) as influenced by extraction conditions. Food Chem 152:276–284. https://doi.org/10.1016/j.foodchem.2013.11.109

Skierka E, Sadowska M, Karwowska A (2007) Optimization of condition for demineralization Baltic cod (Gadus morhua) backbone. Food Chem 105:215–218

Song W-K, Liu D, Sun L-L et al (2019) Physicochemical and biocompatibility properties of type I collagen from the skin of Nile tilapia (Oreochromis niloticus) for biomedical applications. Mar Drugs 17:137

Sousa RO, Martins E, Carvalho DN et al (2020) Collagen from Atlantic cod (Gadus morhua) skins extracted using CO2 acidified water with potential application in healthcare. J Polym Res 27:73. https://doi.org/10.1007/s10965-020-02048-x

Subhan F, Ikram M, Shehzad A, Ghafoor A (2015) Marine collagen: an emerging player in biomedical applications. J Food Sci Technol 52:4703–4707. https://doi.org/10.1007/s13197-014-1652-8

Sukeri N, Sampath Kumar NS, Shaik MI, Sarbon NM (2021) Extractability and physicochemical properties of cobia (Rachycentron canadum) skin collagen as influenced by lactic acid concentration. J Food Process Preserv 45:e15080. https://doi.org/10.1111/jfpp.15080

Tang L, Chen S, Su W et al (2015) Physicochemical properties and film-forming ability of fish skin collagen extracted from different freshwater species. Process Biochem 50:148–155

Tawfik A, Ni S-Q, Awad HM et al (2021) Recent approaches for the production of high value-added biofuels from gelatinous wastewater. Energies 14. https://doi.org/10.3390/en14164936

Tinrat S, Sila-asna M (2017) Optimization of gelatin extraction and physico-chemical properties of fish skin and bone gelatin: its application to Panna cotta formulas. Curr Res Nutr Food Sci J 5:263–273

Tsai M-F, Huang C-Y, Nargotra P et al (2023) Green extraction and purification of chondroitin sulfate from jumbo squid cartilage by a novel procedure combined with enzyme, ultrasound and hollow fiber dialysis. J Food Sci Technol 60:1711–1722. https://doi.org/10.1007/s13197-023-05701-7

Tu Z, Huang T, Wang H et al (2015) Physico-chemical properties of gelatin from bighead carp (Hypophthalmichthys nobilis) scales by ultrasound-assisted extraction. J Food Sci Technol 52:2166–2174

Usman M, Sahar A, Inam-Ur-Raheem M et al (2022) Gelatin extraction from fish waste and potential applications in food sector. Int J Food Sci Technol 57:154–163. https://doi.org/10.1111/ijfs.15286

Ustyuzhanina NE, Bilan MI, Dmitrenok AS et al (2021) Oversulfated dermatan sulfate and heparinoid in the starfish Lysastrosoma anthosticta: structures and anticoagulant activity. Carbohydr Polym 261:117867. https://doi.org/10.1016/j.carbpol.2021.117867

Valcarcel J, Novoa-Carballal R, Pérez-Martín RI et al (2017) Glycosaminoglycans from marine sources as therapeutic agents. Biotechnol Adv 35:711–725. https://doi.org/10.1016/j.biotechadv.2017.07.008

Valcarcel J, Fraguas J, Hermida-Merino C et al (2021a) Production and physicochemical characterization of gelatin and collagen hydrolysates from turbot skin waste generated by aquaculture activities. Mar Drugs 19:491

Valcarcel J, Hermida-Merino C, Piñeiro MM et al (2021b) Extraction and characterization of gelatin from skin by-products of seabream, seabass and rainbow trout reared in aquaculture. Int J Mol Sci 22:12104

Vallet SD, Berthollier C, Ricard-Blum S (2022) The glycosaminoglycan interactome 2.0. Am J Physiol Physiol 322:C1271–C1278. https://doi.org/10.1152/ajpcell.00095.2022

Vázquez JA, Blanco M, Fraguas J et al (2016a) Optimisation of the extraction and purification of chondroitin sulphate from head by-products of Prionace glauca by environmental friendly processes. Food Chem 198:28–35. https://doi.org/10.1016/j.foodchem.2015.10.087

Vázquez JA, Caprioni R, Nogueira M et al (2016b) Valorisation of effluents obtained from chemical and enzymatic chitin production of Illex argentinus pen by-products as nutrient supplements for various bacterial fermentations. Biochem Eng J 116:34–44. https://doi.org/10.1016/j.bej.2015.12.012

Vázquez JA, Noriega D, Ramos P et al (2017a) Optimization of high purity chitin and chitosan production from *Illex argentinus* pens by a combination of enzymatic and chemical processes. Carbohydr Polym 174:262–272. https://doi.org/10.1016/j.carbpol.2017.06.070

Vázquez JA, Ramos P, Mirón J et al (2017b) Production of chitin from penaeus vannamei by-products to pilot plant scale using a combination of enzymatic and chemical processes and subsequent optimization of the chemical production of chitosan by response surface methodology. Mar Drugs 15. https://doi.org/10.3390/md15060180
Vázquez JA, Fraguas J, Novoa-Carvallal R et al (2018a) Isolation and chemical characterization of chondroitin sulfate from cartilage by-products of blackmouth catshark (*Galeus melastomus*). Mar Drugs 16. https://doi.org/10.3390/md16100344
Vázquez JA, Ramos P, Valcarcel J et al (2018b) An integral and sustainable valorisation strategy of squid pen by-products. J Clean Prod 201:207–218. https://doi.org/10.1016/j.jclepro.2018.07.316
Vázquez JA, Durán AI, Menduíña A et al (2019a) Tailor-made process to recover high added value compounds from fishery by-products. In: Green extraction and valorization of by-products from food processing. CRC Press, pp 91–140
Vázquez JA, Fraguas J, Novoa-Carballal R et al (2019b) Optimal isolation and characterisation of chondroitin sulfate from rabbit fish (*Chimaera monstrosa*). Carbohydr Polym 210:302–313. https://doi.org/10.1016/j.carbpol.2019.01.075
Vázquez JA, Fraguas J, González P et al (2020) Optimal recovery of valuable biomaterials, chondroitin sulfate and Bioapatites, from central skeleton wastes of blue shark. Polymers (Basel) 12. https://doi.org/10.3390/polym12112613
Vázquez JA, Hermida-Merino C, Hermida-Merino D et al (2021) Characterization of gelatin and hydrolysates from valorization of farmed Salmon skin by-products. Polymers (Basel) 13:2828
Vázquez JA, Menduíña A, Durán AI et al (2023a) Microbial bioconversion of chemical waste effluents from marine gelatin isolation: production of probiotics under circular economy philosophy. J Clean Prod 416:137952. https://doi.org/10.1016/j.jclepro.2023.137952
Vázquez JA, Pedreira A, Salmerón I et al (2023b) The fermentation of a marine probiotic bacterium on low-cost media formulated with industrial fish gelatin Waterstreams and collagen hydrolysates. PRO 11:2397
Veeruraj A, Arumugam M, Balasubramanian T (2013) Isolation and characterization of thermostable collagen from the marine eel-fish (Evenchelys macrura). Process Biochem 48:1592–1602
Veríssimo NV, Mussagy CU, Oshiro AA et al (2021) From green to blue economy: marine biorefineries for a sustainable ocean-based economy. Green Chem 23:9377–9400. https://doi.org/10.1039/D1GC03191K
Vicente FA, Bradić B, Novak U, Likozar B (2020) A-Chitin dissolution, N-deacetylation and valorization in deep eutectic solvents. Biopolymers 111:e23351. https://doi.org/10.1002/bip.23351
Volpi N (2007) Analytical aspects of pharmaceutical grade chondroitin sulfates. J Pharm Sci 96:3168–3180
Wan M, Qin W, Lei C et al (2021) Biomaterials from the sea: future building blocks for biomedical applications. Bioact Mater 6:4255–4285. https://doi.org/10.1016/j.bioactmat.2021.04.028
Wang L, Liang Q, Chen T et al (2014) Characterization of collagen from the skin of Amur sturgeon (Acipenser schrenckii). Food Hydrocoll 38:104–109
Wang J, Pei X, Liu H, Zhou D (2018) Extraction and characterization of acid-soluble and pepsin-soluble collagen from skin of loach (Misgurnus anguillicaudatus). Int J Biol Macromol 106:544–550
Wang K, Bai F, Zhou X et al (2021) Characterization of chondroitin sulfates isolated from large hybrid sturgeon cartilage and their gastroprotective activity against ethanol-induced gastric ulcers. Food Chem 363:130436. https://doi.org/10.1016/j.foodchem.2021.130436
Weinhold MX, Sauvageau JCM, Keddig N et al (2009) Strategy to improve the characterization of chitosan for sustainable biomedical applications: SAR guided multi-dimensional analysis. Green Chem 11:498–509
Win N, Stevens W (2001) Shrimp chitin as substrate for fungal chitin deacetylase. Appl Microbiol Biotechnol 57:334–341
Xu S, Zhao Y, Song W et al (2023) Improving the sustainability of processing by-products: extraction and recent biological activities of collagen peptides. Food Secur 12:1965

Yamada S, Sugahara K (2008) Potential therapeutic application of chondroitin sulfate/dermatan sulfate. Curr Drug Discov Technol 5:289–301

Yamada S, Sugahara K, Özbek S (2011) Evolution of glycosaminoglycans: comparative biochemical study. Commun Integr Biol 4:150–158

Yang K-R, Tsai M-F, Shieh C-J et al (2021) Ultrasonic-assisted extraction and structural characterization of chondroitin sulfate derived from jumbo squid cartilage. Food Secur 10:2363

Yang G, Hou X, Lu J et al (2022a) Enzymatic modification of native chitin and chitin oligosaccharides by an alkaline chitin deacetylase from microbacterium esteraromaticum MCDA02. Int J Biol Macromol 203:671–678. https://doi.org/10.1016/j.ijbiomac.2022.01.167

Yang H, Wang H, Huang M et al (2022b) Repurposing fish waste into gelatin as a potential alternative for mammalian sources: a review. Compr Rev Food Sci Food Saf 21:942–963. https://doi.org/10.1111/1541-4337.12920

Zeng JN, Jiang BQ, Xiao ZQ, Li SH (2012) Extraction of collagen from fish scales with papain under ultrasonic pretreatment. Adv Mater Res 366:421–424

Zhang Q, Wang Q, Lv S et al (2016) Comparison of collagen and gelatin extracted from the skins of Nile tilapia (Oreochromis niloticus) and channel catfish (Ictalurus punctatus). Food Biosci 13:41–48

Zhang Z, Niu L, Sun L et al (2017) Effects of powdered activated carbon, diatomaceous earth and β-cyclodextrin treatments on the clarity and volatile compounds of tilapia (Oreochromis niloticus) skin gelatin. J Food Meas Charact 11:894–901. https://doi.org/10.1007/s11694-016-9461-6

Zhang H, Wu X, Quan L, Ao Q (2022a) Characteristics of marine biomaterials and their applications in biomedicine. Mar Drugs 20:372

Zhang Z, Wang Y-M, Qiu Y-T et al (2022b) Gelatin from cartilage of Siberian sturgeon (Acipenser baerii): preparation, characterization, and protective function on ultraviolet-A-injured human skin fibroblasts. Front Mar Sci 9:925407

Zhao T, Zhou Y, Mao G et al (2013) Extraction, purification and characterisation of chondroitin sulfate in Chinese sturgeon cartilage. J Sci Food Agric 93:1633–1640

Zhou C, Mi S, Li J et al (2020) Purification, characterisation and antioxidant activities of chondroitin sulphate extracted from Raja porosa cartilage. Carbohydr Polym 241:116306. https://doi.org/10.1016/j.carbpol.2020.116306

Zhu P, Gu Z, Hong S, Lian H (2017) One-pot production of chitin with high purity from lobster shells using choline chloride–malonic acid deep eutectic solvent. Carbohydr Polym 177:217–223. https://doi.org/10.1016/j.carbpol.2017.09.001

# Part III

# Biomedical Applications

# Green Nanomaterials for Biomedical Applications

# 10

Analía Ale, Ayushi G. Patel, Pablo Luis Santo-Orihuela, Rajshree Jotania, and Martin Federico Desimone

**Abstract**

Green synthesis methods for nanoparticles (NPs) have emerged as a promising approach to revolutionize traditional chemistry, offering both environmental and economic advantages. By utilizing green materials rich in polyphenols, proteins, biopolymers, and other biomolecules, this sustainable alternative to chemical reagents enables the synthesis of NPs with superior properties compared to those produced through conventional methods. Nanoparticle characterization is a crucial step in green synthesis, ensuring the quality and efficacy of these nanomaterials. Green-synthesized NPs are finding increasing applications in various fields, including pharmaceutical industries, agriculture, diagnostics, supplements, healthcare products, and targeted drug delivery systems. The convergence of green nanotechnology with drug delivery systems has given rise to a new era of "green nanomedicine", promising sustainable and effective solutions for numerous biomedical challenges. Herein, it is provided an updated overview of the biomedical applications of green nanomaterials, with a particular focus on tissue

A. Ale
Cátedra de Toxicología, Farmacología y Bioquímica Legal (FBCB-UNL), CONICET, Santa Fe, Argentina

A. G. Patel
Department of Physics, Indus Institute of Sciences, Humanities and Liberal Studies, Indus University, Ahmedabad, Gujarat, India

P. L. Santo-Orihuela
Centro de Investigaciones en Plagas e Insecticidas (CIPEIN), Instituto de Investigaciones Científicas y Técnicas para la Defensa (CITEDEF/UNIDEF-CONICET), Buenos Aires, Argentina

Facultad de Farmacia y Bioquímica, Universidad de Buenos Aires (UBA), Buenos Aires, Argentina

 
J. M. Oliveira et al. (eds.), *Sustainable Scaffolds-based Strategies in Tissue Engineering and Regenerative Medicine*, Biomaterials, Bioengineering and Sustainability 8, https://doi.org/10.1007/978-3-031-96274-5_10

engineering and regenerative medicine. We also identified potential gaps and opportunities for future research in this exciting field.

**Keywords**

Green nanotechnology · Synthesis methods · Characterizations techniques · Green nanoparticles · Biomedical applications

## 10.1 Introduction

Green nanomaterials represent an emerging field of nanoscience that focuses on the development of nanoparticles (NPs) through biological pathways (Vijayaram et al. 2024). Its principle is based on "green chemistry" which has been defined as the design of processes and products to reduce (or eliminate) the generation and employment of hazardous substances. Thus, the green synthesis methods of NPs have become a promising way of rethinking the traditional chemistry towards benefiting both the environment and the economy (Jorge de Souza et al. 2019; Singh et al. 2023a). Biological processes related to green synthesis have been stated as suitable to develop NPs ranging from 1 to 100 nm in comparison with other conventional methods. In addition, they are highly recommended given their ecosafety, ecofriendliness, non-toxicity, and cost-effectiveness (Vijayaram et al. 2024). Consequently, green synthesis methods are increasingly favoured over traditional physical and chemical approaches due to concerns about energy consumption, toxic chemical emissions, and complex equipment requirements. For instance, the aerosol method necessitates a high-temperature flame (approximately 2400 K) to produce atomized droplets and metal nanoparticles. Furthermore, chemical methods often involve the use of expensive and toxic reagents like sodium borohydride, as well as dispersants, stabilizers, and organic solvents (Ying et al. 2022).

The biological agents involved in the green synthesis processes could be summarized in three main categories: (i) microorganisms (i.e.: bacteria, fungi, algae); (ii) plants and their extracts (i.e.: leaf, fruits, roots, or stem); (iii) templates (i.e.: biopolymers, membranes, and proteins) (Rafique et al. 2017; Gahlawat and Choudhury 2019) (Fig. 10.1). Green materials, rich in polyphenols, proteins, biopolymers and other biomolecules, offer a sustainable alternative to chemical reagents for nanoparticle synthesis. Indeed, in suitable conditions, these green materials can interact with metal ions to produce nanoparticles with improved properties. Biomolecules can act as

R. Jotania
Department of Physics, School of Sciences, Gujarat University, Ahmedabad, India

M. F. Desimone (✉)
Universidade Federal do Rio Grande (FURG), Instituto de Ciências Biológicas (ICB), Programa de Pós-graduação em Ciências Fisiológicas (PPGCF), Rio Grande, RS, Brazil

Universidad de Buenos Aires (UBA), CONICET, Instituto de Química y Metabolismo del Fármaco (IQUIMEFA), Facultad de Farmacia y Bioquímica, Buenos Aires, Argentina
e-mail: desimone@ffyb.uba.ar

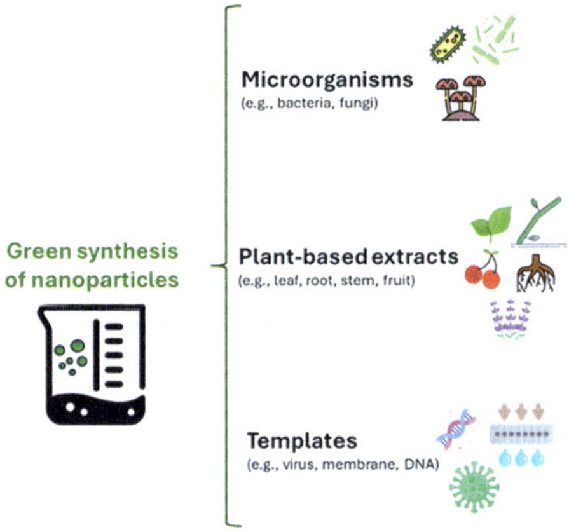


| SYNTHESIS EXAMPLES | REFERENCES |
|---|---|
| Silver NP with Planococcus maritimus MBP-2 | (Pallath et al. 2024) |
| Zinc oxide NP with Bacillus subtilis NH1-8 | (Vosoughian et al. 2023) |
| Copper NP with Aeromonas hydrophila | (Akintelu et al. 2020) |
| Silver NP with Moringa oleifera leaf extract | (Ibrahim 2020) |
| Copper and Zinc NP with Ilex paraguariensis leaves extract | (Galdopórpora et al. 2023) |
| Zinc oxide NP with Vitis vinifera exocarp | (Nivetha et al. 2022) |
| Silver NP with tobacco mosaic virus | (Gubarev et al. 2021) |
| Zinc oxide NP with collagen | (Vijayakumar and Vaseeharan 2018) |
| Graphene oxide-gold NP with hybrid membranes | (Zhang et al. 2013) |

**Fig. 10.1** Main biological agents involved in the green synthesis of nanoparticles and representative examples

reducing agents during the green synthesis of nanomaterials through various mechanisms. These include nucleophilic reactions, ligand binding, transesterification, and chelation. Chelation is particularly important for sequestering metal ions and stabilizing the resulting metal nanoparticles (Acharya et al. 2024). While the reduction of metal ions is the most commonly recognized mechanism, the polyol process and other methods may also contribute to nanoparticle formation (Tenorio Gonzalez et al. 2016).

Notably, green-synthesized NPs would exhibit superior properties compared to those produced through traditional chemical methods. NPs characterization through green synthesis is a key aspect well described in the review carried out by Catalano et al. (2021). The characterization of nano-sized materials is generally done by various physicochemical methods such as X-ray diffraction (XRD), Fourier transform infrared spectroscopy (FT-IR), UV-Vis spectrophotometry (UV-Vis), atomic force microscopy (AFM), scanning electron microscope (SEM), transmission electron microscopy (TEM), dynamic light scattering (DLS), and Inductively coupled plasma atomic emission spectroscopy (ICP-AES) (Table 10.1). According to the aforementioned authors, most of the available publications employed TEM for nanoparticle characterization (~27%), followed by XRD (~18%), SEM (~16%), and DLS (~13%).

Green synthesis has been defined as more beneficial than traditional chemical synthesis in terms of human health safety (Ying et al. 2022). In this sense, green synthesis-based nanoscale compounds are increasingly used in pharmaceutical industries, agriculture, diagnostics, supplements and healthcare products, targeted drug delivery systems, among other fields in biomedicine (Habeeb Rahuman et al. 2022; Singh et al. 2023b). One promising application is the use of green-synthesized NPs as drug delivery systems to enhance the bioavailability of poorly soluble drugs like sorafenib (Khafaga et al. 2023). The convergence of green nanotechnology with drug delivery systems has ushered in a new era of 'green nanomedicine' promising sustainable and effective solutions for various biomedical challenges (Kanwar et al.

**Table 10.1** Main applied methods for nanoparticle characterization

| Method | Main characteristics | Achieved results |
|---|---|---|
| UV-Vis | Extensively employed due to its simplicity, low cost and high versatility. Furthermore, it does not alter the sample and it is less time-consuming | It provides information about nanoparticle size, agglomeration, and concentration |
| DLS | A small sample is suspended in a solution to form a colloidal dispersion. A laser beam is directed through this dispersion, and the resulting scattered light is detected | It measures the size distribution and analyses the behavior of nanoparticle suspensions. It also gives information about the average hydrodynamic diameter of particles. When combined with zeta potential it provides information about the tendency of the nanoparticles to agglomerate |
| FTIR | FTIR spectroscopy utilizes infrared light to generate an interferogram. This raw data is then mathematically processed using a Fourier transform to produce a high-resolution spectrum covering a broad spectral range. The infrared region of the electromagnetic spectrum is characteristic of molecular bond vibrations, enabling the identification and characterization of different substances. It is useful for identification of decomposition or oxidation, detection of contaminants, characterization of unknown materials, analysis of chemical modification or functionalization | Identification of organic, inorganic, and polymeric materials. It allows to identify changes in material composition by alterations in the characteristic patter of the absorption bands |
| SEM | Electron microscopy involves focusing a beam of electrons onto a sample placed in a vacuum chamber. The beam is scanned across the sample, and the emitted signals are collected by detectors. This technique offers high-resolution imaging of nanostructures. However, it can be invasive, as the electron beam can interact with the sample, potentially causing damage. Additionally, sample preparation techniques, such as solvent drying, may introduce artifacts or alter the original nanostructure | Information about morphology and distribution of nanoparticles in cells and other matrices. It also examinates the elemental and chemical composition, size and dimensions |
| TEM | TEM is a powerful technique that utilizes a high-energy electron beam to probe the microstructure of materials. A focused electron beam is transmitted through a thin sample, and the interactions between the electrons and the sample atoms generate various signals, including transmitted electrons. These signals are collected by detectors and converted into an image. TEM can achieve significantly higher spatial resolution, allowing for the visualization of fine details at the nanoscale. However, similar to Scanning Electron Microscopy (SEM), TEM is an invasive technique, as the electron beam can interact with the sample and potentially cause damage, especially in delicate materials | Internal structure images of thin samples (100–300 nm thick) |

(continued)

**Table 10.1** (continued)

| Method | Main characteristics | Achieved results |
|---|---|---|
| XRD | It is considered a versatile, fast, and powerful method by means of X-ray powder diffraction. There are two types: Powder and single crystal. The single crystal diffractometer determines the molecular structure of materials, meanwhile the powder diffractometer is employed for phases' identification | It authenticates the structure of the solid-state materials. It determines the dimensions and shape of the unit cell to identify the detailed structure of the molecules contained withing the unit cell |
| EDX (also known as EDS) | This is a powerful analytical technique that complements electron microscopy by providing elemental information about a sample. When a focused electron beam interacts with the sample, it can excite atoms, causing them to emit X-rays. The energy of these X-rays is specific to each element, allowing for precise identification and quantification. An EDX detector captures and analyzes the emitted X-rays, generating a spectrum that reveals the elemental composition of the sample | This technique offers detailed analytical capabilities, providing both qualitative and quantitative information about selected regions of interest on a sample's surface. By analyzing the spectral data and generating elemental maps, it enables the confirmation of the composition of nanoparticles within the sample |
| AFM | This is a powerful technique for imaging surfaces at the nanoscale. A sharp tip, attached to a flexible cantilever, is brought into close proximity to a sample surface. As the tip scans across the surface, interatomic forces between the tip and the sample cause the cantilever to deflect. A laser beam is reflected off the back of the cantilever onto a detector, and the deflection of the laser spot is measured to determine the forces acting on the tip. This information is used to construct a topographic image of the surface, revealing features with atomic-scale resolution | It creates three-dimensional images of nanoparticles and enables them to determine the size and morphology of them |
| ICP | It is a powerful technique to detect trace-levels concentrations as well as their yield and stability | It provides trace-level concentrations of elements, and allows the detection of multiple elements at the same time |

2019). Green-synthesized nanoparticles (i.e.: Au, Ag, Cu, etc.) have found applications in the development of biosensors for the detection of various molecules, including drugs, metabolites, and other substances (Noah and Ndangili 2022). Furthermore, unique characteristics have been attributed to some NPs such as antifungal, anticancer, antioxidant, anti-inflammatory, and antidiabetic agents (Akintelu and Folorunso 2020; Mitarotonda et al. 2022). In addition, the alarming effect of antibiotic resistance prompted the search for alternative medicine to resolve the microbial resistance conflict whereas nano-based products have also gained importance (Habeeb Rahuman et al. 2022). In this sense, green synthesized silver NPs using an *A. annua* extract demonstrated potent antibacterial activity against various pathogenic bacteria, including *Escherichia coli*, *Staphylococcus aureus*, *Pseudomonas aeruginosa*, *Staphylococcus epidermidis* and *Bacillus subtilis* (Khatoon et al. 2015). Furthermore, concerning microbial resistance, biosynthesized NiO NPs showed inhibition of biofilm formation and antimicrobial activity against methicillin resistant *Staphylococcus aureus* (Saleem et al. 2017). Moreover, copper oxide nanoparticles (CuO-NPs) synthesized using *Acanthospermum hispidum* L. aqueous plant extract demonstrated effective activity against *Plasmodium falciparum* and *Micobacterium tuberculosis* (Pansambal et al. 2017).

A key aspect widely reviewed in literature that involves the green synthesis of NPs (with emphasis on silver NPs) is wound healing. For example, AgNPs were loaded on PF127 (biocompatible-biodegradable polymer) to form a viscous hydrogel that demonstrated enhanced antibacterial properties and significantly increased the wound contraction rate in mice when applied to wound sites (Chinnasamy et al. 2021). Furthermore, the tests using the mouse splint wound model showed that tanic acid modified AgNPs promoted better wound closure, epithelialization, angiogenesis, and formation of the granulation tissue. Additionally, AgNPs elicited expression of VEGF-α, PDGF-β, and TGF-β1 cytokines involved in wound healing (Orlowski et al. 2018). As it is a crucial biological process for tissue repair and regeneration, it needs improvements in the prevention of infection and further complications. In this context, novel therapies are urgently needed to speed up and improve healing processes and, at the same time, cope with antibiotic resistance and prevent wounds from infection; a field of study in which the potential applications of green-synthesized NPs have achieved promising results (Nandhini et al. 2024). Throughout this chapter, we aimed to provide updated information on green nanomaterials biomedical applications, with emphasis on tissue engineering and regenerative medicine, and identify further gaps and potentialities of these novel materials for the near future.

## 10.2 Green Nanomaterials for Tissue Engineering and Regenerative Medicine

Some of the common types of nanomaterials that are clinically used in applications related to healthcare and medicine are shown in Fig. 10.2. Metallic nanoparticles (MNPs), particularly those composed of Ag and Au, have shown substantial potential across various biomedical applications due to their unique properties at the nanoscale, such as high surface area-to-volume ratios and the ability to interact at the cellular level. These nanoparticles demonstrate robust antimicrobial, anticancer, and anti-inflammatory effects, making them ideal for diverse therapeutic applications. For instance, AgNPs have been widely researched for their potent antibacterial effects against a range of pathogenic bacteria, attributed to their ability to disrupt cellular membranes and generate reactive oxygen species (ROS) that damage bacterial cells. In cancer treatment, AuNPs are highly valued for their biocompatibility and functional versatility, serving as carriers for targeted drug delivery, with surface modifications that allow for the attachment of targeting ligands. This specificity not only improves drug accumulation at tumor sites but also minimizes the impact on surrounding healthy tissues. Additionally, due to their optical properties, AuNPs are

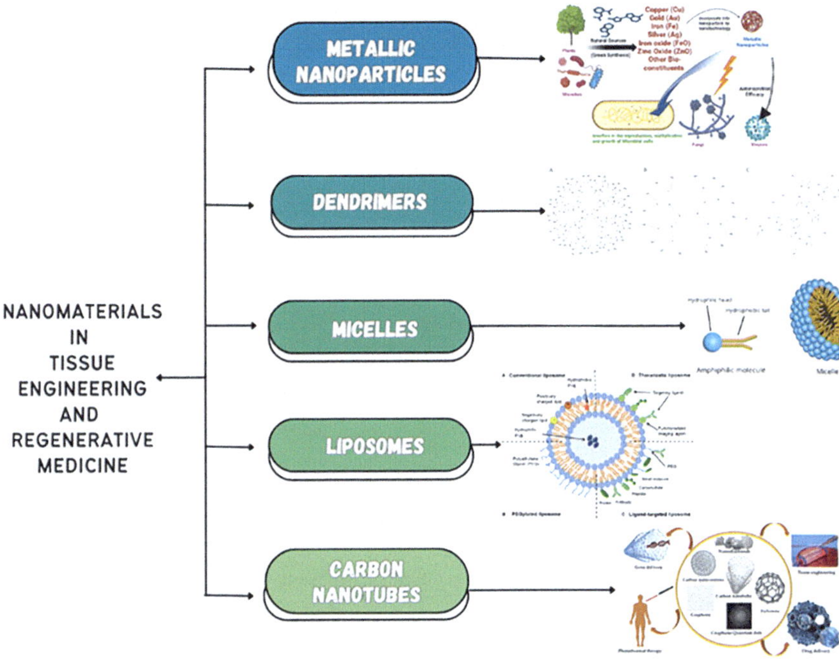


**Fig. 10.2** Types of Nanomaterials for tissue engineering and regenerative medicine applications. Figure developed from (Sim and Wong 2021), (Sercombe et al. 2015; Mendes et al. 2017; Mahor et al. 2021; Patel et al. 2024) under Creative Commons by 4.0

utilized in photothermal therapy, where they convert light into heat to selectively ablate cancer cells. Current research also focuses on enhancing the stability, biocompatibility, and controlled-release capabilities of these nanoparticles, while addressing concerns related to their toxicity and potential environmental impacts (Yaqoob et al. 2020; Burlec et al. 2023).

Secondly, Polymeric nanoparticles, due to their biocompatibility and ability to degrade in the body, are widely utilized in biomedical applications (Fig. 10.3), especially as drug delivery systems. Their nano-scale size and customizable surface properties allow them to carry therapeutic agents with high efficiency, targeting specific tissues or cells. For example, these nanoparticles have shown promise in oncology by encapsulating anticancer drugs, which enhances tumor localization and reduces damage to surrounding healthy cells, thereby mitigating side effects common to conventional chemotherapy (Yaqoob et al. 2020; Gagliardi et al. 2021).

### 10.2.1 Applications in Tissue Regeneration

In tissue engineering, different NPs have diverse roles, enhancing biological, electrical, and mechanical properties of scaffolds for optimized cell growth and tissue repair. AuNPs and titanium dioxide ($TiO_2$NPs), for instance, have demonstrated

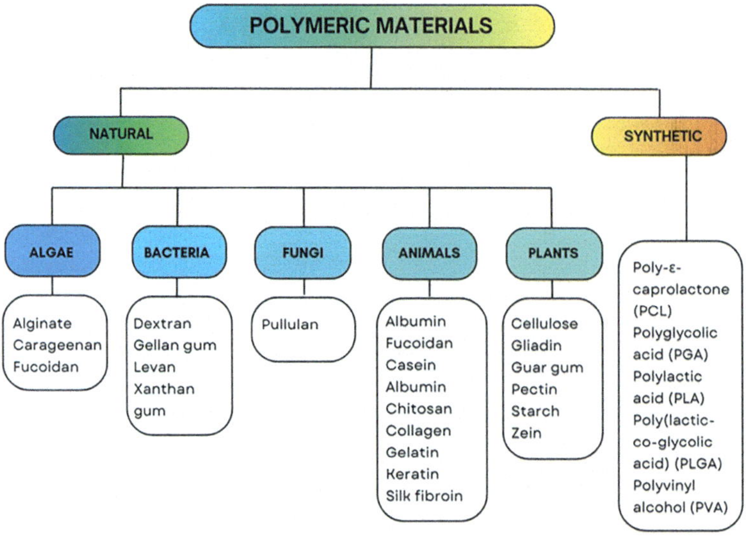


**Fig. 10.3** Various types of polymeric materials employed in bio-nano composites for applications in tissue engineering and regenerative medicine (Geszke-Moritz and Moritz 2024)

efficacy in enhancing cell proliferation rates crucial for bone and cardiac tissue engineering. Specifically, AuNPs are shown to foster osteogenic differentiation of bone precursor cells and modulate osteoclast formation, while also preserving mitochondrial function, making them a promising alternative to bone morphogenetic proteins (BMPs) which, despite their regenerative benefits, are costly and can induce unwanted effects like inflammation (Hasan et al. 2018).

For bone TE, studies highlight the size-dependent efficacy of AuNPs, with certain diameters (20–50 nm) being optimal for enhancing cell differentiation and proliferation. Additionally, combining AuNPs with gelatin in hydrogels has shown positive *in vivo* results, enhancing osteoblast proliferation and differentiation in bone defect sites in animal models. AuNPs and Au-based nanowires incorporated in scaffolds have also contributed to cell proliferation and synapse formation, critical for functional organ transplants and stem cell differentiation without added growth factors (Dong Nyoung Heo et al. 2012). Amino-functionalized mesoporous bioactive glass nanoparticles (MBGNs) with and without Cu ions, integrated into ADA-GEL hydrogels, improve printability, structural fidelity, and bioactivity of 3D-printed constructs for bone tissue engineering. This nanocomposite bioink, with dual enzyme and ion crosslinking, fosters cell proliferation, spreading, osteogenic differentiation, and angiogenesis, making it highly promising for applications in bone repair (Zhu et al. 2022).

Bioactive glass ceramic-based nanoparticles (BGC-NPs) are highly effective materials for bone tissue regeneration. It was found that encapsulating dense BGC-NPs within a gelatin derived from chitosan polymers enhanced alkaline phosphatase activity compared to other mesoporous BGC-NPs. The study indicated that the dense chitosan-gelatin hydrogel containing these bioactive nanoparticles achieved an impressive 80% bone formation rate, outperforming the alternatives (Covarrubias et al. 2018).

Choudhury et al. (2020) made quercetin-loaded chitosan tripolyphosphate nanoparticles using the ionic gelation method, a wound was created on rats and treated with various formulations of quercetin nanoparticles. The 0.03% quercetin nanoparticles significantly reduced tumor necrosis factor-alpha and enhanced healing quality, evidenced by improved granulation tissue, increased blood vessel density, and better collagen organization. Overall, quercetin nanoparticles notably promoted wound healing by modulating relevant cytokines and growth factors (Choudhary et al. 2020). Understanding the interactions of nutraceutical-conjugated metal nanoparticles (MNPs) with the cellular micro-environment is vital for nanomedicine. Another study examined the effects of quercetin-functionalized gold nanoparticles (AuQurNP) on skin fibroblast and keratinocyte migration. AuQurNPs, sized at 47 nm, showed minimal toxicity to fibroblasts and reduced intracellular ROS by 63%. They significantly enhanced cell migration towards wound edges compared to controls. Additionally, AuQurNPs increased TGFβ1 expression, promoting the SMAD signalling pathway, which is crucial for cell migration (Madhyastha et al. 2020). Developing effective tissue adhesives for wet internal tissues remains challenging. Mussel adhesive proteins, especially the catechol group from L-dopa, enhance adhesion when combined with nanoparticles. This study functionalized gelatin with gallic acid and formed a cross-linked network with organosilane to create a bioadhesive. Zinc oxide nanoparticles (ZnO NPs) were incorporated at 1% and 3% to enhance antimicrobial properties. Characterization showed that the addition of ZnO improved mechanical properties and adhesion, especially at 1%, making this Gelatin-Gallic acid/ZnO nanocomposite promising for wound and burn dressings (Ramirez-Barron et al. 2021). A multifunctional chitosan/gelatin@tannic acid cryogel loaded with silver nanoparticles was developed for wound disinfection and healing, demonstrating excellent mechanical properties and compressive repeatability. The *in situ* reduced silver nanoparticles effectively eliminated over 99.9% of *S. aureus* and *E. coli*. The cryogel also reduced TNF-α and ROS levels due to its polyphenol content, showing good biocompatibility and promoting cell proliferation. *In vivo* tests confirmed its efficacy in disinfecting wounds and enhancing skin regeneration without tissue toxicity, making it a promising option for treating infected wounds clinically (Xu et al. 2022).

### 10.2.2 Applications as Antimicrobial Agents

With marine endophytic fungi being a key focus due to their diverse bioactive compounds, a study investigates the biogenic synthesis of AgNPs using *Cladosporium cladosporioides*, isolated from the brown algae *Sargassum wightii*. The AgNPs, measuring under 100 nm, exhibited notable antioxidant and antimicrobial activities. This work reveals the role of NADPH-dependent reductase in AgNPs synthesis, marking it as the first report on this fungal species for AgNP production, with potential for developing broad-spectrum antimicrobial agents (Manjunath Hulikere and Joshi 2019).

MNPs are gaining importance due to their high surface area-to-volume ratio. Plants are vital sources for eco-friendly medicinal compounds and can also facilitate the green synthesis of nanoparticles. This study explored the green synthesis of AgNPs using acetone extract from *Cullen tomentosum* (CTAgNPs) and evaluated their antibacterial activity against *Bacillus cereus* and *Staphylococcus aureus*. The extract demonstrated moderate antibacterial effects, with MIC values of 2.6 mg/mL for *B. cereus* and 3.1 mg/mL for *S. aureus*, while CTAgNPs showed improved activity with MIC values of 1.5 mg/mL and 2.6 mg/mL, respectively. The nanoparticles exhibited a size of 145 nm and a negative surface charge, indicating potential for biopharmaceutical applications (Asong et al. 2023). Green synthesis of AgNPs from *Anthemis pseudocotula* (AP-AgNPs) extract is efficient, eco-friendly, and cost-effective. The AP extract served as both a reducing and stabilizing agent. The AP-AgNPs displayed strong antimicrobial and antibiofilm activity, especially against multidrug-resistant (MDR) *Pseudomonas aeruginosa* (MDR-PA), with significant inhibition in biofilm formation across Gram-negative bacteria like *E. coli* and MDR-AB, as well as Gram-positive MRSA and *S. aureus*. Such study highlights AP-AgNPs as effective agents against select bacterial and fungal strains, promising for antimicrobial applications (Ajlouni et al. 2023). Another study shows that ZnO and ZnO-Ag NPs with two varying concentrations (0.5 and 1 wt%) were embedded in a biodegradable poly(butylene adipate-co-terephthalate) (PBAT) matrix, creating films with improved mechanical, thermal, and crystalline properties. ZnO-Ag NPs particularly enhanced antimicrobial activity due to the synergistic interaction, where ZnO stabilized Ag NPs. These findings support the potential of ZnO-Ag PBAT films in applications like biomedical devices and food packaging (de Souza et al. 2022). Hydroalcoholic extracts from *Linum usitatissimum* were prepared by maceration and tested with CuS nanoparticles for antimicrobial properties using broth macrodilution, agar diffusion, and determination of minimum inhibitory and bactericidal concentrations. The combination effectively combated bacterial infections, offering an alternative to antibiotics. The extracts, particularly from the plant shoots, showed strong antioxidant activities: 128.24 mg GAE/g phenolic content, 30.57% DPPH inhibition, 957.8 μmol Fe(II)/mg in ferric reducing power, and 95.04 mg rutin equivalents/g flavonoid content (Ghaedi et al. 2016). Interest in nanoparticles has surged, especially in microbial synthesis methods, which offer a sustainable alternative to traditional techniques. A recent study used *Bacillus aerius* culture to biosynthesize AgNPs from silver nitrate. The AgNPs exhibited a color change with a plasmon resonance peak at 420 nm, and had a spherical, monodispersed shape (20.12–29.48 nm). FTIR analysis confirmed biomolecular capping, enhancing AgNP stability. These AgNPs showed significant antibacterial and antioxidant properties, highlighting green synthesis as an effective method for creating commercial-grade antioxidant and antibacterial agents (Mahmoud et al. 2017). This study demonstrates a quick, eco-friendly method for synthesizing AgNPs using *Salvia hispanica* (chia) seed extract as a reducing and stabilizing agent. The reduction of silver nitrate at room temperature produced AgNPs with strong antioxidant and antibacterial activities, confirmed by UV-visible absorption at 450 nm, XRD, and SEM, which revealed a spherical morphology with sizes between 40–60 nm. The AgNPs, rich in organic groups from the extract, showed effective catalytic properties and notable antibacterial efficacy at low concentrations, highlighting their potential in biomedical applications (Joshi et al. 2019). Green synthesis of AgNPs using pineapple peel extract yielded monodispersed particles (14–20 nm) without added agents. The concentration ratio of silver ions to peel extract and pH significantly affected the nanoparticle stability. These biogenic AgNPs showed effective photocatalytic degradation of methylene blue dye (98.04% removal under sunlight) and antibacterial activity against *P. aeruginosa* and *B. subtilis*. Optimized pH, dye concentration, and contact time ensured efficient dye removal. This eco-friendly approach leverages sunlight for dye degradation, making it both cost-effective and environmentally sustainable (Agnihotri et al. 2018). This study explores the synthesis of AgNPs using *Nigella sativa* seed extract (NSAgNPs), demonstrating notable antibacterial, antioxidant (89% radical scavenging at 100 µg/ml), and anti-inflammatory (98% RBC membrane stabilization) properties. Additionally, NSAgNPs showed promising antidiabetic potential by inhibiting carbohydrate-hydrolyzing enzymes like α-amylase, α-glucosidase, and DPP-IV. These bioactivities suggest that NSAgNPs, have strong potential for future biomedical applications (Vijayakumar et al. 2021). A novel wound dressing material was developed using cellulosic films loaded with zinc-layered simple hydroxide (allant-ZnLSH) incorporating allantoin. The allant-ZnLSH biohybrid was synthesized by combining zinc chloride, allantoin, and NaOH for controlled pH, then integrated with biocompatible materials like HPMC, CMC, and CNF. These films show strong mechanical properties, optimal water vapor rates for healing, UV-blocking, and transparency for wound monitoring. Antibacterial tests revealed activity against *S. aureus*, indicating the dressing's potential for skin regeneration and antimicrobial protection through zinc and allantoin (Marugan et al. 2023). Beta-glucans, known for their bioactive qualities, are integrated with nanostructured zinc oxide to create flexible, eco-friendly films using *Saccharomyces cerevisiae*-derived glucans. These beta-glucan/zinc oxide composites exhibit strong antibacterial properties, particularly effective against *Staphylococcus epidermidi*, highlighting their promise for advanced wound healing applications (Pino et al. 2023). Cuadra et al. (2022) synthesized ZnO/Ag nanocomposites via a simple chemical method using zinc acetate and silver acetate in ethanol. Different silver concentrations (0, 1, 3, 5, 7, 10%) were tested to examine the influence on bactericidal activity. SEM analysis revealed that adding silver reduced ZnO particle size, while TEM confirmed spherical nanoparticle formation (10–30 nm). Antimicrobial tests showed that ZnO alone was effective against *Staphylococcus aureus* and *Escherichia coli*, with ZnO/Ag composites improving effectiveness by eliminating over 90% of bacteria (Cuadra et al. 2022).

### 10.2.3 Applications as Antioxidant Agents

Bioactive compounds from various organisms, including terpenoids, alkaloids, and polyphenols, exhibit antioxidant properties and can stabilize metallic ions. AgNPs synthesized using these compounds from plant extracts and microorganisms, such as Streptomyces species and fungi, demonstrate significant antioxidant activity, surpassing that of the raw extracts. For instance, AgNPs from *Clerodendrum phlomidis* leaves show improved ferric reducing power and higher radical scavenging activity than the extract alone. Overall, these biogenic AgNPs hold promise for antioxidant applications due to their enhanced properties (Sriranjani et al. 2016). Antioxidant AuNPs are primarily synthesized from plant extracts. For example, *Lactobacillus kimchicus* has been used to produce antioxidant AuNPs with protective properties against cancer cells. Other methods include using proteins from *E. coli* and various fungi and algae to synthesize AuNPs, demonstrating their potential as effective radical scavengers in comparison to traditional gold salts (Markus et al. 2016; Kumar et al. 2020).

A study demonstrates an eco-friendly approach to synthesizing AgNPs from *Streptomyces griseorubens* AU2. The AgNPs, were confirmed by UV–vis spectroscopy at 422 nm, were spherical and crystalline with an average size of 5–20 nm. Their antioxidant properties were assessed via a free radical scavenging assay, showing a dose-dependent increase. This is the first report of AgNP biosynthesis by *S. griseorubens* from soil, indicating its potential for industrial and pharmaceutical applications in producing antioxidant AgNPs (Baygar and Ugur 2017).

Topical use of antioxidants, such as AuNPs, epigallocatechin gallate (EGCG), and α-lipoic acid (ALA), enhances wound healing. This study demonstrated that combinations of EGCG + ALA (EA) and AuNPs + EGCG + ALA (AuEA) promoted cell proliferation and migration in Hs68 and HaCaT cells. AuEA, when applied to mouse skin wounds. Furthermore, they accelerated healing by increasing growth factor expression and antioxidant activity, while reducing inflammation markers. These findings suggest that AuEA's anti-inflammatory and antioxidative effects may aid future antioxidant-based wound treatments (Leu et al. 2012).

### 10.2.4 Applications in Drug Delivery

Nanostructures (made of metals, organic/inorganic, and polymeric materials) are often used for the development of drug delivery systems, research focused on polymeric nanocapsules for pharmaceutical applications presents a dramatic increase in the past two decades (Deng et al. 2020).

Debnath and Srivatstava (2021) discusses carbon-based nanomaterials, such as carbon nanotubes, graphene oxide, and fullerene derivatives, as versatile nanocarriers in drug delivery. These materials are highlighted for their biocompatibility, large surface area, and functionalization flexibility, allowing for controlled drug release, targeting capabilities, and enhanced therapeutic efficacy (Debnath and Srivastava 2021).

Hydrogels made of oxidised starch and CuO have been designed to combat bacteria and respond to colon medication administration stimuli as shown in a study carried out by Namazi et al. (2021) study. This study developed oxidized starch-based bio-nanocomposites (OS/CuO) for colon drug delivery. FTIR and XRD analyses confirmed successful CuO synthesis within the hydrogel network. The impact of $Cu^{2+}$ concentration on swelling, drug loading, and release behavior was evaluated, being possible to select the sample showing the highest values. Additionally, the sample exhibited the strongest antibacterial activity. Overall, the findings suggest that these systems are promising for medical applications (Namazi et al. 2021). Challenges remain in optimizing these nanoparticles for safe clinical applications, including biocompatibility, stability, and effective controlled-release mechanisms.

## 10.3 Conclusion

Green nanomaterials have emerged as a promising and sustainable approach for the development of advanced tissue engineering and regenerative medicine strategies. By leveraging the inherent biocompatibility and multifunctional properties of these nanomaterials, researchers can design innovative scaffolds and delivery systems that promote cell growth, differentiation, and tissue regeneration.

Key advantages of green nanomaterials include their biocompatibility, as green synthesis methods often yield nanomaterials with enhanced biocompatibility and reduced toxicity compared to chemically synthesized counterparts. Additionally, green synthesis methods reduce the reliance on hazardous chemicals and energy-intensive processes, promoting sustainable and eco-friendly practices. Many green nanomaterials are also biodegradable, minimizing potential environmental and health risks. Moreover, green nanomaterials can be engineered to possess multiple functionalities, such as drug delivery, imaging, and cell signalling, enabling targeted and controlled therapies.

While significant progress has been made, further research is needed to fully realize the potential of green nanomaterials in tissue engineering and regenerative medicine. Future studies should focus on developing strategies to precisely control the size, shape, surface chemistry, and porosity of green nanomaterials to optimize their performance. Additionally, rigorous *in vivo* studies are needed to evaluate the long-term safety and efficacy of green nanomaterials. By addressing these challenges and capitalizing on the unique advantages of green nanomaterials, we can pave the way for a sustainable future of tissue engineering and regenerative medicine.

**Acknowledgments** The authors would like to acknowledge grants from the Universidad de Buenos Aires, UBACYT 20020150100056BA and PIDAE2022 and from CONICET PIP 0826 (Martín F. Desimone) which supported this work.

## References

Acharya C, Mishra S, Chaurasia SK, Pandey BK, Dhar R, Pandey JK (2024) Synthesis of metallic nanoparticles using biometabolites: mechanisms and applications. Biometals. https://doi.org/10.1007/s10534-024-00642-w

Agnihotri S, Sillu D, Sharma G, Arya RK (2018) Photocatalytic and antibacterial potential of silver nanoparticles derived from pineapple waste: process optimization and modeling kinetics for dye removal. Appl Nanosci 8(8):2077–2092. https://doi.org/10.1007/s13204-018-0883-9

Ajlouni AW, Hamdan EH, Alshalawi RAE, Shaik MR, Khan M, Kuniyil M, Alwarthan A, Ansari MA, Khan M, Alkhathlan HZ, Shaik JP, Adil SF (2023) Green synthesis of silver nanoparticles using aerial part extract of the Anthemis pseudocotula Boiss. Plant and their biological activity. Molecules 28(1). https://doi.org/10.3390/molecules28010246

Akintelu SA, Folorunso AS (2020) A review on green synthesis of zinc oxide nanoparticles using plant extracts and its biomedical applications. BioNanoScience 10(4):848–863. https://doi.org/10.1007/s12668-020-00774-6

Akintelu SA, Folorunso AS, Folorunso FA, Oyebamiji AK (2020) Green synthesis of copper oxide nanoparticles for biomedical application and environmental remediation. Heliyon 6(7):e04508. https://doi.org/10.1016/j.heliyon.2020.e04508

Asong JA, Frimpong EK, Seepe HA, Katata-Seru L, Amoo SO, Aremu AO (2023) Green synthesis of characterized silver nanoparticle using Cullen tomentosum and assessment of its antibacterial activity. Antibiotics 12(2). https://doi.org/10.3390/antibiotics12020203

Baygar T, Ugur A (2017) Biosynthesis of silver nanoparticles by streptomyces griseorubens isolated from soil and their antioxidant activity. IET Nanobiotechnol 11(3):286–291. https://doi.org/10.1049/iet-nbt.2015.0127

Burlec AF, Corciova A, Boev M, Batir-Marin D, Mircea C, Cioanca O, Danila G, Danila M, Bucur AF, Hancianu M (2023) Current overview of metal nanoparticles' synthesis, characterization, and biomedical applications, with a focus on silver and gold nanoparticles. Pharmaceuticals 16

Catalano PN, Chaudhary RG, Desimone MF, Santo-Orihuela PL (2021) A survey on analytical methods for the characterization of green synthesized nanomaterials. Curr Pharm Biotechnol 22:1–25

Chinnasamy G, Chandrasekharan S, Koh TW, Bhatnagar S (2021) Synthesis, characterization, antibacterial and wound healing efficacy of silver nanoparticles from Azadirachta indica. Front Microbiol 12. https://doi.org/10.3389/fmicb.2021.611560

Choudhary A, Kant V, Jangir BL, Joshi VG (2020) Quercetin loaded chitosan tripolyphosphate nanoparticles accelerated cutaneous wound healing in Wistar rats. Eur J Pharmacol 880. https://doi.org/10.1016/j.ejphar.2020.173172

Covarrubias C, Cádiz M, Maureira M, Celhay I, Cuadra F, von Marttens A (2018) Bionanocomposite scaffolds based on chitosan–gelatin and nanodimensional bioactive glass particles: in vitro properties and in vivo bone regeneration. J Biomater Appl 32(9):1155–1163. https://doi.org/10.1177/0885328218759042

Cuadra JG, Scalschi L, Vicedo B, Guc M, Izquierdo-Roca V, Porcar S, Fraga D, Carda JB (2022) ZnO/Ag nanocomposites with enhanced antimicrobial activity. Appl Sci 12(10). https://doi.org/10.3390/app12105023

de Souza AG, Komatsu LGH, Barbosa RFS, Parra DF, Rosa DS (2022) The effect of ZnO nanoparticles as Ag-carrier in PBAT for antimicrobial films. Polym Bull 79(6):4031–4048. https://doi.org/10.1007/s00289-021-03681-2

Debnath SK, Srivastava R (2021) Drug delivery with carbon-based nanomaterials as versatile nanocarriers: progress and prospects. Front, Nanotechnol,vol 3. https://doi.org/10.3389/fnano.2021.644564

Deng S, Gigliobianco MR, Censi R, Di Martino P (2020) Polymeric nanocapsules as nanotechnological alternative for drug delivery system: current status, challenges and opportunities. Nano 10

Gagliardi A, Giuliano E, Venkateswararao E, Fresta M, Bulotta S, Awasthi V, Cosco D (2021) Biodegradable polymeric nanoparticles for drug delivery to solid tumors. Front Pharmacol:12

Gahlawat G, Choudhury AR (2019) A review on the biosynthesis of metal and metal salt nanoparticles by microbes. RSC Adv 9(23):12944–12967. https://doi.org/10.1039/c8ra10483b

Galdopórpora MJ, Municoy S, Ibarra F, Puente V, Antezana EP, Echazú IAM, Desimone FM, Galdopórpora JM, Municoy S, Ibarra F, Puente V, Antezana PE, Echazú MIA, Desimone MF (2023) A green synthesis method to tune the morphology of CuO and ZnO nanostructures

Geszke-Moritz M, Moritz M (2024) Biodegradable polymeric nanoparticle-based drug delivery systems: comprehensive overview, perspectives and challenges. Polymers (Basel):16

Ghaedi M, Yousefi-Nejad M, Safarpoor M, Khafri HZ, Tyagi I, Agarwal S, Gupta VK (2016) Synthesis of CuS nanoparticles and evaluation of its antimicrobial properties in combination with linum usitatissimum root and shoot extract. Desalin Water Treat 57(51):24456–24466. https://doi.org/10.1080/19443994.2016.1138896

Gubarev AS, Lezov AA, Mikhailova ME, Senchukova AS, Mikusheva NG, Sobolev VY, Arkhipenko MV, Karpova OV, Tsvetkov NV (2021) Green synthesis of silver nanoparticles with the tobacco mosaic virus. Rev Adv Chem 11(3):189–196. https://doi.org/10.1134/S207997802103002X

Habeeb Rahuman HB, Dhandapani R, Narayanan S, Palanivel V, Paramasivam R, Subbarayalu R, Thangavelu S, Muthupandian S (2022) Medicinal plants mediated the green synthesis of silver nanoparticles and their biomedical applications. IET Nanobiotechnol 16(4):115–144. https://doi.org/10.1049/nbt2.12078

Hasan A, Morshed M, Memic A, Hassan S, Webster TJ, Marei HES (2018) Nanoparticles in tissue engineering: applications, challenges and prospects. Int J Nanomedicine 13:5637–5655

Heo DN, Ko W-K, Bae MS, Lee JB, Lee D-W, Byun W, Lee CH, Kim E-C, Jung B-Y, Kwon IK (2012) Enhanced bone regeneration with a gold nanoparticle-hydrogel complex. J Mater Chem B 2019(November). https://doi.org/10.1039/C3TB21246G

Ibrahim ATA (2020) Toxicological impact of green synthesized silver nanoparticles and protective role of different selenium type on Oreochromis niloticus: hematological and biochemical response. J Trace Elem Med Biol 61:126507. https://doi.org/10.1016/j.jtemb.2020.126507

Jorge de Souza TA, Rosa Souza LR, Franchi LP (2019) Silver nanoparticles: an integrated view of green synthesis methods, transformation in the environment, and toxicity. Ecotoxicol Environ Saf 171:691–700. https://doi.org/10.1016/j.ecoenv.2018.12.095

Joshi N, Pathak A, Anupam R, Jain N, Singh J, Upadhyaya CP (2019) A rapid and efficient biosynthesis of metallic nanoparticles using aqueous extract of chia (Salvia hispanica L.) seeds. BioNanoScience 9(4):893–902. https://doi.org/10.1007/s12668-019-00672-6

Kanwar R, Rathee J, Salunke DB, Mehta SK (2019) Green nanotechnology-driven drug delivery assemblies. ACS Omega 4(5):8804–8815. https://doi.org/10.1021/acsomega.9b00304

Khafaga DSR, El-Khawaga AM, Elfattah Mohammed RA, Abdelhakim HK (2023) Green synthesis of nano-based drug delivery systems developed for hepatocellular carcinoma treatment: a review. Mol Biol Rep 50(12):10351–10364. https://doi.org/10.1007/s11033-023-08823-5

Khatoon N, Ahmad R, Sardar M (2015) Robust and fluorescent silver nanoparticles using Artemisia annua: biosynthesis, characterization and antibacterial activity. Biochem Eng J 102:91–97. https://doi.org/10.1016/j.bej.2015.02.019

Kumar H, Bhardwaj K, Nepovimova E, Kuča K, Dhanjal DS, Bhardwaj S, Bhatia SK, Verma R, Kumar D (2020) Antioxidant functionalized nanoparticles: a combat against oxidative stress. Nano 10:1–31

Leu JG, Chen SA, Chen HM, Wu WM, Hung CF, Der YY, Tu CS, Liang YJ (2012) The effects of gold nanoparticles in wound healing with antioxidant epigallocatechin gallate and α-lipoic acid. Nanomed Nanotechnol Biol Med 8(5):767–775. https://doi.org/10.1016/j.nano.2011.08.013

Madhyastha H, Halder S, Queen IN, Madhyastha R, Mohanapriya A, Sudhakaran R, Sajitha LS, Banerjee K, Bethasiwi P, Daima H, Navya PN, Maruyama M, Nakajima Y (2020) Surface refined AuQuercetinnanoconjugate stimulates dermal cell migration: possible implication in wound healing. RSC Adv 10(62):37683–37694. https://doi.org/10.1039/d0ra06690g

Mahmoud W, Elazzazy AM, Danial EN (2017) In vitro evaluation of antioxidant, biochemical and antimicrobial properties of biosynthesized silver nanoparticles against multidrug-resistant bacterial pathogens. Biotechnol Biotechnol Equip 31(2):373–379. https://doi.org/10.1080/13102818.2017.1282836

Mahor A, Singh PP, Bharadwaj P, Sharma N, Yadav S, Rosenholm JM, Bansal KK (2021) Carbon-based nanomaterials for delivery of biologicals and therapeutics: a cutting-edge technology. C 7(1):19. https://doi.org/10.3390/c7010019

Manjunath Hulikere M, Joshi CG (2019) Characterization, antioxidant and antimicrobial activity of silver nanoparticles synthesized using marine endophytic fungus – Cladosporium cladosporioides. Process Biochem 82:199–204. https://doi.org/10.1016/j.procbio.2019.04.011

Markus J, Mathiyalagan R, Kim YJ, Abbai R, Singh P, Ahn S, Perez ZEJ, Hurh J, Yang DC (2016) Intracellular synthesis of gold nanoparticles with antioxidant activity by probiotic Lactobacillus kimchicus DCY51T isolated from Korean kimchi. Enzym Microb Technol 95:85–93. https://doi.org/10.1016/j.enzmictec.2016.08.018

Marugan E, Rebitski EP, Darder M, Balestra SRG, del Real G, Aranda P (2023) Allantoin-zinc layered simple hydroxide biohybrid as antimicrobial active phase in cellulosic bionanocomposites as potential wound dressings. Appl Clay Sci 241. https://doi.org/10.1016/j.clay.2023.107002

Mendes LP, Pan J, Torchilin VP (2017) Dendrimers as nanocarriers for nucleic acid and drug delivery in cancer therapy. Molecules:22

Mitarotonda R, Giorgi E, Eufrasio-da-Silva T, Dolatshahi-Pirouz A, Mishra YK, Khademhosseini A, Desimone MF, De Marzi M, Orive G (2022) Immunotherapeutic nanoparticles: from autoimmune disease control to the development of vaccines. Biomater Adv 212726. https://doi.org/10.1016/j.bioadv.2022.212726

Namazi H, Pooresmaeil M, Hasani M (2021) Oxidized starch/CuO bio-nanocomposite hydrogels as an antibacterial and stimuli-responsive agent with potential colon-specific naproxen delivery. Int J Polym Mater Polym Biomater 70(18):1296–1305. https://doi.org/10.1080/00914037.2020.1798431

Nandhini J, Karthikeyan E, Elizabeth Rani E, Karthikha VS, Sakthi Sanjana D, Jeevitha H, Rajeshkumar S, Venugopal V, Priyadharshan A (2024) Advancing engineered approaches for sustainable wound regeneration and repair: harnessing the potential of green synthesized silver nanoparticles. Eng Regen 5(3):306–325. https://doi.org/10.1016/j.engreg.2024.06.004

Nivetha K, Vinotha V, Albeshr MF, Mahboob S, Manzoor I, Govindarajan M, Vaseeharan B (2022) Synthesis and characterization of Vitis vinifera exocarp-mediated ZnO nanoparticles: an evaluation of biological potential and ecotoxicity. J Drug Deliv Sci Technol 77:103846. https://doi.org/10.1016/j.jddst.2022.103846

Noah NM, Ndangili PM (2022) Green synthesis of nanomaterials from sustainable materials for biosensors and drug delivery. Sensors Int 3:100166. https://doi.org/10.1016/j.sintl.2022.100166

Orlowski P, Zmigrodzka M, Tomaszewska E, Ranoszek-Soliwoda K, Czupryn M, Antos-Bielska M, Szemraj J, Celichowski G, Grobelny J, Krzyzowska M (2018) Tannic acid-modified silver nanoparticles for wound healing: the importance of size. Int J Nanomedicine 13:991–1007. https://doi.org/10.2147/IJN.S154797

Pallath N, Francis B, Devanesan S, Asemi N, AlSalhi MS, Natarajan S (2024) Synthesis of silver nanoparticles from marine bacteria and evaluation of antimicrobial, antifungal and cytotoxic effects. J King Saud Univ Sci 36(2):103073. https://doi.org/10.1016/j.jksus.2023.103073

Pansambal S, Deshmukh K, Savale A, Ghotekar S, Pardeshi O, Jain G, Aher Y, Pore D (2017) Phytosynthesis and biological activities of fluorescent CuO nanoparticles using acanthospermum hispidum L. extract. J Nanostruct 7(3):165–174. https://doi.org/10.22052/jns.2017.03.001

Patel J, Kumar GS, Roy H, Maddiboyina B, Leporatti S, Bohara RA (2024) From nature to nanomedicine: bioengineered metallic nanoparticles bridge the gap for medical applications. Discov Nano 19

Pino P, Pellegrino G, Ronchetti S, Mollea C, Bosco F, Onida B (2023) Antibacterial β-glucan/zinc oxide nanocomposite films for wound healing. BioNanoScience 13(2):426–435. https://doi.org/10.1007/s12668-023-01079-0

Rafique M, Sadaf I, Rafique MS, Tahir MB (2017) A review on green synthesis of silver nanoparticles and their applications. Artif Cells Nanomed Biotechnol 45(7):1272–1291. https://doi.org/10.1080/21691401.2016.1241792

Ramirez-Barron SN, Sanchez-Valdes S, Betancourt R, Gallardo CA, Puente-Urbina B, Rodriguez-Fernández OS, Carneiro-da Cunha MG, dos Santos-Correia MT, Sanchez-Martinez ZV (2021) Preparation and characterization of gelatin-gallic acid/ZnO nanocomposite with antibacterial properties as a promising multi-functional bioadhesive for wound dressing applications. Int J Adhes Adhes 104. https://doi.org/10.1016/j.ijadhadh.2020.102749

Saleem S, Ahmed B, Khan MS, Al-Shaeri M, Musarrat J (2017) Inhibition of growth and biofilm formation of clinical bacterial isolates by NiO nanoparticles synthesized from Eucalyptus globulus plants. Microb Pathog 111:375–387. https://doi.org/10.1016/j.micpath.2017.09.019

Sercombe L, Veerati T, Moheimani F, Wu SY, Sood AK, Hua S (2015) Advances and challenges of liposome assisted drug delivery. Front Pharmacol:6

Sim S, Wong NK (2021) Nanotechnology and its use in imaging and drug delivery (review). Biomed Rep 14(5). https://doi.org/10.3892/br.2021.1418

Singh H, Desimone MF, Pandya S, Jasani S, George N, Adnan M, Aldarhami A, Bazaid AS, Alderhami SA (2023a) Revisiting the green synthesis of nanoparticles: uncovering influences of plant extracts as reducing agents for enhanced synthesis efficiency and its biomedical applications. Int J Nanomedicine 18(null):4727–4750. https://doi.org/10.2147/IJN.S419369

Singh NB, Chaudhary RG, Desimone MF, Anupam A, Saroj SK (2023b) Green synthesized nanomaterials for safe technology in sustainable agriculture. Curr Pharm Biotechnol 24:61–85

Sriranjani R, Srinithya B, Vellingiri V, Brindha P, Anthony SP, Sivasubramanian A, Muthuraman MS (2016) Silver nanoparticle synthesis using Clerodendrum phlomidis leaf extract and preliminary investigation of its antioxidant and anticancer activities. J Mol Liq 220:926–930. https://doi.org/10.1016/j.molliq.2016.05.042

Tenorio Gonzalez FN, Bolarín Miró AM, Sánchez De Jesús F, Cortés Escobedo CA, Ammar S (2016) Mechanism and microstructural evolution of polyol mediated synthesis of nanostructured M-type SrFe12O19. J Magn Magn Mater 407:188–194. https://doi.org/10.1016/j.jmmm.2016.01.068

Vijayakumar S, Vaseeharan B (2018) Antibiofilm, anti cancer and ecotoxicity properties of collagen based ZnO nanoparticles. Adv Powder Technol 29(10):2331–2345. https://doi.org/10.1016/j.apt.2018.06.013

Vijayakumar S, Divya M, Vaseeharan B, Chen J, Biruntha M, Silva LP, Durán-Lara EF, Shreema K, Ranjan S, Dasgupta N (2021) Biological compound capping of silver nanoparticle with the seed extracts of Blackcumin (Nigella sativa): a potential antibacterial, antidiabetic, anti-inflammatory, and antioxidant. J Inorg Organomet Polym Mater 31(2):624–635. https://doi.org/10.1007/s10904-020-01713-4

Vijayaram S, Razafindralambo H, Sun Y-Z, Vasantharaj S, Ghafarifarsani H, Hoseinifar SH, Raeeszadeh M (2024) Applications of green synthesized metal nanoparticles – a review. Biol Trace Elem Res 202(1):360–386. https://doi.org/10.1007/s12011-023-03645-9

Vosoughian N, Asadbeygi M, Mohammadi A, Soudi MR (2023) Green synthesis of zinc oxide nanoparticles using novel bacterium strain (Bacillus subtilis NH1-8) and their in vitro antibacterial and antibiofilm activities against Salmonella typhimurium. Microb Pathog 185:106457. https://doi.org/10.1016/j.micpath.2023.106457

Xu N, Yuan Y, Ding L, Li J, Jia J, Li Z, He D, Yu Y (2022) Multifunctional chitosan/gelatin@tannic acid cryogels decorated with in situ reduced silver nanoparticles for wound healing. Burn Trauma 10. https://doi.org/10.1093/burnst/tkac019

Yaqoob AA, Ahmad H, Parveen T, Ahmad A, Oves M, Ismail IMI, Qari HA, Umar K, Mohamad Ibrahim MN (2020) Recent advances in metal decorated nanomaterials and their various biological applications: a review. Front Chem:8

Ying S, Guan Z, Ofoegbu PC, Clubb P, Rico C, He F, Hong J (2022) Green synthesis of nanoparticles: current developments and limitations. Environ Technol Innov 26:102336. https://doi.org/10.1016/j.eti.2022.102336

Zhang P, Zhang X, Zhang S, Lu X, Li Q, Su Z, Wei G (2013) One-pot green synthesis, characterizations, and biosensor application of self-assembled reduced graphene oxide–gold nanoparticle hybrid membranes. J Mater Chem B 1(47):6525–6531. https://doi.org/10.1039/C3TB21270J

Zhu H, Monavari M, Zheng K, Distler T, Ouyang L, Heid S, Jin Z, He J, Li D, Boccaccini AR (2022) 3D bioprinting of multifunctional dynamic nanocomposite bioinks incorporating cu-doped mesoporous bioactive glass nanoparticles for bone tissue engineering. Small 18(12). https://doi.org/10.1002/smll.202104996

# Nature-Inspired Engineered Biomaterials to Guide Cell Function

# 11

Jennifer A. Westphal, Andrew E. Bryan, and Greg M. Harris

**Abstract**

The innovative design of natural scaffolds and biomaterials to drive a desired cellular response is of great importance for the advancement of present-day tissue engineering. Decellularized extracellular matrices, biopolymer scaffolds, and extracellular matrix-inspired polymers have shown promising potential for directing cellular behaviors to promote regeneration by means such as cellular alignment, enhanced proliferation, and directed differentiation. A variety of naturally derived and inspired polymeric scaffoldings fabricated in combination with organic materials that include a multitude of micropatterning processes to direct cell function are outlined herein. The current methods for the development of micropatterned biomatrices are specific to the materials at hand and can be categorized as follows: (1) additive manufacturing, (2) instability-induced surface micropatterning, (3) patterning by dewetting, (4) patterning by evaporation, and (5) patterning by electric field and thermal gradient. Although these

J. A. Westphal
Department of Biomedical Engineering, University of Cincinnati, Cincinnati, OH, USA
e-mail: westphja@mail.uc.edu

A. E. Bryan
Department of Chemical and Environmental Engineering, University of Cincinnati, Cincinnati, OH, USA
e-mail: bryanae@mail.uc.edu

G. M. Harris (✉)
Department of Biomedical Engineering, University of Cincinnati, Cincinnati, OH, USA

Department of Chemical and Environmental Engineering, University of Cincinnati, Cincinnati, OH, USA

Neuroscience Graduate Program, University of Cincinnati College of Medicine, Cincinnati, OH, USA
e-mail: gregory.harris@uc.edu



J. M. Oliveira et al. (eds.), *Sustainable Scaffolds-based Strategies in Tissue Engineering and Regenerative Medicine*, Biomaterials, Bioengineering and Sustainability 8, https://doi.org/10.1007/978-3-031-96274-5_11

techniques appear to be mainly tailored toward polymeric substances, the science behind these applied processes is rooted within nature. The responsive cellular performances and cell-material interactions are also explored, focusing on the topography of the material's surface for cellular guidance and the resulting cellular behaviors specifically cell morphology and gene and protein expression.

**Keywords**

Micropatterning · Tissue engineering · Tissue scaffolds · Extracellular matrix · Biomaterials

## 11.1 Introduction

Physical cues derived from biomaterial surface characteristics have shown tremendous promise as a method to direct cell growth, behavior, morphology, and differentiation (Hu et al. 2016). These physical factors are specific to the native tissue microenvironment and can include stiffness, permeability, and roughness. In bioengineered materials, the scaffold stiffness, mechanical performance under stress, and two- (2D) and three-dimensional (3D) terrain are important to mimic such that they positively impact the structure and function of the cell for tissue regeneration purposes (Hu et al. 2016; Cady et al. 2020; Xu et al. 2020a). Other types of biomaterial properties arise from the composition of electrical and chemical microenvironments which surround the cell, and which can activate cellular signaling pathways to alter cell morphology and behavior. Environmental electrical cues such as bioelectricity in excitable tissues can be replicated with the use of conductive or piezoelectric biomaterials (Zhao et al. 2022). Chemical cues within a native cell microenvironment can range from surface chemistry to sustained growth factor release in biomaterials to complex extracellular matrix (ECM)-induced signaling via integrins. These mechanisms can be programmed into biomaterials with the incorporation of growth factors, ECM proteins, proteoglycans, carbohydrates, and their derivatives. The ECM constituents create diverse mechano-chemical properties within cellular microenvironments, rendering cell-ECM signaling both chemical and mechanical in nature (Kshitiz et al. 2016).

To comprehensively impact cellular function *in vitro*, the mechanical properties and surface chemistry of bioengineered materials are designed to promote distinct functions such as optimized cell-scaffold adhesion (Hu et al. 2016; Lee et al. 2017), cellular alignment (Hu et al. 2016; Wang et al. 2007; Atmanli and Domian 2013), and cytoskeletal elongation (Hu et al. 2016). Cellular alignment and orientation in fibers of natural tissues, for example, is especially important for countless biological processes and is therefore imperative to mimic when fabricating biomaterials (Hu et al. 2016; Papadaki et al. 2001; Wigmore and Dunglison 1998; Chiu et al. 2004; Vunjak-Novakovic et al. 2004). As seen in nature, alignment and directionality of tissues and fibrous materials is widespread, ranging from algae (de Grooth et al. 1985) to aligned spider silk (Salehi et al. 2020) to cell-secreted factors that form the architecture of the ECM. Cell-derived materials such as micropatterned decellularized ECM (Cady et al. 2020; Harris et al. 2017; Takeda and Xu 2014) have recently become popular in an effort to engineer more biocompatible, nature-inspired biomaterials. The ability to guide cell morphology and behavior with engineered tissues holds promise in a variety of applications including the study of cell-cell and cell-substrate interactions (Chen et al. 2005; McBeath et al. 2004), tissue regeneration (Griffith and Naughton 2002; Langer and Vacanti 1993), and regenerative medicine (Ghosh and Ingber 2007), as well as drug screening (Griffith and Naughton 2002; Whitesides et al. 2001) and the development of biosensors (Kane et al. 1999). Many different methods are currently used to pattern biomaterials for cell guidance, including soft lithography (Xu et al. 2020a; Kane et al. 1999; Xia and Whitesides 1998; Yap and Zhang 2007), photolithography (Yap and Zhang 2007; Britland et al. 1992a, b; Lom et al. 1993; Healy 1996), microcontact printing (Yap and Zhang 2007; Mrksich et al. 1997; Xia et al. 1995; Csucs et al. 2003), microfluidic patterning (Folch and Toner 1998; Folch et al. 1999), stencil patterning (Folch et al. 2000; Ostuni et al. 2000), and robotic printing (Yap and Zhang 2007; Xu et al. 2004; Pardo et al. 2003; Roth et al. 2004). These techniques provide micropatterns, or grooves, necessary for guiding cell growth through directionality, alignment, and elongation (Kane et al. 1999). Contact guidance of cells from this physical topography occurs when cells align and elongate on patterned surfaces, and the elongated cells can promote directional cell migration, which in turn can lead to alterations in gene expression and facilitate tissue healing (Cady et al. 2020; Xu et al. 2020a).

Biomaterials can interact with cells and direct behaviors as they are sensitive to environmental factors and respond to changes thereby making the interplay between cells and the biomechanical and biochemical cues imposed by the surrounding ECM crucial to mediating many vital cell functions (Hu et al. 2016; Cady et al. 2020; Xu et al. 2020a). In an *in vitro* environment, the tunability of cell directionality, organization, and alignment depends on vigorous nano- and microfiber fabrication methods to assemble the artificial extracellular environments that mimic the native, *in vivo* cell environments (Hu et al. 2016). To address the need for improved nature-based biomaterials that are capable of guiding cell function, techniques such as micropatterning and bioprinting biomaterials including silk fibroin (Sun et al. 2021), decellularized ECM (Cady et al. 2020; Féréol et al. 2011), and specialized hydrogels have been developed. These techniques, along with a variety of nature-derived biomaterials for an improved cell response, are highlighted in Fig. 11.1. Engineering the matrix niche introduces a modern approach in studying and manipulating cell fate. Fabricated ECM not only provides cells with a 3D scaffold capable of storing soluble factors to enhance their activities, but also supplies cells with the structural support needed for attachment, resilience, and remodeling for tissue homeostasis (Cheng et al. 2020). This review focuses on exploring these nature-based materials, their associated micropatterning techniques, and their applications in the field of tissue engineering and regenerative medicine.

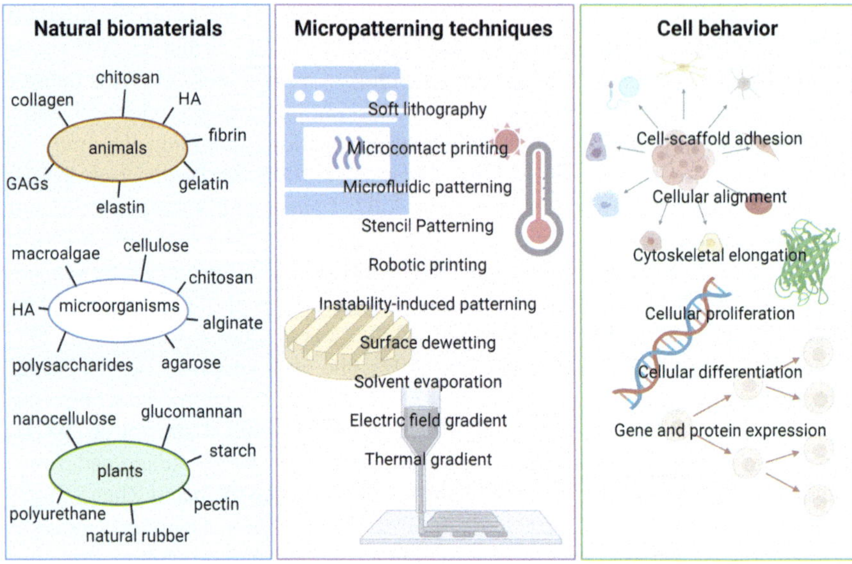


**Fig. 11.1** Schematic outlining the various examples of natural biomaterials, micropatterning approaches for fabricating organized surface topographies, and resulting cell behavior and function that can be directed through biomaterials

## 11.2 Nature-Inspired, Engineered Materials for Regenerative Medicine

### 11.2.1 Material Sources

#### 11.2.1.1 Biomaterial Sources

**Animal Based**

Natural biomaterials used in regenerative medicine originate from animals, plants, and microorganisms and can span from bioinert to bioactive as well as bio-inspired materials (Insuasti-Cruz et al. 2022). Bird waste from the poultry industry, for example, is a great source of biomaterials including collagen, gelatin, chondroitin sulfate, bioceramics, and hyaluronic acid (Tarafdar et al. 2021). Similarly, the molted skin of reptiles provides keratin, collagen-I, hydroxyproline, and fibrin. Waste products from fish remain a dependable source of biopolymers which include elastin peptides, glycosaminoglycans, lipids, chitosan, fibronectin, fibrinogen, fibrin, lectin, thrombin, and chondroitin sulfate (Insuasti-Cruz et al. 2022; Thirukumaran et al. 2022). Proteins that are extracted from many different fresh- and saltwater fish species exhibit unique properties as a result of the biodiversity in ecosystems and evolved traits such as intraspecific recognition and communication, bioluminescence, magnetoreception, and ability to change sex (Insuasti-Cruz et al. 2022; Wainwright and Longo 2017; Treberg and Speers-Roesch 2016). Many biomaterials are also extracted from insect exoskeletons which consist of monosaccharides, collagen, melanin, chitosan, elastin, silk, and polylactic acid. Recently, spiders and scorpions have gained much interest for their protein-rich secretions from their glands including cobwebs and venom (Rodríguez de la Vega et al. 2010; Erdeş et al. 2014; Liu et al. 2019; Vassilevski et al. 2009). The outer shell layer of crustaceans also provides valuable matrix biomaterials such as crustacyanin, chitosan, chitin, collagen, mussel adhesive protein, exopolysaccharides, heparin, lectin, aneroin, and melanin (Insuasti-Cruz et al. 2022; El Knidri et al. 2018). These extracted proteins are easily incorporated into biomaterials for micropatterning techniques, as they render bioactive and cell-favorable microenvironments. The various micropatterning methods are covered in a following section.

#### Plant Based

Plant-derived materials for micropatterning tissue scaffolds can originate from trees, ferns, mosses, and lichens. Cellulose, nanocellulose, pectin, natural rubber, zein protein, polyurethane, glucomannan and starch are the most common plant based materials frequently used in bioengineering (Insuasti-Cruz et al. 2022; Jovic et al. 2019). The natural biopolymer nanocellulose, for instance, is abundant, versatile, and cost-effective and has shown much promise in the field of tissue regeneration due to its biocompatibility, strong mechanical properties, biodegradability, and unique surface chemistry (Jovic et al. 2019). In particular, nanocellulose has been used in vascular grafts (Jacob et al. 2022; Ghorbani et al. 2020; Pooyan et al. 2013), bio-scaffolds for skin tissue engineering (Jacob et al. 2022; Priya et al. 2008; Groeber et al. 2011), grafts for nerve regeneration (Jacob et al. 2022), and for cross-linking and building a hydroxyapatite membrane for bone reconstruction (Jacob et al. 2022; Huang et al. 2019). Moreover, cellulose also has bacterial origins from microorganisms such as algae, differing from plant-derived cellulose by possessing greater chemical purity, high crystallinity, high water-holding capacity (100%), and increased tensile strength (Naomi et al. 2020). Polyurethane, synthesized from vegetable oils, is another popular biopolymer frequently used in biomedical applications.

Similar to animal-derived biomaterials, plants also provide many proteins and substances popular in the field of tissue engineering. Micropatterned biopolymers from these materials can typically be fabricated via soft lithography or photografting. Porous biomaterials such as many plant-derived materials are difficult to incorporate into scaffolds with a 2D or 3D structure as a result of the size, shape, and fabrication methods associated with creating cellular layers (Ueno et al. 2019). Ueno et al. developed a new micropatterning technique which does not require high temperatures, a vacuum, or high voltages for patterning porous materials capable of containing cells. That method was proven to achieve various patterned microstructures with a wide range of scales, regardless of the material used (Ueno et al. 2019).

Further, Watson et al. recently developed a subcellular micropatterning method of proteins which utilizes fibrinogen anchors to allow the micropatterning of any protein at high homogeneity and specificity while maintaining its activity. This method enhances the micropatterning outcomes on numerous substrates, facilitates multiplexed micropatterning processes, and improves the on-pattern activity of fragile proteins, especially molecular motors found in plant tissues that function to hydrolyze ATP and utilize the energy in a variety of cellular functions (Watson et al. 2021). As an alternative to applying micropatterning techniques, many decellularized plant tissues already hold desirable properties such as micropatterns, which are embodied in the structure and function of plants. Fontana et al. found human mesenchymal stem cells and human dermal fibroblasts were able to conform to the microstructure of natural plant frameworks, resulting in cell-pattern registration and cellular alignment. Each plant leaf had a unique topography as a result of the natural patterning of microgrooves on its surface and cells responded positively to topographical cues by aligning with the existing structural patterns of the leaves (Fontana et al. 2017).

#### Microorganism Based

Many natural biomaterials are synthesized by microorganisms such as bacteria, fungi, and algae. Based on their physical characteristics, most microbiologists argue that algae, except for certain species of macroalgae, are classified as microorganisms (Lee and Ryu 2021). Microalgae can either be eukaryotic or prokaryotic. Algae synthesizes polysaccharides through the process of photosynthesis, which accounts for a main product harvested for biomedical applications. Macroalgae, commonly known as seaweed, is the most utilized in engineering biomaterials. Based on their pigment, these macroalgae are divided into three categories: Chlorophyta (green-), Rhodophyta (red-), and Phaeophyta (brown macroalgae). Red and brown macroalgae are rich in arachidonic and eicosapentaenoic acids, while green algae contain high levels of linoleic, oleic, palmitic, and hexadecatetraenoic acids (Biancarosa et al. 2018; Tiwari and Troy 2015). Alginate composites have recently gained much popularity in tissue engineering for their biologically favorable properties including biocompatibility, ease of gelation, porous and comb-like architecture, low cost, and environmental advantages (Bar-Shai et al. 2021; Popa et al. 2015; Liberski et al. 2016). Further, algal polysaccharides are either naturally produced or derived from genetically engineered microorganisms and possess many desirable properties for biomaterials. These include carrageenan, cellulose, chitosan, fucoidan, laminarin, alginate, agarose, and agra (Insuasti-Cruz et al. 2022; Khrunyk et al. 2020). Likewise, fungal species including molds and yeasts produce an unlimited number of polysaccharides, particularly protein complexes and $\beta$-glucans, alongside chitosan and chitin (Insuasti-Cruz et al. 2022). Different species of bacteria also produce biomaterials such as alginate, cellulose, xanthan, polyglutamic acid, polylysine, polyphosphates, and hyaluronate (Hinchliffe et al. 2021). Bacteria are prokaryotes and can be genetically modified or regulated by antibiotics for safer and easier manipulation in bioengineering (Plavec and Berlec 2020). Lately, bacteria have been used in synthetic biology to fabricate innovative biofilms for their ability to secrete extracellular polymer-like materials (Flores Bueso et al. 2018).

These microorganism-derived biomaterials also require their own methods of micropatterning and handling to be fabricated into scaffolds with defined surface topography for favorable cell behavior. Unlike with animal- or plant-derived biomaterials, most microorganisms require a humid environment to synthesize products for bioengineering. The presence of moisture challenges material integrity as well as the microstructure morphology of the secreted biomaterial. Liu et al. optimized these surface-engineered micropatterns on bacterial cellulose for application in scar-free wound healing of the skin by utilizing low-energy $CO_2$ laser etching to establish different micropatterns on the cellulose surface. Grooves exhibited a width of ~100 μm and depth of ~300 μm, displaying distinct hydrophilic and hydrophobic properties, and maintained their structure under a humid environment. Both *in vitro* and *in vivo* analysis on cellular migration, collagen deposition, and histology showed micropatterns significantly impacted wound healing progress compared to surfaces without micropatterns (Liu et al. 2023a). Another technique developed by Kimura et al. utilizes bacterial cellulose and a filtration-assisted magnetic micropatterning method which forms a magnetic trap by a modulated field such that floating particles in medium were able to be patterned. The problem with field modulation is that floating particles beyond the distance of the modulator surface cannot be patterned, and to solve this, the liquid medium was evaporated such that particles reach within the distance to be trapped. However, this evaporation caused a change in medium composition which led to coagulation of particles and damage to cells. Here, bacterial cellulose was chosen because it floats in medium and therefore a conventional magnetic patterning technique was unable to be applied successfully. However, the combined magnetic technique with filtration enabled reliable micropatterning of bacterial cellulose through formation of a line pattern from the magnetic effect (Kimura et al. 2007).

#### 11.2.1.2 Cell-Derived Materials and Decellularized Extracellular Matrices

Many of the natural biomaterials used in medical applications or engineering are derived from cells. Cell-derived matrices (CDM) and decellularized ECM (dECM) are common cell-free scaffoldings. CDM is commonly referred to as the extracellular matrix remaining after a decellularization process of lab-grown cell cultures whereas dECM is generally referred to as being derived from decellularizing whole tissue, although strict definitions may vary. Like dECM, CDM provides a tunable, bioactive culturing surface that specifically mimics properties of natural tissue while eliminating the use of animal tissue as a source (Cheng et al. 2020; Chan et al. 2021). The main difference, however, lies within the quantity produced and physiochemical characteristics of the ECM (Chan et al. 2021). CDM can possess a very similar chemical composition to dECM of the same tissue origin, but it is difficult to fabricate CDM with the same physical properties of whole tissue due to a lack of three-dimensional structural organization found in CDM (Chan et al. 2021). Often, CDM does not hold any mechanical or structural characteristics apart from its chemical composition. Because of this, CDM is generally more suitable for applications where ECM macrostructure on the tissue level is not needed (Chan et al. 2021).

The tunable composition of CDM allows these matrices to be optimized with various cell types, culture conditions, decellularization techniques, and micropatterning methods (Chan et al. 2021; Sart et al. 2020) for studying cell proliferation, migration, differentiation, and cell-ECM signaling (Harris et al. 2018). CDM fabricated from various cell types have been utilized in scaffold applications both *in vitro* and *in vivo*. For example, fibroblast-secreted CDM has been shown to promote formation of capillary-like tubules of endothelial cells *in vitro* (Berthod et al. 2006; Soucy and Romer 2009). Similarly, CDM produced from fibroblasts originating from tumor-associated stroma matched the physical and chemical characteristics seen *in vivo* (Amatangelo et al. 2005). Tumor-affiliated CDM successfully promoted the desmoplastic differentiation of normal fibroblasts and stimulated the invasive response of breast cancer cells (Amatangelo et al. 2005; Castelló-Cros et al. 2009; Hoshiba and Tanaka 2013), which suggests that CDM provides necessary physio-chemical cues to initiate the transformations during stromatogenesis *in vivo* (Kutys et al. 2013).

On the contrary to CDM, dECM is more robust structurally in its natural state. dECM can be processed for scaffold applications by removing cells, however, dECM can lose its original structure and integrity during handling and intense processing to incorporate into different scaffold types. To fabricate dECM suitable for tissue culture, ECM is first decellularized followed by lyophilization and finally re-solubilized in enzymes for the digestion of cell-derived proteins. While the overall 3D structure of dECM is lost, new processes can be applied to produce micropatterned scaffolds. Fabrication methods of dECM for micropatterned surfaces include electrospinning (Cady et al. 2020; Xu et al. 2020a; Orkwis et al. 2020, 2022) and 3D microfilament bioprinting using dECM-bioink (Min et al. 2022; Ma et al. 2018; Mao et al. 2020). CDM and dECM in its digested form are both desirable for tissue culture and hold similar features such that their terms are used interchangeably. In this chapter moving forward, the term dECM will be used for all decellularized matrices, regardless of cell or tissue origin. The biomimetic micro- and macrostructure of electrospun or bioprinted dECM presents seeded cells with structure, guidance, and orientation, which induce cellular alignment, cytoskeletal elongation, and gene and protein expression. For example, researchers have demonstrated the ability of dECM to positively impact cell migration, proliferation, matrix remodeling, and other cellular mechanisms (Cady et al. 2020; Harris et al. 2017; Mao and Schwarzbauer 2005; Cukierman et al. 2001; Hellewell et al. 2017), however, the critical gap of these matrices really lies within the alignment and organization of the dECM which is imperative to the function of the tissue. Cady et al. fabricated micropatterned polydimethylsiloxane (PDMS) substrates to guide the alignment of cells and the resulting dECM for mimicking the alignment found within native tissues. In this work, soft lithography was utilized to pattern the PDMS and cell morphology was found to be significantly more elongated than that of cells cultured on nonpatterned PDMS. Further, fibroblasts seeded on micropatterned PDMS secreted an aligned dECM oriented in the direction of the grooves as opposed to the unaligned matrix secreted on nonpatterned PDMS and coverslips (Cady et al. 2020). Xu et al. further analyzed protein expression on these micropatterned PDMS substrates, concluding that proteins c-Jun and p75NTR showed higher expression in Schwann cells seeded on micropatterned surfaces compared to nonpatterned surfaces (Xu et al. 2020b).

Similarly, Xu et al. utilized micropatterned coverslips as well as micropatterned laminin-, collagen-I, and fibronectin-coated PDMS stamps to induce cellular behaviors such as elongation, spreading, and protein expression in Schwann cells. Using microcontact printing, islands of distinct area, shape, and aspect ratio were printed onto PDMS substrates for which the cell shape, spreading, nuclear elongation, and c-Jun expression were analyzed. Schwann cells on islands patterned with increasing aspect ratio possessed elongated nuclei, indicating that cell and nuclear shape could serve as potential regulators for c-Jun expression (Xu et al. 2020a). c-Jun was upregulated on higher aspect ratio islands, however, c-Jun expression from Schwann cells on nonpatterned surfaces varied in response to cell spreading changes over several substrate stiffnesses (Xu et al. 2020a). An explanation for this could be when Schwann cells are seeded on micropatterned surfaces, upstream Hippo signaling may be blocked by the obstruction of cell-cell contact by grooves, which activates YAP/TAZ signaling. Therefore, it may be possible that YAP/TAZ was activated in Schwann cells seeded on the stiffer, nonpatterned surfaces as well as on the softer, micropatterned surfaces. Dupont et al. and Grove et al. have also demonstrated that YAP/TAZ is inactivated as cell spreading is obstructed, resulting in an increase in cell dedifferentiation (Dupont et al. 2011; Grove et al. 2017).

Towards a nature-inspired perspective are microalgal biofilms consisting of extracellular polymeric substances (EPS). Similar to dECM, these microalgal biofilms are comprised of phospholipids, proteins, polysaccharides, and uronic acids as well as a few functional groups including carboxylic, hydroxyl, phosphoric, and amino groups (Christenson and Sims 2011; Cheah and Chan 2021). Biofilms are made up of approximately 10% algal cells and 90% EPS and due to their robust 3D matrix and biochemical properties, have potential as biomaterials (Cheah and Chan 2021; Brackman and Coenye 2016). For instance, biofilm wound healing models have shown success in expediting wound closure of the skin and antimicrobial efficacy after burn injuries in several *in vitro* and *in vivo* models (Brackman and Coenye 2016). Living micro-ecological hydrogels (LMH) containing functionalized algae and probiotic bacteria have also been engineered for continuous delivery of oxygen and antibacterial substances into chronic, difficult-to-heal wound sites. LMHs were capable of continuously releasing oxygen to prevent hypoxia while simultaneously knocking out colonized pathogenic bacteria which resulted in significant healing of infected diabetic wounds (Chen et al. 2023a).

Examples of the above-mentioned biomaterials, their origin, and type, along with their specific micropatterning techniques are provided in Table 11.1. These scaffold examples are not limited and represent current, state-of-the-art developments in tissue engineering and regeneration.

Table 11.1: Overview of natural and nature-inspired biomaterials in state-of-the-art tissue engineering applications along with their unique micropatterning techniques and resulting cell performance. Table sections are classified into natural (animal-, plant-, and microorganism-derived) and nature-inspired (bioactive and composites/mesh fibers) materials.

**Table 11.1** Summary of nature-inspired, micropatterned biomaterial scaffolds

| Material classification, type | Description | Micropatterning technique | Cell behavior | Reference |
|---|---|---|---|---|
| *Natural, animal derived* | 3D cell-laden porous gelatin hydrogel for soft tissue reconstruction | Photolithography, layered manufacturing | Increased cellular viability and proliferation | Chu et al. (2023) |
| | Ovalbumin scaffold for modulating Schwann cell behavior | Patterned by micro-molding and surface-biomodification with peptides | Increased Schwann cell proliferation, migration, and elongation. Upregulated gene expression | Gao et al. (2023) |
| | Fibrin scaffolds for cardiomyocyte alignment and contractility | Micropatterning with PDMS photolithography | Improved iPSC alignment, elongation, and contractile properties with electrical stimulation | English et al. (2023) |
| | dECM bioink scaffolds for bioprinting spinal cord tissue | 3D bioprinting | Enhanced PC12 adhesion, proliferation, long term survival, and upregulated laminin-1 and neurofilament-M production | Garrido Dos Santos et al. (2023) |
| | Hydroxyapatite/collagen scaffold in gelation bath for bone tissue engineering | 3D bioprinting | Increased BMSC adhesion and proliferation. Upregulated alkaline phosphate expression | Guo et al. (2023) |
| *Natural, plant derived* | Elastic wood-derived scaffolds with anisotropic channels for bone reconstruction | Simple delignification treatment of natural wood | Promoted cellular polarization along direction of channels, increased BMSC focal adhesion. Upregulated osteogenic factors through an increase in YAP expression | Liu et al. (2023b) |
| | Decellularized leaf scaffold for aligned skeletal muscle assembly | Innate uniaxial micropatterned topography, biofunctionalized with copolymer to prevent non-specific protein adsorption | Increased adhesion, growth, and alignment of various human cells including hESC-derived muscle cells. Cells adopted parallel orientation and elongated along leaf topography | Yun et al. (2023) |
| | Graphene-cellulose acetate-sodium alginate scaffold for formation of ovarian cancer spheroid and drug assessment | Drop-casting method | Increased NIH-3T3 proliferation and aggregation. Observed hypoxia induced expression of HIF-$\alpha$ and VEGF | Suryavanshi et al. (2023) |
| | Surface-phosphorylated cellulose nanofiber scaffolds for bone cell differentiation and calcification | Surface phosphorylation reaction for uniform and stable scaffolds | Increased osteoblast adhesion, proliferation, and osteogenic differentiation | Liu et al. (2023c) |
| *Natural, microorganism derived* | 3D filaments from polyhydroxy butyrate-micronized bacterial cellulose | 3D printing for a micropatterned lace architecture | Supported L929 cell viability and adhesion | Celestino et al. (2023) |

| | | | | |
|---|---|---|---|---|
| *Nature-inspired, bioactive* | Nanofibers with polydopamine-copper ion complexes and peptides | Electrospinning with stepwise modification | Increased cellular adhesion, endothelial formation, and showed positive vascular remodeling | Yi et al. (2023) |
| | PDMS hydrogels with PDA and RGA to induce a regulatory macrophage-like phenotype | Photo- and soft lithography | Enhanced cell adhesion, elongation, and morphology. Upregulated Mreg markers expression | Pham et al. (2023) |
| | Gelatin-PCL/silk fibroin composite membrane for accelerated wound healing | Photolithography with PDMS stamp | Guided oriented cell growth, increased angiogenesis, upregulated actin genes and proteins, and wound contraction | Li et al. (2023a) |
| | Surface-modified aligned PCL/silk fibroin scaffolds for corneal stromal regeneration | Fiber alignment with electrospinning | Increased adhesion, proliferation, and gene expression in corneal stromal keratocytes | Bhattacharjee et al. (2023) |
| | Diels-Alder photoclick patterning of extracellular matrix for controlling cell behavior | Light-guided pattering platform | Spatially controlled ERK activity and migration in mammalian cells | Bailey et al. (2023) |
| *Nature-inspired, composite/mesh* | Nanofibrous PCL/PPDO sponges to regulate neural stem cell fate for spinal cord regeneration | Directional electrospinning and subsequent gas-foaming treatment | Enhanced neural stem cell infiltration, alignment, and differentiation. Upregulated molecular pathways | Li et al. (2023b) |
| | Micropatterned PMMA hydrogel for stem cell culture | Deep UV lithography | Enhanced stem cell proliferation, migration, and differentiation | Samal et al. (2023) |
| | Microcellular thermoplastic polyurethane/nanoclay scaffold for tissue engineering applications | Aligned scaffolds through microcellular injection molding | Incorporation of nanoclays increased MG63 cell dispersion morphology, adhesion, viability, and proliferation | Geng et al. (2023) |
| | Ingenious multicomponent nanoyarn scaffolds from PetOx and PCL to mimic fibrillar and tubular structure of tendons | Electrospinning and bundling of nanoyarns | Nanoyarns guided growing cells along nanofiber direction. Increased cytocompatibility with mouse and equine tendon cells | Schynkel et al. (2023) |

## 11.2.2 Polymer Based Biomaterials

### 11.2.2.1 Extracellular Matrix-Inspired Biopolymers

ECM is composed of a complex network of extracellular macromolecules secreted by tissue-specific cells and contains water, proteins, and polysaccharides and can be tuned with unique topography and composition depending on tissue type (Spiller et al. 2021). Biopolymers such as ECM complemented with certain synthetic polymers can produce stable scaffolds with tunable material properties, including stiffness, degradation rate, and porosity (Xing et al. 2020). These can then be additionally tuned through alterations of chemical crosslinking. Polymer-ECM coupled fibers, with the biocompatibility of ECM proteins and the chemical stability of synthetic polymers, have shown to promote long-term cellular programming (Shu et al. 2013). Altering the density and fibril tracks of ECM contributes to the microenvironmental cues which guide sprouting trajectories of pericytes, vascular endothelial cells, and forming vasculature of angiogenic neovessels (Rauff et al. 2023). dECM, for example, has recently become popular for combination with synthetic polymers to render more bioactive, acellular surfaces with greater cell affinity, improved host response, and better integration with native tissue. With the addition of dECM, selective polymers can be functionalized and properties of both the polymer and ECM can be preserved. For example, synthetic polymers such as conductive polycaprolactone (PCL) and piezoelectric polyvinylidene fluoride-trifluoroethylene (PVDF-TrFE) provide a favorable, bioactive cell microenvironment when combined with dECM (Cady et al. 2020; Xu et al. 2020a; Orkwis et al. 2020, 2022). Other selective polymers for combination with ECM include poly(lactic acid) (PLA) (Patil et al. 2023), poly(L-lactide) (PLLA) (Ravindran Girija et al. 2023), polyvinyl alcohol (PVA) (Nuttelman et al. 2001), and poly(desamino tyrosyl-tyrosine carbonate) (PDTEC) with poly(ethylene glycol) (PEG) (Goyal et al. 2017).

Fibronectin domains coupled to PEG scaffolds showed an increase in fibroblast migration in skin wound models when compared to unmodified or synthetic RGD (arginylglycylaspartic acid) functionalized polymer scaffolds (Ghosh et al. 2006). Similarly, fibronectin-coated PVA gels provided a favorable surface topography for enhanced fibroblast adhesion, proliferation, and migration (Nuttelman et al. 2001). In a muscle injury model, laminin conjugated to PEG hydrogels showed greater cell proliferation and myogenic differentiation (Ziemkiewicz et al. 2018). Further, in bone defect models, a PLGA/PLA-based mesh scaffold coated with dECM was shown to deliver bone morphogenic protein (BMP-2) and generate an osteogenic microenvironment, nearly completely healing the defect in a mouse ectopic and rat calvarial model compared to the control and fibronectin coated scaffolds (Kim et al. 2015).

Collagen, the largest component of ECM proteins, has also been combined with natural and synthetic polymers including alginate, bioactive glass nanoparticles, and PCL. Lee et al. demonstrated that a greater collagen content in a collagen-alginate composite scaffold increased the material's elasticity and amplified the osteogenic differentiation of MSCs (Lee et al. 2018). Likewise, collagen was incorporated into a PCL-glass nanoparticle hybrid material as a biocompatible delivery device for endothelial progenitor cells in skin wound healing (Wang et al. 2018). These studies support that ECM-derived materials complement synthetic polymers to create chemically and mechanically robust, tunable materials for tissue engineering. Stemming from the properties of native ECM, it is anticipated that hybrid scaffoldings exhibit intensified biorecognition and reduce the likelihood of cellular rejection.

#### 11.2.2.2 Biomaterial Surface Modifications

Surface modification for bioengineered materials can be categorized as physical or chemical. Physical alterations change the surface topography or roughness with little or no change in chemical composition. Examples include machining, etching, and grit-blasting (Bose et al. 2018). Chemical methods on the other hand, can oxidize a surface by application of surface functionalization, single layer coatings, or coatings of multiple layers with varying chemical compositions (Roy et al. 2012; Vahabzadeh et al. 2015). These can be achieved by techniques such as electrochemical deposition, atomic layer deposition, and plasma and chemical vapor deposition (Stallard et al. 2016; O'Shaughnessy et al. 2006; Roy et al. 2011; Bandyopadhyay et al. 2017).Various surface treatments and additive manufacturing methods for micropatterning surfaces are outlined in Fig. 11.2. These topics are further explored below.

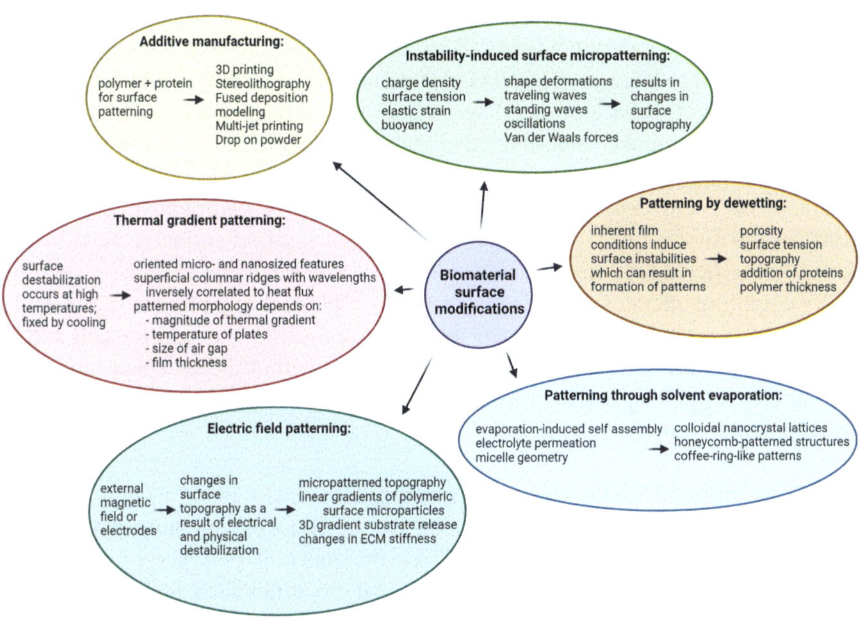


**Fig. 11.2** Visual outline of micropatterning methods for biomaterial fabrication in tissue engineering

### Additive Manufacturing

Protein coatings or surface treatments have recently been implemented for fabricating polymers with greater cell affinity through enhanced biocompatibility and cell adherence. Biomaterial surfaces can be modified with the use of smart polymers and additive manufacturing (AM), a process in which precise control over a surface's chemical and physical properties is able to produce highly customizable materials during the manufacturing process (Bose et al. 2018). AM can also be used as a post-processing tool for site-specific surface modification. AM techniques such as 3D printing, stereolithography, fused deposition modeling, drop on powder, and multi-jet printing have been used to modify material nano and micro topography to induce desirable cell responses. For example, Lee et al. introduced hydroxyapatite to polymer constituents prior to stereolithography. This method produced polymer-based scaffoldings with fully crystallized granules of ceramic which protruded from the surface, providing surface chemistry, topography, and crystallography similar to that of bone (Lee et al. 2008). This surface treatment resulted in increased hydrophilicity and serum protein adsorption, leading to greater cell attachment, spreading, and proliferation compared to non-treated surfaces. Pérez-Davila et al. also applied AM by fused deposition modeling to achieve polylactic acid-hydroxyapatite scaffolds, which provided surface roughness and greater biocompatibility compared to scaffolds without hydroxyapatite. Young's moduli of AM fabricated hydroxyapatite scaffolds increased up to ~50% compared to scaffolds without hydroxyapatite and were able to support proliferation of MG63 cells for up to 21 days (Pérez-Davila et al. 2023). Traditional techniques only coat outer surfaces, whereas AM offers new approaches to alter the chemistry of exposed surfaces throughout the life of the material i.e., the degradation process (Bose et al. 2018). Likewise, a study by Chen et al. applied AM to fabricate a novel biodegradable polymer, poly (ethylene glycol)-co-poly (glycerol sebacate) acrylate (PEGSA), blended with hydroxyapatite nanoparticles to produce osteo-inductive and photocurable resins for 3D printing. Osteogenesis was promoted in 3D printed scaffolds and confirmed *in vivo*, with mesenchymal stem cells attached and differentiated on scaffolds. Interestingly, SEM images showed unexpected grated pattens on scaffold surfaces, most likely due to the restricted printing precision along the z-axis since digital light processing additive printing operates through a layer-by-layer photocuring process (Chen et al. 2023b). As expected, these patterned microgrooves were beneficial for cell adhesion as their width ranged from 5–10 μm, which is within the accepted range for cellular contact guidance. This further suggests AM processes have high potential for biopolymer scaffold fabrication.

### Instability-Induced Surface Micropatterning

Surface instability-induced micropatterning is a rather novel patterning method that unlike other patterning techniques, utilizes the inherent polymer properties to deform material surfaces. Electrical and chemical properties such as charge density, surface tension, elastic strain, and buoyancy can deform a material's surface by capillary waves which localize material flow through feedback loops created by reaction-diffusion dynamics. This phenomenon causes shape deformations, including traveling waves, standing waves, and oscillations. Additional instabilities that induce micropatterns include steric, polar, and electrostatic interactions as well as Van der Waals forces (Fukami et al. 2015; Jacobs and Whitesides 2001; Würthner et al. 2023; Nemani et al. 2018) resulting from liquid-polymer or polymer-polymer instabilities. An inherent competition between these destabilizing forces exists and is responsible for the size and topography of the material surface patterning. The two main types of polymer films for applying instability-induced patterning consist of unstable and metastable films. Unstable materials have their instability attributed to the inherent topographical chemistry and metastable materials are instable due to the stimulation of external thermal, electrical, or mechanical forces (Nemani et al. 2018). For instance, Junisu et al. showed that poly(2-vinyl pyridine) (P2VP) films display significant surface patterns when dried after immersing in an aqueous solution of $HAuCl_4$ with low contents of $K_2CO_3$, having crater morphologies which resulted from capillary flows, Marangoni flows, and pinning and depinning events within the polymer (Junisu et al. 2022). Likewise, Wang et al. optically induced electrohydrodynamic instabilities in PDMS which resulted in microscale patterning of the polymer surface. Projected light patterns induced pillar formation in PDMS substrates within 5 to 90 s in dark regions and subsequentially spread to illuminated regions with thickness less than 2 μm. Pillars, or grooves, had diameters between 15 and 30 μm, which is appropriate for cellular contact guidance. Inter-pillar spacing was controlled by adjusting the applied DC voltage (110 and 210 V) of the two electrodes between which the PDMS substrate was sandwiched (Wang et al. 2014). With voltage control of pillar diameter and pillar position via projected light patterns, this efficient and low cost optically-induced electrohydrodynamic instability method revels new opportunities for fabricating micro- and nanoscale polymer structures for use in many biomedical applications.

### Dewetting

Dewetting is a process which also utilizes surface instabilities to achieve deformations resulting in morphological development and self-organization and consequently in patterning (Müller-Buschbaum 2003; Xing et al. 2007; Bäumchen and Jacobs 2010). The dewetting technique is especially sought after for use in polymers due to a polymers' ability to change between states of matter (solid and liquid) by exposure to temperatures above and below the glass transition temperature (Nemani et al. 2018). For patterning surfaces, dewetting is driven by the inherent film conditions which exist initially such as thickness and tensions between individual fibers and the polymer sheet. The motive to pattern these surfaces, although typically synthetic using the dewetting technique is rooted in nature by mimicking the remarkable wetting and adhesion properties of many micro- and nanopatterned materials (Naffouti et al. 2017; Xue and Han 2011). For example, in nature it is observed that Locus leaves do not get wet in the rain, geckos can locomote on a ceiling, and springtails can breathe underwater (Chudak et al. 2020). Likewise, this phenomenon is seen in many organic and inorganic films including metals, liquids, and semiconductors, and is propelled by mass transport and surface tension (Naffouti et al. 2017).

Polymer surface patterning via dewetting has shown great success in achieving desired cellular functions. For example, a 2019 study shows that biocompatible polymer PLGA was patterned via thin film dewetting and functionalized with antibodies to target the epithelial cell adhesion molecules of cancer cells, demonstrating the binding of cancer cells to the modified patterned surfaces. This simple, inexpensive method delineates the fabrication of functionalized, patterned PLGA surfaces for the efficient isolation of cancer cells from blood samples (Jaiswal et al. 2019). Tronser et al. investigated nanoporous superhydrophobic–hydrophilic micropatterns for inhibiting the differentiation of mouse embryonic stem cells (mESCs) and simultaneously enabling the formation of arrays of microdroplets via discontinuous dewetting. This method allowed high-throughput screening of undifferentiated mESCs. After observing the inhibition of spontaneous mESC differentiation on the droplet microarray for 72 h, a fourfold increase in cell growth rate of mESCs cultured on the microarrays was seen. The behavior of mESCs was attributed to the porosity and topography of their polymer surfaces, achieved through the dewetting process (Tronser et al. 2017). Similarly, elliptical ring arrays of proteins fabricated via a combination of colloidal lithography dewetting and surface initiated atom-transfer radical polymerization showed to promote mouse MC3T3-E1 osteoblast adhesion and enhance the symmetry of the actin cytoskeleton (Liu et al. 2013). Other techniques such as photolithography, microcontact printing, and vapor deposition can be applied alongside dewetting to achieve specific patterns and cellular responses (Nemani et al. 2018).

**Evaporation**

The evaporation of solvents from the surfaces of polymers can leave behind non-volatile solutes in their assembled form, which in turn causes unique surface patterning. Patterning by evaporation is a process similar to surface dewetting, however, the evaporation technique may create an asymmetrical surface anatomy which is attributed to variable convection (Nemani et al. 2018). To avoid this nonuniformity in surface topography, confined geometry methods have been applied for producing patterns of high regularity (Tao et al. 2008; Lin and Granick 2005). Incorporating biopolymers for a more nature-inspired material, bacterial cellulose humidity sensors have been fabricated via solvent evaporation-induced self-assembly and electrolyte permeation to display laminar alignment (Ma et al. 2022). Further, the evaporation method has been used to engineer hierarchically assembled copolymer micelles with desirable shapes and internal structures. Controlled evaporative self-assembly was able to aid in the spontaneous assembly of micelle geometry, creating nanometer-sized microscopic stripes ideal for applications in tissue eingineering (Han et al. 2012; Wu et al. 2017). Honeycomb-patterned microporous structures, colloidal nanocrystal superlattices, and coffee-ring-like patterns are additional structures with more complex features which can be fabricated by interfacial evaporation-induced self-assembly on different length scales (Ma and Hao 2011). Self-assembly has become a practical approach for creating a variety of nanostructures and hierarchical structures because of its simple and effective technique to prepare patterned and mesoporous films (Grabowska et al. 2018). The topographical microstructure of honeycomb films, for example, can promote cell adhesion and control cellular morphology (Li et al. 2023c; Valsesia et al. 2008; Kawano et al. 2012) as well as increase cell proliferation and viability (Gao et al. 2022; Huber et al. 2022).

**Electric Field and Thermal Gradient**

External gradients such as heat flux or electric fields applied to polymer surfaces can cause chemical and physical destabilization which result in formation of patterns (Nemani et al. 2018). For different polymer surfaces, controlled electric fields can be tuned by applying patterned top electrodes for generating ordered arrays of pillars (Dwivedi et al. 2018). Electric fields and thermal gradients often go hand in hand. For example, studies show that pattern formation in polymer films takes place at high temperatures and is then fixed by cooling the films to room temperature. Localized variations in the electric field then cause the destabilization of the air-polymer interface to be concentrated along the direction of the greatest electric field, leaving the film surface with oriented micro- and nanosized features (Schaffer et al. 2000; Kunti et al. 2020). Similar to electric fields, solely applying thermal gradients can also be used to destabilize polymer surfaces for patterning. In this technique, the polymer film is placed between two plates with different temperatures such that a small air gap remains, separating the material from the top plate. The temperature gradient across the air-polymer interface creates superficial columnar ridges with wavelengths which are inversely correlated to the initial heat flux (Nemani et al. 2018; Ciftci et al. 2019; Schäffer et al. 2003; Mukherjee and Sharma 2015). The remaining morphology of the film depends on the magnitude of the thermal gradient, temperatures of the plates, the size of the air gap, and the film thickness (Mukherjee and Sharma 2015).

Enhanced cellular performance on surfaces patterned with electric fields and thermal gradients has been shown to have promising potential. Huang et al. investigated a magnetically controllable 3D-gradient release of nerve growth factor for enhancing topographical nerve guidance after a peripheral nerve injury. The model consisted of magnetic poly(lactic-co-glycolic acid) microcapsules with a self-owned reversible thermo-responsive sol-to-gel phase transformation ability, stimulated via multiple magnetic guidance for micropatterning the microcapsules. Material topography resulted in linear axonal alignment in more than 90% of cells and neurite extension was almost two-fold greater compared with controls. *In vivo*, this patterned implant demonstrated accelerated restoration of motor function within 14 days and the model further contributed to axonal outgrowth and remyelination after 28 days (Huang et al. 2022). Likewise, Herath et al. embedded magnetic beads into ECM for external magnetic field stimulation. Their viscoelastic model of modified ECM allowed for the design of a stretch test to quantify the changes in stiffness induced by the magnetic field. Their future work will involve studying varying stiffness gradients in ECM to influence cell behavior *in vitro* (Herath et al. 2014). Finally, Zhou et al. controlled the distribution of an electrostatic field for generating linear gradients of polymeric microparticles on surfaces. They showed that such gradients promoted the migration of human neuroblastoma cells, concluding that this method will be applicable to research in nerve repair and wound closure (Zhou et al. 2022).

### 11.2.3 Advancements of Patterned Surfaces in Biological Applications

Recent developments in engineering biomaterials for tissue regeneration focus not only on the material itself, but also, on the fabrication process to render microenvironments that mimic native tissue to ultimately guide cell function and regeneration. Fabrication methods of these biomaterials include 3D microcontact printing (Yap and Zhang 2007; Mrksich et al. 1997; Xia et al. 1995; Csucs et al. 2003), electrospinning (Orkwis et al. 2020, 2022), lyophilization, phase separation, soft lithography (Xu et al. 2020a; Kane et al. 1999; Xia and Whitesides 1998; Yap and Zhang 2007) and photolithography (Yap and Zhang 2007; Britland et al. 1992a, b; Lom et al. 1993; Healy 1996), solvent casting, material injection, microfluidic patterning (Folch and Toner 1998; Folch et al. 1999), robotic printing (Yap and Zhang 2007; Xu et al. 2004; Pardo et al. 2003; Roth et al. 2004), and stencil patterning (Folch et al. 2000; Ostuni et al. 2000). Most of these processes leave the materials with groves or ridges that can align and elongate cells with directionality towards a wound gap in the context of injury. For example, in many different tissues including nerves (Cady et al. 2020; Xu et al. 2020a; Orkwis et al. 2020, 2022), skin (Salehi et al. 2020; Sun et al. 2021; Priya et al. 2008; Groeber et al. 2011; Amatangelo et al. 2005; Ghosh et al. 2006; Wang et al. 2018), cardiac (Atmanli and Domian 2013; Papadaki et al. 2001; Liberski et al. 2016; Xing et al. 2020; Echave et al. 2017), and muscle (Atmanli and Domian 2013; Papadaki et al. 2001; Wigmore and Dunglison 1998; Chiu et al. 2004; Ziemkiewicz et al. 2018; Ye et al. 2023), it is desired that biomaterials provide contact guidance cues to the cells in order to promote cellular alignment, directional migration, cytoskeletal elongation, and gene and protein expression (Cady et al. 2020; Xu et al. 2020a). Contact guidance is typically achieved through the use of micropatterned surfaces which possess high cell affinity. The grooves or ridges allow cells to attach in a way that guides cells with direction and parallel alignment, promoting cytoskeletal elongation and morphological changes through a regenerative cell phonotype. For instance, a current study by Han et al. reported the regulatory effects of scaffold surface micropatterns on the myofibroblastic differentiation of fibroblasts for inhibiting the onset of fibrosis (Han et al. 2023). *In vitro* and *in vivo* studies show PCL surfaces with micro-columns triggered F- to G-actin transition, inhibiting the nuclear transfer of myocardin-related transcription factor-A and suppressing the α-smooth muscle actin marker of myofibroblasts, which hindered the overall formation of peri-implant fibrotic capsules in all patterned scaffolds (Han et al. 2023). Further, Lu et al. investigated the performance of tendon stem/progenitor cells (TSPCs) on micropatterned silk fibroin films in a rat Achilles tendon defect model (Lu et al. 2023). Immunohistochemistry and western blot results show tendon marker genes were significantly higher in micropatterned silk fibroin film repair groups, elucidating that micropatterned films promoted remodeling of injured tendons and regulated TSPC differentiation through the $\alpha 2\beta 1$/FAK/PI3K/AKT signaling pathway (Lu et al. 2023).

## 11.3 Conclusion

Nature-inspired biomaterials have gained immense popularity for their biocompatibility and improved cellular behaviors. Here, we overviewed the realm of biomaterials from both synthetic and natural origins and introduced a variety of micropatterning techniques which in literature, have enhanced cellular adhesion, proliferation, differentiation, and protein and gene expression. Further, we explored extracellular matrix-based scaffolds including decellularized matrices and extracellular matrix-inspired polymers for improved cellular biocompatibility and scaffold tunability. Several biomaterial fabrication techniques in which patterning occurs naturally were also discussed with a focus placed on resulting cellular responses.

**Acknowledgements** The authors gratefully acknowledge support from the Department of Defense (DM190692 to GMH and AEB), the University of Cincinnati (GMH), and the National Science Foundation (DM190692 to GMH and JW). The authors would also like to thank Katelyn Jansen and Josh Long for their assistance in performing related research.

## References

Amatangelo MD, Bassi DE, Klein-Szanto AJP, Cukierman E (2005) Stroma-derived three-dimensional matrices are necessary and sufficient to promote desmoplastic differentiation of normal fibroblasts. Am J Pathol 167:475–488

Atmanli A, Domian IJ (2013) Generation of aligned functional myocardial tissue through microcontact printing. J Vis Exp e50288. https://doi.org/10.3791/50288

Bailey SJ et al (2023) Diels-Alder photoclick patterning of extracellular matrix for spatially controlled cell behaviors. Adv Mater:e2303453. https://doi.org/10.1002/adma.202303453

Bandyopadhyay A et al (2017) In vivo response of laser processed porous titanium implants for load-bearing implants. Ann Biomed Eng 45:249–260

Bar-Shai N, Sharabani-Yosef O, Zollmann M, Lesman A, Golberg A (2021) Seaweed cellulose scaffolds derived from green macroalgae for tissue engineering. Sci Rep 11:11843

Bäumchen O, Jacobs K (2010) Slip effects in polymer thin films. J Phys Condens Matter 22:033102

Berthod F, Germain L, Tremblay N, Auger FA (2006) Extracellular matrix deposition by fibroblasts is necessary to promote capillary-like tube formation in vitro. J Cell Physiol 207:491–498

Bhattacharjee P, Madden PW, Patriarca E, Ahearne M (2023) Optimization and evaluation of oxygen-plasma-modified, aligned, poly (Є-caprolactone) and silk fibroin nanofibrous scaffold for corneal stromal regeneration. Biomater Biosyst 12:100083

Biancarosa I et al (2018) Chemical characterization of 21 species of marine macroalgae common in Norwegian waters: benefits of and limitations to their potential use in food and feed. J Sci Food Agric 98:2035–2042

Bose S, Robertson SF, Bandyopadhyay A (2018) Surface modification of biomaterials and biomedical devices using additive manufacturing. Acta Biomater 66:6–22

Brackman G, Coenye T (2016) In vitro and in vivo biofilm wound models and their application. Adv Exp Med Biol 897:15–32

Britland S et al (1992a) Micropatterning proteins and synthetic peptides on solid supports: a novel application for microelectronics fabrication technology. Biotechnol Prog 8:155–160

Britland S, Clark P, Connolly P, Moores G (1992b) Micropatterned substratum adhesiveness: A model for morphogenetic cues controlling cell behavior. Exp Cell Res 198:124–129

Cady E et al (2020) Micropatterning decellularized ECM as a bioactive surface to guide cell alignment, proliferation, and migration. Bioengineering (Basel) 7:102

Castelló-Cros R, Khan DR, Simons J, Valianou M, Cukierman E (2009) Staged stromal extracellular 3D matrices differentially regulate breast cancer cell responses through PI3K and beta1-integrins. BMC Cancer 9:94
Celestino MF et al (2023) 3D filaments based on polyhydroxy butyrate-micronized bacterial cellulose for tissue engineering applications. J Funct Biomater 14:464
Chan WW et al (2021) Towards biomanufacturing of cell-derived matrices. Int J Mol Sci 22:11929
Cheah YT, Chan DJC (2021) Physiology of microalgal biofilm: a review on prediction of adhesion on substrates. Bioengineered 12:7577–7599
Chen CS, Jiang X, Whitesides GM (2005) Microengineering the environment of mammalian cells in culture. MRS Bull 30:194–201
Chen G, Wang F, Zhang X, Shang Y, Zhao Y (2023a) Living microecological hydrogels for wound healing. Sci Adv 9:eadg3478
Chen Y-T, Chuang Y-H, Chen C-M, Wang J-Y, Wang J (2023b) Development of hybrid scaffolds with biodegradable polymer composites and bioactive hydrogels for bone tissue engineering. Biomater Adv 153:213562
Cheng HW, Yuan MT, Li CW, Chan BP (2020) Cell-derived matrices (CDM)-methods, challenges and applications. Methods Cell Biol 156:235–258. Elsevier
Chiu J-J, Chen L-J, Chen C-N, Lee P-L, Lee C-I (2004) A model for studying the effect of shear stress on interactions between vascular endothelial cells and smooth muscle cells. J Biomech 37:531–539
Christenson L, Sims R (2011) Production and harvesting of microalgae for wastewater treatment, biofuels, and bioproducts. Biotechnol Adv 29:686–702
Chu W-S, Park H, Moon S (2023) Novel fabrication of 3-D cell laden micro-patterned porous structure on cell growth and proliferation by layered manufacturing. Bioengineering (Basel) 10:1092
Chudak M, Chopra V, Hensel R, Darhuber AA (2020) Elastohydrodynamic dewetting of thin liquid films: elucidating underwater adhesion of topographically patterned surfaces. Langmuir 36:11929–11937
Ciftci HT, Van LP, Koopmans B, Kurnosikov O (2019) Polymer patterning with self-heating atomic force microscope probes. J Phys Chem A 123:8036–8042
Csucs G, Michel R, Lussi JW, Textor M, Danuser G (2003) Microcontact printing of novel co-polymers in combination with proteins for cell-biological applications. Biomaterials 24:1713–1720
Cukierman E, Pankov R, Stevens DR, Yamada KM (2001) Taking cell-matrix adhesions to the third dimension. Science 294:1708–1712
de Grooth BG, Geerken TH, Greve J (1985) The cytodisk: a cytometer based upon a new principle of cell alignment. Cytometry 6:226–233
Dupont S et al (2011) Role of YAP/TAZ in mechanotransduction. Nature 474:179–183
Dwivedi S, Narayanan R, Chaudhary R, Mukherjee R, Atta A (2018) Controlled nanoscale electrohydrodynamic patterning using mesopatterned template. ACS Omega 3:9781–9789
Echave MC, Saenz del Burgo L, Pedraz JL, Orive G (2017) Gelatin as biomaterial for tissue engineering. Curr Pharm Des 23:3567–3584
El Knidri H, Belaabed R, Addaou A, Laajeb A, Lahsini A (2018) Extraction, chemical modification and characterization of chitin and chitosan. Int J Biol Macromol 120:1181–1189
English EJ, Samolyk BL, Gaudette GR, Pins GD (2023) Micropatterned fibrin scaffolds increase cardiomyocyte alignment and contractility for the fabrication of engineered myocardial tissue. J Biomed Mater Res A 111:1309–1321
Erdeş E et al (2014) Characterization of Leiurus abdullahbayrami (Scorpiones: Buthidae) venom: peptide profile, cytotoxicity and antimicrobial activity. J Venom Anim Toxins Incl Trop Dis 20:48
Féréol S et al (2011) Micropatterned ECM substrates reveal complementary contribution of low and high affinity ligands to neurite outgrowth. Cytoskeleton (Hoboken) 68:373–388
Flores Bueso Y, Lehouritis P, Tangney M (2018) In situ biomolecule production by bacteria; a synthetic biology approach to medicine. J Control Release 275:217–228

Folch A, Toner M (1998) Cellular micropatterns on biocompatible materials. Biotechnol Prog 14:388–392

Folch A, Ayon A, Hurtado O, Schmidt MA, Toner M (1999) Molding of deep polydimethylsiloxane microstructures for microfluidics and biological applications. J Biomech Eng 121:28–34

Folch A, Jo B-H, Hurtado O, Beebe DJ, Toner M (2000) Microfabricated elastomeric stencils for micropatterning cell cultures. J Biomed Mater Res 52:346–353

Fontana G et al (2017) Biofunctionalized plants as diverse biomaterials for human cell culture. Adv Healthc Mater 6

Fukami K et al (2015) Spontaneous formation of microgroove arrays on the surface of p-type porous silicon induced by a turing instability in electrochemical dissolution. ChemPhysChem 16:1613–1618

Gao W et al (2022) The hollow porous sphere cell carrier for the dynamic three-dimensional cell culture. Tissue Eng Part C Methods 28:610–622

Gao H et al (2023) Surface topologized ovalbumin scaffolds containing YIGSR peptides for modulating Schwann cell behavior. Int J Biol Macromol 253:127015

Garrido Dos Santos M et al (2023) Production of a bioink containing decellularized spinal cord tissue for 3D bioprinting. Tissue Eng Part A 30:61–74. https://doi.org/10.1089/ten.TEA.2023.0078

Geng T et al (2023) The study on the morphology and compression properties of microcellular TPU/nanoclay tissue scaffolds for potential tissue engineering applications. Polymers (Basel) 15:3647

Ghorbani M, Nezhad-Mokhtari P, Sohrabi H, Roshangar L (2020) Electrospun chitosan/nanocrystalline cellulose-graft-poly(N-vinylcaprolactam) nanofibers as the reinforced scaffold for tissue engineering. J Mater Sci 55:2176–2185

Ghosh K, Ingber DE (2007) Micromechanical control of cell and tissue development: implications for tissue engineering. Adv Drug Deliv Rev 59:1306–1318

Ghosh K, Ren X-D, Shu XZ, Prestwich GD, Clark RAF (2006) Fibronectin functional domains coupled to hyaluronan stimulate adult human dermal fibroblast responses critical for wound healing. Tissue Eng 12:601–613

Goyal R et al (2017) Development of hybrid scaffolds with natural extracellular matrix deposited within synthetic polymeric fibers. J Biomed Mater Res A 105:2162–2170

Grabowska E, Marchelek M, Paszkiewicz-Gawron M, Zaleska-Medynska A (2018) Metal oxide photocatalysts. In: Metal oxide-based photocatalysis. Elsevier, pp 51–209. https://doi.org/10.1016/B978-0-12-811634-0.00003-2

Griffith LG, Naughton G (2002) Tissue engineering – current challenges and expanding opportunities. Science 295:1009–1014

Groeber F, Holeiter M, Hampel M, Hinderer S, Schenke-Layland K (2011) Skin tissue engineering – in vivo and in vitro applications. Adv Drug Deliv Rev 63:352–366

Grove M et al (2017) YAP/TAZ initiate and maintain Schwann cell myelination. elife 6:e20982

Guo C, Wu J, Zeng Y, Li H (2023) Construction of 3D bioprinting of HAP/collagen scaffold in gelation bath for bone tissue engineering. Regen Biomater 10:rbad067

Han W, Byun M, Li B, Pang X, Lin Z (2012) A simple route to hierarchically assembled micelles and inorganic nanoparticles. Angew Chem Int Ed Engl 51:12588–12592

Han W et al (2023) Modulating myofibroblastic differentiation of fibroblasts through actin-MRTF signaling Axis by micropatterned surfaces for suppressed implant-induced fibrosis. Research (Wash D C) 6:0049

Harris GM et al (2017) Nerve guidance by a decellularized fibroblast extracellular matrix. Matrix Biol 60–61:176–189

Harris GM, Raitman I, Schwarzbauer JE (2018) Cell-derived decellularized extracellular matrices. Methods Cell Biol 143:97–114. Elsevier

Healy K (1996) Kinetics of bone cell organization and mineralization on materials with patterned surface chemistry. Biomaterials 17:195–208

Hellewell AL, Rosini S, Adams JC (2017) A rapid, scalable method for the isolation, functional study, and analysis of cell-derived extracellular matrix. J Vis Exp 55051. https://doi.org/10.3791/55051

Herath SCB et al (2014) Characterization of uniaxial stiffness of extracellular matrix embedded with magnetic beads via bio-conjugation and under the influence of an external magnetic field. J Mech Behav Biomed Mater 30:253–265
Hinchliffe JD, Parassini Madappura A, Syed Mohamed SMD, Roy I (2021) Biomedical applications of bacteria-derived polymers. Polymers (Basel) 13:1081
Hoshiba T, Tanaka M (2013) Breast cancer cell behaviors on staged tumorigenesis-mimicking matrices derived from tumor cells at various malignant stages. Biochem Biophys Res Commun 439:291–296
Hu Y, You J-O, Aizenberg J (2016) Micropatterned hydrogel surface with high-aspect-ratio features for cell guidance and tissue growth. ACS Appl Mater Interfaces 8:21939–21945
Huang C et al (2019) Bio-inspired nanocomposite by layer-by-layer coating of chitosan/hyaluronic acid multilayers on a hard nanocellulose-hydroxyapatite matrix. Carbohydr Polym 222:115036
Huang W-C, Lin C-C, Chiu T-W, Chen S-Y (2022) 3D gradient and linearly aligned magnetic microcapsules in nerve guidance conduits with remotely spatiotemporally controlled release to enhance peripheral nerve repair. ACS Appl Mater Interfaces 14:46188–46200
Huber F et al (2022) Influence of 3D printing parameters on the mechanical stability of PCL scaffolds and the proliferation behavior of bone cells. Materials (Basel) 15:2091
Insuasti-Cruz E et al (2022) Natural biomaterials from biodiversity for healthcare applications. Adv Healthc Mater 11:e2101389
Jacob S et al (2022) Nanocellulose in tissue engineering and bioremediation: mechanism of action. Bioengineered 13:12823–12833
Jacobs HO, Whitesides GM (2001) Submicrometer patterning of charge in thin-film electrets. Science 291:1763–1766
Jaiswal N, Hens A, Chatterjee M, Mahata N, Chanda N (2019) Ethylenediamine assisted functionalization of self-organized poly (d, l-lactide-co-glycolide) patterned surface to enhance cancer cell isolation. J Colloid Interface Sci 534:122–130
Jovic TH, Kungwengwe G, Mills AC, Whitaker IS (2019) Plant-derived biomaterials: a review of 3D bioprinting and biomedical applications. Front Mech Eng 5:19
Junisu BA, Ching-Ya Chang I, Sun Y-S (2022) Film instability induced by swelling and drying. Langmuir 38:13009–13020
Kane RS, Takayama S, Ostuni E, Ingber DE, Whitesides GM (1999) Patterning proteins and cells using soft lithography. Biomaterials 20:2363–2376
Kawano T, Iwama N, Ishihata H, Shimauchi H, Shimomura M (2012) Preparation and biomedical application of self-organized honeycomb-patterned polymer films. In: Sasaki K, Suzuki O, Takahashi N (eds) Interface oral health science 2011. Springer, pp 22–26. https://doi.org/10.1007/978-4-431-54070-0_4
Khrunyk Y, Lach S, Petrenko I, Ehrlich H (2020) Progress in modern marine biomaterials research. Mar Drugs 18:589
Kim IG et al (2015) Bioactive cell-derived matrices combined with polymer mesh scaffold for osteogenesis and bone healing. Biomaterials 50:75–86
Kimura T, Kamioka T, Kuga S (2007) Filtration-assisted magnetic micropatterning of bacterial cellulose. Polym J 39:1199–1201
Kshitiz AJ, Chang H, Goyal R, Levchenko A (2016) Mechanics of microenvironment as instructive cues guiding stem cell behavior. Curr Stem Cell Rep 2:62–72
Kunti G, Agarwal T, Bhattacharya A, Maiti TK, Chakraborty S (2020) On-chip concentration and patterning of biological cells using interplay of electrical and thermal fields. Anal Chem 92:838–844
Kutys ML, Doyle AD, Yamada KM (2013) Regulation of cell adhesion and migration by cell-derived matrices. Exp Cell Res 319:2434–2439
Langer R, Vacanti JP (1993) Tissue engineering. Science 260:920–926
Lee S-M, Ryu C-M (2021) Algae as new kids in the beneficial plant microbiome. Front Plant Sci 12:599742

Lee K-W, Wang S, Yaszemski MJ, Lu L (2008) Physical properties and cellular responses to crosslinkable poly(propylene fumarate)/hydroxyapatite nanocomposites. Biomaterials 29:2839–2848

Lee YB et al (2017) Microcontact printing of polydopamine on thermally expandable hydrogels for controlled cell adhesion and delivery of geometrically defined microtissues. Acta Biomater 61:75–87

Lee H, Woo H-M, Kang B-J (2018) Impact of collagen-alginate composition from microbead morphological properties to microencapsulated canine adipose tissue-derived mesenchymal stem cell activities. J Biomater Sci Polym Ed 29:1042–1052

Li J et al (2023a) Micropatterned composite membrane guides oriented cell growth and vascularization for accelerating wound healing. Regen Biomater 10:rbac108

Li Z et al (2023b) Three-dimensional nanofibrous sponges with aligned architecture and controlled hierarchy regulate neural stem cell fate for spinal cord regeneration. Theranostics 13:4762–4780

Li Z et al (2023c) Selenium-decorated biocompatible honeycomb films with redox-switchable surface for controlling cell adhesion/detachment. J Colloid Interface Sci 635:503–513

Liberski A, Latif N, Raynaud C, Bollensdorff C, Yacoub M (2016) Alginate for cardiac regeneration: from seaweed to clinical trials. Glob Cardiol Sci Pract 2016:e201604

Lin Z, Granick S (2005) Patterns formed by droplet evaporation from a restricted geometry. J Am Chem Soc 127:2816–2817

Liu W et al (2013) Elliptical polymer brush ring array mediated protein patterning and cell adhesion on patterned protein surfaces. ACS Appl Mater Interfaces 5:12587–12593

Liu K et al (2019) Structural foundation for insect-selective activity of acylpolyamine toxins from spider Araneus ventricosus. Chem Res Toxicol 32:659–667

Liu H, Hu Y, Wu X, Hu R, Liu Y (2023a) Optimization of surface-engineered micropatterns on bacterial cellulose for guided scar-free skin wound healing. Biomol Ther 13:793

Liu K et al (2023b) Highly anisotropic and elastic cellulosic scaffold guiding cell orientation and osteogenic differentiation via topological and mechanical cues. Carbohydr Polym 321:121292

Liu Q, Li Q, Hatakeyama M, Kitaoka T (2023c) Proliferation and differential regulation of osteoblasts cultured on surface-phosphorylated cellulose nanofiber scaffolds. Int J Biol Macromol 253:126842

Lom B, Healy KE, Hockberger PE (1993) A versatile technique for patterning biomolecules onto glass coverslips. J Neurosci Methods 50:385–397

Lu K et al (2023) Micropattern silk fibroin film facilitates tendon repair in vivo and promotes tenogenic differentiation of tendon stem/progenitor cells through the α2β1/FAK/PI3K/AKT signaling pathway in vitro. Stem Cells Int 2023:2915826

Ma H, Hao J (2011) Ordered patterns and structures via interfacial self-assembly: superlattices, honeycomb structures and coffee rings. Chem Soc Rev 40:5457–5471

Ma X et al (2018) Rapid 3D bioprinting of decellularized extracellular matrix with regionally varied mechanical properties and biomimetic microarchitecture. Biomaterials 185:310–321

Ma H et al (2022) Strong bacterial cellulose-based films with natural laminar alignment for highly sensitive humidity sensors. ACS Appl Mater Interfaces 14:3165–3175

Mao Y, Schwarzbauer JE (2005) Fibronectin fibrillogenesis, a cell-mediated matrix assembly process. Matrix Biol 24:389–399

Mao Q et al (2020) Fabrication of liver microtissue with liver decellularized extracellular matrix (dECM) bioink by digital light processing (DLP) bioprinting. Mater Sci Eng C Mater Biol Appl 109:110625

McBeath R, Pirone DM, Nelson CM, Bhadriraju K, Chen CS (2004) Cell shape, cytoskeletal tension, and RhoA regulate stem cell lineage commitment. Dev Cell 6:483–495

Min K et al (2022) Three-dimensional microfilament printing of a decellularized extracellular matrix (dECM) bioink using a microgel printing bath for nerve graft fabrication and the effectiveness of dECM graft combined with a polycaprolactone conduit. ACS Appl Bio Mater 5:1591–1603

Mrksich M, Dike LE, Tien J, Ingber DE, Whitesides GM (1997) Using microcontact printing to pattern the attachment of mammalian cells to self-assembled monolayers of alkanethiolates on transparent films of gold and silver. Exp Cell Res 235:305–313
Mukherjee R, Sharma A (2015) Instability, self-organization and pattern formation in thin soft films. Soft Matter 11:8717–8740
Müller-Buschbaum P (2003) Dewetting and pattern formation in thin polymer films as investigated in real and reciprocal space. J Phys Condens Matter 15:R1549–R1582
Naffouti M et al (2017) Complex dewetting scenarios of ultrathin silicon films for large-scale nanoarchitectures. Sci Adv 3:eaao1472
Naomi R, Bt Hj Idrus R, Fauzi MB (2020) Plant- vs. bacterial-derived cellulose for wound healing: a review. Int J Environ Res Public Health 17:6803
Nemani SK et al (2018) Surface modification of polymers: methods and applications. Adv Mater Interfaces 5:1801247
Nuttelman CR, Mortisen DJ, Henry SM, Anseth KS (2001) Attachment of fibronectin to poly(vinyl alcohol) hydrogels promotes NIH3T3 cell adhesion, proliferation, and migration. J Biomed Mater Res 57:217–223
O'Shaughnessy WS, Gao M, Gleason KK (2006) Initiated chemical vapor deposition of trivinyltrimethylcyclotrisiloxane for biomaterial coatings. Langmuir 22:7021–7026
Orkwis JA et al (2020) Development of a piezoelectric PVDF-TrFE fibrous scaffold to guide cell adhesion, proliferation, and alignment. Macromol Biosci 20:e2000197
Orkwis JA et al (2022) Mechanical stimulation of a bioactive, functionalized PVDF-TrFE scaffold provides electrical signaling for nerve repair applications. Biomater Adv 140:213081
Ostuni E, Kane R, Chen CS, Ingber DE, Whitesides GM (2000) Patterning mammalian cells using elastomeric membranes. Langmuir 16:7811–7819
Papadaki M et al (2001) Tissue engineering of functional cardiac muscle: molecular, structural, and electrophysiological studies. Am J Physiol Heart Circ Physiol 280:H168–H178
Pardo L, Wilson WC, Boland T (2003) Characterization of patterned self-assembled monolayers and protein arrays generated by the ink-jet method. Langmuir 19:1462–1466
Patil D, Kumari S, Chatterjee K (2023) Bioinspired nanotopography on 3D printed tissue scaffold to impart mechanobactericidal and osteogenic activities. Colloids Surf B Biointerfaces 228:113401
Pérez-Davila S et al (2023) Physicochemical properties of 3D-printed polylactic acid/hydroxyapatite scaffolds. Polymers (Basel) 15:2849
Pham HL et al (2023) PDMS micropatterns coated with PDA and RGD induce a regulatory macrophage-like phenotype. Micromachines (Basel) 14:673
Plavec TV, Berlec A (2020) Safety aspects of genetically modified lactic acid bacteria. Microorganisms 8:297
Pooyan P, Kim IT, Jacob KI, Tannenbaum R, Garmestani H (2013) Design of a cellulose-based nanocomposite as a potential polymeric scaffold in tissue engineering. Polymer 54:2105–2114
Popa EG, Reis RL, Gomes ME (2015) Seaweed polysaccharide-based hydrogels used for the regeneration of articular cartilage. Crit Rev Biotechnol 35:410–424
Priya SG, Jungvid H, Kumar A (2008) Skin tissue engineering for tissue repair and regeneration. Tissue Eng Part B Rev 14:105–118
Rauff A et al (2023) Dynamic biophysical cues near the tip cell microenvironment provide distinct guidance signals to angiogenic neovessels. Ann Biomed Eng 51:1835–1846. https://doi.org/10.1007/s10439-023-03202-4
Ravindran Girija A et al (2023) Collagen functionalization of polymeric electrospun scaffolds to improve integration into full-thickness wounds. Pharmaceutics 15:880
Rodríguez de la Vega RC, Schwartz EF, Possani LD (2010) Mining on scorpion venom biodiversity. Toxicon 56:1155–1161
Roth EA et al (2004) Inkjet printing for high-throughput cell patterning. Biomaterials 25:3707–3715
Roy M, Bandyopadhyay A, Bose S (2011) Induction plasma sprayed Nano hydroxyapatite coatings on titanium for orthopaedic and dental implants. Surf Coat Technol 205:2785–2792
Roy M, Fielding GA, Beyenal H, Bandyopadhyay A, Bose S (2012) Mechanical, in vitro antimicrobial, and biological properties of plasma-sprayed silver-doped hydroxyapatite coating. ACS Appl Mater Interfaces 4:1341–1349

Salehi S, Koeck K, Scheibel T (2020) Spider silk for tissue engineering applications. Molecules 25:737
Samal P et al (2023) Direct deep UV lithography to micropattern PMMA for stem cell culture. Mater Today Bio 22:100779
Sart S, Jeske R, Chen X, Ma T, Li Y (2020) Engineering stem cell-derived extracellular matrices: decellularization, characterization, and biological function. Tissue Eng Part B Rev 26:402–422
Schaffer E, Thurn-Albrecht T, Russell TP, Steiner U (2000) Electrically induced structure formation and pattern transfer. Nature 403:874–877
Schäffer E, Harkema S, Roerdink M, Blossey R, Steiner U (2003) Morphological instability of a confined polymer film in a thermal gradient. Macromolecules 36:1645–1655
Schynkel L et al (2023) Cell guiding multicomponent nanoyarn tendon scaffolds with tunable morphology and flexibility. ACS Appl Mater Interfaces 15:42241–42250
Shu JY, Panganiban B, Xu T (2013) Peptide-polymer conjugates: from fundamental science to application. Annu Rev Phys Chem 64:631–657
Soucy PA, Romer LH (2009) Endothelial cell adhesion, signaling, and morphogenesis in fibroblast-derived matrix. Matrix Biol 28:273–283
Spiller S, Clauder F, Bellmann-Sickert K, Beck-Sickinger AG (2021) Improvement of wound healing by the development of ECM-inspired biomaterial coatings and controlled protein release. Biol Chem 402:1271–1288
Stallard CP, Solar P, Biederman H, Dowling DP (2016) Deposition of non-fouling PEO-like coatings using a low temperature atmospheric pressure plasma jet: deposition of non-fouling PEO-like coatings…. Plasma Process Polym 13:241–252
Sun W, Gregory DA, Tomeh MA, Zhao X (2021) Silk fibroin as a functional biomaterial for tissue engineering. Int J Mol Sci 22:1499
Suryavanshi P, Kudtarkar Y, Chaudhari M, Bodas D (2023) Fabricating a low-temperature synthesized graphene-cellulose acetate-sodium alginate scaffold for the generation of ovarian cancer spheriod and its drug assessment. Nanoscale Adv 5:5045–5053
Takeda YS, Xu Q (2014) Fabrication of 2D and 3D constructs from reconstituted decellularized tissue extracellular matrices. J Biomed Nanotechnol 10:3631–3637
Tao AR, Huang J, Yang P (2008) Langmuir-Blodgettry of nanocrystals and nanowires. Acc Chem Res 41:1662–1673
Tarafdar A et al (2021) Advances in biomaterial production from animal derived waste. Bioengineered 12:8247–8258
Thirukumaran R et al (2022) Resource recovery from fish waste: prospects and the usage of intensified extraction technologies. Chemosphere 299:134361
Tiwari BK, Troy DJ (2015) Seaweed sustainability – food and nonfood applications. In: Seaweed sustainability. Elsevier, pp 1–6. https://doi.org/10.1016/B978-0-12-418697-2.00001-5
Treberg JR, Speers-Roesch B (2016) Does the physiology of chondrichthyan fishes constrain their distribution in the deep sea? J Exp Biol 219:615–625
Tronser T, Popova AA, Jaggy M, Bastmeyer M, Levkin PA (2017) Droplet microarray based on patterned superhydrophobic surfaces prevents stem cell differentiation and enables high-throughput stem cell screening. Adv Healthc Mater 6:1700622
Ueno H, Sato K, Yamada K, Suzuki T (2019) Micropatterning method for porous materials using the difference of the glass transition temperature between exposed and unexposed areas of a thick-photoresist. Micromachines 11:54
Vahabzadeh S, Roy M, Bandyopadhyay A, Bose S (2015) Phase stability and biological property evaluation of plasma sprayed hydroxyapatite coatings for orthopedic and dental applications. Acta Biomater 17:47–55
Valsesia A et al (2008) Use of nanopatterned surfaces to enhance immunoreaction efficiency. Anal Chem 80:1418–1424
Vassilevski AA, Kozlov SA, Grishin EV (2009) Molecular diversity of spider venom. Biochemistry (Mosc) 74:1505–1534
Vunjak-Novakovic G, Altman G, Horan R, Kaplan DL (2004) Tissue engineering of ligaments. Annu Rev Biomed Eng 6:131–156

Wainwright PC, Longo SJ (2017) Functional innovations and the conquest of the oceans by acanthomorph fishes. Curr Biol 27:R550–R557
Wang D-Y, Huang Y-C, Chiang H, Wo AM, Huang Y-Y (2007) Microcontact printing of laminin on oxygen plasma activated substrates for the alignment and growth of Schwann cells. J Biomed Mater Res B Appl Biomater 80:447–453
Wang F et al (2014) Optically induced electrohydrodynamic instability-based micro-patterning of fluidic thin films. Microfluid Nanofluid 16:1097–1106
Wang C et al (2018) Highly efficient local delivery of endothelial progenitor cells significantly potentiates angiogenesis and full-thickness wound healing. Acta Biomater 69:156–169
Watson JL et al (2021) High-efficacy subcellular micropatterning of proteins using fibrinogen anchors. J Cell Biol 220:e202009063
Whitesides GM, Ostuni E, Takayama S, Jiang X, Ingber DE (2001) Soft lithography in biology and biochemistry. Annu Rev Biomed Eng 3:335–373
Wigmore PM, Dunglison GF (1998) The generation of fiber diversity during myogenesis. Int J Dev Biol 42:117–125
Wu Y, Wang K, Tan H, Xu J, Zhu J (2017) Emulsion solvent evaporation-induced self-assembly of block copolymers containing pH-sensitive block. Langmuir 33:9889–9896
Würthner L, Goychuk A, Frey E (2023) Geometry-induced patterns through mechanochemical coupling. Phys Rev E 108:014404
Xia Y, Whitesides GM (1998) Soft lithography. Annu Rev Mater Sci 28:153–184
Xia Y, Mrksich M, Kim E, Whitesides GM (1995) Microcontact printing of octadecylsiloxane on the surface of silicon dioxide and its application in microfabrication. J Am Chem Soc 117:9576–9577
Xing R, Luo C, Wang Z, Han Y (2007) Dewetting of polymethyl methacrylate on the patterned elastomer substrate by solvent vapor treatment. Polymer 48:3574–3583
Xing H, Lee H, Luo L, Kyriakides TR (2020) Extracellular matrix-derived biomaterials in engineering cell function. Biotechnol Adv 42:107421
Xu T et al (2004) Construction of high-density bacterial colony arrays and patterns by the ink-jet method. Biotechnol Bioeng 85:29–33
Xu Z, Orkwis JA, DeVine BM, Harris GM (2020a) Extracellular matrix cues modulate Schwann cell morphology, proliferation, and protein expression. J Tissue Eng Regen Med 14:229–242
Xu Z, Orkwis JA, Harris GM (2020b) Preparation of tunable extracellular matrix microenvironments to evaluate Schwann cell phenotype specification. JoVE 61496. https://doi.org/10.3791/61496
Xue L, Han Y (2011) Pattern formation by dewetting of polymer thin film. Prog Polym Sci 36:269–293
Yap FL, Zhang Y (2007) Protein and cell micropatterning and its integration with micro/nanoparticles assembly. Biosens Bioelectron 22:775–788
Ye K et al (2023) LIPUS-SCs-Exo promotes peripheral nerve regeneration in cavernous nerve crush injury-induced ED rats via PI3K/Akt/FoxO signaling pathway. CNS Neurosci Ther 29:3239–3258. https://doi.org/10.1111/cns.14256
Yi B et al (2023) Step-wise CAG@PLys@PDA-Cu2+ modification on micropatterned nanofibers for programmed endothelial healing. Bioact Mater 25:657–676
Yun J et al (2023) Aligned skeletal muscle assembly on a biofunctionalized plant leaf scaffold. Acta Biomater S1742-7061(23):00550–0. https://doi.org/10.1016/j.actbio.2023.09.016
Zhao G et al (2022) Rational design of electrically conductive biomaterials toward excitable tissues regeneration. Prog Polym Sci 131:101573
Zhou Z et al (2022) Manipulating electrostatic field to control the distribution of bioactive proteins or polymeric microparticles on planar surfaces for guiding cell migration. Colloids Surf B Biointerfaces 209:112185
Ziemkiewicz N et al (2018) Laminin-111 functionalized polyethylene glycol hydrogels support myogenic activity in vitro. Biomed Mater 13:065007

# Sustainable Ceramic-Based Biomaterials for Bone Tissue Engineering

# 12

Sandra Pina, Rui L. Reis, and J. Miguel Oliveira

**Abstract**

In today's world, the need for sustainable development has reached an unprecedented level of urgency. Our planet is grappling with numerous environmental crises, including the escalating threat of climate change, pervasive pollution, and the rapid depletion of natural resources. In response to these challenges, a promising path forward involves utilizing renewable resources as primary sources for biomaterials and tissue engineering applications. This approach not only mitigates environmental impacts but also advances medical technologies. Bioceramics, particularly calcium phosphates, have gained significant traction as synthetic bone graft substitutes due to their remarkable similarity to the mineral component of natural bone tissue. In line with sustainability and the principles of a circular economy, these bioceramics can be derived from various waste materials, such as fish scales, fish bones, eggshells, and seashells. Additionally, marine sponges and agricultural waste products (e.g., rice straw and rice husk) present a sustainable and economical source of silica with applications in biomedicine, targeted bioglasses synthesis, and drug delivery. As research in sustainable bioceramics advances, further improvements in material properties, processing techniques, and clinical outcomes are anticipated, potentially revolutionizing the field of bone tissue engineering and regenerative medicine.

S. Pina (✉) · R. L. Reis · J. M. Oliveira
3B's Research Group, I3Bs – Research Institute on Biomaterials, Biodegradables and Biomimetics, University of Minho, Headquarters of the European Institute of Excellence on Tissue Engineering and Regenerative Medicine, AvePark - Parque de Ciência e Tecnologia, Rua Ave 1, Edifício 1 (Sede), Barco, Guimarães, Portugal

ICVS/3B's–PT Government Associate Laboratory, Braga/Guimarães, Portugal
e-mail: sandra.pina@i3bs.uminho.pt



J. M. Oliveira et al. (eds.), *Sustainable Scaffolds-based Strategies in Tissue Engineering and Regenerative Medicine*, Biomaterials, Bioengineering and Sustainability 8, https://doi.org/10.1007/978-3-031-96274-5_12

**Keywords**

Bioceramics · Biowaste · Sustainability · Eco-friendly · Bone · Tissue engineering

## 12.1 Introduction

Innovative biomedical implants have transformed the lives of numerous patients by acting as substitutes or promoting the regeneration of damaged tissues in various tissue engineering (TE) and regenerative medicine applications. Ideal biomaterials should be capable of releasing therapeutic agents, such as pro-angiogenic, anti-resorptive, recombinant growth factors, and anabolic molecules, in a controlled manner over time and space, thereby promoting bone healing and regeneration.

Among the wide array of biomaterials employed, bioceramics stand out as the most used types for bone and dental implants. These materials find extensive use in diverse medical disciplines, including but not limited to artificial total hip, knee, and shoulder replacements, as well as in plates, screws, vertebrae spacers, alveolar bone replacements, mandibular reconstruction, spinal fusion, end osseous tooth replacement implants, and orthodontic anchors.

Bioceramics encompass a variety of inorganic non-metallic biomaterials, such as calcium phosphates (CaPs), calcium carbonates, calcium silicates, and bioglass. Renowned for their exceptional biocompatibility, osteoconductivity, hardness, brittleness, and resistance to corrosion, these materials offer significant advantages (Pina et al. 2015). However, limitations include low fracture toughness, brittleness, elasticity, and high stiffness, which can be overcome by combining them with different polymers to create biocomposite materials. These biomaterials can substitute the commonly employed metallic implants, eliminating the requirement for a subsequent surgical procedure for implant removal, avoiding corrosion or metal buildup in tissues, and minimizing stress-shielding (Montazerian et al. 2022).

The increasing focus on eco-friendly materials is driving the adoption of sustainable biomaterial precursors, which are often disposed of in the environment without proper scientific oversight, posing socio-economic and environmental challenges. Consequently, the implementation of effective waste management practices is imperative to harness the potential of these waste resources responsibly. Biowaste materials obtained from marine sources (e.g., mollusk shells, shellfish carapaces, cuttlefish bone, fishbone, and sponges) and mammalian sources (e.g., poultry, bovine, camel, horse, and pig bones) has emerged as extensively researched candidates for bioceramic-based materials towards bone TE, while minimizing their environmental footprint (Tarafdar et al. 2021). Additionally, agricultural waste products (e.g., rice straw, rice husk, rice ash, sugarcane leave ash, and corn cob) represent valuable sources of silica, that can be converted into biogenic silica nanoparticles and bioglass, used in biomedical implants applications (Ismail and Mohamad 2021; Azeena et al. 2017). They can also act as carriers for gene and medication delivery, with the nanoparticle pores serving as storage spaces for medicinal compounds. It has been demonstrated that each component can function as a standalone biomaterial; however, these components can be modified and combined with other biomaterials to enhance bioactivity, improve mechanical strength, and/or enable controlled release capabilities.

In parallel, utilizing eco-friendly methods such as sol-gel processing, chemical or hydrothermal pretreatments, and pyrolysis is crucial to enhance the sustainability, biocompatibility, mechanical properties, conversion rate, biodegradability, and digestibility of the bone substitutes. These methods also reduce process energy consumption and increase production volume. Lately, additive manufacturing (AM), exemplified by 3D printing, has been used as a novel frontline technology for the versatile multimaterial fabrication of scaffolds using sustainable resources (Decante et al. 2024). It allows the control of fine structural features such as porosity, pore size, and interconnectivity, which conventional techniques cannot achieve, as well as, the incorporation of cells/bioactive agents into the scaffolds is possible during printing (Costa et al. 2019; Decante et al. 2021). A suitable porous structure, featuring high porosity and an open structure, can warrant optimal osteogenic effects by facilitating the ingrowth of bone and blood vessels (Zhou et al. 2021).

The present chapter presents a comprehensive overview of sustainable bioceramics derived from waste sources, covering their features, processing, mechanical and biological properties for applications in bone tissue repair and regeneration.

## 12.2 Sustainable Natural Resources

### 12.2.1 Natural-Derived Calcium Phosphates

#### 12.2.1.1 Animal Residues

CaPs derived from biowaste have garnered significant attention over the synthetic ones, primarily because of their economic and environmental benefits aligning with the global trend of the circular economy (Carella et al. 2021). Animal bone biowaste, which composition is rich in CaPs, namely hydroxyapatite (HAp) and/or β-tricalcium phosphate (β-TCP), offers a sustainable and cost-effective source of materials, with promising results in terms of biocompatibility, bioactivity, and osteoconductivity, for applications in bone TE (Table 12.1) (Shen et al. 2019). Besides, the presence of a variety of minerals (e.g., Zn, Mg, Na, K, Fe, etc.) in the biowaste improves the biological performance of the biogenic HAp. The significant benefit of animal bone-derived biomaterials is related with the preservation of the natural structure and inherent 3D structure of bone, inorganic components and after the elimination of immunogenic substances, like collagen and protein, through processes such as calcination or other techniques (Zhu et al. 2022). A myriad of studies has been exploring the potential of different types of natural-derived biomaterials. For instance, chicken bone-derived HAp demonstrated considerable feasibility as an osteogenic filler in chitosan membrane for guided bone regeneration (Bee and Hamid 2019; Luna-Domínguez et al. 2018; Shah et al. 2022). The chicken-bone derived biomaterials demonstrated a greater ability to form apatite layer when immersed in simulated body fluid (SBF) compared to pure HAp (Yadav et al. 2022).

**Table 12.1** Summary of the most recent studies on bioceramics-based biomaterials obtained from diverse biowaste sources for TE applications

| Biowaste | Derived biomaterials | Processing technology | Main properties | Application | Reference |
|---|---|---|---|---|---|
| Chicken bone | HAp/chitosan membrane | Solvent casting and evaporation method | Improved bioactivity and decrease of degradation rate with HAp loading increase | Guided bone regeneration | Bee and Hamid (2019) |
| | HAp/whitlockite biphasic powder combined with chitosan membrane | Biphasic powder was obtained from chicken bones heat-treated at 1000 °C | *In vivo* studies showed the acceleration of new bone formation in rabbit calvarial defects, by an enhancement of osteogenic activity by increasing the expression of the biomarker osteonectin, the early formation of mineralized ECM, and by fully developed osteons | Guided bone regeneration | Luna-Domínguez et al. (2018) |
| Chicken bone and eggshell | HAp | 3D printing | Eggshell was used as a filler aiming to increase the mechanical properties (fracture strain (~62%) and toughness (~0.52 $MJ/m^3$)) | Orthopedic and bone TE | Das et al. (2023) |
| Fish bone | Gelatin methacrylate/nano fish bone hybrid hydrogel | Photo-cross-linking | Immune microenvironment modulation by inducing the secretion of cytokines and differentiation of hDPSCs for osteogenesis | Bone regeneration | Huang et al. (2020) |
| Fish scale | Biogenic HAp combined with yttria-stabilized zirconia (YSZ) powder and coating on a Ti–6Al–4 V alloy | Powders prepared by spray-drying; plasma coating | Well-grown apatite spherulite layers on the coatings surface; good L929 cell viability (95%) at the highest concentration on the coated specimens | n.d. | Anene et al. (2023) |
| | Decellularized and decollagenized fish scales | Freeze-drying | Decellularized fish scales showed more prominent in anti-inflammatory osteoimmune microenvironment regulation and enhanced bone regeneration *in vivo* and *in vitro* | Bone defect regeneration | Qin et al. (2023) |

| | | | | | |
|---|---|---|---|---|---|
| Eggshell | CaO nanoparticles incorporated with PLA | Electrospinning | 20% of CaO resulted in increased young modulus (138%), 30% reduction in bacterial viability against *S. aureus* and 11% against *E. coli* after 6 h of bacterial exposure, and high capacity of osteogenic differentiation | Bone TE | Canales et al. (2023) |
| | Mesoporous carbonated HAp | Synthesis in a custom-made pilot scale microwave reactor | Presence of Na, Mg and Si trace elements was confirmed; excellent biocompatibility and good cell attachment | Bone fillers, drug/protein delivery carriers, and TE scaffolds | Muthu et al. (2022) |
| | $CaCO_3$ to create biogenic bioglass | Sol-gel method | PCL/Zein scaffold containing biogenic bioglass powder improved the thermal stability of the scaffolds | Bone TE in non-load bearing | El-Korashy et al. (2024) |
| | Eggshell-derived CaP and chitosan composite | Coagulation-assisted extrusion printing | Average fiber diameter of ~500 μm; pore sizes of 3–20 μm; ectopic bonelike tissue formation | Bone-graft application | Dadhich et al. (2016) |

(continued)

**Table 12.1** (continued)

| Biowaste | Derived biomaterials | Processing technology | Main properties | Application | Reference |
|---|---|---|---|---|---|
| Seashell | Fe-doped HAp nanoparticles | Microwave-meadiated method | High biocompatibility and excellent antibacterial performance toward *E. coli* and *S. pneumoniae* | Drug targeting, hyperthermia cancer therapy, and magnetic resonance imaging | Karunakaran et al. (2019) |
| | HAp nanoparticles-based scaffolds | Direct foaming method | Open interconnected porosity (87–89%), pore size 50–300 μm, and good mechanical properties | Bone replacement | Scialla et al. (2020) |
| | HAp synthesized from mussel shells | Dissolution-precipitation synthesis | Hap/chitosan composites presented flexural strength and hardness of the composite material in the range of the compact bone | Bone TE | Galotta et al. (2023) |
| | HAp synthesized from sand lobster shells | Precipitation method | HAp/15 wt.% polyethylene oxide/chitosan scaffolds demonstrated good porosity, micropores and macropores to supply nutrients to cells for bone tissue growth | Bone TE | Kadek Hariscandra Dinatha et al. (2024) |
| | Carbonate-substituted Hap nanorods | Dissolution-precipitation synthesis | High antibacterial activity against *P. aeruginosa* and *S. aureus* | Alveolar bone TE | Jamilludin et al. (2023) |
| | Multiphasic CaP/HAp | Coagulation-assisted extrusion 3D printing | Significant *in vitro* cytocompatibility and *in vivo* biocompatibility | Bone TE | Dadhich et al. (2021) |
| Eggshell and seashell | 70% HAp/30% PCL | Extrusion 3D printing | Human osteoblast cells attachment and maintenance | TE | Hembrick-Holloman et al. (2020) |

| | | | | | |
|---|---|---|---|---|---|
| Marine organisms | Biosilica extracted from *Dragmacidon reticulatum* | 3D printing | Increased MC3T3-E1 and L929 cell viability; promote bone healing in the cranial critical defects after 45 days of implantation | Bone TE | dos Santos Jorge Sousa et al. (2023) |
| | Biosilica extracted from *Tedania ignis* | Solid-liquid extraction using sodium hypochlorite | Biosilica presented better material degradation and granulation tissue formation, in comparison to bioglass | Bone repair | Cruz et al. (2020) |
| | Biosilica isolated from the *Axinella infundibuliformis* combined with collagen composites | Crosslinking with 1-ethyl-3-(3-dimethylaminopropyl)-carbodiimide hydrochloride followed by freeze-drying | Increased mechanical properties and high cell viability | Bone regeneration | Martins et al. (2022) |
| | Diatomite/chitosan composite scaffolds | Freeze-drying | Increased water uptake capacity with diatomite presence; 82–90% porosity; compression moduli of 67.3–90.1 kPa; enhanced ALP activity and mineral formation on scaffold surface | Bone TE | Tamburaci and Tihminlioglu (2018) |
| | Gelatin/chitosan/hyaluronic acid loaded with diatom and β-sitosterol | Freeze-drying | Porosity of 53–90% with interconnected pore sizes of 240 μm; hemolytic index 0.11–1.02%; mechanical strength of 19.3 MPa | Bone regeneration | Mohammadi et al. (2024) |
| | Diatoms/silk fibroin | Solvent casting-particulate leaching method | Porosity of 85–87%; diatoms improved cell activity and proliferation; boosted the osteogenic potential of MG63 osteoblast-like cells | Bone regeneration | Le et al. (2018) |

(continued)

**Table 12.1** (continued)

| Biowaste | Derived biomaterials | Processing technology | Main properties | Application | Reference |
|---|---|---|---|---|---|
| Agricultural waste | Recycled rice husk and eggshell | Precipitation method | Compressive strength of 23 MPa; blood vessels formation after 24, 48, and 96 h; antibacterial activity against *S. aureus* and *E. coli* | Bone healing | Shendage et al. (2024) |
| | Silica extracted from rice husk- based composites | Chemical synthesis without any harsh chemical reagents | Silica average particle size of 200–230 nm; pore size of 95–350 nm; efficacy against *S. aureus* and *E. coli* growth inhibition | Dental and drug release | Varshney et al. (n.d.) |
| | Silica nanoparticles extracted from rice and wheat husk | Synthesis by sol-gel process | High-efficiency anticancer effect on the MCF-7 cell line | Drug release and cancer treatment | Porrang et al. (2021) |
| | Silica nanoparticles extracted from rice husk systems | The ashes were calcined at 700 °C | Significant tumor inhibition in nude mice after treatment; the system loaded with doxorubicin showed a higher toxicity than without in cancerous cells | Cancer treatment | Espinoza et al. (2023) |
| | Rice husk ash reinforced HAp | Sintering | Porosities of 39 ± 0.1% to 56.53 ± 0.14% range; wear rate of the composites significantly reduced with the addition of the rice husk | Bone repair | Adamu et al. (2024) |
| | Silica foam monolith using waste rice husk ash | 3D printing and burning the binder and template. | Porosity of ~89% with different shapes of pores of 2 nm–450 μm | Drug release | Hossain et al. (2023) |

*CaP* calcium phosphates, *ECM* extracellular matrix, *HAp* hydroxyapatite, *n.d.* not defined, *PCL* polycaprolactone, *TE* tissue engineering

Bioceramics obtained from bovine bone have also proven their competence for xenograft development for applications such as orthopedics, dentistry, wound dressing, and aesthetic surgery (Ratnayake et al. 2017; Han et al. 2022).

Indeed, some natural-derived CaP-based biomaterials are already available in the market, namely Endobon® Xenograft and Straumann® Xenograft derived from bovine bone, and Zcore™ Porcine Xenograft derived from porcine cancellous bone. They are characterized by their intricate interconnecting macroscopic and microscopic porous structures, both biomaterials serve as robust scaffolds, fostering the formation and ingrowth of new bone tissue with remarkable efficacy.

CaPs extracted from fish residues (bones and scales) have been used as an implantable material for tooth restoration, beauty, bone tissue regeneration, pharmaceutical, and chemical industries owing to its large resources, good biocompatibility, and superior biological activity and safety (Table 12.1) (Terzioğlu et al. 2018; Duta and Dorcioman 2021). The micro-patterned and hierarchical structures of fish scales serve as an external mechanical stimulus that effectively promotes a range of cell activities, such as migration, proliferation, and differentiation, with a particular emphasis on fostering osteogenesis (Bas et al. 2020). Fish bones consist predominantly of ~60% calcium orthophosphate, whereas fish scales contain a mixture of calcium-deficient HAp, CaPs, and various trace elements, comprising about 38–46% of their structure. The development of fish residues-derived HAp and/or β-TCP powders or scaffolds involves tailored processing conditions, such as calcination temperatures and times, to optimize their physicochemical and biological performance (Fig. 12.1a–c) (Qin et al. 2022; Putra et al. 2024). Fish bone-based hydrogel showed enhanced mechanical properties and immune microenvironment modulation when combined 5 wt% fish bone with gelatin methacrylate (GelMA) (Fig. 12.1d) (Huang et al. 2020). On the other hand, fish scales exhibit a notable increase in anti-inflammatory and pro-healing cytokines expressions, within 6 weeks of

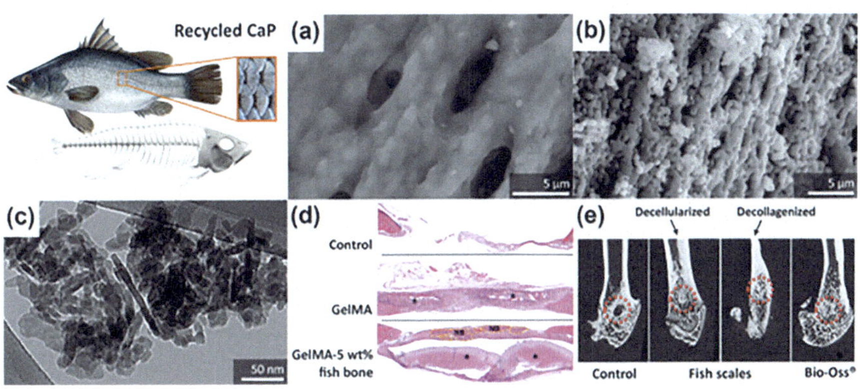


**Fig. 12.1** Fish residues for CaPs-based biomaterials production: (**a**) pristine fish bone, (**b**) HAp obtained from fish bone, (**c**) nano-HAp obtained from fish scale, (**d**) H&E staining of fish bone/Gelatin methacryloyl (GelMA) composites after 4 weeks implanted in rat cranial bone defect, and (**e**) micro-computed tomography (mCT) images of decellularized and decollagenized fish scales after 6 weeks of implantation in a rat femoral defect model. (Reprinted with permission from Putra et al. 2024)

implantation in a rat femoral defect model, consequently fostering bone healing as the commercial Bio-Oss® (Fig. 12.1e) (Qin et al. 2023).

#### 12.2.1.2 Eggshells and Seashells Sources

Eggshells and seashells are other valuable sources that offer significant benefits as bone substitute materials (Table 12.1) (Waheed et al. 2020; Azarian and Sutapun 2022). These sources are rich in calcium carbonate ($CaCO_3$), with eggshells typically containing around 96% $CaCO_3$, 1% magnesium carbonate, and 1% CaP, whereas seashells are composed of 95–97% $CaCO_3$ (Oliveira et al. 2013; Summa et al. 2022). The remaining components consist of small amounts of organic matter. Eggshells or seashells typically go through an initial thermal treatment to convert $CaCO_3$ into CaO, followed by reaction with a phosphorus source, using chemical precipitation, solid-state, or hydrothermal treatment, to obtain CaPs powders/particles (Figs. 12.2a–d) (Carella et al. 2021; Borciani et al. 2023). Recycled Ca-rich biomaterials have been explored to combine with hydrogel composites showing improved osteogenic expression (Fig. 12.2e) and to favor de novo bone generation *in vivo* (Fig. 12.2f). Additionally, seashells biowaste of black *Chlamys varia* were used to synthesize mesoporous superparamagnetic HAp nanoparticles showing high biocompatibility and excellent antibacterial performance (Karunakaran et al.

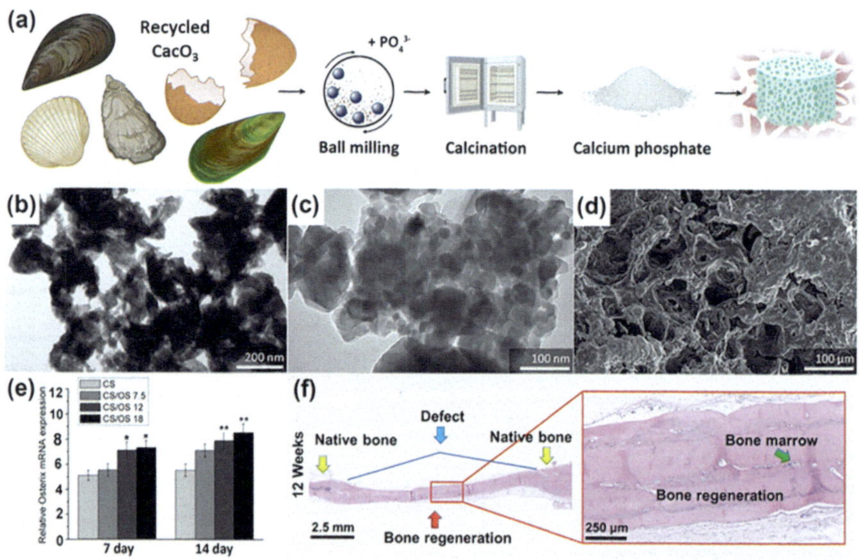


**Fig. 12.2** Eggshell and seashells sources: (**a**) Illustration of the processing steps to obtain CaP powders and 3D scaffolds; (**b**) HAp particles obtained by solid-state sintering; (**c**) Aragonite nanoparticles obtained from bivalve mollusk shells; (**d**) Aragonite nanoparticles composites containing gelatin, dextran, and dextrin; (**e**) Relative mRNA expression of MG-63 cells exposed to calcium sulfate/oyster shells composites for 7 and 14 days; and (**f**) Hematoxylin and eosin (H&E) staining on eggshell particles/GelMA scaffolds showing bone regeneration on critical-sized cranial rat bone defects after 12 weeks implantation. (Reprinted with permission from Putra et al. 2024)

2019). Hembrick-Holloman et al. (Hembrick-Holloman et al. 2020) have used nano-HAp derived from eggshell and clam seashell, in combination with polycaprolactone (PCL), to print complex scaffolds with a HAp weight percentage of 70%. The scaffolds demonstrated high bioactivity, enabling osteoblast cells to adhere and proliferate effectively.

Eggshells biowaste have also been used as a filler aiming to enhance the properties of the final product (Kumar et al. 2022; Das et al. 2023; Ganesan et al. 2020). For example, Manojit Das et al. (2023) used eggshell as a reinforcement in a chicken bone extract producing printed structures with enhanced mechanical and biological properties holding potential for applications in orthopedic and bone TE. Elsewhere, Canales et al. (2023) developed nanocomposite electrospun fibers of polylactic acid incorporating CaO nanoparticles derived from eggshells for bone regeneration. A substantial increase in Young's modulus was observed with the addition of 20 wt.% CaO, along with a significant reduction in the viability of *S. aureus* and *E. coli* after 6 h of exposure.

## 12.2.2 Natural-Derived Silica

### 12.2.2.1 Marine Organism Sources

Marine organisms, such as sponges, are among the most fascinating organisms that have garnered significant interest for their production of biosilica, which is a key component of their skeletal structures. Biosilica is obtained from calcination of the sponges usually at 800 °C.

Silica ranks as the third most common biogenic mineral, following $CaCO_3$ and CaP (Ravera et al. 2016). This natural form of silica is highly valued in biomedical science due to its biocompatibility and ability to promote bone growth. Silicon-containing materials, including bio-glass and amorphous silica particles, have indeed been proposed for bone TE (Table 12.1) (Gaharwar et al. 2013). The beneficial effect of silicon on bone formation is well-established, particularly demonstrated by the enhanced differentiation of mesenchymal stem cells induced by silicon-doped β-tricalcium phosphate (Camiré et al. 2006).

Different species of marine sponges, including *Geodia barretti*, *Geodia macandrewii*, *Geodia atlantica*, *Stelletta normani*, *Axinella infundibuliformis*, and *Phakellia ventilabrum*, have demonstrated bioactivity that is dependent on their biosilica content (Kaya et al. 2021; Dudik et al. 2021; Ehrlich et al. 2010). For instance, a study by Dudik et al. (2021) demonstrated that, after 28 days of immersion in SBF, the biosilica obtained from *Geodia atlantica* exhibited the highest bioactivity, and capabilities for bone TE applications. Martins et al. (2022) developed different biocomposites based on silica (bioglass (BG), Diatomaceous Earth (DE), and biosilica (BS)) derived from the marine sponge *Axinella infundibuliformis* and marine collagen, with enhanced mechanical properties and ability to support adhesion and proliferation of L929 fibroblast-like cells, more evidenced in the BS composition, for non-load-bearing bone tissue regeneration. The biomaterials were characterized by porosity higher than 85% and 70% interconnectivity, and particular morphologies. BG showed uniform particles, while DE exhibited hierarchical structures

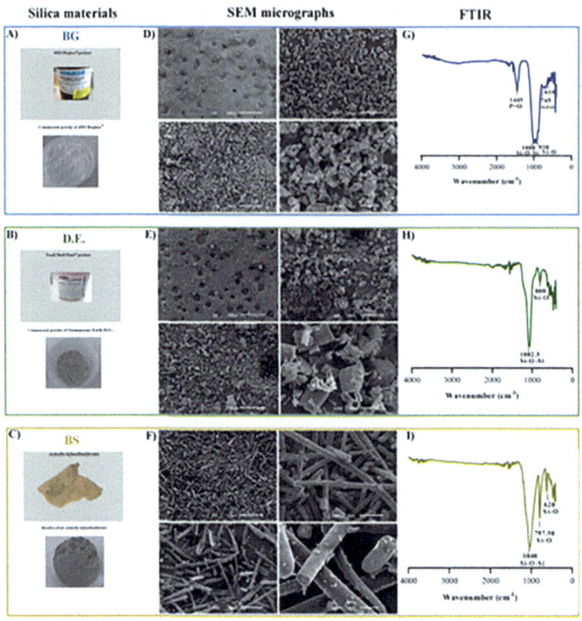


J) EDS

| Composition (Wt) | Silica-derived materials (%) | | |
|---|---|---|---|
| | BG | D.E | BS |
| O | 40.0 | 44.9 | 44.7 |
| C | 30.5 | 34.2 | 24.6 |
| Si | 10.8 | 19.3 | 27.0 |
| Ca | 10.1 | 0.1 | 0.9 |
| Na | 7.3 | - | 0.8 |
| P | 1.3 | - | 0.5 |
| Al | - | 1 | 0.2 |
| Fe | - | 0.4 | 0.3 |
| K | - | - | 0.5 |
| Mg | - | - | 0.4 |
| S | - | - | 0.2 |

**Fig. 12.3** Silica-based materials derived from the marine sponge *Axinella infundibuliformis*, (**a**) Bioglass® 45S5 (BG); (**b**) Diatomaceous Earth (DE); and (**c**) Biosilica (BS). (**d–f**) SEM images of the silica-based structures. (**g–i**) FTIR spectra of the silica-based materials. (**j**) Chemical composition of the materials obtained by EDS analysis. (Reprinted with permission from Martins et al. 2022)

characteristic of the diatom species, featuring tube-like porous formations (Fig. 12.3). In contrast, BS was composed of needle-like particles, typical of a specific type of sponge spicules.

Another significant natural source of porous silica is diatom biosilica, which has been extensively studied for the development of advanced biomaterials, including those used in bone regeneration and implantable medical devices (Şan et al. 2009). Diatom biosilica is distinguished by its highly porous morphology and hierarchical micro- to nanoscale structures known as frustules. Diatom frustules are mainly composed of amorphous clusters of very small silica particles, typically measuring a few tens of nanometers. This form of $SiO_2$ can react with $Ca(OH)_2$ to produce calcium silicate hydrates. Diatomite structures offer a large surface area, easily modifiable surface chemistry, good permeability (0.1–10 mD), porosity (35–65%), non-toxicity, and high biocompatibility, making them ideal for medical applications (Kastis et al. 2006). Its structure features well-organized pores on both outer and inner surfaces, with numerous sieve pores approximately 40 nm in size that facilitate the exchange of nutrients and gases.

A few studies have proposed the use of diatoms combined with different organic/inorganic biomaterials. Among them, Hahn et al. (2018) have demonstrated that incorporating diatom particles into silk fibroin sponges enhances cell activity and proliferation, as well as the osteogenic potential of MG63 osteoblast-like cells. The

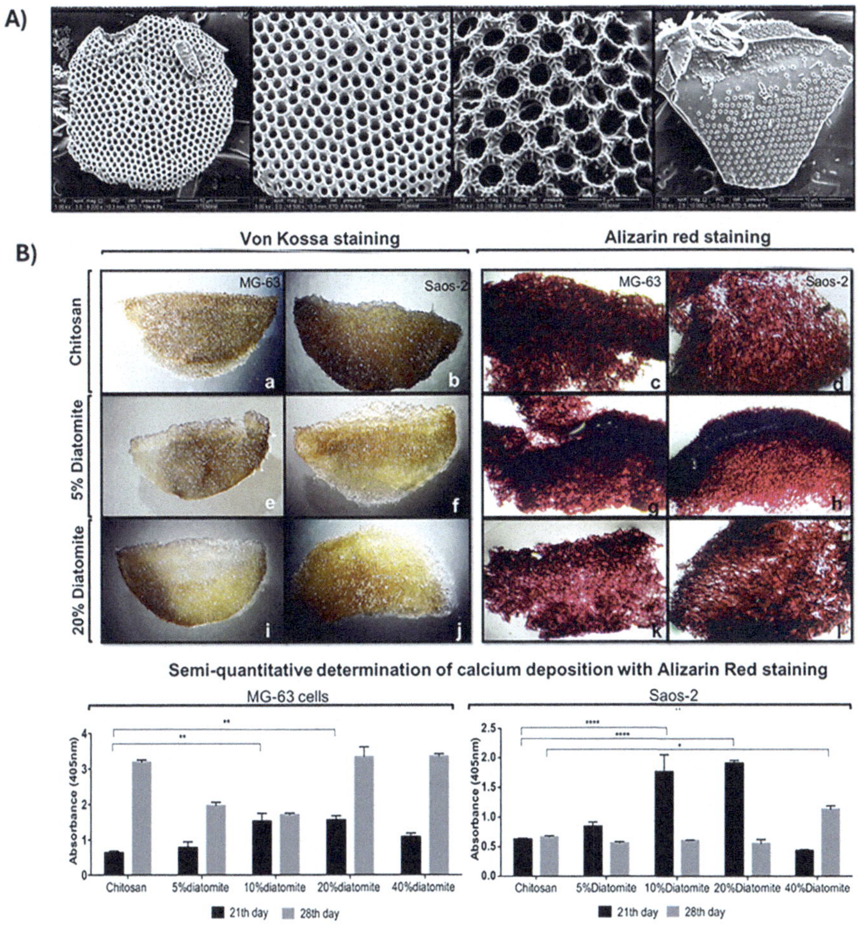


**Fig. 12.4** Chitosan and chitosan/diatomite composite scaffolds: Microstructure images of diatom frustules presenting morphologies with 5, 10 and 20 μm scales (**a**); and von Kossa and alizarin red staining (2× magnification) and semi-quantitative determination of calcium deposition with alizarin red extraction (**b**). (Reprinted with permission from Tamburaci and Tihminlioglu 2018. © 2018 Elsevier B.V.)

silicon released from the diatom particles significantly promoted cell proliferation and increased the expression of early bone markers, including alkaline phosphatase, osterix, and collagen type I. Tamburaci and Tihminlioglu (2018) developed composites using natural silica source diatomite combined with chitosan for bone TE scaffolding applications. Chitosan/diatomite scaffolds exhibited an average pore size of 218–319 μm range and porosity of 82–90%, and diatom frustules presented a pore size range of 0.3–5 μm (Fig. 12.4A). *In vitro* studies using MG-63 and Saos-2 cells demonstrated mineralization on chitosan and chitosan/diatomite scaffolds (5–20%) over 28 days incubation period, which is favorable for osteoblast proliferation (Fig. 12.4B). The presence of diatom frustules positively influenced cell attachment by modifying the surface topography. Incorporating diatomite resulted in high ALP activity and enhanced biomineralization on the scaffold surface.

Recently, incorporating diatom and β-sitosterol into composites made of gelatin, chitosan, and hyaluronic acid has also demonstrated significant improvements in osteoblast proliferation, and promoting angiogenesis and leverages the anti-inflammatory properties of β-sitosterol (Mohammadi et al. 2024). The scaffolds exhibited an open porosity ranging from 53% to 90% (depending on the percentage of the diatom in the structure) with interconnected pore sizes of 240 μm (Fig. 12.5I). An increase in compressive strength from 1.9 MPa to 11.2 MPa and Young's modulus from 5.6 MPa to 19.3 MPa was observed with the incorporation of diatom. Pure diatom and scaffolds demonstrated excellent hemocompatibility, with a hemolytic index of 0.11–1.02% range, well below the allowed level of 2% for biomaterials (Fig. 12.5II). Additionally, the scaffolds showed promise as a carrier for therapeutic agents, demonstrating the ability to locally release β-sitosterol to utilize its anti-inflammatory and angiogenic properties.

#### 12.2.2.2 Agricultural Wastes

Reutilization of agricultural waste (e.g., rice husk, bagasse, and wheat husk) is an important, environmentally friendly, and cost-effective strategy for the extraction of silica, over traditional silica compounds based on chemical precursors (e.g., sodium silicate and tetraethyl orthosilicate) (Karande et al. 2021). Typically, the structure, composition, and morphology of silica derived from agricultural waste are greatly influenced by the extraction method used, which involves green synthesis methods that minimize environmental impact (Karumanchi et al. 2022). High-purity silica particles are produced through a pyrolysis process, followed by acid leaching to remove unwanted metallic elements, and then calcined at temperatures ranging from 500 °C to 700 °C.

In the past decade, substantial biomedical advancements have been achieved in the preparation of silica nanoparticles (SNPs), which can be extracted from rice husk and rice husk ash, the most advantageous waste materials (Athinarayanan et al. 2015). Thanks to their beneficial structural and chemical properties, SNP have been employed as drug delivery systems, in TE, and in cancer treatment (Table 12.1) (Porrang et al. 2021; Battegazzore et al. 2014). As an example, Espinoza et al. (2023) developed a multifunctional system using SNP derived from rice husk, coated with chitosan (CH) and loaded with doxorubicin (DOX), followed by functionalization with cell membrane (CM), aiming to improve the side effects of the cancer drug and promote good recovery for the patient post-treatment. The cytotoxicity results demonstrated that SNP–CH–DOX–CM exhibited significantly higher efficacy in treating the HepG2 cancer cell line compared to free DOX, indicating superior cell inhibition capability. The anticancer effect of the nanoparticles on cancer cells showed that the tumor volume and weight decreased in nude mice treated with SNP–CH–DOX–CM, with slow tumor progression (Fig. 12.6).

Various types of bioceramics, with rice straw ash as silica precursor, have been also processed for bone TE, including bioglass, wollastonite ($CaSiO_3$), forsterite ($Mg_2SiO_4$), and diopside (Choudhary et al. 2021). For instance, wollastonite can be

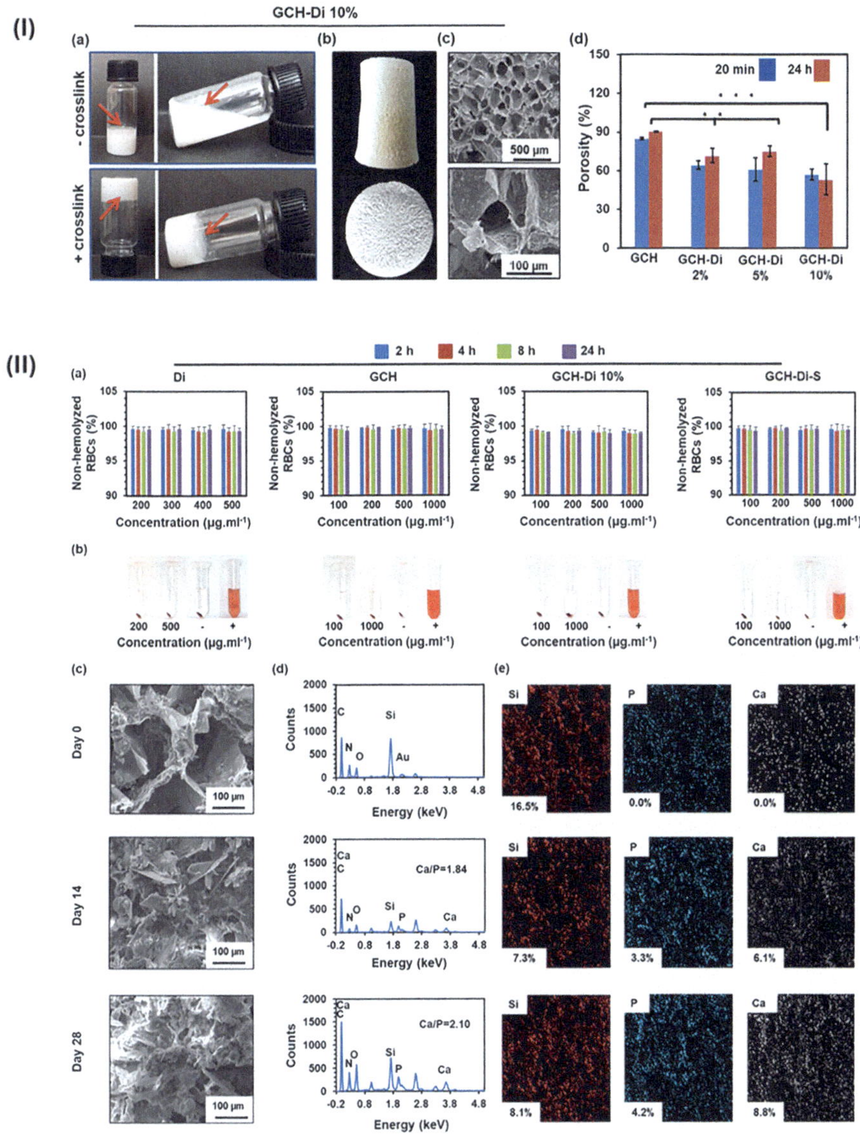


**Fig. 12.5** (**I**) Scaffolds preparation (**a**), macroscopic images (**b** and **c**), and porosity analysis with different diatom concentrations (**d**); $^{*}p < 0.05$, $^{**}p < 0.01$, $^{**}p < 0.001$. GCH (−): scaffold non-crosslinked; GCH (+): final scaffold with crosslinked; GCH-Di 2%: scaffold with incorporate 2% diatom (Di); GCH-Di 5%: scaffold with incorporate 5% Di; GCH-Di 10%: scaffold with incorporated 10% Di; GCH-Di-S: GCH scaffold with incorporated 10% Di, BS. (**II**) Non-hemolyzed red blood cells (RBCs) percentage of diatom and the scaffolds (**a** and **b**), $^{*}$ $p < 0.05$, $^{**}p < 0.01$, $^{***}p < 0.001$; SEM images, energy-dispersive X-ray spectroscopy analysis, and elemental mapping analysis for biomineralization of the GCH-Di-S scaffolds after 0, 14, and 28 days immersed in simulated body fluid (**c**–**e**). Data are represented as the mean ± SD (N = 3). (Reprinted with permission from Mohammadi et al. 2024)

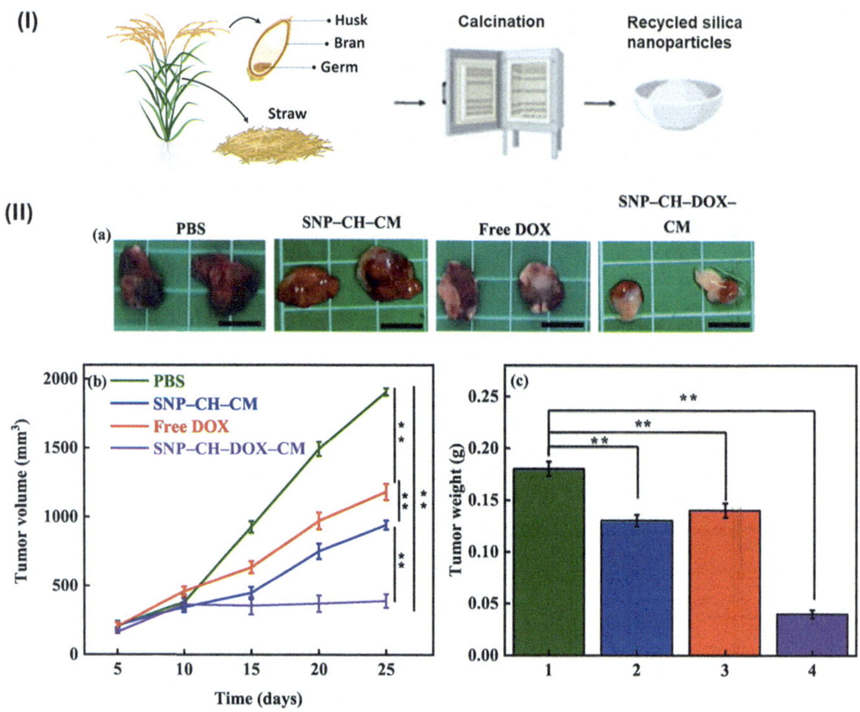


**Fig. 12.6** (**I**) Silica nanoparticles extracted from rice husk. (**II**) Representative images of tumors formed in nude mouse models (scale bar: 1 cm) (**a**), tumor volumes (**b**), and average tumor weights after treatment with PBS, SNP–CH–CM, free DOX, and SNP–CH–DOX–CM nanocarriers (**c**). $_{**}$p < 0.01. (Reprinted with permission from Espinoza et al. 2023)

synthesized by mixing silica derived from rice husk with CaO in an equimolar ratio, while forsterite can be obtained by combining MgO with rice husk-derived silica (Palakurthy et al. 2019). On the other hand, a mixture of eggshell-derived CaO, MgO, and rice husk-derived silica results in diopside (Choudhary et al. 2021). Those bioceramics have demonstrated in vitro biodegradability and the capability to induce apatite formation. Additionally, synthesized rare-earth ion-integrated SNPs have demonstrated potential for developing cost-effective bioimaging contrast agents (Araichimani et al. 2020).

## 12.3 Conclusions and Future Perspectives

The development of sustainable ceramic-based biomaterials from biowaste and renewable sources represents a significant leap forward in tissue engineering. This progress is driven by the urgent need for environmentally friendly, cost-effective solutions for tissue repair and regeneration, aligning seamlessly with the global shift towards a circular economy. Bioceramics stand out for their unique combination of high intrinsic strength, exceptional biocompatibility, and remarkable versatility, making them invaluable across a wide spectrum of applications including orthopedics, dentistry, bone tissue engineering, and drug delivery systems. These innovative biomaterials can be precisely engineered and customized to enhance specific properties, offering unprecedented control over their performance. Furthermore, they can incorporate dopant ions, growth factors, and/or therapeutic drugs to significantly improve their biological efficacy, opening new avenues for personalized medicine and targeted treatments. Despite the promising progress in utilizing biowaste for sustainable bioceramics and medical devices, several key issues must be addressed. While various biowaste materials (e.g., eggshells, seashells, and fish residues) have demonstrated biocompatibility, long-term *in vivo* studies are necessary to evaluate the performance and safety of waste-derived bioceramics for bone implant applications. Scaling up production from laboratory to industrial levels while navigating complex regulatory landscapes is crucial for the successful clinical translation of these sustainable biomaterials. Additionally, further research is needed to fully exploit additive manufacturing technologies, particularly in maximizing the potential of waste-derived bone substitutes. Such integration strategy could lead to more precise and complex geometries, patient-specific implants and scaffolds. Likewise, developing hybrid materials that combine biogenic silica with other sustainable biomaterials could lead to improved properties, potentially addressing current limitations in mechanical strength or biodegradation rates. Lastly, exploring new agricultural waste sources and optimizing extraction and purification processes could significantly broaden the range of available sustainable bioceramics, potentially uncovering materials with unique and beneficial properties.

**Acknowledgments** This study was funded by the Portuguese Foundation for Science and Technology (FCT) for the distinctions attributed to SP (CEECIND/03673/2017 and 2023.07400.CEECIND).

## References

Adamu MA et al (2024) Rice husk ash filled with hydroxyapatite based composite for biomedical application. In: School of engineering and engineering technology annual conference (SEET conference)

Anene FA et al (2023) Biomechanical properties and corrosion resistance of plasma-sprayed fish scale hydroxyapatite (FsHA) and FsHA-doped Yttria-stabilized zirconia coatings on Ti–6Al–4V alloy for biomedical applications. Coatings 13(1):199

Araichimani P et al (2020) Rare-earth ions integrated silica nanoparticles derived from rice husk via microwave-assisted combustion method for bioimaging applications. Ceram Int 46(11, Part A):18366–18372

Athinarayanan J et al (2015) Synthesis of biogenic silica nanoparticles from rice husks for biomedical applications. Ceram Int 41(1, Part A):275–281

Azarian MH, Sutapun W (2022) Biogenic calcium carbonate derived from waste shells for advanced material applications: a review. Front Mater:9

Azeena S et al (2017) Antibacterial activity of agricultural waste derived wollastonite doped with copper for bone tissue engineering. Mater Sci Eng C 71:1156–1165

Bas M et al (2020) Mechanical and biocompatibility properties of calcium phosphate bioceramics derived from salmon fish bone wastes. 21(21)

Battegazzore D et al (2014) Rice husk as bio-source of silica: preparation and characterization of PLA-silica bio-composites. RSC Adv 4(97):54703–54712

Bee S-L, Hamid ZAA (2019) Characterization of chicken bone waste-derived hydroxyapatite and its functionality on chitosan membrane for guided bone regeneration. Compos Part B 163:562–573

Borciani G et al (2023) Marine biological waste as a source of hydroxyapatite for bone tissue engineering applications. Ceram Int 49(2):1572–1584

Camiré CL et al (2006) Material characterization and *in vivo* behavior of silicon substituted α-tricalcium phosphate cement. J Biomed Mater Res Part B Appl Biomater 76B(2):424–431

Canales D et al (2023) Preparation and characterization of novel poly (lactic acid)/calcium oxide nanocomposites by electrospinning as a potential bone tissue scaffold. Biomater Adv 153:213578

Carella F et al (2021) The use of calcium phosphates in cosmetics, state of the art and future perspectives. Materials 14(21):6398

Choudhary R et al (2021) Biomineralization, dissolution and cellular studies of silicate bioceramics prepared from eggshell and rice husk. Mater Sci Eng C Mater Biol Appl 118:111456

Costa JB et al (2019) Indirect printing of hierarchical patient-specific scaffolds for meniscus tissue engineering. Bio-Des Manuf 2(4):225–241

Cruz MDA et al (2020) In vivo biological effects of marine biosilica on a tibial bone defect in rats. Braz Arch Biol Technol 63:63

Dadhich P et al (2016) A simple approach for an eggshell-based 3D-printed osteoinductive multiphasic calcium phosphate scaffold. ACS Appl Mater Interfaces 8(19):11910–11924

Dadhich P et al (2021) Direct 3D printing of seashell precursor toward engineering a multiphasic calcium phosphate bone graft. ACS Biomater Sci Eng 7(8):3806–3820

Das M et al (2023) 3D printing of tough nature inspired hierarchical architecture using chicken bone and eggshell biowaste for biomedical applications. Ceram Int 49(17, Part B):29274–29287

Decante G et al (2021) Engineering bioinks for 3D bioprinting. Biofabrication 13(3):032001

Decante G et al (2024) Sustainable highly stretchable and tough gelatin-alkali lignin hydrogels for scaffolding and 3D printing applications. Mater Today Commun 39:108875

dos Santos Jorge Sousa K et al (2023) 3D printed scaffolds manufactured with biosilica from marine sponges for bone healing in a cranial defect in rats. Mar Biotechnol 25(2):259–271

Dudik O et al (2021) Bioactivity of biosilica obtained from North Atlantic deep-sea sponges. Front Mar Sci:8

Duta L, Dorcioman G (2021) A review on biphasic calcium phosphate materials derived from fish discards. 11(11)

Ehrlich H et al (2010) Modern views on desilicification: biosilica and abiotic silica dissolution in natural and artificial environments. Chem Rev 110(8):4656–4689

El-Korashy AS et al (2024) An in vitro evaluation of the biological activity of biogenic eggshell-derived nanosized bioglass, poly (ε-caprolactone), and zein protein 3D composite scaffolds for bone tissue engineering. Catrina Int J Environ Sci:1–19

Espinoza MJC et al (2023) Synthesis and characterization of silica nanoparticles from rice ashes coated with chitosan/cancer cell membrane for hepatocellular cancer treatment. Int J Biol Macromol 228:487–497

Gaharwar AK et al (2013) Bioactive silicate nanoplatelets for osteogenic differentiation of human mesenchymal stem cells. Adv Mater 25(24):3329–3336

Galotta A et al (2023) Dissolution-precipitation synthesis and cold sintering of mussel shells-derived hydroxyapatite and hydroxyapatite/chitosan composites for bone tissue engineering. Open Ceramics 15:100418

Ganesan K et al (2020) A new assessment on mechanical properties of jute fiber mat with egg shell powder/nanoclay-reinforced polyester matrix composites. J Nat Fibers 17(4):482–490

Han K-S et al (2022) Wound healing efficacy of biocompatible hydroxyapatite from bovine bone waste for bone tissue engineering application. J Environ Chem Eng 10(1):106888
Hembrick-Holloman V et al (2020) Ecofriendly production of bioactive tissue engineering scaffolds derived from egg- and sea-shells. J Mater Res Technol 9(6):13729–13739
Hossain SS et al (2023) Development of tri-modal-pore architecture of silica monolith via extrusion-based 3D printing and employing rice husk ash. Mater Today Commun 34:105246
Huang L et al (2020) Biomimetic gelatin methacrylate/nano fish bone hybrid hydrogel for bone regeneration via osteoimmunomodulation. ACS Biomater Sci Eng 6(6):3270–3274
Ismail H, Mohamad H (2021) Bioactivity and biocompatibility properties of sustainable wollastonite bioceramics from rice husk ash/rice straw ash: a review. Materials (Basel) 14(18)
Jamilludin MA et al (2023) Functionalized cellulose nanofibrils in carbonate-substituted hydroxyapatite nanorod-based scaffold from long-spined sea urchin (Diadema setosum) shells reinforced with polyvinyl alcohol for alveolar bone tissue engineering. RSC Adv 13(46):32444–32456
Kadek Hariscandra Dinatha I et al (2024) Porous scaffold hydroxyapatite from sand lobster shells (Panulirus homarus) using polyethylene oxide/chitosan as polymeric porogen for bone tissue engineering. J Biomed Mater Res Part B Appl Biomater 112(1):e35341
Karande SD et al (2021) Green and sustainable synthesis of silica nanoparticles. Nanotechnol Environ Eng 6(2):29
Karumanchi M, Nerella R, Mikkili I (2022) An integrated approach on extraction methodologies of nanosilica from cultivated agricultural wastes and microstructural characteristics – a review. Mater Today Proc 62:2883–2891
Karunakaran G et al (2019) Sodium dodecyl sulfate mediated microwave synthesis of biocompatible superparamagnetic mesoporous hydroxyapatite nanoparticles using black Chlamys varia seashell as a calcium source for biomedical applications. Ceram Int 45(12):15143–15155
Kastis D et al (2006) Properties and hydration of blended cements with calcareous diatomite. Cem Concr Res 36(10):1821–1826
Kaya M et al (2021) Sponge-derived natural bioactive glass microspheres with self-assembled surface channel arrays opening into a hollow core for bone tissue and controlled drug release applications. Chem Eng J 407:126667
Kumar N, Walia RS, Angra S (2022) An interactive study on wear behaviour and mechanical properties of carbonized eggshells filler loaded glass-jute reinforced polyester hybrid biocomposites. Int J Interact Des Manuf (IJIDeM) 16(3):913–925
Le TDH et al (2018) Enhancing bioactive properties of silk fibroin with diatom particles for bone tissue engineering applications. J Tissue Eng Regen Med 12(1):89–97
Luna-Domínguez JH et al (2018) Development and in vivo response of hydroxyapatite/whitlockite from chicken bones as bone substitute using a chitosan membrane for guided bone regeneration. Ceram Int 44(18):22583–22591
Martins E et al (2022) 3D biocomposites comprising marine collagen and silica-based materials inspired on the composition of marine sponge skeletons envisaging bone tissue regeneration. Mar Drugs 20(11):718
Mohammadi M et al (2024) Diatom-guided bone healing via a hybrid natural scaffold. Heliyon 10(4):e25878
Montazerian M et al (2022) Bioceramic coatings on metallic implants: an overview. Ceram Int 48(7):8987–9005
Muthu D et al (2022) Rapid synthesis of eggshell derived hydroxyapatite with nanoscale characteristics for biomedical applications. Ceram Int 48(1):1326–1339
Oliveira DA, Benelli P, Amante ER (2013) A literature review on adding value to solid residues: egg shells. J Clean Prod 46:42–47
Palakurthy S et al (2019) In vitro bioactivity and degradation behaviour of β-wollastonite derived from natural waste. Mater Sci Eng C 98:109–117
Pina S, Oliveira JM, Reis RL (2015) Natural-based nanocomposites for bone tissue engineering and regenerative medicine: a review. Adv Mater 27(7):1143–1169

Porrang S et al (2021) Preparation and in-vitro evaluation of mesoporous biogenic silica nanoparticles obtained from rice and wheat husk as a biocompatible carrier for anti-cancer drug delivery. Eur J Pharm Sci 163:105866
Putra NE, Zhou J, Zadpoor AA (2024) Sustainable sources of raw materials for additive manufacturing of bone-substituting biomaterials. Adv Healthc Mater 13(1):2301837
Qin D et al (2022) Development and application of fish scale wastes as versatile natural biomaterials. Chem Eng J 428:131102
Qin D et al (2023) Natural micropatterned fish scales combing direct osteogenesis and osteoimmunomodulatory functions for enhancing bone regeneration. Compos Part B 255:110620
Ratnayake JTB et al (2017) Development and characterization of a xenograft material from New Zealand sourced bovine cancellous bone. J Biomed Mater Res B Appl Biomater 105(5):1054–1062
Ravera E et al (2016) Biosilica and bioinspired silica studied by solid-state NMR. Coord Chem Rev 327-328:110–122
Şan O, Gören R, Özgür C (2009) Purification of diatomite powder by acid leaching for use in fabrication of porous ceramics. Int J Miner Process 93(1):6–10
Scialla S et al (2020) Mussel shell-derived macroporous 3D scaffold: characterization and optimization study of a bioceramic from the circular economy. Mar Drugs 18(6):309
Shah S et al (2022) Microstructural analysis of biowaste-derived hydroxyapatite-chitosan nanocomposites. Micro Nano Lett 17(13):369–376
Shen X et al (2019) Novel technologies in utilization of byproducts of animal food processing: a review. Crit Rev Food Sci Nutr 59(21):3420–3430
Shendage SS et al (2024) Sustainable development of a bioactive material from recycled rice husk and eggshell for bone regeneration application. ACS Sustain Chem Eng 12(7):2598–2610
Summa D et al (2022) Trends and opportunities of bivalve shells' waste valorization in a prospect of circular blue bioeconomy. Resources 11(5):48
Tamburaci S, Tihminlioglu F (2018) Biosilica incorporated 3D porous scaffolds for bone tissue engineering applications. Mater Sci Eng C 91:274–291
Tarafdar A et al (2021) Advances in biomaterial production from animal derived waste. Bioengineered 12(1):8247–8258
Terzioğlu P, Öğüt H, Kalemtaş A (2018) Natural calcium phosphates from fish bones and their potential biomedical applications. Mater Sci Eng C 91:899–911
Varshney S et al (n.d.) Silica from waste rice husk and its polymeric hydrogel-based composite filler: synthesis, characterisation, and antibacterial activity. Adv Mater Process Technol:1–16
Waheed M et al (2020) Channelling eggshell waste to valuable and utilizable products: a comprehensive review. Trends Food Sci Technol 106:78–90
Yadav MK et al (2022) A low-cost approach to develop silica doped Tricalcium Phosphate (TCP) scaffold by valorizing animal bone waste and rice husk for tissue engineering applications. Ceram Int 48(17):25335–25345
Zhou C et al (2021) Biofabrication (3D bioprinting) Laboratory at Sichuan University. Bio-Des Manuf 4(2):432–439
Zhu Q et al (2022) Preparation and characterization of two novel osteoinductive fishbone-derived biphasic calcium phosphate bone graft substitutes. J Biomater Appl 37(4):600–613

# Sustainable Scaffold Therapies for Craniofacial Tissue Engineering

13

Uyen M. N. Cao, Jan C. Kwan, Yuli Zhang, and Simon D. Tran

**Abstract**

The loss of craniofacial tissue due to congenital and developmental conditions can lead to various negative impacts such as aesthetic, functional, and physiological aspects of one's life. Conventional reconstruction methods have been characterized by limited availability, risk of resorption, and immune reaction. Advances in tissue engineering, including the fabrication of biomaterials, have been developed to overcome these challenges and allow the reconstruction and long-term rehabilitation of craniofacial defects. This task is complicated and time-consuming, owing to the intricate nature of the craniofacial region, which includes both highly vascularized and innervated soft and hard tissues with varying growth rates. Although some of these biomaterials are well-developed, their sustainability is often overlooked and is rarely the topic of discussion. Recently, greater attention has been placed on green development within tissue engineering, leading to a more prominent emphasis on the production of sustainable scaffolding materials. Here, we aim to discuss different sustainability concepts, as well as review current examples of sustainable biomaterials including synthetic materials such as PEEK, PEG, PLGA and natural-derived materials such as alginate, chitosan, and soy protein used in scaffolding therapies. The clinical applications of these materials for craniofacial tissue regeneration are also overviewed.

**Keywords**

Scaffolding therapies · Sustainable biomaterials · Craniofacial tissue engineering · Tissue engineering

U. M. N. Cao · J. C. Kwan · Y. Zhang · S. D. Tran (✉)
McGill Craniofacial Tissue Engineering and Stem Cells Laboratory, Faculty of Dental Medicine and Oral Health Sciences, McGill University, Montreal, QC, Canada
e-mail: uyen.cao@mail.mcgill.ca; jan.kwan@mail.mcgill.ca; yuli.zhang@mail.mcgill.ca; simon.tran@mcgill.ca



J. M. Oliveira et al. (eds.), *Sustainable Scaffolds-based Strategies in Tissue Engineering and Regenerative Medicine*, Biomaterials, Bioengineering and Sustainability 8, https://doi.org/10.1007/978-3-031-96274-5_13

## 13.1 Introduction

Approximately 15% of newborns have abnormalities every year, and many of these involve craniofacial problems, resulting in functional impairments and other medical complications (Forrest and Hopper 2013). Other than congenital abnormalities, various craniofacial defects could be postnatally caused by trauma, inflammation, or cancer surgeries (Miura et al. 2006; Zhang and Yelick 2018). Given the complex geometry and architecture, craniofacial bone reconstruction is considered crucial for its function of supporting adjacent soft tissues and providing anchorage for dental structures. Without timely reconstruction, craniofacial defects can lead to cerebral atrophy, neuropsychiatric disorders, malocclusions, and aesthetic problems (Brown and Shaw 2010; Zhang and Yelick 2018; Gu et al. 2023). Traditional craniofacial reconstruction methods such as autologous, allogenic, or xenogeneic grafts, however, can face limitations such as limited availability (Henkel et al. 2013), risk of site resorption (Lee et al. 2014), weakened mechanical properties (Dimitriou et al. 2011), increased risks of disease transmission or immune rejection (Aghali 2021). These concerns have driven researchers to develop alternatives that are more target specific and sustainable (Fig. 13.1) (Datta et al. 2017).

These alternative strategies include tissue engineering and the use of biomaterials, providing biocompatible scaffolds for cell proliferation, differentiation and tissue regeneration, which is considered one of the most promising strategies for craniofacial reconstruction (Datta et al. 2017; Zhang and Yelick 2018). Several biomaterials have been developed, which could be categorized into two groups according to their origins: synthetic or natural. Synthetical scaffolds include polyether ether ketone (PEEK), polyethylene glycol (PEG), poly(lactic-co-glycolic acid (PLGA), calcium phosphate, bio-glass, polymethyl methylacrylate (PMMA); and those bio-matrix derived from plants or animals, such as alginate, gelatin, soybean proteins, hyaluronic acid (HA) and basement membrane matrix. Although these

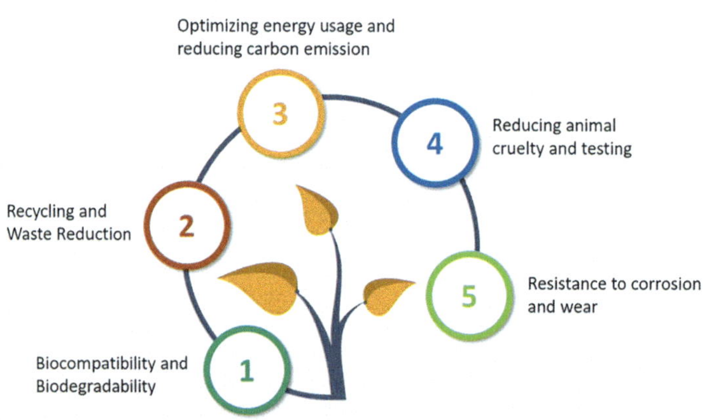


**Fig. 13.1** Schematic representation of major sustainable criteria in scaffolding therapies for tissue engineering

biomaterials could be considered well-developed, and with some having been applied commercially, pharmaceutically, or even clinically, discussions regarding their sustainability are rare. However, on a more positive note, researchers are becoming more aware of this issue and are making efforts to optimize the fabrication method to be less harmful to the environment, especially for animals. For example, the use of Matrigel® has faced criticism regarding its product fabrication process, particularly the induction of tumors in mice followed by their subsequent sacrifice, which raises ethical concerns and questions regarding sustainability (Hughes et al. 2010; Hayashi et al. 2004). In contrast, those biomaterials extracted from plants could be more sustainable, plants tend to grow much faster than animals, and extraction could be completed without the need to sacrifice. Progress has been made in developing greener technologies, such as enzyme-assisted, ultrasound-assisted, and extrusion-assisted extraction applied to enhance extraction yield, increase production, and while achieving desired properties (Saji et al. 2022). Furthermore, another concept of biomaterial sustainability is the ability to support and maintain the reconstruction process over long-term in the human body. It requires biomaterials to be biocompatible, resistant to corrosion and wear, and able to support osseointegration in craniofacial bone reconstruction (Rajan et al. 2021). An additional criterion to evaluate their sustainability is their ability to support the reconstruction without needing additional surgical intervention. This demands that the scaffold exhibit high biocompatibility, biodegradability, or bioactivity, thereby minimizing prognostic inflammation and pain (Amirthalingam et al. 2021; Rizwan et al. 2017b). Herein, we will briefly summarize some of the current sustainable biomaterials (Table 13.1) for scaffold therapies and their applications in the craniofacial region.

## 13.2 Common Biomaterial for Scaffolding Therapies in Head and Neck Region

### 13.2.1 Synthetic Materials

#### 13.2.1.1 Calcium Phosphate (CaP)

Calcium phosphates are the major inorganic components of bones. Since the early 1900s, man-made calcium phosphate has been extensively studied and clinically applied in patients for bone regeneration (Jeong et al. 2019). Nowadays, the material is still widely used, serving numerous functionalities such as bone cement (Liu et al. 2022a), bone scaffolds (Wattanaanek et al. 2022; Schulz et al. 2023), coatings and implants (Feng et al. 2023; Arunjaroensuk et al. 2023), due to its excellent mechanical properties, biocompatibility and osseointegration capacity (Jeong et al. 2019). Different types of calcium phosphate have been commercialized, such as hydroxyapatite (HAp), tricalcium phosphate (TCP), brushite or dicalcium phosphate dihydrate (DCPD), and whitlockite (WH) (Jeong et al. 2019). Among those common types of calcium phosphate, HAp constitutes the highest amount of calcium and possesses the most thermodynamically stable form, exhibiting low

**Table 13.1** Some current biomaterials used in the field of craniofacial regeneration

| Scaffolds | Advantages | Disadvantages | Applications in craniofacial and oral regions | References |
|---|---|---|---|---|
| Calcium phosphate | Bioactivity<br>Biocompatibility<br>Osteoconductivity<br>Low immunogenicity | Brittleness<br>Slow resorption rate | Bone cement<br>Bone graft/substitute<br>Maxillary/mandibular augmentation | Sun and Yang (2015), Gali et al. (2017, 2018), van Oirschot et al. (2023), Knabe et al. (2023), and Liu et al. (2022a) |
| Bioglass | High bioactivity<br>Biocompatibility<br>Antibacterial property<br>Good thermal stability | Brittleness<br>Low fracture resistance<br>Low bending strength | Bone graft material<br>Endosseous/dental implants<br>Drug carrier<br>Wound dressing | Rizwan et al. (2017a), Amirthalingam et al. (2021), Tu et al. (2023), Wang and Li (2016), Dey et al. (2023), Sharifi et al. (2023), and Yang et al. (2023) |
| PMMA | Biocompatibility<br>Lightweight<br>Good impact resistance<br>Low thermal conductivity | Non-biodegradability<br>Limited integration with tissues<br>Potential release of non-reacted monomers | Bone cement<br>Bone scaffolds<br>Cranial implants | Miao et al. (2023), Huang et al. (2015), Ghasemi et al. (2023), and Abdulhussain Kadhum and Nassir (2023) |
| PEEK | Excellent chemical resistance<br>High melting temperature<br>High modulus of elasticity and tensile strength<br>High load capacity<br>Great biocompatibility | High cost of manufacturing<br>Lack of thermoformability | Dental implants<br>Craniofacial implants<br>PEEK-loaded hydrogel for bone regeneration<br>Drug carrier | Najeeb et al. (2016), Ouldyerou et al. (2022), Behnecke and Petersen (2020), Alipour et al. (2023), and Rammos et al. (2015) |

| | | | | |
|---|---|---|---|---|
| PEG | Low immunogenicity<br>Non-toxic<br>Adjustable mechanical properties<br>Resistance to protein degradation | Non-biodegradability<br>Limited bioactivity<br>Lack of cell-specific adhesion<br>Relatively high cost of fabrication | Drug carrier for bone regeneration<br>Scaffolds for gene delivery<br>Microgels as bone scaffolds | Sun et al. (2023), Wehrhan et al. (2012), Kamalakar et al. (2023), and Stengelin et al. (2020) |
| PLGA | Biodegradability<br>High bioavailability<br>Tunable mechanical properties<br>Controlled drug release | Slow resorption rate<br>Hydrophobicity<br>Relatively high cost | Drug carrier<br>Scaffolds for periodontal tissue/bone regeneration | Kadlecova et al. (2023), Gentile et al. (2014), Chachlioutaki et al. (2022), and Qian et al. (2019) |
| Collagen | Excellent biocompatibility<br>Biodegradability<br>Versatility<br>Low antigenicity<br>Similarity to natural ECM | Poor mechanical strength and structural stability<br>Rapid degradation rate | Collagen oral wound dressings<br>Bone substitutions/grafts<br>Collagen scaffolds/matrices<br>Adhesive membranes for soft tissue regeneration<br>Collagen plugs for hemostasis | Wu et al. (2022), Chamieh et al. (2016), Correia et al. (2022), Samiei et al. (2022), and Kao and Scott (2007) |
| Gelatin | Biocompatibility<br>Biodegradability<br>Low toxicity<br>Cost-effectiveness<br>Ease of manufacturing and modifications | Poor mechanical strength and structural stability<br>Potential allergic reactions when derived from animal sources | Bone scaffolds<br>Microparticles for cell/drug carriers<br>Gelatin sponges for wound dressings/skin substitutes<br>TMJ cartilage scaffolds<br>Microdroplets for endodontic regeneration<br>Microfilms for corneal regeneration | Tu et al. (2023), Liu et al. (2022a), Sharifi et al. (2023), Chen et al. (2023), Hu et al. (2023a), Wang et al. (2023b), Hinkelmann et al. (2022), and Farasatkia et al. (2021) |

(continued)

**Table 13.1** (continued)

| Scaffolds | Advantages | Disadvantages | Applications in craniofacial and oral regions | References |
|---|---|---|---|---|
| Alginate | Biocompatibility<br>Biodegradability<br>Versatility<br>Low immunogenicity | Poor mechanical strength<br>Lack of cell adhesion sites<br>Challenges for cell migration in dense matrices | Scaffolds for skin regeneration<br>Bone/cartilage scaffolds<br>Microtubes for salivary gland regeneration<br>Drug carriers | Jorgensen et al. (2022), Paul et al. (2023), Chen et al. (2017), Xu et al. (2023), Han et al. (2017), and Heo et al. (2017) |
| Cellulose | Renewability<br>Biodegradability<br>Non-toxicity<br>Source abundance<br>Low production cost | Low non-specific protein binding affinity<br>Excessive water absorption | Skin and wound dressings<br>Drug carriers<br>Bone/tendon/ligament/cartilage scaffolds<br>Cellulose scaffolds for neural regeneration<br>Vascular grafts | Busuioc et al. (2022), Wu et al. (2020), Hosny et al. (2021), Ahn et al. (2018), Yang et al. (2018), Geisel et al. (2016), Fusco et al. (2022), Ashley et al. (1960), and Muller et al. (2006) |
| Silk fibroin | Biocompatibility<br>Biodegradability<br>Robust mechanical strength<br>High tensile strength and elasticity | Slow degradation rate<br>High brittleness<br>Relatively high cost | Silk membranes/sponges/scaffolds for bone regeneration<br>Silk coatings for dental implants<br>Silk fibroin films for dental barrier membranes<br>Silk scaffolds for wound dressings | Smeets et al. (2017), Han et al. (2015), Rameshbabu et al. (2020), and Rider et al. (2018) |

| | | | | |
|---|---|---|---|---|
| Chitosan | Biocompatibility<br>Antimicrobial properties<br>Hemostatic properties<br>Low immunogenicity<br>Non-toxicity | Low mechanical strength and resistance<br>Difficulty in hydrogel's pore size control<br>Insolubility in aqueous solutions | Hydrogel for wound dressings<br>Bone scaffolds<br>Drug carriers/delivery systems<br>Coatings for dental implants<br>Membranes for dental pulp regeneration<br>Microtubes for nerve-vascular regeneration | Masood et al. (2019), Şenel (2010), Da Cunha et al. (2023), Giunchedi et al. (2002), Iqbal et al. (2017), Sanjaya IGPH (2022), Mu et al. (2022), Samiei et al. (2023), and Hsu et al. (2012) |
| Soy proteins | Tailorable biodegradability<br>Renewability and sustainability<br>Low immunogenicity<br>Cost-effectiveness | Limited mechanical strength<br>High surface hydrophobicity<br>Sensitivity to environmental factors (pH and temperature) | Wound dressings<br>Porous scaffolds for bone regeneration<br>Soy hydrogel as drug carriers | Xue et al. (2023), Las Heras et al. (2023), Ahn et al. (2018), Liu et al. (2020), and Chien et al. (2014) |

solubility when exposed to tissue fluid (Nithya and Meenakshi Sundaram 2015; Hanh et al. 2021). HAp has also been known to be osteoconductive, meaning that it stimulates osteoblast adherence, proliferation, and deposition of calcium to form new bone. Clinical studies by Gali and his team were conducted to evaluate the regeneration potential of bone marrow aspirate (BMA)-coated HAp for the reconstruction of maxilla-mandibular osseous defects. Clinical and radiological evaluations of the studies after 6 months revealed desirable osseointegration of BMA-HAp scaffolds with adjacent bone and no complications, including immunoreaction or infection, were reported (Gali et al. 2017, 2018).

#### 13.2.1.2 Bioglass (BG)

There are three major types of bioactive glasses: silicate glass (45S5), glass-ceramic (BonAlive or S53P4), and silica-based glass (13-93 and 13-93B1 bioactive glass). The main components of all BG include silicon dioxide ($SiO_2$), sodium dioxide ($Na_2O$), calcium dioxide (CaO), and phosphorus pentoxide ($P_2O_5$) (Huang 2017). When introduced to the body, BG possesses anti-inflammatory and antibacterial properties, induces angiogenesis, and promotes bone healing through osteoconductivity, osteoinductivity and osteointegration (O'Donnell and Hill 2010). The early generations of BG had high brittleness, making them less suitable for bone integration and maturation. To overcome this limitation, manipulations of BG composition have been made, including the addition of polymers into BG, such as collagen, silk, fibrin, hyaluronic acid, and alginate (Tu et al. 2023; Sharifi et al. 2023; Yasaswini et al. 2017; Yang et al. 2023; Xu et al. 2023; Hu et al. 2023b; Lee et al. 2022). Since then, the applications of BG have been more extensive, the material is not only used for bone regeneration but also can be applied in drug delivery systems (Yao et al. 2022; Wang and Li 2016), tendon healing (Xu et al. 2023) and dermal repair (Sharifi et al. 2023).

#### 13.2.1.3 Titanium

Titanium and its alloys have been used for hard tissue replacements such as in bone or dental implants for decades (Bothe 1940). Nowadays, the commercially pure titanium (CP-Ti) and its alloys (such as Ti−15Mo−5Zr (Rao and Houska 1979), Ti−15Mo−5Zr−3Al (Matsuda et al. 1997) or Ti−15Zr−4Nb−4Ta alloy (Okazaki 2001)) have been increasingly popular. Owing to their superior inertness, corrosion resistance, low immunogenicity and excellent osseointegration, titanium implants are generally considered sustainable for various medical applications (Hanawa 2019). Titanium is also a recyclable material and recycling processes have been well-established for many years, which minimize the environmental impact of producing new materials (Lee and Chang 2001). In the craniofacial region, metal has been used with great success as bone fixators (Zhao et al. 2012), bone screws (Varitimidis et al. 2021), and most particularly, dental implants for the replacement of lost teeth (Van Velzen et al. 2015). Modern CP-Ti and titanium alloy dental implants have remarkably high success rates of at least 89% over a 10-year lifespan (Van Velzen et al. 2015; Jung et al. 2008). Thus, thanks to the many above-mentioned advantages, titanium will continue to be one of the most popular materials used in medical and dental fields for future years.

#### 13.2.1.4 Polymethyl Methacrylate (PMMA)

The synthetic nondegradable polymer is fabricated by the polymerization of methyl methacrylate. PMMA was first introduced as a cement in the dental industry by Otto Rohm in 1901 (Rohm 1901), but it was not until the 1950s that the polymer was first applied in orthopedic surgeries with the establishment of prosthesis anchorage for the femoral head (Charnley 1960). As the material is rigid, lightweight and durable, it is commonly used for orthopedic bone scaffolds or bone cement (Miao et al. 2023; Moncayo-Matute et al. 2023; Abdulhussain Kadhum and Nassir 2023). One of the most popular applications of PMMA is craniofacial implants (Huang et al. 2015; Rammos et al. 2015; Fiaschi et al. 2016). Huang et al. (2015) conducted a cohort study on 20 patients receiving 22 PMMA implants for the management of acquired cranial deformities. Even though two patients experienced persistent temporal hollowing and facial asymmetric dissatisfaction, the overall complication rate was quite low (14%) without any signs of infection, hematoma or cerebrospinal fluid leak. The findings indicated that PMMA could be a supreme material for customized craniofacial implants (Huang et al. 2015). However, further research needs to be done to reduce the complication rate and enhance the satisfaction of the patients.

#### 13.2.1.5 Polyetheretherketone (PEEK)

This polymer is a semicrystalline thermoplastic material, synthesized by the stepwise dialkylation of ether ketone monomers (Najeeb et al. 2016). PEEK has been used as a prosthetic biomaterial for craniofacial implants and prosthetics (Najeeb et al. 2016), supportive scaffolds for bone regeneration (Dondani et al. 2023) or drug delivery systems (Behnecke and Petersen 2020), since it possesses attractive properties such as superior inertness, great thermal stability even at the temperature of 200–300 °C, excellent mechanical strength, high fatigue and abrasion resistance as well as allied elasticity to the natural bone (Ponnappan et al. 2009; Punchak et al. 2017). A recent study by Alipour et al. was conducted to develop an injectable PEEK hydrogel for craniofacial bone regeneration. The research team administered Aldehyde-cellulose nanocrystalline/silk fibroin (ADCNCs/SF) hydrogel into critical-sized cranial defects in rats. Their findings through radiographs and histology showed that there was greater bone formation in the defect area, which indicated that the novel material had a promising capacity for osteogenic induction (Alipour et al. 2023). However, further investigation should be done in the load-bearing regions to optimize this novel scaffold before it can be commercially ready for the patients.

#### 13.2.1.6 Poly(Ethylene Glycol) (PEG)

The primary structure of PEG consists of repeating ethylene oxide monomers connected in the end with different functional groups, which can be fabricated through the process of photopolymerization (Peppas et al. 1999). This hydrophilic material possesses many advantages such as high biocompatibility, low immunogenicity, resistance to protein degradation, and controllable mechanical properties and chemical compositions. PEG also gains attention for its hydrophilic nature, versatility in developing hydrogels, controllable release of bioactive molecules and promoting cell adhesion (Zhu 2010; Kadlecova et al. 2023; Jiang et al. 2023). Thanks to those beneficial characteristics, PEG has been extensively applied in many drug delivery systems, hydrogel and supportive 3D scaffolds for tissue regeneration (Xin et al. 2019; Ghuman et al. 2021; Jiang et al. 2023; Peppas et al. 1999). For example, a study reported a silver-nanoparticle-encapsulated chitosan-PEG hydrogel in the management of wound healing in diabetic patients. After the implantation of the new hydrogel into the lesions, the composite hydrogel gradually released silver particles into the surrounding environment and improved the wound healing rate. The authors indicated that this well-developed hydrogel could be a potential for soft tissue healing, especially in chronic diabetic lesions (Masood et al. 2019).

#### 13.2.1.7 Poly Lactic-Co-Glycolic Acid (PLGA)

PLGA is a combination of two polymers including poly lactic acid (PLA) and polyglycolic acid (PGA), which can be fabricated from renewable sources such as corn starch or sugarcane (Loureiro and Pereira 2020). This material is among very few FDA-approved polymers and has been extensively investigated as a potential carrier for drugs, proteins and genes into the human body, owing to the facts that it has excellent biocompatibility, biodegradability, adjustable mechanical properties and the ability to encapsulate and release drugs in a controlled manner (Bouissou et al. 2006; Ruhe et al. 2003). PLGA has also been tested as a scaffold for orofacial tissue regeneration. In 2019, Qian and his team fabricated a silver-modified/collagen-coated electrospun PLGA/PCL to treat alveolar bone defects in mouse periodontal disease models. Their findings indicated that the composite scaffold enhanced the alveolar/craniofacial bone regeneration by 31.8% compared to the control group and was an effective treatment for periodontitis (Qian et al. 2019). While PLGA shows promise in the context of tissue engineering and sustainability, ongoing research and further development are crucial to address challenges and improve its environmental impacts.

### 13.2.2 Naturally Derived Materials

#### 13.2.2.1 Collagen

Collagen is among the most essential proteins in the connective tissues of animals and humans. Twenty-nine types of collagen have been identified, among which collagen types I, II and III are the most abundant in the body (Van Der Rest and Garrone 1991). Collagen can be derived from different sources, including bovine, porcine, or fish. Collagen obtained from by-products of the food industry or sustainably harvested can enhance the sustainability of the material (Cao et al. 2022). In the field of oral and craniofacial regeneration, owing to its excellent biological properties, it has been employed as a supportive scaffold for oral mucosa ulcers and keratinized gingiva (Suebsamarn et al. 2022), membranes for guided bone regeneration (Elangovan 2018; Wang et al. 2018b), matrix for dentin remineralization (Wu et al. 2015), and hydrogel for promoting pulp-dentin complex revitalization (Holiel et al. 2021). Moreover, this natural material has also been used for the management of temporomandibular joint dysfunction. Wang et al. constructed an injectable bi-layer hyaluronic-acid/type I collagen (HA-SH/Col I) scaffold embedded with bone mesenchymal stem cells and chondrocytes for the regeneration of fibrocartilage and subchondral bone. The results indicated that the hydrogel scaffold could promote an appropriate microenvironment for the proliferation and chondrogenic matrix secretion of the cells, which suggested that this novel collagen-based scaffold can be a prospective treatment for mandibular condylar cartilage regeneration (Wang et al. 2021).

#### 13.2.2.2 Gelatin

Gelatin is acquired by the process of denaturation and partial hydrolysis of collagen, which is derived from animal connective tissues including bone, cartilage, and tendon (Xing et al. 2014). Gelatin exhibits a characteristic sol-gel transition in response to physical factors such as temperature, and pH, or chemical factors such as formaldehyde, and glutaraldehyde (Xing et al. 2014). Since gelatin possesses multiple cell adhesion motifs, excellent biocompatibility and biodegradability, it has been widely utilized in tissue engineering as 3D scaffolds (Tu et al. 2023; Sharifi et al. 2023; Wang et al. 2023b), drug carriers (Madkhali 2023; Hinkelmann et al. 2022), hydrogels (Tan et al. 2023), sponges and films for wound healing applications (Hu et al. 2023a; Kozlowska et al. 2023). In the orofacial region, it has been used as a bioink for 3D-printed scaffolds for the management of temporomandibular cartilage degeneration (Helgeland et al. 2021). The authors reported that the 3D-printed gelatin-based scaffolds were able to promote the proliferation and chondrogenic differentiation of mesenchymal stem cells, which could have great potential in cartilage regeneration (Helgeland et al. 2021). In addition, Yang et al. (2021) fabricated cell-laden Gelatin Methacrylate (GelMa) microspheres using the electrostatic microdroplet method in the application of endodontic regeneration. The *in vitro* experiments demonstrated that human dental pulp stem cells (hDPSC) encapsulated in the GelMa were able to migrate, proliferate and secrete extracellular matrix proteins. *In vivo* experiments also confirmed the positive results for generating a vascularized pulp-like tissue in mouse subcutaneous implantation, showing that GelMa possesses a significant prospect for pulp tissue regeneration.

#### 13.2.2.3 Alginate

Alginate is a natural polysaccharide extracted from a renewable source of the *Phaeophycean* brown seaweed family. As there is increasing attention on the green movement, scientists are now working on increasing the efficiency of alginate extraction, minimizing energy consumption and waste production. Hence, various sustainable approaches have been developed in the alginate extraction process such as ultrasound-assisted (Youssouf et al. 2017), microwave-assisted (Yuan and Macquarrie 2015) and extrusion-assisted methods (Sugiono et al. 2019). The material has been proven to be biocompatible, biodegradable, and non-toxicity, and has been extensively used in food, pharmaceutical, and bioengineering industries. In the oral and craniofacial regions, a study on drug-loaded alginate hydrogel has been conducted for the reparation of oral bone tissue (Chen et al. 2017). Experiments of human periodontal ligament cells and bone marrow stromal cells embedded in alginate hydrogels exhibited significant proliferation *in vitro*. Positive *in vivo* results also confirmed favourable biocompatibility and less inflammation. Furthermore, alginate hydrogels expressed osteogenesis-related genes and had superior osteoinductive bone capacity compared to the control group (Chen et al. 2017). Another study by Jorgensen et al. (2022) developed biocompatible alginate microtubes to promote epithelial and mesenchymal cell interactions for salivary gland regeneration (Jorgensen et al. 2022). The findings indicated the capacity of alginate hydrogel to facilitate cell-cell interactions, self-organization and expression of epithelial markers after, which could be a promising co-culture system for salivary gland engineering in the future (Jorgensen et al. 2022).

#### 13.2.2.4 Cellulose

Cellulose is among the most sustainable and abundant polysaccharides in nature. There are two main types of cellulose: plant-based cellulose produced by plants' cell walls and bacterial cellulose produced by certain strains of microorganisms such as *Agrobacterium, Rhizobium, Sarcina* and *Gluconacetobacter* (Zhang et al. 2022). Cellulose has gained attention as a renewable biomaterial owing to its biocompatibility, biodegradability and structural properties; however, its insolubility in aqueous environment makes it burdensome to fabricate. To overcome this problem, esterification, enzymatic hydrolysis, and high-pressure homogenization can be utilized to enhance solubility. In oral health research, as cellulose nanoparticles and nanofibers can enhance site-specific targeting of the tissues and antimicrobial activity against pathogens, the products have been used extensively as carriers in multiple drug delivery systems, especially in the management of periodontal diseases, dental caries and endodontic diseases (Hosny et al. 2021; Johnson et al. 2020; Paul et al. 2023). Furthermore, numerous studies have reported the use of cellulose nanofibrils for the regeneration of alveolar bone, showing that the cellulose-based scaffolds could induce excellent antibacterial levels, suitable macropores for nutrient supply to encourage cell growth, and new blood vessel formation (Jamilludin et al. 2023; Feroz and Dias 2021).

#### 13.2.2.5 Silk Fibroin

Silk fibroin, a natural protein extracted from the silkworm of *Bombyx mori* moth, has gained a great deal of attention in biomedical research for the regeneration of bone, cartilage, and tendons, due to its excellent mechanical strength, flexibility, lightweight and biocompatibility (Sun et al. 2021). The material can be chemically adjusted to enhance the effectiveness of growth factors and adhesion molecules, as well as fabricated into multiple forms including hydrogels (Li et al. 2024), sponges (Brown et al. 2017), films (Song et al. 2019), and membranes (Wang et al. 2023d) for different applications. In the dental implant industry, Wang et al. (2023d) developed a silk fibroin hydrogel membrane for guided bone regeneration (Wang et al. 2023d). The membranes were formed by photo-crosslinking technique and ethanol treatment resulting in the coexistence of the network structure and beta-sheet structure, which remarkably improved the strength and toughness of the novel membrane. The author demonstrated that the membrane was biocompatible and more resistant to degradation than other natural-derived membranes such as collagen-based or chitosan-based counterparts. *In vivo* experiments showed that the silk fibroin membranes had an outstanding capacity for spatial maintenance and soft-tissue barrier for bone regeneration (Wang et al. 2023d).

#### 13.2.2.6 Chitosan

Chitosan is a natural polysaccharide derived from the alkaline deacetylation of chitin—a component found mostly in the exoskeletons of crustaceans such as crabs or shrimps. This is among the most sustainable polysaccharides as tons of crustaceans' shelves are discarded every year from the seafood industry. The shell waste can be recycled and fabricated for various applications in biomedical industries (Gholap et al. 2024). Owing to its hydrophilic, non-immunogenic, biocompatible, biodegradable, permeation-enhancing properties, chitosan is frequently produced in the forms of hydrogels or nanoparticles for various drug delivery systems (Cai et al. 2022; Masood et al. 2019). In the oral cavity, chitosan-based gels have been proven to reduce the discomfort of inflammation and ulcers on the oral mucosa, especially in immunosuppressive patients. Hence, it is a subline candidate for the management of oral mucositis, as the gels can act as a carrier for therapeutic medication, inhibit bacterial growth and stimulate tissue reorganization (Şenel 2010). Chitosan microspheres containing chlorhexidine diacetate were reported to effectively treat buccal infections, particularly caused by *Candida albicans*. The *in vivo* findings showed the capacity of chitosan-based microspheres to offer a prolonged release of medication on the buccal mucosa (Giunchedi et al. 2002). Another application of chitosan is a bone scaffold for craniofacial regeneration. Cao et al. reported that mineralized tissues around chitosan-scaffold implantation showed a higher level of osteoid formation with complete bone mineralization compared to the control groups; they also expressed various osteogenic-related genes including alkaline phosphatase (ALP), runt-related transcription factor (RUNX2) and bone morphogenetic proteins 2 (BMP-2) (Cao et al. 2018). Hence, chitosan could be a promising therapy in the future for the regeneration of bone and hard tissues, especially in the craniofacial area.

#### 13.2.2.7 Soy Protein

Soy-based products have long been utilized in tissue engineering and pharmaceutical industries, owing to their cost-effectiveness, environmental sustainability, biocompatibility and biodegradability (Ramji and Shah 2014). Soybean is rich in proteins (mostly glycinin and beta-conglycinin) and isoflavones—a bioactive and antioxidant molecule, and are known for their anti-inflammatory, wound healing, cholesterol serum lowering effects and cancer protection abilities (Distefano 2023; Van Der Eecken et al. 2023). Owing to its multi-benefits, numerous studies have been conducted to fabricate soy bioproducts, such as hydrogels, injectable paste, thin films, and microcapsules (Liu et al. 2022b; Wu et al. 2020; Ramji and Shah 2014; Chien and Shah 2012; Lin et al. 2016; Alesaeidi et al. 2023; Phelan et al. 2020; Ahn et al. 2018; Chien et al. 2013). Alesaedi et al. recently reported to formulate soy protein isolate (SPI)/alginate (A)/hydroxyapatite (HA) composite hydrogel for bone regeneration. The authors realized that although soy protein isolate possesses high biocompatibility, tailorable biodegradability, and provides excellent support for cell attachment and growth, the material has poor mechanical properties. Therefore, the incorporation of SPI with other biomaterials such as alginate and hydroxyapatite would help to overcome the issues of weak mechanical performance (Alesaeidi et al. 2023). Recently, Liu et al. (2022b) incorporated soy protein and soy peptide powder into alginate/gelatin-based hydrogel to develop a bio-ink for 3D printing scaffolds. The authors reported that the integration of soy protein and soy peptide into the hydrogel accelerated the degradation rate, and the effect of soy protein on degradation was more remarkable. The i*n vitro* findings showed that the addition of soy-based powder into the composite hydrogel induced better cell attachment, migration and proliferation. Further *in vivo* results on chicken embryos and rat subcutaneous implants suggested that the 3D printed scaffolds could promote angiogenesis as neo-vessels sprouted more in groups with soy protein and soy peptide compared to the control group (Liu et al. 2022b). In summary, although these results are still preliminary, soy-based materials possess great potential for scaffolding therapies. More aspects of soy-based scaffolds should be profoundly investigated before they can be applied widely in patients and other clinical settings.

## 13.3 Applications in Craniofacial Soft and Hard Tissue Engineering

Scaffold-based regenerative strategies, incorporating principles of tissue engineering and regenerative medicine, along with the integration of cell and drug-based technologies, have gained widespread attention, as they aim at the reconstruction and long-term rehabilitation of craniofacial defects (Fig. 13.2). This task is complicated by the intricate nature of the craniofacial region, which includes both highly vascularized and innervated soft and hard tissues with varying growth rates in complex three-dimensional (3D) anatomical defects (Zhang and Yelick 2018). This complexity has been previously reported as the "biological boundary," when unaddressed, results in the growth of soft connective tissues in areas needing bone repair (Whitaker 1989). Techniques like guided tissue regeneration (GTR) (Dahlin et al. 1988) and guided bone regeneration (GBR) (Kaigler et al. 2013) have been consistently used to control the ingrowth of specific tissue types in craniofacial defects. These scaffold-based techniques provide a physical framework for new tissue or bone growth and can be made from a variety of biomaterials, each with its own set of limitations based on intended clinical use and surrounding microenvironment. Additionally, the use of craniofacial scaffolds for tissue engineering, often porous can facilitate the delivery of biofactors and natural tissue regeneration, while

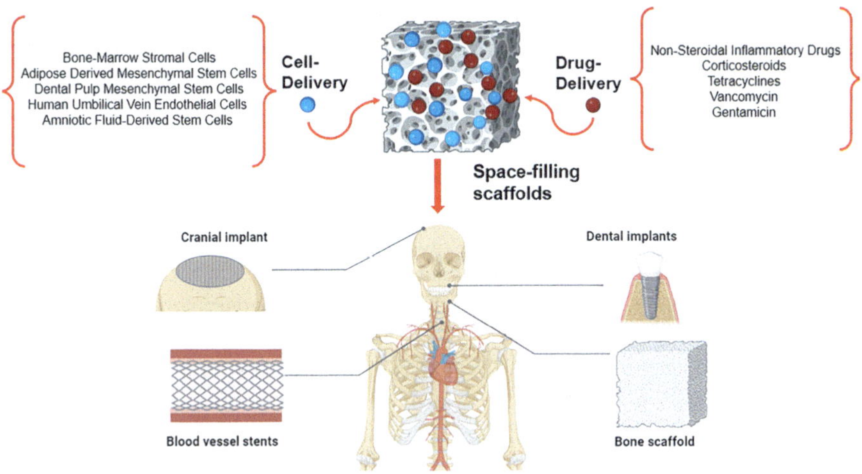


**Fig. 13.2** Schematic representation of the three major applications of scaffolding therapies: cell-delivery, drug-delivery and space-filling approaches for craniofacial soft and hard tissue engineering

maintaining shape and resisting functional loads in the complex 3D craniofacial anatomy (Tollemar et al. 2016). The choice of scaffold material—be it synthetic, natural, or a combination—further influences these properties, adding to the complexity of craniofacial tissue engineering.

The design of the ideal craniofacial scaffold is complex; however, an often-overlooked factor is the goal of creating sustainable scaffold therapies. Several studies have reported on novel approaches that utilize advanced materials and have embodied principles of green chemistry. In this section, we review recent advances and practical implementations of sustainable scaffold-based therapies for craniofacial soft and hard tissue engineering, with a specific focus on their effectiveness in tissue regeneration.

### 13.3.1 Structural and Space-Filling Scaffolds

Both structural and space-filling scaffolds have specialized functions for reconstructive surgery in the craniofacial region. Structural scaffolds are designed to provide mechanical support and maintain the overall structural integrity of the craniofacial region, typically these types of scaffolds are often made from materials with high mechanical strength such as Titanium (Ti) (Wang et al. 2023c), and certain ceramics (Wang et al. 2019) and polymer composites (Hollister et al. 2005). Structural scaffolds are mainly required for high load-bearing capacity in cases where reconstruction of major bone defects in the jaw or skull is required and can mimic the mechanical properties of the surrounding bone being replaced. While, space-filling scaffolds are primarily utilized in scenarios where the objective is to promote the growth of new bone or soft tissue, rather than to offer immediate structural support (Wu et al. 2021). These scaffolds are typically employed to fill smaller craniofacial defects or voids. In general, space-filling scaffolds are composed of materials that are softer and more pliable compared to those used in structural scaffolds. Examples of such materials include collagen (Chamieh et al. 2016), hydrogels (Drury and Mooney 2003), and various polymers that facilitate cell attachment and growth (Bashir et al. 2023). Owing to their inherent objectives, space-filling scaffolds are also designed to be biodegradable and bioresorbable (Lu et al. 2001). This means they gradually break down and are absorbed by the body as new tissue grows concurrently, thereby eliminating the need for secondary surgery to remove the scaffold.

In the field of craniofacial tissue engineering, the use of sustainable scaffolds is not explicitly stated, however, much of the general trend within the field is moving in that direction (Xie et al. 2021; Biswal et al. 2020; Baranwal et al. 2022). One example is a study by Han and colleagues that reported on the use of epigenetics, where the authors targeted and overexpressed an epigenetic regulator plant homeodomain finger protein 8 (PHF8) that specifically removes and targets methyl groups from specific histone sites (Han et al. 2015). The overexpression of PHF8, found in pre-osteoblasts and bone marrow stromal cells, enhanced key osteogenic genes and proteins only when seeded onto silk fibroin scaffolds in a serum-free medium, which led to the promotion of bone formation and regeneration in 8-week-old mice with 4-mm-diameter calvarial critical-sized defects. Silk fibroin is an example of a natural polymer and represents a shift to more sustainable resources (Sun et al. 2021). The effectiveness of silk fibroin has also been exemplified in a study by Smeets et al. assessing different silk-based barrier membranes, their results demonstrated that β-tricalcium phosphate (β-TCP)-modified silk showed the most promising results in terms of supporting bone formation, defect bridging, and restoration of original bone microarchitecture (Smeets et al. 2017). When compared to a commercially available collagen membrane there was lower resorbability suggesting the advantage of β-TCP-modified silk in longer-term bone regeneration and a benefit to guided bone regeneration. Collagen is another example of a protein-based polymer that has been reported as a sustainable resource (Rezvani Ghomi et al. 2021; Coppola et al. 2020). Momose et al. investigated the use of a collagen hydrogen scaffold combined with fibroblast growth factor-2 (FGF-2) implanted into class II furcation defects in dogs (Momose et al. 2016). The combination of collagen type I and FGF2 promoted soft tissue ingrowth, in addition to reconstruction of the alveolar bone. Periodontal attachment was also observed, with the authors suggesting that FGF-2 could have facilitated guided self-assembly. Similarly, Wang and colleagues took an alternative approach by developing resveratrol (RSV) containing collagen scaffolds and investigating its effect on soft tissue wound healing and bone regeneration (Wang et al. 2018a). RSV is a polyphenolic compound known to be anti-inflammatory and an antioxidant, which has been shown to create a conducive environment for cell proliferation and differentiation. The authors observed the proliferation and differentiation of human adipose stem cells (hASCs) on collagen/RSV scaffolds and effective wound closure and contraction when compared to the collagen scaffold alone. It was also observed that collagen scaffolds containing RSV were more effective at inducing hASC differentiation.

The use of natural polymers in craniofacial tissue engineering has gained considerable attention and many researchers have made significant strides in harnessing these polymers for engineering both soft and hard tissues (Thrivikraman et al. 2017; Chen et al. 2022). One effective strategy involves combining these natural polymers with other bio-based materials (Krishani et al. 2023). This approach aims to enhance certain qualities that may be lacking in a single polymer. Such advancements are particularly crucial when focusing on cell delivery. The success of scaffold-based therapies often hinges on cell behavior during initial seeding or encapsulation of cells before implantation. This step becomes even more challenging when incorporating biomaterials from sustainable sources.

### 13.3.2 Cell Delivery

Cell delivery, in the context of tissue engineering, primarily refers to the introduction or integration of living cells into a damaged or diseased tissue site to promote regeneration and healing, which has been widely investigated within the field of craniofacial tissue engineering (Wang et al. 2023a). The type of cells can consist of stem cells or specific tissue cells that are introduced into the target area. Typically, before implantation, the selected cells are often seeded or encapsulated onto scaffolds which helps support the cells and allow them to attach, grow, and form new tissue. This is generally defined as a three-dimensional (3D) framework to mimic the natural ECM of tissues (Zhang and Yelick 2018).

An important consideration of scaffold therapies is their ability to similarly replicate key ECM components that can promote cell behavior. A benefit is the use of natural polymers which tend to be well tolerated *in vivo* and with safer degradation by-products as compared to synthetic polymers (Reddy et al. 2021; Samir et al. 2022). Consequently, there is an increasing shift among researchers towards the use of polymer scaffolds that are naturally and indirectly sustainably derived, especially in the field of craniofacial tissue engineering (Brovold et al. 2018). Wang et al. developed an injectable photo-crosslinked hydrogel composed of methacrylate silk fibroin (RSFMA) and methacrylate hyaluronic acid (MeHA) for dental pulp regeneration (Wang et al. 2022). Upon encapsulation of human dental pulp stem cells (hDPSCs) within a hydrogel ratio of 1% MeHA to 3% RSFMA, led to a very rapid proliferation of cells, growth, and osteogenic differentiation ability that was significantly different from other ratios tested. A similar study also investigated the use of hDPSCs for endodontic regeneration but instead used gelatin methacryloyl (GelMA) microspheres as an injectable vehicle for hDPSCs *in vivo* (Yang et al. 2021). Chen and colleagues investigated the use of Arg-Gly-Asp (RGD) sequences grafted onto chitosan mixed with calcium phosphate cement (CPC) to create a biofunctionalized scaffold (Chen et al. 2014). The authors discovered *in vitro* that seeding a coculture of human umbilical vein endothelial cells (HUVEC) and human osteoblasts (HOB) onto CPC-RGD scaffolds had shown evidence of cell-synthesized bone mineral content and microcapillary-like structures when compared to the CPC control. Yousefiasl et al. also investigated the use of chitosan through the 3D printing of a chitosan-alginate-HAp nanocomposite scaffold. Scaffolds with 5% HAp amplified apatite formation, increased cell viability, and no toxicity was observed when seeded with mesenchymal stem cells (Yousefiasl et al. 2023b). Despite the certain physiochemical limitations of naturally derived polymers, the use of functionalization to further enhance and improve the efficacy of scaffolds for craniofacial tissue engineering has been extensively studied in many recent *in vitro* and *in vivo*.

To further add to the complexity of this concept, natural biopolymers alone often exhibit a range of poor physiochemical properties, such as biodegradability, poor mechanical properties, or a lack of bioactivity (Puertas-Bartolomé et al. 2021). Alginate is a good example as it comes from a sustainable source and has the advantage of a low cost of production (Sahoo and Biswal 2021). However, when it is stand-alone it exhibits several poor physiochemical properties such as a lack of bioactive sites for cell adhesion, low mechanical strength, and a high degradation rate (Lee and Mooney 2012). Zhang and colleagues proposed the use of egg whites in conjunction with alginate in developing a hydrogel that can grow human salivary spheroids as an alternative 3D culture scaffold for the differentiation of salivary ductal cells into acinar-like cells (Zhang et al. 2020). Using a cross-linking technique via a frozen calcium chloride disk technique and using egg white-alginate at 1.5% alginate concentration, exhibited the highest cell viability. In contrast, EWA with a 3% alginate concentration was more effective in promoting the formation of larger salivary spheroid-like structures. Susanto et al. investigated using the waste from fisheries specifically barramundi scales as a source of collagen combined with chitosan to develop a collagen-chitosan membrane for GTR (Susanto et al. 2019). The *in vitro* results indicate the presence of a fibrous membrane and ideal pore size needed for GTR, although further studies are required for its *in vivo* use there is great potential and a step in the direction of sustainably derived materials. Da Cunha and colleagues investigated a similar combination of collagen with chitosan, *in vivo,* using a 3:1 ratio as a scaffold for repair in rat mandibular defects (Da Cunha et al. 2023). Collagen and chitosan scaffolds seeded with and without dental pulp mesenchymal stem cells (DPSCs) observed evidence of bone formation but when compared to defects seeded only with DPSCs there was an observed greater osteoregenerative capacity. Both stem cell and scaffold choice are important factors that constantly need to be addressed dependent on the desired tissue to be engineered. Instead of examining DPSCs, Mohammed et al. considered the use of second trimester human amniotic fluid-derived stem cells (hAF-MSCs) seeded onto nano-hydroxyapatite (nHAp) and chitosan scaffolds containing 30% nHAp for bone healing in induced tibia defects of rabbits (Mohammed et al. 2019). The authors reported that seeded scaffolds showed new bone formation at 2–3 weeks post-transplantation, in addition, to complete bone healing at the defect site at 4 weeks. Whereas in non-seeded scaffolds bone regeneration was limited and at the same 4-week mark defects were only semi-healed. The objective of cell delivery in scaffold-based therapies for tissue engineering extends beyond simply selecting the appropriate stem cells for seeding or encapsulation within scaffolds, which aid in cell attachment, proliferation, and differentiation. Incorporating a drug into the scaffold can further enhance its efficacy, facilitating an even more customized approach.

### 13.3.3 Drug Delivery

A critical factor in drug delivery is controlled and sustained drug release while maintaining therapeutic levels both over the short and long term (Adepu and Ramakrishna 2021). For craniofacial tissue engineering, it is common for drugs to be encapsulated or incorporated within nanoparticles, hydrogels, or membranes (Moioli et al. 2007). Applications of drug delivery scaffolds are typically used to further enhance tissue regeneration and healing, in addition to reducing inflammation and infection (Ekenseair et al. 2013). It is not uncommon to see a combination of both synthetic and natural polymers used in the design composition of drug-delivery-based scaffolds with promising results while still being sustainably derived (Donnaloja et al. 2020; Bharadwaz and Jayasuriya 2020).

Infection at the site of bone or tissue repair is a likely possibility and, in many cases, studies have addressed this limitation by incorporating existing cell-delivery-based scaffolds to co-deliver specific antibiotics locally. Hassani Besheli et al. investigated contaminated bone injuries through the combination of vascular endothelial growth factor (VEGF) loaded on silk fibroin nanoparticles embedded within a silk scaffold containing vancomycin, *in vitro* (Hassani Besheli et al. 2018). Another study investigated *de novo* bone formation in critical-sized mandibular continuity defects in *Macaca fascicularis*, reporting increased bone density and volume in defects with a type I absorbable collagen sponge scaffold functionalized with chimeric anti-bone morphogenetic protein-2 (BMP-2) monoclonal antibodies (Xie et al. 2017). Similarly for bone repair in the alveolar region, Calasans-Maia and colleagues developed alginate microspheres encapsulated with minocycline-loaded nanocrystalline carbonated hydroxyapatite and tested its efficacy *in vivo*, which exhibited both antibacterial activity against *Enterococcus faecalis* and cytocompatibility and osteoconduction properties (Calasans-Maia et al. 2019). Petit et al. synthesized an injectable thermosensitive chitosan hydrogel (ChiG) loaded with statin, atorvastatin and lovastatin, nano-emulsions for the treatment of periodontitis lesions (Petit et al. 2020). To enhance the solubility and bioavailability of the statins, D-α-Tocopherol polyethylene glycol succinate (TPGS) was used to create the nano emulsion before mixing with the chitosan hydrogel. The *in vitro* results demonstrated decreased gene expression of pro-inflammatory markers, TNF-α, IL-1β, and RANKL, while mice with calvarial bone defects treated with both atorvastatin TPGS nano-emulsions ChiG and lovastatin TPGS nano-emulsions ChiG observed the highest bone healing effect and decreased inflammation of soft tissue surrounding treated lesions.

The concept of sustainable scaffold therapies for tissue engineering often can be misinterpreted to only use natural or sustainably derived advanced materials (Papageorgiou 2018). However, the idea of sustainable materials can also extend into the use and reuse of materials that possess an overall minimal impact on health and the environment (Jahangirian et al. 2018; Papageorgiou 2018). For example, synthetic polymers can be sustainable with an additional advantage of several synthetic polymers sanctioned for clinical use by various regulatory authorities and have also been classified as green materials (Gentile et al. 2014; Babak and Hadi 2013). Rahmati et al. characterized, *in vitro*, a core-shell nano-fibrous membrane capable of supporting both hard and soft tissue in the periodontium (Rahmati et al. 2022). This was achieved by embedding various bioactive glass nanoparticle concentrations within a polycaprolactone (PCL)/gelatin core and a polyvinyl alcohol (PVA)/polyvinylidene fluoride (PVDF) shell, utilizing electrospinning. Whereas Chachlioutaki et al. similarly investigated periodontal tissue engineering but instead electrospinning a scaffold made out of silk sericin, poly lactide-co-glycolic acid (PLGA), and ketoprofen, a nonsteroidal anti-inflammatory drug (Chachlioutaki et al. 2022). Another example specific to oral and craniofacial bone regeneration is an *in vitro* study that reported the use of mesoporous silica nanoparticles (MSN) at a ratio of 30% incorporated into an alginate/chitosan-based porous composite scaffold demonstrating biomineralization properties and increased cell viability when compared to the alginate-chitosan scaffold without MSN and sample groups with MSN ratios of 10% and 20% (Yousefiasl et al. 2023a). Although significant advances have been made in drug delivery-based scaffolds, the ability to tune the physiochemical properties of a scaffold for a desired drug or antibiotic to possess sustained and therapeutic release over the short and long term remains a complex issue in the field of drug delivery systems.

## 13.4 Conclusion and Future Perspectives

Both congenital and acquired craniofacial defects have long been one of the major issues affecting millions of people worldwide. Some conventional materials for craniofacial reconstruction are associated with a lack of quantity, and poor biocompatibility, of which the by-products often result in negative environmental consequences. As awareness about ecological impacts emphasized in the last decade, more and more sustainable alternative biomaterials have been developed and applied in the field. Several methods can be incorporated within the workflow of the development of sustainable scaffold therapies, especially in the field of craniofacial tissue engineering. It is not uncommon for craniofacial defects, regardless of size, to be anatomically complex and to require scaffolds that precisely match the shape of the defect. In addition, to the potential biomechanical stressors that the respective scaffold may experience based upon its location. Thus, the process of customization and precise fabrications are methods that can address these challenges and improve the desired therapeutic effect. Recently, the use of additive manufacturing, particularly 3D printing, bio-printing and newly developed 4D printing technologies have become much more advanced while also being inexpensive and accessible in the market. Thanks to state-of-the-art technology, 4D printing has allowed researchers to fabricate biomedical models or scaffolds that are dynamic, viable and stimuli-responsive systems. Existing printing technologies in several capacities are sustainable by developing scaffolds to minimize waste, re-usability of materials, and improved energy efficiency. Additionally, the use of novel bioinks combined with specific antibodies or proteins can further improve and contribute to the sustainable workflow in scaffold therapies for craniofacial soft and hard tissue engineering. Regardless, the process of developing ideal biomaterials that can match requirements to conserve the environment and tackle the complexity of human organs is still a long journey. It could take years of relentless work and effort until researchers can achieve that aim. Yet, hopefully, the employment of green biomaterials in tissue engineering, particularly in craniofacial regeneration can be more attainable shortly, which, therefore, can improve the quality of lives of many patients worldwide.

## References


Abdulhussain Kadhum S, Nassir NA (2023) Synthesis of bioactive heat cured PMMA/PEKK blend reinforced by nano titanium dioxide for bone scaffold applications. J Mech Behav Biomed Mater 150:106258

Adepu S, Ramakrishna S (2021) Controlled drug delivery systems: current status and future directions. Molecules 26:5905

Aghali A (2021) Craniofacial bone tissue engineering: current approaches and potential therapy. Cells 10:2993

Ahn S, Chantre CO, Gannon AR, Lind JU, Campbell PH, Grevesse T, O'Connor BB, Parker KK (2018) Soy protein/cellulose nanofiber scaffolds mimicking skin extracellular matrix for enhanced wound healing. Adv Healthc Mater 7:e1701175

Alesaeidi S, Kahrizi MS, Ghorbani Tajani A, Hajipour H, Ghorbani M (2023) Soy protein isolate/sodium alginate hybrid hydrogel embedded with hydroxyapatite for tissue engineering. J Polym Environ 31:396–405

Alipour M, Ghorbani M, Johari Khatoonabad M, Aghazadeh M (2023) A novel injectable hydrogel containing polyetheretherketone for bone regeneration in the craniofacial region. Sci Rep 13:864

Amirthalingam S, Lee SS, Pandian M, Ramu J, Iyer S, Hwang NS, Jayakumar R (2021) Combinatorial effect of nano whitlockite/nano bioglass with FGF-18 in an injectable hydrogel for craniofacial bone regeneration. Biomater Sci 9:2439–2453

Arunjaroensuk S, Thunyakitpisal P, Nampuksa K, Monmaturapoj N, Mattheos N, Pimkhaokham A (2023) Stability of guided bone regeneration with two ratios of biphasic calcium phosphate at implant sites in the esthetic zone: a randomized controlled clinical trial. Clin Oral Implants Res 34:850–862

Ashley FL, Stone RS, Edwards JW, Sloan R (1960) Further studies on the application of monomolecular cellulose filter tubes to create artificial tendon sheaths in the hand and wrist. West J Surg Obstet Gynecol 68:156–161

Babak G, Hadi A (2013) Biodegradable polymers. In: Rolando C, Francisca R (eds) Biodegradation. IntechOpen, Rijeka

Baranwal J, Barse B, Fais A, Delogu GL, Kumar A (2022) Biopolymer: a sustainable material for food and medical applications. Polymers 14:983

Bashir MH, Korany NS, Farag DBE, Abbass MMS, Ezzat BA, Hegazy RH, Dörfer CE, Fawzy El-Sayed KM (2023) Polymeric nanocomposite hydrogel scaffolds in craniofacial bone regeneration: a comprehensive review. Biomol Ther 13:205

Behnecke M, Petersen S (2020) Establishment of PEEK-associated drug delivery systems—limits and perspectives. J Mater Sci Chem Eng 8:32–41

Bharadwaz A, Jayasuriya AC (2020) Recent trends in the application of widely used natural and synthetic polymer nanocomposites in bone tissue regeneration. Mater Sci Eng C 110:110698

Biswal T, Badjena SK, Pradhan D (2020) Sustainable biomaterials and their applications: a short review. Mater Today: Proc 30:274–282

Bothe RT (1940) Reaction of bone to multiple metallic implants. Surg Gynecol Obstet 71:59–602

Bouissou C, Rouse J, Price R, Van Der Walle C (2006) The influence of surfactant on PLGA microsphere glass transition and water sorption: remodeling the surface morphology to attenuate the burst release. Pharm Res 23:1295–1305

Brovold M, Almeida JI, Pla-Palacín I, Sainz-Arnal P, Sánchez-Romero N, Rivas JJ, Almeida H, Dachary PR, Serrano-Aulló T, Soker S, Baptista PM (2018) Naturally-derived biomaterials for tissue engineering applications. Adv Exp Med Biol 1077:421–449

Brown JE, Moreau JE, Berman AM, Mcsherry HJ, Coburn JM, Schmidt DF, Kaplan DL (2017) Shape memory silk protein sponges for minimally invasive tissue regeneration. Adv Healthc Mater 6:1600762

Brown JS, Shaw RJ (2010) Reconstruction of the maxilla and midface: introducing a new classification. Lancet Oncol 11:1001–1008

Busuioc C, Isopencu G, Banciu A, Banciu DD, Oprea O, Mocanu A, Deleanu I, Zaulet M, Popescu L, Tanasuica R, Vasilescu M, Stoica-Guzun A (2022) Bacterial cellulose hybrid composites with calcium phosphate for bone tissue regeneration. Int J Mol Sci 23:16180

Cai S, Lei T, Bi W, Sun S, Deng S, Zhang X, Yang Y, Xiao Z, Du H (2022) Chitosan hydrogel supplemented with metformin promotes neuron-like cell differentiation of gingival mesenchymal stem cells. Int J Mol Sci 23:3276

Calasans-Maia MD, Barboza Junior CAB, Soriano-Souza CA, Alves A, Uzeda MJP, Martinez-Zelaya VR, Mavropoulos E, Rocha Leão MH, De Santana RB, Granjeiro JM, Rossi AM (2019) Microspheres of alginate encapsulated minocycline-loaded nanocrystalline carbonated hydroxyapatite: therapeutic potential and effects on bone regeneration. Int J Nanomedicine 14:4559–4571

Cao C, Xiao Z, Ge C, Wu Y (2022) Animal by-products collagen and derived peptide, as important components of innovative sustainable food systems-a comprehensive review. Crit Rev Food Sci Nutr 62:8703–8727

Cao D, Xu Z, Chen Y, Ke Q, Zhang C, Guo Y (2018) Ag-loaded MgSrFe-layered double hydroxide/chitosan composite scaffold with enhanced osteogenic and antibacterial property for bone engineering tissue. J Biomed Mater Res B Appl Biomater 106:863–873

Chachlioutaki K, Karavasili C, Adamoudi E, Bouropoulos N, Tzetzis D, Bakopoulou A, Fatouros DG (2022) Silk sericin/PLGA electrospun scaffolds with anti-inflammatory drug-eluting properties for periodontal tissue engineering. Biomater Adv 133:112723

Chamieh F, Collignon AM, Coyac BR, Lesieur J, Ribes S, Sadoine J, Llorens A, Nicoletti A, Letourneur D, Colombier ML, Nazhat SN, Bouchard P, Chaussain C, Rochefort GY (2016) Accelerated craniofacial bone regeneration through dense collagen gel scaffolds seeded with dental pulp stem cells. Sci Rep 6:38814

Charnley J (1960) Anchorage of the femoral head prosthesis to the shaft of the femur. J Bone Joint Surg Br 42-B:28–30

Chen L, Shen R, Komasa S, Xue Y, Jin B, Hou Y, Okazaki J, Gao J (2017) Drug-loadable calcium alginate hydrogel system for use in oral bone tissue repair. Int J Mol Sci 18:989

Chen L, Ye J, Gao C, Deng F, Liu W, Zhang Q (2023) Design and fabrication of gelatin-based hydrogel loaded with modified amniotic extracellular matrix for enhanced wound healing. Heliyon 9:e20521

Chen M, Jiang R, Deng N, Zhao X, Li X, Guo C (2022) Natural polymer-based scaffolds for soft tissue repair. Front Bioeng Biotechnol 10:954699

Chen W, Thein-Han W, Weir MD, Chen Q, Xu HHK (2014) Prevascularization of biofunctional calcium phosphate cement for dental and craniofacial repairs. Dent Mater 30:535–544

Chien KB, Chung EJ, Shah RN (2014) Investigation of soy protein hydrogels for biomedical applications: materials characterization, drug release, and biocompatibility. J Biomater Appl 28:1085–1096
Chien KB, Makridakis E, Shah RN (2013) Three-dimensional printing of soy protein scaffolds for tissue regeneration. Tissue Eng Part C Methods 19:417–426
Chien KB, Shah RN (2012) Novel soy protein scaffolds for tissue regeneration: material characterization and interaction with human mesenchymal stem cells. Acta Biomater 8:694–703
Coppola D, Oliviero M, Vitale GA, Lauritano C, D'Ambra I, Iannace S, De Pascale D (2020) Marine collagen from alternative and sustainable sources: extraction, processing and applications. Mar Drugs 18:214
Correia C, Sousa RO, Vale AC, Peixoto D, Silva TH, Reis RL, Pashkuleva I, Alves NM (2022) Adhesive and biodegradable membranes made of sustainable catechol-functionalized marine collagen and chitosan. Colloids Surf B Biointerfaces 213:112409
Da Cunha MR, Maia FLM, Iatecola A, Massimino LC, Plepis AMG, Martins V, Da Rocha DN, Mariano ED, Hirata MC, Ferreira JRM, Teixeira ML, Buchaim DV, Buchaim RL, De Oliveira BEG, Pelegrine AA (2023) In Vivo evaluation of collagen and chitosan scaffold, associated or not with stem cells, in bone repair. J Funct Biomater 14:357
Dahlin C, Linde A, Gottlow J, Nyman S (1988) Healing of bone defects by guided tissue regeneration. Plast Reconstr Surg 81:672–676
Datta P, Ozbolat V, Ayan B, Dhawan A, Ozbolat IT (2017) Bone tissue bioprinting for craniofacial reconstruction. Biotechnol Bioeng 114:2424–2431
Dey N, Mohny FP, Betsy Reshma G, Rao D, Ganguli M, Santhiya D (2023) Bioinspired synthesis of bioactive glass nanocomposites for hyaluronic acid delivery to bone and skin. Int J Biol Macromol 253:127262
Dimitriou R, Jones E, Mcgonagle D, Giannoudis PV (2011) Bone regeneration: current concepts and future directions. BMC Med 9:66
Distefano JK (2023) The role of choline, soy Isoflavones, and probiotics as adjuvant treatments in the prevention and management of NAFLD in postmenopausal women. Nutrients 15:2670
Dondani JR, Iyer J, Tran SD (2023) Surface treatments of PEEK for osseointegration to bone. Biomol Ther 13:464
Donnaloja F, Jacchetti E, Soncini M, Raimondi MT (2020) Natural and synthetic polymers for bone scaffolds optimization. Polymers 12:905
Drury JL, Mooney DJ (2003) Hydrogels for tissue engineering: scaffold design variables and applications. Biomaterials 24:4337–4351
Ekenseair AK, Kasper FK, Mikos AG (2013) Perspectives on the interface of drug delivery and tissue engineering. Adv Drug Deliv Rev 65:89–92
Elangovan S (2018) Dental implants placed in alveolar ridge augmented using guided bone regeneration procedure performed using resorbable collagen membranes and particulate bone grafts using simultaneous or staged approach exhibit a high survival rate. J Evid Based Dent Pract 18:173–175
Farasatkia A, Kharaziha M, Ashrafizadeh F, Salehi S (2021) Transparent silk/gelatin methacrylate (GelMA) fibrillar film for corneal regeneration. Mater Sci Eng C Mater Biol Appl 120:111744
Feng Y, Wu D, Knaus J, Kessler S, Ni B, Chen Z, Avaro J, Xiong R, Colfen H, Wang Z (2023) A bioinspired gelatin-amorphous calcium phosphate coating on titanium implant for bone regeneration. Adv Healthc Mater 12:e2203411
Feroz S, Dias G (2021) Hydroxypropylmethyl cellulose (HPMC) crosslinked keratin/hydroxyapatite (HA) scaffold fabrication, characterization and in vitro biocompatibility assessment as a bone graft for alveolar bone regeneration. Heliyon 7:e08294
Fiaschi P, Pavanello M, Imperato A, Dallolio V, Accogli A, Capra V, Consales A, Cama A, Piatelli G (2016) Surgical results of cranioplasty with a polymethylmethacrylate customized cranial implant in pediatric patients: a single-center experience. J Neurosurg Pediatr 17:705–710
Forrest CR, Hopper RA (2013) Craniofacial syndromes and surgery. Plast Reconstr Surg 131:86e–109e

Fusco D, Meissner F, Podesser BK, MarsanO A, Grapow M, Eckstein F, Winkler B (2022) Small-diameter bacterial cellulose-based vascular grafts for coronary artery bypass grafting in a pig model. Front Cardiovasc Med 9:881557

Gali RS, Chinnaswamy R, Devireddy SK, Shaik MV, Kumar RVK, Kanubaddy SR, Vaka RB, Harish YS, Pathapati RM (2018) Concentrated bone marrow aspirate-coated hydroxyapatite for reconstruction of small-to-moderate-sized mandibular defects caused by the removal of benign pathologies. Contemp Clin Dent 9:535–540

Gali RS, Devireddy SK, Mohan Rao N, Kishore Kumar RV, Kanubaddy SR, Dasari M, Sowjanya K, Pathapati RM (2017) Autogenous bone marrow aspirate coated synthetic hydroxyapatite for reconstruction of Maxillo-mandibular osseous defects: a prospective study. J Maxillofac Oral Surg 16:71–78

Geisel N, Clasohm J, Shi X, Lamboni L, Yang J, Mattern K, Yang G, Schafer KH, Saumer M (2016) Microstructured multilevel bacterial cellulose allows the guided growth of neural stem cells. Small 12:5407–5413

Gentile P, Chiono V, Carmagnola I, Hatton PV (2014) An overview of poly(lactic-co-glycolic) acid (PLGA)-based biomaterials for bone tissue engineering. Int J Mol Sci 15:3640–3659

Ghasemi F, Jahani A, Moradi A, Ebrahimzadeh MH, Jirofti N (2023) Different modification methods of poly methyl methacrylate (PMMA) bone cement for orthopedic surgery applications. Arch Bone Jt Surg 11:485–492

Gholap AD, Rojekar S, Kapare HS, Vishwakarma N, Raikwar S, Garkal A, Mehta TA, Jadhav H, Prajapati MK, Annapure U (2024) Chitosan scaffolds: expanding horizons in biomedical applications. Carbohydr Polym 323:121394

Ghuman H, Matta R, Tompkins A, Nitzsche F, Badylak SF, Gonzalez AL, Modo M (2021) ECM hydrogel improves the delivery of PEG microsphere-encapsulated neural stem cells and endothelial cells into tissue cavities caused by stroke. Brain Res Bull 168:120–137

Giunchedi P, Juliano C, Gavini E, Cossu M, Sorrenti M (2002) Formulation and in vivo evaluation of chlorhexidine buccal tablets prepared using drug-loaded chitosan microspheres. Eur J Pharm Biopharm 53:233–239

Gu L, Huang R, Ni N, Gu P, Fan X (2023) Advances and prospects in materials for craniofacial bone reconstruction. ACS Biomater Sci Eng 9:4462–4496

Han Q, Yang P, Wu Y, Meng S, Sui L, Zhang L, Yu L, Tang Y, Jiang H, Xuan D, Kaplan DL, Kim SH, Tu Q, Chen J (2015) Epigenetically modified bone marrow stromal cells in silk scaffolds promote craniofacial bone repair and wound healing. Tissue Eng Part A 21:2156–2165

Han Y, Li Y, Zeng Q, Li H, Peng J, Xu Y, Chang J (2017) Injectable bioactive akermanite/alginate composite hydrogels for in situ skin tissue engineering. J Mater Chem B 5:3315–3326

Hanawa T (2019) Titanium-tissue Interface reaction and its control with surface treatment. Front Bioeng Biotechnol 7:170

Hanh L, Van Hai L, The Hoang N, Hanh TH, D., Minh Hai, L. & Viet Nam, N. (2021) In vitro biodegradation behavior of biodegradable hydroxyapatite coated AZ31 alloy treated at various pH values. J Appl Biomater Funct Mater 19:22808000211010037

Hassani Besheli N, Damoogh S, Zafar B, Mottaghitalab F, Motasadizadeh H, Rezaei F, Shokrgozar MA, Farokhi M (2018) Preparation of a codelivery system based on vancomycin/silk scaffold containing silk nanoparticle loaded VEGF. ACS Biomater Sci Eng 4:2836–2846

Hayashi Y, Emoto T, Futaki S, Sekiguchi K (2004) Establishment and characterization of a parietal endoderm-like cell line derived from Engelbreth-Holm-Swarm tumor (EHSPEL), a possible resource for an engineered basement membrane matrix. Matrix Biol 23:47–62

Helgeland E, Rashad A, Campodoni E, Goksoyr O, Pedersen TO, Sandri M, Rosen A, Mustafa K (2021) Dual-crosslinked 3D printed gelatin scaffolds with potential for temporomandibular joint cartilage regeneration. Biomed Mater 16:035026

Henkel J, Woodruff MA, Epari DR, Steck R, Glatt V, Dickinson IC, Choong PFM, Schuetz MA, Hutmacher DW (2013) Bone regeneration based on tissue engineering conceptions — a 21st century perspective. Bone Res 1:216–248

Heo EY, Ko NR, Bae MS, Lee SJ, Choi B-J, Kim JH, Kim HK, Park SA, Kwon IK (2017) Novel 3D printed alginate–BFP1 hybrid scaffolds for enhanced bone regeneration. J Ind Eng Chem 45:61–67

Hinkelmann S, Springwald AH, Schulze S, Hempel U, Mitrach F, Wolk C, Hacker MC, Schulz-Siegmund M (2022) Mineralizing gelatin microparticles as cell carrier and drug delivery system for siRNA for bone tissue engineering. Pharmaceutics 14:548

Holiel AA, Mahmoud EM, Abdel-Fattah WM, Kawana KY (2021) Histological evaluation of the regenerative potential of a novel treated dentin matrix hydrogel in direct pulp capping. Clin Oral Investig 25:2101–2112

Hollister S, Lin C, Saito E, Lin C, Schek R, Taboas J, Williams J, Partee B, Flanagan C, Diggs A, Wilke E, Van Lenthe G, Müller R, Wirtz T, Das S, Feinberg S, Krebsbach P (2005) Engineering craniofacial scaffolds. Orthod Craniofac Res 8:162–173

Hosny KM, Alkhalidi HM, Alharbi WS, Md S, Sindi AM, Ali SA, Bakhaidar RB, Almehmady AM, Alfayez E, Kurakula M (2021) Recent trends in assessment of cellulose derivatives in designing novel and nanoparticulate-based drug delivery systems for improvement of oral health. Polymers (Basel) 14:92

Hsu SH, Huang GS, Feng F (2012) Isolation of the multipotent MSC subpopulation from human gingival fibroblasts by culturing on chitosan membranes. Biomaterials 33:2642–2655

Hu XM, Wang CC, Xiao Y, Jiang P, Liu Y, Qi ZQ (2023a) Enhanced wound healing and hemostasis with exosome-loaded gelatin sponges from human umbilical cord mesenchymal stem cells. World J Stem Cells 15:947–959

Hu Z, Lv X, Zhang H, Zhuang S, Zheng K, Zhou T, Cen L (2023b) An injectable gel based on photo-cross-linkable hyaluronic acid and mesoporous bioactive glass nanoparticles for periodontitis treatment. Int J Biol Macromol 257:128596

Huang GJ, Zhong S, Susarla SM, Swanson EW, Huang J, Gordon CR (2015) Craniofacial reconstruction with poly(methyl methacrylate) customized cranial implants. J Craniofac Surg 26:64–70

Huang J (2017) Chapter 20 – Design and development of ceramics and glasses. In: Vishwakarma A, Karp JM (eds) Biology and engineering of stem cell niches. Academic Press, Boston

Hughes CS, Postovit LM, Lajoie GA (2010) Matrigel: a complex protein mixture required for optimal growth of cell culture. Proteomics 10:1886–1890

Iqbal H, Ali M, Zeeshan R, Mutahir Z, Iqbal F, Nawaz MAH, Shahzadi L, Chaudhry AA, Yar M, Luan S, Khan AF, Rehman IU (2017) Chitosan/hydroxyapatite (HA)/hydroxypropylmethyl cellulose (HPMC) spongy scaffolds-synthesis and evaluation as potential alveolar bone substitutes. Colloids Surf B Biointerfaces 160:553–563

Jahangirian H, Lemraski EG, Rafiee-Moghaddam R, Webster TJ (2018) A review of using green chemistry methods for biomaterials in tissue engineering. Int J Nanomedicine 13:5953–5969

Jamilludin MA, Dinatha IKH, Supii AI, Partini J, Kusindarta DL, Yusuf Y (2023) Correction: functionalized cellulose nanofibrils in carbonate-substituted hydroxyapatite nanorod-based scaffold from long-spined sea urchin (Diadema setosum) shells reinforced with polyvinyl alcohol for alveolar bone tissue engineering. RSC Adv 13:34755

Jeong J, Kim JH, Shim JH, Hwang NS, Heo CY (2019) Bioactive calcium phosphate materials and applications in bone regeneration. Biomater Res 23:4

Jiang Z, Jiang K, Si H, Mcbride R, Kisiday J, Oakey J (2023) One step encapsulation of mesenchymal stromal cells in PEG Norbornene microgels for therapeutic actions. ACS Biomater Sci Eng 9:6322–6332

Johnson A, He JL, Kong F, Huang YC, Thomas S, Lin HV, Kong ZL (2020) Surfactin-loaded k-carrageenan oligosaccharides entangled cellulose nanofibers as a versatile vehicle against periodontal pathogens. Int J Nanomedicine 15:4021–4047

Jorgensen M, Ramesh P, Toro M, Evans E, Moskwa N, Zhang X, Sharfstein ST, Larsen M, Xie Y (2022) Alginate hydrogel microtubes for salivary gland cell organization and cavitation. Bioengineering (Basel) 9:38

Jung RE, Pjetursson BE, Glauser R, Zembic A, Zwahlen M, Lang NP (2008) A systematic review of the 5-year survival and complication rates of implant-supported single crowns. Clin Oral Implants Res 19:119–130

Kadlecova Z, Sevriugina V, Lysakova K, Rychetsky M, Chamradova I, Vojtova L (2023) Liposomes affect protein release and stability of ITA-modified PLGA-PEG-PLGA hydrogel carriers for controlled drug delivery. Biomacromolecules 25:67–76

Kaigler D, Pagni G, Park CH, Braun TM, Holman LA, Yi E, Tarle SA, Bartel RL, Giannobile WV (2013) Stem cell therapy for craniofacial bone regeneration: a randomized, controlled feasibility trial. Cell Transplant 22:767–777

Kamalakar A, Tobin B, Kaimari S, Toma AI, Moriarity I, Gautam S, Bhattaram P, Abramowicz S, Drissi H, García AJ, Wood LB, Goudy SL (2023) Delivery of a jagged1-PEG-MAL hydrogel with pediatric human bone cells regenerates critically-sized craniofacial bone defects. bioRxiv

Kao ST, Scott DD (2007) A review of bone substitutes. Oral Maxillofac Surg Clin North Am 19:513–521

Knabe C, Stiller M, Kampschulte M, Wilbig J, Peleska B, Gunster J, Gildenhaar R, Berger G, Rack A, Linow U, Heiland M, Rendenbach C, Koerdt S, Steffen C, Houshmand A, Xiang-Tischhauser L, Adel-Khattab D (2023) A tissue engineered 3D printed calcium alkali phosphate bioceramic bone graft enables vascularization and regeneration of critical-size discontinuity bony defects in vivo. Front Bioeng Biotechnol 11:1221314

Kozlowska J, Skopinska-Wisniewska J, Kaczmarek-Szczepanska B, Grabska-ZielinskA S, Makurat-Kasprolewicz B, Michno A, Ronowska A, Wekwejt M (2023) Gelatin and gelatin/starch-based films modified with sorbitol for wound healing. J Mech Behav Biomed Mater 148:106205

Krishani M, Shin WY, Suhaimi H, Sambudi NS (2023) Development of scaffolds from bio-based natural materials for tissue regeneration applications: a review. Gels 9:100

Las Heras K, Garcia-Orue I, Aguirre JJ, De La Caba K, Guerrero P, Igartua M, Santos-Vizcaino E, Hernandez RM (2023) Soy protein/beta-chitin sponge-like scaffolds laden with human mesenchymal stromal cells from hair follicle or adipose tissue promote diabetic chronic wound healing. Biomater Adv 155:213682

Lee KY, Mooney DJ (2012) Alginate: properties and biomedical applications. Prog Polym Sci 37:106–126

Lee PS, Heinemann C, Zheng K, Appali R, Alt F, Krieghoff J, Bernhardt A, Boccaccini AR, Van Rienen U, Hintze V (2022) The interplay of collagen/bioactive glass nanoparticle coatings and electrical stimulation regimes distinctly enhanced osteogenic differentiation of human mesenchymal stem cells. Acta Biomater 149:373–386

Lee SH, Chang YI (2001) Effects of recycling on the mechanical properties and the surface topography of nickel-titanium alloy wires. Am J Orthod Dentofacial Orthop 120:654–663

Lee SH, Yoo CJ, Lee U, Park CW, Lee SG, Kim WK (2014) Resorption of autogenous bone graft in cranioplasty: resorption and reintegration failure. Korean J Neurotrauma 10:10–14

Li H, Tong Z, Fang Y, Liu F, He F, Teng C (2024) Biomimetic injectable hydrogel based on methacrylate-modified silk fibroin embedded with Kartogenin for superficial cartilage regeneration. ACS Biomater Sci Eng 10:507–514

Lin HH, Hsieh FY, Tseng CS, Hsu SH (2016) Preparation and characterization of a biodegradable polyurethane hydrogel and the hybrid gel with soy protein for 3D cell-laden bioprinting. J Mater Chem B 4:6694–6705

Liu F, Liu C, Zheng B, He J, Liu J, Chen C, Lee IS, Wang X, Liu Y (2020) Synergistic effects on incorporation of beta-tricalcium phosphate and graphene oxide nanoparticles to silk fibroin/soy protein isolate scaffolds for bone tissue engineering. Polymers (Basel) 12:69

Liu N, Huang S, Guo F, Zhai S, Wang D, Li F, Liu C (2022a) Calcium phosphate cement with icariin-loaded gelatin microspheres as a local drug delivery system for bone regeneration. Biomed Eng Online 21:89

Liu Y, Hu Q, Dong W, Liu S, Zhang H, Gu Y (2022b) Alginate/gelatin-based hydrogel with soy protein/peptide powder for 3D printing tissue-engineering scaffolds to promote angiogenesis. Macromol Biosci 22:e2100413

Loureiro JA, Pereira MC (2020) PLGA based drug carrier and pharmaceutical applications: the most recent advances. Pharmaceutics 12:903

Lu L, Zhu X, Valenzuela RG, Currier BL, Yaszemski MJ (2001) Biodegradable polymer scaffolds for cartilage tissue engineering. Clin Orthop Relat Res 391:S251–S270

Madkhali OA (2023) Drug delivery of gelatin nanoparticles as a biodegradable polymer for the treatment of infectious diseases: perspectives and challenges. Polymers (Basel) 15:4327

Masood N, Ahmed R, Tariq M, Ahmed Z, Masoud MS, Ali I, Asghar R, Andleeb A, Hasan A (2019) Silver nanoparticle impregnated chitosan-PEG hydrogel enhances wound healing in diabetes induced rabbits. Int J Pharm 559:23–36

Matsuda Y, Nakamura T, Ido K, Oka M, Okumura H, Matsushita T (1997) Femoral component made of Ti-15Mo-5Zr-3Al alloy in total hip arthroplasty. J Orthop Sci 2:166–170

Miao X, Yang S, Zhu J, Gong Z, Wu D, Hong J, Cai K, Wang J, Fang X, Lu J, Jiang G (2023) Bioactive mineralized small intestinal submucosa acellular matrix/PMMA bone cement for vertebral bone regeneration. Regen Biomater 10:rbad040

Miura M, Miura Y, Sonoyama W, Yamaza T, Gronthos S, Shi S (2006) Bone marrow-derived mesenchymal stem cells for regenerative medicine in craniofacial region. Oral Dis 12:514–522

Mohammed EEA, Beherei HH, El-Zawahry M, Farrag ARH, Kholoussi N, Helwa I, Gaber K, Allam MA, Mabrouk M, Aleem AKA (2019) Combination of human amniotic fluid derived-mesenchymal stem cells and nano-hydroxyapatite scaffold enhances bone regeneration. Open Access Maced J Med Sci 7:2739–2750

Moioli EK, Clark PA, Xin X, Lal S, Mao JJ (2007) Matrices and scaffolds for drug delivery in dental, oral and craniofacial tissue engineering. Adv Drug Deliv Rev 59:308–324

Momose T, Miyaji H, Kato A, Ogawa K, Yoshida T, Nishida E, Murakami S, Kosen Y, Sugaya T, Kawanami M (2016) Collagen hydrogel scaffold and fibroblast growth Factor-2 accelerate periodontal healing of class II furcation defects in dog. Open Dent J 10:347–359

Moncayo-Matute FP, Torres-Jara PB, Vazquez-Silva E, Pena-Tapia PG, Moya-Loaiza DP, Abad-Farfan G (2023) Finite element analysis of a customized implant in PMMA coupled with the cranial bone. J Mech Behav Biomed Mater 146:106046

Mu X, Liu H, Yang S, Li Y, Xiang L, Hu M, Wang X (2022) Chitosan tubes inoculated with dental pulp stem cells and stem cell factor enhance facial nerve-vascularized regeneration in rabbits. ACS Omega 7:18509–18520

Muller FA, Muller L, Hofmann I, Greil P, Wenzel MM, Staudenmaier R (2006) Cellulose-based scaffold materials for cartilage tissue engineering. Biomaterials 27:3955–3963

Najeeb S, Zafar MS, Khurshid Z, Siddiqui F (2016) Applications of polyetheretherketone (PEEK) in oral implantology and prosthodontics. J Prosthodont Res 60:12–19

Nithya R, Meenakshi Sundaram N (2015) Biodegradation and cytotoxicity of ciprofloxacin-loaded hydroxyapatite-polycaprolactone nanocomposite film for sustainable bone implants. Int J Nanomedicine 10(Suppl 1):119–127

O'Donnell MD, Hill RG (2010) Influence of strontium and the importance of glass chemistry and structure when designing bioactive glasses for bone regeneration. Acta Biomater 6:2382–2385

Okazaki Y (2001) A new Ti–15Zr–4Nb–4Ta alloy for medical applications. Curr Opinion Solid State Mater Sci 5:45–53

Ouldyerou A, Merdji A, Aminallah L, Roy S, Mehboob H, Ozcan M (2022) Biomechanical performance of Ti-PEEK dental implants in bone: an in-silico analysis. J Mech Behav Biomed Mater 134:105422

Papageorgiou GZ (2018) Thinking green: sustainable polymers from renewable resources. Polymers (Basel) 10:952

Paul M, Pramanik SD, Sahoo RN, Dey YN, Nayak AK (2023) Dental delivery systems of antimicrobial drugs using chitosan, alginate, dextran, cellulose and other polysaccharides: a review. Int J Biol Macromol 247:125808

Peppas NA, Keys KB, Torres-Lugo M, Lowman AM (1999) Poly(ethylene glycol)-containing hydrogels in drug delivery. J Control Release 62:81–87

Petit C, Batool F, Stutz C, Anton N, Klymchenko A, Vandamme T, Benkirane-Jessel N, Huck O (2020) Development of a thermosensitive statin loaded chitosan-based hydrogel promoting bone healing. Int J Pharm 586:119534

Phelan MA, Kruczek K, Wilson JH, Brooks MJ, Drinnan CT, Regent F, Gerstenhaber JA, Swaroop A, Lelkes PI, Li T (2020) Soy protein nanofiber scaffolds for uniform maturation of human induced pluripotent stem cell-derived retinal pigment epithelium. Tissue Eng Part C Methods 26:433–446

Ponnappan RK, Serhan H, Zarda B, Patel R, Albert T, Vaccaro AR (2009) Biomechanical evaluation and comparison of polyetheretherketone rod system to traditional titanium rod fixation. Spine J 9:263–267

Puertas-Bartolomé M, Mora-Boza A, García-Fernández L (2021) Emerging biofabrication techniques: a review on natural polymers for biomedical applications. Polymers (Basel) 13:1209

Punchak M, Chung LK, Lagman C, Bui TT, Lazareff J, Rezzadeh K, Jarrahy R, Yang I (2017) Outcomes following polyetheretherketone (PEEK) cranioplasty: systematic review and meta-analysis. J Clin Neurosci 41:30–35

Qian Y, Zhou X, Zhang F, Diekwisch TGH, Luan X, Yang J (2019) Triple PLGA/PCL scaffold modification including silver impregnation, collagen coating, and electrospinning significantly improve biocompatibility, antimicrobial, and osteogenic properties for orofacial tissue regeneration. ACS Appl Mater Interfaces 11:37381–37396

Rahmati S, Labbaf S, Yaghini J, Talebi A (2022) Polymeric nanocomposite multifunctional core-shell membrane for periodontal repair and regeneration applications. Mater Today Chem 26:101097

Rajan ST, Subramanian B, Arockiarajan A (2021) A comprehensive review on biocompatible thin films for biomedical application. Ceram Int 48:4377–4400

Rameshbabu AP, Bankoti K, Datta S, Subramani E, Apoorva A, Ghosh P, Jana S, Manchikanti P, Roy S, Chaudhury K, Dhara S (2020) Bioinspired 3D porous human placental derived extracellular matrix/silk fibroin sponges for accelerated bone regeneration. Mater Sci Eng C Mater Biol Appl 113:110990

Ramji K, Shah RN (2014) Electrospun soy protein nanofiber scaffolds for tissue regeneration. J Biomater Appl 29:411–422

Rammos CK, Cayci C, Castro-Garcia JA, Feiz-Erfan I, Lettieri SC (2015) Patient-specific polyetheretherketone implants for repair of craniofacial defects. J Craniofac Surg 26:631–633

Rao VB, Houska CR (1979) Kinetics of the phase transformation in a Ti-15Mo-5Zr-3AI alloy as studied by X-ray diffraction. Metall Trans A 10:355–358

Reddy MSB, Ponnamma D, Choudhary R, Sadasivuni KK (2021) A comparative review of natural and synthetic biopolymer composite scaffolds. Polymers (Basel) 13:1105

Rezvani Ghomi E, Nourbakhsh N, Akbari Kenari M, Zare M, Ramakrishna S (2021) Collagen-based biomaterials for biomedical applications. J Biomed Mater Res B Appl Biomater 109:1986–1999

Rider PM, Brook IM, Smith PJ, Miller CA (2018) Reactive inkjet printing of regenerated silk fibroin films for use as dental barrier membranes. Micromachines (Basel) 9:46

Rizwan M, Hamdi M, Basirun WJ (2017a) Bioglass(R) 45S5-based composites for bone tissue engineering and functional applications. J Biomed Mater Res A 105:3197–3223

Rizwan M, Hamdi M, Basirun WJ (2017b) Bioglass® 45S5-based composites for bone tissue engineering and functional applications. J Biomed Mater Res A 105:3197–3223

Rohm O (1901) On the polymerization products of acrylic acid. Chemistry. Dissertation, University of Tubingen, Tubingen

Ruhe PQ, Hedberg EL, Padron NT, Spauwen PH, Jansen JA, Mikos AG (2003) rhBMP-2 release from injectable poly (DL-lactic-co-glycolic acid)/calcium-phosphate cement composites. JBJS 85:75–81

Sahoo DR, Biswal T (2021) Alginate and its application to tissue engineering. SN Appl Sci 3:30

Saji S, Hebden A, Goswami P, Du C (2022) A brief review on the development of alginate extraction process and its sustainability. Sustainability 14:5181

Samiei M, Alipour M, Khezri K, Saadat YR, Forouhandeh H, Abdolahinia ED, Vahed SZ, Sharifi S, Dizaj SM (2022) Application of collagen and mesenchymal stem cells in regenerative dentistry. Curr Stem Cell Res Ther 17:606–620

Samiei M, Dalir Abdollahinia E, Amiryaghoubi N, Fathi M, Barar J, Omidi Y (2023) Injectable thermosensitive chitosan/gelatin hydrogel for dental pulp stem cells proliferation and differentiation. Bioimpacts 13:63–72

Samir A, Ashour FH, Hakim AAA, Bassyouni M (2022) Recent advances in biodegradable polymers for sustainable applications. npj Mater Degrad 6:68

Sanjaya Igph MS (2022) Chitosan as bone scaffold for craniofacial bone regeneration: a systematic review. Open Access Maced J Med Sci 10:705–711

Schulz MC, Holtzhausen S, Nies B, Heinemann S, Muallah D, Kroschwald L, Paetzold-Byhain K, Lauer G, Sembdner P (2023) Three-dimensional plotted calcium phosphate scaffolds for bone defect augmentation-A new method for regeneration. J Pers Med 13:464

Şenel S (2010) Potential applications of chitosan in oral mucosal delivery. J Drug Deliv Sci Tech 20:23–32

Sharifi E, Yousefiasl S, Laderian N, Rabiee N, Makvandi P, Pourmotabed S, Ashrafizadeh M, Familsattarian F, Fang W (2023) Cell-loaded genipin cross-linked collagen/gelatin skin substitute adorned with zinc-doped bioactive glass-ceramic for cutaneous wound regeneration. Int J Biol Macromol 251:125898

Smeets R, Knabe C, Kolk A, Rheinnecker M, Gröbe A, Heiland M, Zehbe R, Sachse M, Große-Siestrup C, Wöltje M, Hanken H (2017) Novel silk protein barrier membranes for guided bone regeneration. J Biomed Mater Res B Appl Biomater 105:2603–2611

Song JE, Sim BR, Jeon YS, Kim HS, Shin EY, Carlomagno C, Khang G (2019) Characterization of surface modified glycerol/silk fibroin film for application to corneal endothelial cell regeneration. J Biomater Sci Polym Ed 30:263–275

Stengelin E, Kuzmina A, Beltramo GL, Koziol MF, Besch L, Schroder R, Unger RE, Tremel W, Seiffert S (2020) Bone scaffolds based on degradable Vaterite/PEG-composite microgels. Adv Healthc Mater 9:e1901820

Suebsamarn O, Kamimura Y, Suzuki A, Kodama Y, Mizuno R, Osawa Y, Komatsu T, Sato T, Haga K, Kobayashi R, Naito E, Kida M, Kishimoto K, Mizuno J, Hayasaki H, Izumi K (2022) In-process monitoring of a tissue-engineered oral mucosa fabricated on a micropatterned collagen scaffold: use of optical coherence tomography for quality control. Heliyon 8:e11468

Sugiono S, Masruri M, Estiasih T, Widjanarko SB (2019) Optimization of extrusion-assisted extraction parameters and characterization of alginate from brown algae (Sargassum cristaefolium). J Food Sci Technol 56:3687–3696

Sun H, Yang HL (2015) Calcium phosphate scaffolds combined with bone morphogenetic proteins or mesenchymal stem cells in bone tissue engineering. Chin Med J 128:1121–1127

Sun S, Cui Y, Yuan B, Dou M, Wang G, Xu H, Wang J, Yin W, Wu D, Peng C (2023) Drug delivery systems based on polyethylene glycol hydrogels for enhanced bone regeneration. Front Bioeng Biotechnol 11:1117647

Sun W, Gregory DA, Tomeh MA, Zhao X (2021) Silk fibroin as a functional biomaterial for tissue engineering. Int J Mol Sci 22:1499

Susanto A, Satari MH, Abbas B, Koesoemowidodo RSA, Cahyanto A (2019) Fabrication and characterization of chitosan-collagen membrane from barramundi (Lates Calcarifer) scales for guided tissue regeneration. Eur J Dent 13:370–375

Tan Y, Fan S, Wu X, Liu M, Dai T, Liu C, Ni S, Wang J, Yuan X, Zhao H, Weng Y (2023) Fabrication of a three-dimensional printed gelatin/sodium alginate/nano-attapulgite composite polymer scaffold loaded with leonurine hydrochloride and its effects on osteogenesis and vascularization. Int J Biol Macromol 249:126028

Thrivikraman G, Athirasala A, Twohig C, Boda SK, Bertassoni LE (2017) Biomaterials for craniofacial bone regeneration. Dent Clin N Am 61:835–856

Tollemar V, Collier ZJ, Mohammed MK, Lee MJ, Ameer GA, Reid RR (2016) Stem cells, growth factors and scaffolds in craniofacial regenerative medicine. Genes Dis 3:56–71

Tu X, Guo L, Li Y, Tan G, Chen R, Wu J, Miao G, Guo L, Zhang C, Zou T, Zhang Y, Jiang Q (2023) 3D-printed gelatin/sodium alginate/58S bioactive glass scaffolds promote osteogenesis in vitro and in vivo. J Biomater Appl 37:1758–1766
Van Der Eecken H, Joniau S, Berghen C, Rans K, De Meerleer G (2023) The use of soy Isoflavones in the treatment of prostate cancer: a focus on the cellular effects. Nutrients 15:4856
Van Der Rest M, Garrone R (1991) Collagen family of proteins. FASEB J 5:2814–2823
Van Oirschot B, Mikos AG, Liu Q, Van Den Beucken J, Jansen JA (2023) Fast degradable calcium phosphate cement for maxillofacial bone regeneration. Tissue Eng Part A 29:161–171
Van Velzen FJ, Ofec R, Schulten EA, Ten Bruggenkate CM (2015) 10-year survival rate and the incidence of peri-implant disease of 374 titanium dental implants with a SLA surface: a prospective cohort study in 177 fully and partially edentulous patients. Clin Oral Implants Res 26:1121–1128
Varitimidis S, Dailiana Z, Agorastakis D, Fyllos A, Zibis A, Hantes M, Malizos K (2021) Long-term functional results of intra- and extra-articular hand fractures treatment with titanium miniature plates and screws with minimum follow-up of 24 months. Cureus 13:e15438
Wang CC, Wang CH, Chen HC, Cherng JH, Chang SJ, Wang YW, Chang A, Yeh JZ, Huang YH, Liu CC (2018a) Combination of resveratrol-containing collagen with adipose stem cells for craniofacial tissue-engineering applications. Int Wound J 15:660–672
Wang F, Cai X, Shen Y, Meng L (2023a) Cell-scaffold interactions in tissue engineering for oral and craniofacial reconstruction. Bioact Mater 23:16–44
Wang H, Xu Y, Wang P, Ma J, Wang P, Han X, Fan Y, Bai D, Sun Y, Zhang X (2021) Cell-mediated injectable blend hydrogel-BCP ceramic scaffold for in situ condylar osteochondral repair. Acta Biomater 123:364–378
Wang L, Zhang Y, Xia Y, Xu C, Meng K, Lian J, Zhang X, Xu J, Wang C, Zhao B (2022) Photocross-linked silk fibroin/hyaluronic acid hydrogel loaded with hDPSC for pulp regeneration. Int J Biol Macromol 215:155–168
Wang MM, Flores RL, Witek L, Torroni A, Ibrahim A, Wang Z, Liss HA, Cronstein BN, Lopez CD, Maliha SG, Coelho PG (2019) Dipyridamole-loaded 3D-printed bioceramic scaffolds stimulate pediatric bone regeneration in vivo without disruption of craniofacial growth through facial maturity. Sci Rep 9:18439
Wang S, Chen H, Huang J, Shen S, Tang Z, Tan X, Lei D, Zhou G (2023b) Gelatin-modified 3D printed PGS elastic hierarchical porous scaffold for cartilage regeneration. APL Bioeng 7:036105
Wang T, Yu Z, Song Y, Yang L, Chen L, Dang J, Zeng X, Guo S, Su Y, Song B (2023c) Dura mater assists 3D printed porous titanium scaffolds in healing rabbit calvarial defects. Mater Today Commun 34:105111
Wang X, Li W (2016) Biodegradable mesoporous bioactive glass nanospheres for drug delivery and bone tissue regeneration. Nanotechnology 27:225102
Wang Y, Hua Y, Zhang Q, Yang J, Li H, Li Y, Cao M, Cai Q, Yang X, Zhang X, Li C (2018b) Using biomimetically mineralized collagen membranes with different surface stiffness to guide regeneration of bone defects. J Tissue Eng Regen Med 12:1545–1555
Wang Y, Yang Z, Chen X, Jiang X, Fu G (2023d) Silk fibroin hydrogel membranes prepared by a sequential cross-linking strategy for guided bone regeneration. J Mech Behav Biomed Mater 147:106133
Wattanaanek N, Suttapreyasri S, Samruajbenjakun B (2022) 3D printing of calcium phosphate/calcium sulfate with alginate/cellulose-based scaffolds for bone regeneration: multilayer fabrication and characterization. J Funct Biomater 13:47
Wehrhan F, Amann K, Molenberg A, Lutz R, Neukam FW, Schlegel KA (2012) PEG matrix enables cell-mediated local BMP-2 gene delivery and increased bone formation in a porcine critical size defect model of craniofacial bone regeneration. Clin Oral Implants Res 23:805–813
Whitaker LA (1989) Biological boundaries: a concept in facial skeletal restructuring. Clin Plast Surg 16:1–10
Wu DT, Munguia-Lopez JG, Cho YW, Ma X, Song V, Zhu Z, Tran SD (2021) Polymeric scaffolds for dental, oral, and craniofacial regenerative medicine. Molecules 26:7043

Wu M, Wu P, Xiao L, Zhao Y, Yan F, Liu X, Xie Y, Zhang C, Chen Y, Cai L (2020) Biomimetic mineralization of novel hydroxyethyl cellulose/soy protein isolate scaffolds promote bone regeneration in vitro and in vivo. Int J Biol Macromol 162:1627–1641

Wu SS, Wells M, Ascha M, Duggal R, Gatherwright J, Chepla K (2022) Head and neck wound reconstruction using biodegradable temporizing matrix versus collagen-chondroitin silicone bilayer. Eplasty 22:e31

Wu XT, Mei ML, Li QL, Cao CY, Chen JL, Xia R, Zhang ZH, Chu CH (2015) A direct electric field-aided biomimetic mineralization system for inducing the remineralization of dentin collagen matrix. Materials (Basel) 8:7889–7899

Xie W, Li T, Tiraferri A, Drioli E, Figoli A, Crittenden JC, Liu B (2021) Toward the next generation of sustainable membranes from green chemistry principles. ACS Sustain Chem Eng 9:50–75

Xie Y, Su Y, Min S, Tang J, Goh BT, Saigo L, Ansari S, Moshaverinia A, Zhang C, Wang J, Liu Y, Khojasteh A, Zadeh HH, Wang S (2017) Collagen sponge functionalized with chimeric anti-BMP-2 monoclonal antibody mediates repair of critical-size mandibular continuity defects in a nonhuman primate model. Biomed Res Int 2017:8094152

Xin S, Chimene D, Garza JE, Gaharwar AK, Alge DL (2019) Clickable PEG hydrogel microspheres as building blocks for 3D bioprinting. Biomater Sci 7:1179–1187

Xing Q, Yates K, Vogt C, Qian Z, Frost MC, Zhao F (2014) Increasing mechanical strength of gelatin hydrogels by divalent metal ion removal. Sci Rep 4:4706

Xu H, Zhu Y, Xu J, Tong W, Hu S, Chen YF, Deng S, Yao H, Li J, Lee CW, Chan HF (2023) Injectable bioactive glass/sodium alginate hydrogel with immunomodulatory and angiogenic properties for enhanced tendon healing. Bioeng Transl Med 8:e10345

Xue Y, Li Y, Zhang D, Xu W, Ning C, Han D (2023) Calcium phosphate silicate microspheres with soybean lecithin as a sustained-release bone morphogenetic protein-delivery system for bone tissue regeneration. ACS Biomater Sci Eng 9:2596–2607

Yang J, Du M, Wang L, Li S, Wang G, Yang X, Zhang L, Fang Y, Zheng W, Yang G, Jiang X (2018) Bacterial cellulose as a supersoft neural interfacing substrate. ACS Appl Mater Interfaces 10:33049–33059

Yang M, Yu S, Zhao P, Shi G, Guo Y, Xie L, Lyu G, Yu J (2023) Fabrication of biologically inspired electrospun collagen/silk fibroin/bioactive glass composited nanofibrous to accelerate the treatment efficiency of wound repair. Int Wound J 20:687–698

Yang T, Zhang Q, Xie L, Zhang R, Qian R, Tian Y, Chen G, Tian W (2021) hDPSC-laden GelMA microspheres fabricated using electrostatic microdroplet method for endodontic regeneration. Mater Sci Eng C 121:111850

Yao H, Luo J, Deng Y, Li Z, Wei J (2022) Alginate-modified mesoporous bioactive glass and its drug delivery, bioactivity, and osteogenic properties. Front Bioeng Biotechnol 10:994925

Yasaswini MS, Prabhakara Rao KV, Tanuja P, Motakatla NR (2017) Comparison of marginal periosteal pedicle graft and bioactive glass with platelet-rich fibrin and bioactive glass in the treatment of Intrabony defects – a clinicoradiographic study. J Pharm Bioallied Sci 9:S96–S102

Yousefiasl S, Manoochehri H, Makvandi P, Afshar S, Salahinejad E, Khosraviyan P, Saidijam M, Soleimani Asl S, Sharifi E (2023a) Chitosan/alginate bionanocomposites adorned with mesoporous silica nanoparticles for bone tissue engineering. J Nanostructure Chem 13:389–403

Yousefiasl S, Sharifi E, Salahinejad E, Makvandi P, Irani S (2023b) Bioactive 3D-printed chitosan-based scaffolds for personalized craniofacial bone tissue engineering. Eng Regen 4:1–11

Youssouf L, Lallemand L, Giraud P, Soulé F, Bhaw-Luximon A, Meilhac O, D'Hellencourt CL, Jhurry D, Couprie J (2017) Ultrasound-assisted extraction and structural characterization by NMR of alginates and carrageenans from seaweeds. Carbohydr Polym 166:55–63

Yuan Y, Macquarrie DJ (2015) Microwave assisted step-by-step process for the production of fucoidan, alginate sodium, sugars and biochar from Ascophyllum nodosum through a biorefinery concept. Bioresour Technol 198:819–827

Zhang W, Yelick PC (2018) Craniofacial tissue engineering, vol 8. Cold Spring Harb Perspect Med, p a025775

Zhang X, Wang D, Liu S, Tang J (2022) Bacterial cellulose nanofibril-based pickering emulsions: recent trends and applications in the food industry. Food Secur 11:4064

Zhang Y, Pham HM, Munguia-Lopez JG, Kinsella JM, Tran SD (2020) The optimization of a novel hydrogel—egg white-alginate for 2.5D tissue engineering of salivary spheroid-like structure. Molecules 25:5751

Zhao X, Zhu ZA, Sun YH, Wang Y, Zhao J, Zhang YJ, Dai KR (2012) Nickel-titanium shape-memory sawtooth-arm embracing fixator for periprosthetic femoral fractures. Int Orthop 36:619–626

Zhu J (2010) Bioactive modification of poly(ethylene glycol) hydrogels for tissue engineering. Biomaterials 31:4639–4656

# Sustainable Electrical Responsive Scaffolds

# 14

Clara Piccirillo and Robert C. Pullar

**Abstract**

Tridimensional (3D) porous structures, referred to as scaffolds, are widely used in biomedicine, as support to cell adhesion and growth, leading to cell differentiation and tissue regeneration; electrically conductive scaffolds have specific applications, as some tissues (musculoskeletal, neural and cardiac tissues) are positively stimulated by electricity. Generally conductive 3D scaffolds are made of synthetic polymers (PPy, PEDOT, PANI); there is however growing interest in developing materials from natural sources, for them to have reduced impact on the environment. Here, we discuss the most important reports in electrically conductive scaffolds prepared from natural and/or renewable sources; the natural polymers include chitosan, cellulose, hyaluronic acid, collagen, alginate and silk-derived materials. As these compounds are not conductive, they were modified with the addition of either the conductive polymers mentioned above, conductive carbon in different forms (graphite, graphene, etc.) or metallic nanoparticles; in some cases, a combination of these additives was explored, to study possible synergistic effects. The literature available on this topic confirms the potential of these sustainable materials for specific biomedical applications. Future investigations should focus on the use of sustainable materials also as conductive additives. Moreover, applications in the field of 4D printing should also be considered.

C. Piccirillo (✉)
CNR NANOTEC, Institute of Nanotechnology, Campus Ecoteckne, Lecce, Italy
e-mail: clara.piccirillo@nanotec.cnr.it

R. C. Pullar
Dipartimento di Scienze Molecolari e Nanosistemi (DSMN), Università Ca' Foscari Venezia, Venice, VE, Italy
e-mail: robertcarlyle.pullar@unive.it



J. M. Oliveira et al. (eds.), *Sustainable Scaffolds-based Strategies in Tissue Engineering and Regenerative Medicine*, Biomaterials, Bioengineering and Sustainability 8, https://doi.org/10.1007/978-3-031-96274-5_14

**Keywords**

Sustainable materials · Electrically conductive polymers · Biomedical applications · Tissue regeneration · Natural polymers

## 14.1 Introduction

Tri-dimensional (3D) porous structures are acquiring an increasing importance in tissue engineering. The main use of these structures, generally referred to as scaffolds, is to provide a support for cell adhesion and growth/proliferation, which can then lead to tissue regeneration. For 3D scaffolds to be effective in this role, they must be biocompatible and biodegradable (Nikolova and Chavali 2019); in addition to this, however, they must also possess other properties, which can favour the attachment of the cells and promote their growth. A key issue is the large-scale porosity of the pores or voids in the scaffold matrix material (as opposed to any small-scale porosity within the matrix material itself), i.e. the size of the pores/voids and their interconnection. These parameters are, in fact, crucial for the cells to diffuse into the structure, together with the necessary nutrients, oxygen, etc. (Loh and Choong 2013), and are usually on the scale of hundreds of microns in diameter. Moreover, particular selected molecules could be added to the scaffolds, to improve their performance: these could be drugs, DNA or RNA based molecules for gene therapy, MRI contrast agents or compounds for theranostic treatments, anti-bacterial substances to prevent infections, or growth factors which can further promote cellular growth.

In addition to this, according to the kind of tissue they are employed on, specific properties may favour tissue formation and regeneration. Electrical conductivity, either intrinsic or induced by an external stimulus, is a characteristic which can be important for some 3D scaffold applications; more specifically, musculoskeletal, neural and cardiac tissues, amongst others, can be positively affected by an electrical stimulation (Dixton and Gormillon 2022; Yi et al. 2023; Mehrotra et al. 2023). It has to be highlighted that, as some tissues are themselves electrically conductive, the use of a scaffold with comparable electrical conductivity can already be beneficial and correspond to higher cellular regeneration, even in the absence of an external electrical stimulus (Martins et al. 2014).

The most common conductive materials employed to fabricate 3D scaffolds are conductive polymers such as polypyrrole (PPy), poly(3,4-ethylenedioxythiophene) (PEDOT), polythiophene (PTh) and polyaniline (PANI); the electrical conductivity of these polymers is due to the electron delocalisation characteristic of these aromatic compounds (Schöbel and Boccaccini 2023). Indeed, several studies showed their conductivity corresponding to enhanced cellular growth (Kanaan and Piedade 2022 and references therein). It has to be highlighted, however, that all these polymers are synthetic materials, whose preparation can have a significant impact on the environment, both in terms of use of non-renewable resources and pollution associated with the preparation itself. Both PEDOT and PANI, for instance, are generally prepared using acid solutions (Cameron and Skabara 2020; Mantione et al. 2017).

In recent years there has been a growing interest in the development of materials of natural origin, i.e. obtained from renewable sources, to minimise the impact on the environment (Batool et al. 2024), especially biopolymers. In some cases, materials were also prepared from the residue of some industrial process, thereby reducing the wastes and converting them into more valuable compounds.

The 3D electro-responsive scaffolds made from natural/sustainable sources will be described herein.

## 14.2 Scaffolds Made from Natural Polymers

Several polymers of natural origin (biopolymers) are employed in biomedicine. These are macromolecules extracted from biomass, or synthesised by a living organism; they include starches and polysaccharides (often glucose-based) and polypeptides/proteins. However, all of these compounds, which will be described in detail in the sections below, are not inherently electrically conductive; therefore, they have to be mixed with different conductive compounds, to form composites. In this work, we will consider the studies in which the natural polymers constitute the main component and/or the framework of the composites used for the scaffold structure, and not those in which they are the dopant and/or the minor components. The natural biopolymers discussed in this review are shown in Fig. 14.1.

Different natural materials have been employed; for the majority of them, however, the conductive additives used could be divided into 3 groups: (1) the conductive polymers normally used to prepare the scaffolds mentioned above were employed, not as the main components, but as dopants; (2) graphitic carbon in different forms (i.e. carbon nanotubes, graphene and/or its derivatives, carbon from natural sources) is added; and (3) metallic nanoparticles are added. Some cases include a combination of these various approaches.

### 14.2.1 Chitin and Chitosan

Chitosan (Cs) is a natural polysaccharide derived from chitin (Ch), which after cellulose is the second most abundant polysaccharide in the biosphere; Ch forms the hard exoskeletons of crustaceans and insects, and also part of the cell wall of fungi. In crustaceans and insects, Ch is a hierarchical mix of fibrils / fibres surrounded by crystals of calcium carbonate—a natural composite material. Ch is the polysaccharide 2-acetamido-2-deoxy-b-d-glucose, and is identical to cellulose (see below) but with a hydroxyl group replaced by an acetamido group; Cs is obtained by the deacetylation of chitin (Fig. 14.1). It can be extracted from the shells of crustacean species (i.e. shrimps, molluscs, etc.) as a residue of the food industry (Oladzadabbasabadi et al. 2022).

Cs is one of the most used natural polymers in biomedicine, due to its properties, such as biocompatibility, antibacterial and anti-oxidant activities (Oladzadabbasabadi et al. 2022); due to its positive charge, it can also interact with anionic biomolecules

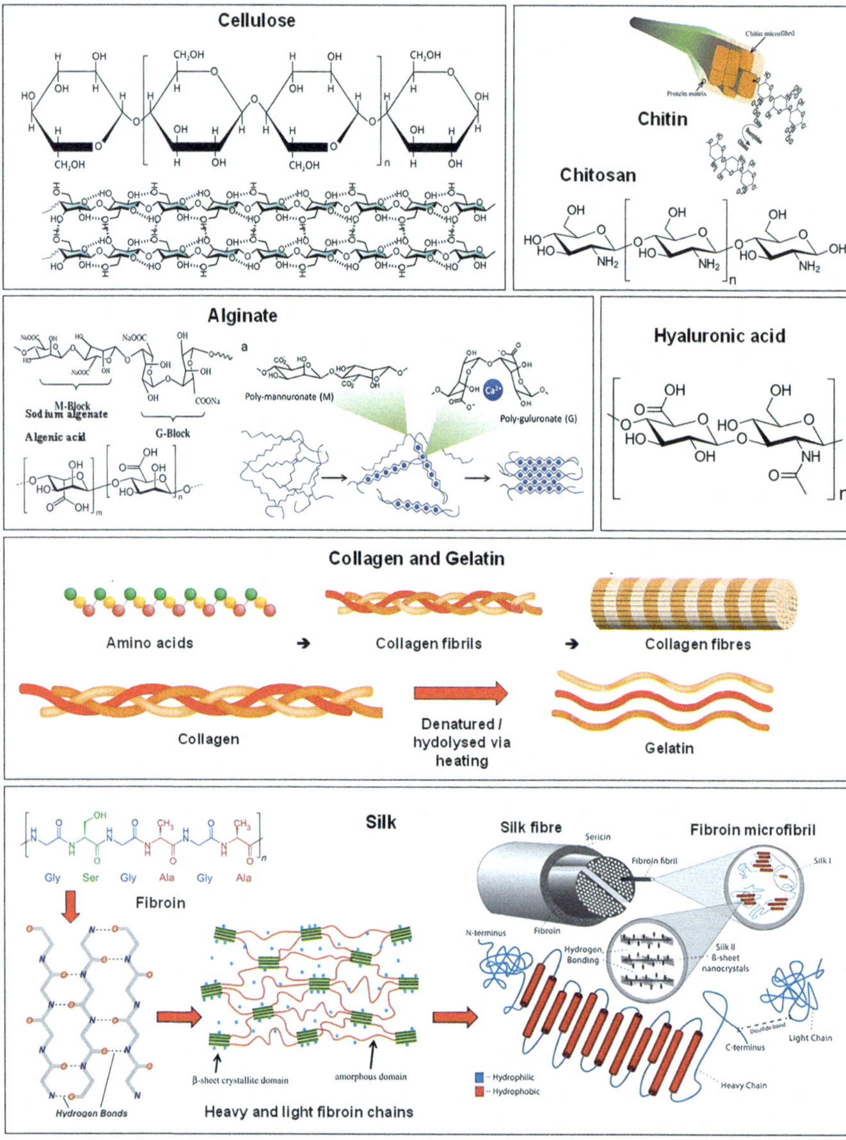


**Fig. 14.1** Examples of various natural biopolymers for use in conductive scaffolds: cellulose, chitosan, alginate, hyaluronic acid, collagen and gelatin, and silk

such as phospholipids. It is not, however, electrically conductive. 3D Cs scaffolds can be prepared with a cross-linking process, i.e. forming covalent bonds between the functional groups of two different Cs chains and an appropriate molecule (cross-linking agent); this allows one to obtain a 3D structure with higher stability and improved mechanical properties (Levengood and Zhang 2014; Józwiack et al.

2017). Cs scaffolds can also be prepared in form of hydrogel, which will become solid at a suitable temperature (generally 37 °C) (Morello et al. 2021).

Literature reports several Cs-based composites made with conductive polymers. In a study by Huang et al. (2010), a composite with PPy was investigated, with a PPy concentration of 2.5%wt (Cs 97.5%wt), leading to a conductivity of about $10^{-3}$ S/cm; the authors performed tests with Schwann cells both in the absence and presence of an applied electric field. Results showed that an enhanced cellular adhesion and expression was observed thanks to the electrical conductivity in both conditions, although the increase was more significant when the field was applied (Huang et al. 2010).

In another work, an oligopyrrole was employed instead of the polymer PPy, i.e. an additive with a shorter chain was considered; this was done to enhance the biodegradability of the composite, without affecting the electrical conductivity. The composites, in the form of a hydrogel, indeed showed significant electrical conductivity (about $10^{-3}$ S/cm for all oligopyrrole concentrations) while their degradation was comparable to that of the unmodified Cs. Cellular growth, on the other hand, was higher for the samples with higher levels of conductive dopant (>0.1% w/w to Cs) (Kashi et al. 2018).

Scaffolds made from Cs nanofibers were modified with PEDOT:PSS (polystyrene sulfonate) in concentrations up to 1% wt (Adedi et al. 2019). It was observed that, along with higher electrical conductivity (almost $10^{-3}$ S/cm), the addition of the dopant corresponded to fibres with smaller diameters; in fact, the average value decreased from 144 to 84 nm. Cellular growth tests, performed on mesenchymal stem cells, showed higher proliferation and cell adhesion on the 1% composite.

Cs was also made conductive with graphitic carbon in different forms. One of the most interesting works was performed by Scalera et al. (2021), as they prepared a composite made of Cs and graphitic carbon obtained from cork powder with a pyrolysis process. Cork powder, a natural compound, was a waste residue of the cork stopper industry, its use resulting in an entirely sustainable conducting scaffold material, as also the conductive additive was from a renewable source. Results showed that the addition of 1%wt/v of pyrolysed cork powder led to a significant increase in the electrical conductivity, up to about $10^{-5}$ S/cm. Although this value is lower than that obtained in many other studies, this conductivity is comparable to that of some cardiac tissues (Stinstra et al. 2005), making these sustainable scaffolds suitable for specific applications.

The use of carbon nanofibers was also reported for conductive Cs scaffolds (Martins et al. 2014). The nanofibers, in a concentration of 1% w/v, corresponded to a conductivity of about $10^{-3}$ S/cm. The composites were shown to favour the cellular activity of cardiomyocytes, as a significantly higher value was registered; it has to be highlighted that these results were achieved without applying an electrical stimulus, confirming the higher stimulation to cellular growth was induced by the electrical conductivity itself. Cs was also made conductive using graphene oxide (GO) (Jiang et al. 2019); with such additive (concentration 0.75% w/w), the conductivity reached values of about $10^{-3}$ S/cm. Cellular tests performed on heart H9C2 cells showed that GO's presence led to an increased expression of specific cardiac proteins. A similar approach was employed by Gholizadeh et al. (2017), but as conductive agent carbon nanotubes (CNTs) were considered.

In all the studies mentioned above, in addition to the electrical conductivity, other properties were all positively affected by the presence of the additives; an improvement in the mechanical behaviour, for instance, was registered (i.e. higher Young's modulus, tensile strength, etc.). Often the overall porosity did not change significantly, but different additive concentrations corresponded to different pore size.

Considering the modification with conductive NPs, gold (Au) NPs were considered in different studies; this is due to both their high conductivity for the presence of surface plasmon resonance (SPR) and to their use in biomedicine. Baei et al. (2016), for instance, report Cs-based hydrogel scaffolds with Au NPs included in their structure, which led to significant electrical conductivity (about $10^{-3}$ S/cm). The tests on cellular growth showed that the cellular activity was the comparable; however, when Au NPs were employed, enhanced cardiac differentiation of mesenchymal stems cells was observed. In a successive study, Cs was used together with k-carrageenan (kCA), a negatively charged sulfonated polysaccharide obtained from red algae (see Sect. 2.5 below). In this work, Au NPs were added to the Cs-kCA composite (Pourjavadi et al. 2019); cell proliferation and attachment were significantly increased by the NPs presence.

An application of conductive scaffold on a different sector of biomedicine, i.e. bone regeneration, was explored by Radwan-Pragłowska et al. (2020). They employed fungal Cs mixed with 10 wt% hydroxyapatite NPs and cross-linked to form an aerogel scaffold, which was then covered with a thin layer of poly (lactic acid) (PLA) fibres loaded with NPs of either $TiO_2$, Au or Pt. The hydroxyapatite NPs were needles 72 nm long and 20 nm wide, the low-density PLA fibres were between 0.61–1.56 μm wide (average diameter <1 μm), while the $TiO_2$, Au and Pt NPs were 95 nm, 20–40 nm and 10–20 nm, respectively.The resulting hydrogels had a swelling capacity in water of 211–232%, above the 200% level suitable for bone recovery (Turnbull et al. 2018). Under standard conditions in simulated body fluid (SBF) for 7 days only $TiO_2$ NPs increased biomineralisation to form hydroxyapatite by 17–18%, but with an applied current (5 V DC in SBF), scaffolds with Au and Pt NPs increased this by 22–24% and 16–19%, respectively, while it had no effect on the nonconductive $TiO_2$ NPs with no increase. All were *in vitro* tested for cytotoxicity with MG-63 bone cells, and all showed >100% cell viability.

Ch itself was used to make conductive plasma-treated chitin—multi-walled carbon nanotube (MWCNT) composite scaffolds showing very good neuron adhesion, and an ability to support their synaptic function (Singh et al. 2016), with the potential to be an implantable electrode for the stimulation and repair of neurons. 0.07 wt% MWCNTs coated with carboxymethyl cellulose were added to the chitin precursor solution, $O_2$ plasma treatment was carried out on the scaffold in a radio frequency plasma reactor, and this was then coated with lysine before cell attachment. The plasma-treated chitin-CNT composite not only an had an electrical conductivity of 0.0289 S/cm (compared to 0.0073 S/cm without plasma treatment), but the $O_2$ plasma treatment also introduced more oxygen species onto the scaffold surface, and it became significantly more hydrophilic, with a greater Young's modulus. Neuron viability experiments showed excellent neuron attachment onto the plasma-treated composite scaffolds after 7–21 days, and neuron adhesion was significantly higher after 14 days on plasma-treated chitin-MWCNT scaffolds compared to chitin alone, showing the advantage of the induced conductivity.

### 14.2.2 Cellulose

Cellulose is a natural polysaccharide constituted of D-glucose monomers linked by the b(1–4) glycosidic bond, giving long chains of 9000–15,000 units in plants—the monomer is disaccharide D-glucose $(C_{12}H_{20}O_{10})_n$ units linked by the C-O-C glycosidic bond (Fig. 14.1). Moreover, the different cellulose chains interact with each other through electrostatic interactions such as hydrogen bond and other weaker Van der Waals forces, leading to a 3D networked structure. Cellulose is the most abundant biopolymer n the biosphere, and can be extracted from plants, trees and wood. However, often for technological applications cellulose produced from selected bacterial strains is used (Lahiri et al. 2021). The most common bacterium for its production is the Gram negative *Gluconacetobacter xylinus*, which produce filaments of the polysaccharide. Bacterial cellulose (BC) has higher purity than that extracted from plants; moreover, changing the bacterial culture conditions, it is possible to control some of the features of the produced BC. Some intrinsic BC properties such as biocompatibility and anti-bacterial activity make it suitable for biomedicine applications.

Several studies were reported on modified BC to make it conductive and widen its biomedical uses. In situ polymerisation of PPy on the surface of nanocelluose filaments was investigated by Mueller et al. (2013); a similar approach was considered also by Liang et al. (2015) on cellulose crystalline structure. In both cases a significant increase in the conductivity was measured (more than 3 orders of magnitude). In the case of nanocellulose, cellular studies showed that cells adhesion was higher for the PPy-containing composites in the first 1–3 days but at the seventh day the activity was comparable to that of the non-modified cellulose. In a more recent work, a scaffold made of BC was decorated with PPy NPs (Srinivasan et al. 2023); with this method the authors fabricated a 3D flexible scaffold usable as a cardiac patch. The bioactivity tests showed an improved cellular differentiation towards cardiomyocytes, due to the enhanced electrical conductivity.

The opposite approach was employed by Guo et al. (2021), where a conductive 3D network made of graphene was covered with bacterial cellulose by culturing the *Acetobacter xylinum* strain on its surface; this led to a uniform and compact coating. This 3D system showed to promote the growth and adhesion of neural stem cells, and favour their differentiation into neurons, leading to a 3D neural network.

BC was also employed with a combination of conductive polymer (PPy) and a form of carbon (CNTs) for a hydrogel formulation (Wang et al. 2020); in this work the synergistic effect between the two conductive additives was shown, as the hydrogel reached conductivity values of about $10^{-3}$ S/cm. The authors performed tests with NIH3T3 cells, both in absence and in presence of an electric field. With no electric field, a better cell spreading was registered, together with enhanced cellular growth. This effect was even more significant when an electric field was applied.

Electrospun cellulose fibres were used to make conductive nanofibrous scaffolds, able to mimic the neural extracellular matrix (ECM). These were made conductive either by addition of 0.2 wt% MWCNTs, or by transformation into carbon nanofibers (CNFs) via carbonisation (Kuzmenko et al. 2016) (Fig. 14.2). The cellulose scaffold was made of fibres between 300–1500 nm, and became densely covered by MWCNTs, with a conductivity of 0.18 S/cm. When carbonised, the CNFs were reduced to 50–250 nm, but had a much higher conductivity of 11.2 S/cm. The CNF was also much less hydrophilic with a contact angle of 87.2° compared to 34.3° for the MWCNT coated cellulose fibres. However, the CNF showed no cytotoxicity with cell viability of 95%, while the MWCNT coated fibres were potentially cytotoxic with a cell viability of only 38%. Despite this, after testing with SH-SY5Y human neuroblastoma cells, differentiation towards mature neurons was distinctively greater on both electrically conductive scaffolds within only 10 days of differentiation, compared to the cellulose fibres alone (Fig. 14.2), indicating the advantages of these conductive cellulose derived scaffolds.

A conductive 3D printing ink was made from cellulose nanofibrils (CNF) with single-walled carbon nanotubes (SWCNTs), using alginate to promote the cross-linking process. Two conductive bioinks with varying CNT levels were produced, with ratios of NFC/CNT/alginate equal to 60/20/20 and 70/10/20. This was 3D printed using freeform reversible embedding hydrogel (FRESH) bioprinting, which employs low-viscosity liquids in a supporting bath of gelatin that can be easily separated from the printed scaffold, and then rapidly crosslinked with viscous polymer gels. Results showed that the cross-linking had a significant detrimental effect on

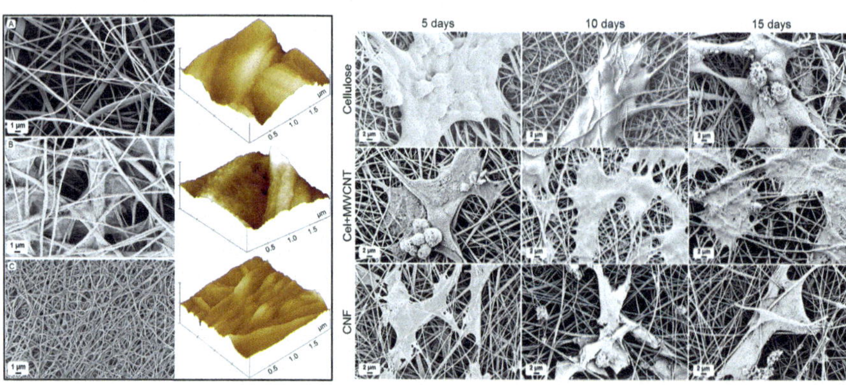


**Fig. 14.2** Left: SEM and AFM images of: (**a**) the electrospun cellulose fibre, (**b**) cellulose fibre + MWCNT, and (**c**) carbonised cellulose carbon nanofibre (CNF) scaffold materials. Right: SEM images of SH-SY5Y neuroblastoma cells on the scaffolds after 5, 10, and 15 days of culturing. (Reprinted from Kuzmenko et al. (2016), copyright © 2015 Elsevier B.V, used with permission from Elsevier)

the electrical conductivity of the 3D scaffolds, rendering them virtually non-conductive. However, non-crosslinked scaffolds showed good electrical conductivity of 0.205 S/cm with 10 wt% SWCNTs, and 2.13 S/cm with 20 wt% SWCNTs, reaching the percolation threshold. Conductive scaffolds were prepared that could promote the differentiation of SH-SY5Y cells, suggesting that these NFC-SWCNT conductive inks could 3D print scaffolds to differentiate neural cells. Cells cultivated on non-conductive NFC/alginate scaffolds formed neural cells after 10 days, but failed to generate a realistic neural network, and had poor differentiation, showing the importance of a conductive scaffold being employed for this application. It was shown that 10 wt% CNTs provided sufficient conductive properties to allow communication between separated neural cells, the conductivity further promoting maturation of the neural cells (Bordoni et al. 2020). The more conductive scaffold with 20 wt% CNTs had no attached living cells, however, suggesting cytotoxity from too-higher concentration of CNTs had a very negative effect. This effect, already seen in previous studies with CNTs (Sachar and Saxena 2011), highlights the fact that often in the best performing composite a compromise between different properties has to be considered; indeed, some additives can improve selected functional properties, but negatively affect others.

### 14.2.3 Hyaluronic Acid

Hyaluronic acid (HA) is another natural disaccharide biopolymer, found throughout the human body's connective, epithelial (skin) and neural tissue. It is an anionic, non-sulphated, glycosaminoglycan (long unbranched carbohydrate polysaccharide), consisting of up to 20,000 disaccharide monomers, and acts as a lubricant. HA's full IUPAC name is (1 → 4)-(2-Acetamido-2-deoxy-D--gluco)-(1 → 3)-D-glucuronoglycan, and it is D-glucose with attached carboxylic acid and acetamido groups (Fig. 14.1). It is one of the main components of extracellular matrices (ECMs), has a significant influence cell proliferation and migration, is involved in the growth of many malignant tumours, and is found in the extracellular capsule of streptococcal bacteria contributing to their virulence (Wessels et al. 1991). However, HA also has a key role in wound repair via tissue regeneration, inflammation response and angiogenesis (Shaharudin and Aziz 2016). As such, it has great potential as a biopolymer for use in biomedical scaffolds. Indeed, crosslinked HA can be employed for the fabrication of 3D scaffolds, also in form of hydrogels, which can be employed in biomedicine for tissue regeneration, for instance for cartilage tissue (Xue et al. 2020; He et al. 2021).

HA was combined with conductive polymers to make conductive composite scaffolds. Yang et al. (2016) reported a HA hydrogel with PPy incorporated into its structure; to prepare it, pyrrole was placed into a HA solution, to form first a hydrogel in which HA was conjugated to pyrrole. Successively the monomer was polymerised, to form PPy; this gave enhanced stability and mechanical properties to the composite hydrogel. In that study, different pyrrole concentrations in solution were tested to optimise the performance of the materials in terms of electrical conductivity, which reached the range $10^{-3}$–$10^{-2}$ S/cm. Cellular tests, performed on 3 T3 fibroblasts, showed improved adhesion in comparison with the unmodified hydrogel.

Conductive polymers in form of NPs were also employed within a HA matrix; either PANI or PPy NPs were studied by Truong et al. (2022). In both cases, a suspension of polymer NPs was added to a HA solution in different concentrations (up to 0.108%); the homogeneous suspensions were then crosslinked and freeze-dried to form solid scaffolds. The addition of the polymers led to a decrease in the average pore size and porosity, while an increase in the elasticity modulus was seen—the latter being higher for the scaffolds made with PPy. Cellular tests performed on fibroblast cell lines showed that the reduced porosity did not affect cellular adhesion; for PPy and PANI highest concentrations, however, decreased cellular activity was observed.

A combination of PPy and CNTs was considered in the study by Jisoo et al. (2017) within a HA matrix modified with catechol; both additives were tested in different concentrations, i.e. 0.5% and 1% for CNTs and 30 and 50 mM for PPy. Results showed that CNT 1% and PPy 30 mM gave the best results in terms of elastic modulus and stability. A significant decrease in the impedance was observed, indicating a synergistic effect between the two components. Cellular studies on neural stem cells showed that the conductive hydrogels were more effective in promoting cellular differentiation; moreover, hydrogels modified with just CNTs gave better results than those modified with PPy, possibly because of the difference in porosity and mechanical behaviour of the two systems. This comparison is interesting, as it shows that the performance of these composite scaffolds may be affected by different properties/parameters, and that several issues have to be considered in their formulation.

An interesting investigation was performed by Rastin et al. (2020b), as they employed a different material as conductive phase, i.e. nanosheets of a MXene. MXenes are a family of 2D transition metal carbide/nitride nanosheets discovered in 2011 (Naguib et al. 2011), analogous to graphene, which have been extensively explored due to their outstanding properties such as high specific surface area, high electrical conductivity, relatively low toxicity and biodegradability, and they can be functionalised to be hydrophilic. They have also been explored for several biomedical applications, such as bioimaging and biosensing, cancer therapies and drug delivery, but their potential in tissue engineering has been mostly overlooked. In this work, an electroconductive cell-laden bioink composed of $Ti_3C_2$ MXene nanosheets dispersed homogeneously within HA and a small amount of alginate (HA/Alg) to form a hydrogel was 3D bioprinted, and its potential for tissue engineering and capability to assemble functional scaffolds to regenerate damaged tissue was explored. The introduction of 1 wt% Alg to the HA hydrogel was necessary to increase the yield stress and complex modulus to give adequate mechanical strength after printing. The inks had excellent rheological properties, allowing the creation of multilayered 3D-structures that maintained their shape with high resolution, but with added electrical conductivity from the MXene nanosheets. The electrical conductivity of the nanocomposite bioink containing 1 and 5 mg/ml MXene nanosheets was 0.55 and 0.72 S/m, respectively, which was significantly higher than that of the pure HA/Alg hydrogel, with enhanced mechanical properties. With encapsulated human embryonic kidney 293 cells (HEK-293) they displayed a high cell viability of >95% in both bulk hydrogel and 3D bioprinted structures.

### 14.2.4 Collagen and Gelatin

Collagen is the most abundant protein in the human body, accounting for 25–35% of the mammalian body's total proteins It is the main structural protein in the extracellular matrix, and is the primary building block of skin, ligaments, tendons, muscles and other connective tissues, as well as occurring in organs, bones, blood vessels and the lining of the intestine. It is responsible for healthy joints and skin elasticity, and can help repair damaged skin, joints and connective tissue. Collagen consists of three alpha (α)-chains of peptides that self-assemble into a left-handed triple helix structure; these individual collagen triple helixes being called tropocollagen. There are two basic forms of collagen: type I collagen is formed of two α1-chains and one α2-chain, while type II collagen is formed of three α1-chains. These triple helix chains are cross-linked at the end telopeptide regions of adjacent collagen helixes, forming larger collagen fibrils that in turn bind together to form collagen fibres (Fig. 14.1).

Gelatin is a protein derived usually from degraded animal collagen (often cows and pigs, which can cause issues with some religions) through heat treatment, although it can also be extracted from fish and any other animal. It is essentially denatured or hydrolysed collagen separated into separate, shorter protein strands, and is commonly used to make drug capsules, cosmetics, ointments, and is an additive in many foods.

Type-I collagen scaffolds made by cryogelation were made conductive with the addition of 0.1, 0.5 and 1 w/v% amino-functionalised graphene before gelation, the amino-functionalised graphene aiding the crosslinking of the collagen (Agarwal et al. 2021) (Fig. 14.3). The nerve conduit scaffold had sufficiently high porosity (88.3–93.6%, 60–200 μm) with interconnectivity between the pores, along with a Young's modulus of 100–347 kPa, but a conductivity of only 0.0029–0.0038 S/cm. Nevertheless, it showed an enhanced expression of CD90 and CD73 genes upon an electrical stimulation of 100 mV/mm with rat bone-marrow derived mesenchymal stem cells (BM-MSCs), which proliferated over between 3–5 days. Electrical stimulus during the neuronal differentiation stage also increased the gene expression of MAP-2 kinase and β-tubulin III, markers of neuronal lineage. Furthermore, an increased secretion of ATP (an energy source or cofactor for cells) from the BM-MSCs was also observed upon an electrical stimulation of 1 V for 5 min, and this could help in the regeneration of neurons and immuno-modulation. Upon clinical transplantation of 0.5% graphene collagen cryogels into the injured spinal cord of an adult rat, it supported cellular growth and neuronal cellular migration, and had potential for suppressing the occurrence of neuro-inflammation (Fig. 14.3), a major barrier in neural tissue engineering. These results show that such conducting

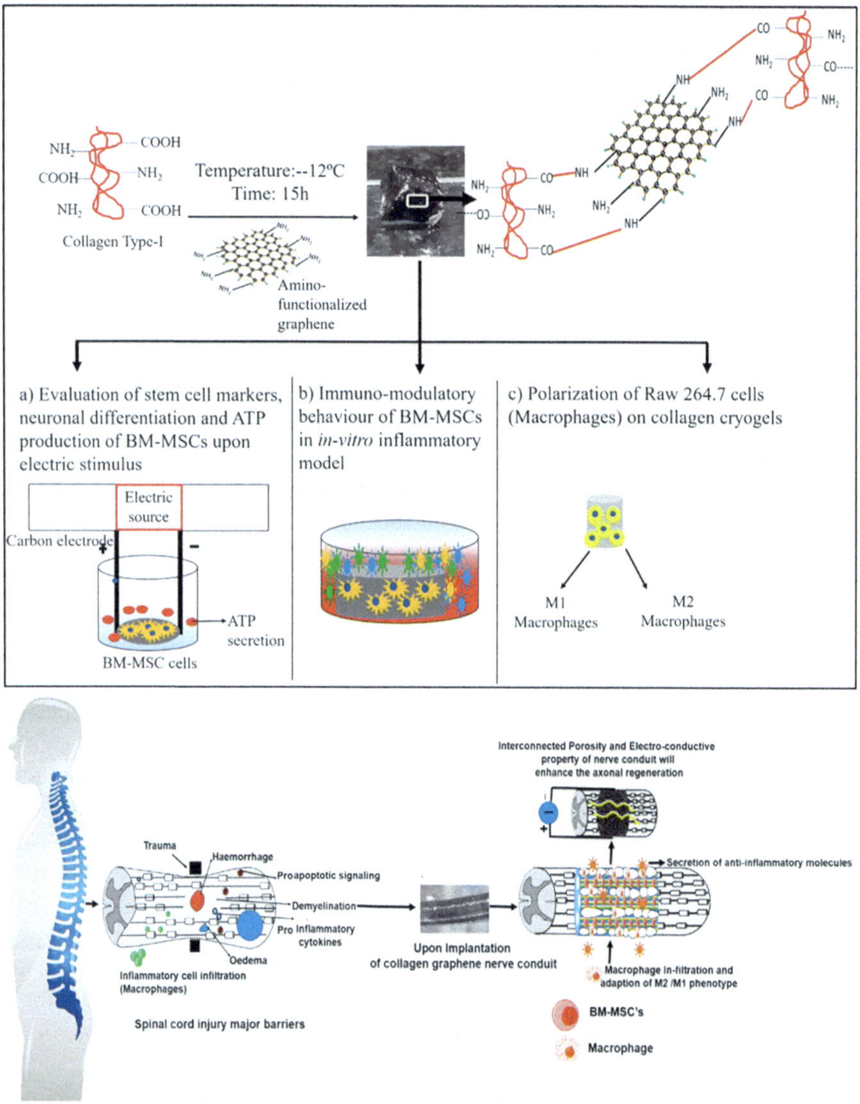


**Fig. 14.3** Top: Fabrication of collagen-graphene composite cryogels with crosslinking by the amino-functionalised graphene, with (**a**) electrical stimulation for the proliferation of BM-MSCs, (**b**) reduction of neuro-inflammation, and (**c**) evaluation of macrophages on collagen cryogels. Bottom: Illustration of the effects of conductive collagen-graphene cryogels on spinal cord injury, reducing neuro-inflammation, promoting BM-MSC proliferation and secreting anti-inflammatory molecules. (Reprinted from Agarwal et al. (2021), copyright © 2020 Elsevier B.V, used with permission from Elsevier)

scaffolds under electrical stimulation may aid the establishment of neural connections, promote cell migration and proliferation which is key for spinal cord regeneration, and also increase the immune-modulatory secretions that may prevent cell inflammation.

A hydrogel of Type-I collagen with an unstated amount of alginate added for crosslinking was loaded with 0.26 and 0.52 wt% PEDOT:PSS. The hydrogel itself had an ionic conductivity of 0.12 S/m, that increased to 0.27 and 0.35 S/m in gels containing 0.26 wt% and 0.52 wt% PEDOT:PSS, respectively. There was no significant change in cytotoxicity in 2D gels, but 0.52 wt% PEDOT:PSS showed some evidence of toxicity in 3D structures. The 0.26 wt% conductive hydrogels, on the other hand, improved the maturation and beating of cardiomyocytes–derived from stem cells, with enhanced beating frequency, along with an increased amplitude and speed of contraction. This could enable the fabrication of engineered cardiac tissues that beat and maturate on their own without external mechanical or electrical stimulations, leading to improvements in tissue engineering approaches to treat heart failure, arrhythmia and possibly diseases of other electrically sensitive tissues (Roshanbinfar et al. 2018).

Gelatin needs to be thermally bioprinted to cross link, and this temperature-induced gelation is often slow and unstable. A way to avoid this is to modify gelatin with photopolymerisable methacryloyl groups via covalent cross-linking under UV or visible light. This functionalised gelatin methacryloyl (GelMA) is a promising bioink as the cross-linking density can be easily controlled (Gungor-Ozkerim et al. 2018).

A biocompatible conductive GelMA hydrogel (using gelatin from cold water fish skin) with added 0.1–0.3 wt/v% PEDOT:PSS was made via UV photopolymerisation, with mechanical, degradation and swelling properties, as well as conductivity, being tuneable by changing the PEDOT:PSS concentration (Spencer et al. 2018). The impedance of the hydrogels decreased from 449.0 kOhm for pure GelMA to 281.2 and 261.0 kOhm for hydrogels containing 0.1% and 0.3% PEDOT:PSS at a frequency of 1 Hz, and *ex vivo* experiments demonstrated that the threshold voltage necessary to stimulate contraction in explanted abdominal tissue connected by the composite hydrogels decreased from 9.3 V for GelMA to 6.7 and 4.0 V for the two PEDOT:PSS hydrogels. *In vitro* studies showed that composite hydrogels with 0.1% PEDOT:PSS were biocompatible, with viability and spreading of C2C12 myoblasts comparable to pure GelMA controls. These GelMA PEDOT:PSS hydrogels were then bioprinted to form complex 3D cell-laden structures (Spencer et al. 2019). They were formed by an initial cross-linking step of the PEDOT:PSS with $Ca^{2+}$ ions and a subsequent photopolymerisation step using visible light to cross-link the GelMA hydrogels, enabling the tuning of mechanical properties with Young's modulus of ~40–150 kPa, as well as enabling conductivity to be tuned by varying the PEDOT:PSS concentration. The hydrogels degraded *in vivo* with no significant inflammatory response, and high cytocompatibility and cell spreading were demonstrated in both wet-spinning and 3D bioprinting of cell-laden hydrogels.

Interestingly, in a study GelMA 3D fibre scaffolds (porcine skin gelatin, UV crosslinking) laden with C2C12 myoblast cells *without* any electrically conductive additives were exposed to electrical stimulation during bioprinting, inducing cell alignment and differentiation (Yang et al. 2021). Cell viability, cell proliferation, cell guidance and myogenic differentiation in GelMA were investigated in response to applied electric fields. The microfibrous cells were randomly oriented when present at low and high densities with various electric field conditions, but cells at medium density were highly aligned with an electric field >0.4 kV/cm for >60 s. The study found that the most appropriate were an electric field strength of 0.8 kV/mm applied for 120 s with a medium cell density of $15 \times 10^6$ cells/mL. Three different cell densities were investigated ($7 \times 10^6$, $15 \times 10^6$ and $30 \times 10^6$ cells/mL) and a low and medium density relatively high cell alignment and reasonable cell viability (<80%) were obtained. However, low cell viability (<75%) was observed at high cell density, and in particular, an electric field of 1.2 kV/cm along with 120 s stimulation time seemed to result in significantly high cell death. In addition, 4D bioprinting of the gelatin hydrogel was used to print a cell-laden fibrous bundle, designed to mimic the structure of skeletal muscle (perimysium). The 4D printing technique employs a time element as the fourth dimension to vary the structures produced, to mimic those in the human body. By varying the crosslinking times the degree of swelling and, hence, the structure produced could be varies and controlled. 30 and 60 min gave perfectly rolled, tubular structures, and structures were produced mimicking that of skeletal muscle tissue such as deltoid and biceps brachii.

### 14.2.5 Alginate

Alginates are polysaccharide biopolymers derived from algae and seaweed, and are a series of linear, unbranched polysaccharides consisting of two groups: b-(1–4)-D-mannuronic acid (M) and a-(1–4)-L-guluronic acid (G). This is not a random copolymer, but contains regions of alternating G and M, the G:M ratio depending on the seaweed source, and varying from 30% to 70% G (Fig. 14.1). It contains flattened 'sheet-like' sections and buckled chains, and they are anionic polysaccharides, so they can bind well to metal cations, and can also form 3D gel networks with metallic NPs and with metal cations. As well as its high biocompatibility and stiffness, alginate can be used to model neural tissue in 3D bioprinting (Fantini et al. 2019), implant stem cells (Chung et al. 2014) and can stimulate the metabolism for regenerative medicine (Klontzas et al. 2019).

In some sections above, the use of alginate was reported within other matrices to favour the cross-linking process of the scaffolds. However, alginate, and another related algae-derived polysaccharide k-carrageenan, has also been employed as the main component of conductive scaffolds, as described below.

Alginate (Alg), alginate-graphene oxide (Alg-GO) composite microspheres were made with calcium ion cross-linking, as an alternative bone graft substitute (Devi et al. 2021). The GO was 300–800 nm × 0.7–1.2 nm, and 3.3 wt% was added to the alginate. Moreover, composites with dexamethasone (Dex) were also prepared; this is a corticosteroid used to treat many illnesses including rheumatic problems, skin diseases, severe allergies, asthma, and inflammation. 1.5 mm microspheres had a

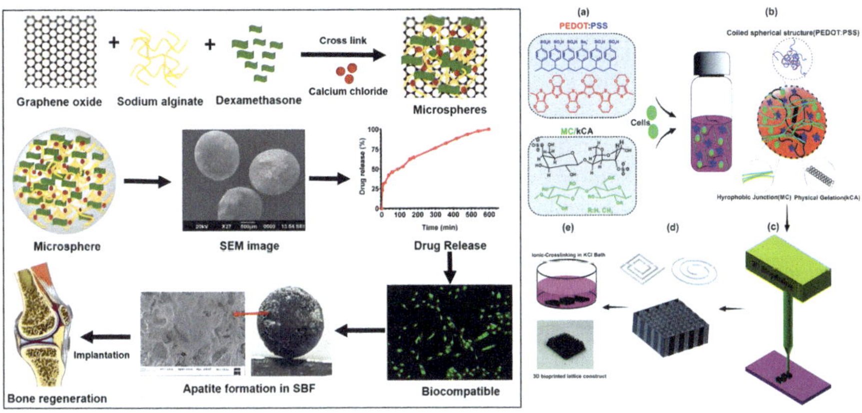


**Fig. 14.4** Left: Conductive alginate-graphene oxide-dexamethasone (Alg-GO-Dex) composite microspheres, Dex drug release, and their biocompatibility and bioactivity. (Reprinted from Devi et al. (2021), copyright © 2021 Elsevier B.V, used with permission from Elsevier). Right: Conductive k-Carrageenan-methyl cellulose PEDOT:PSS bioink, cell inclusion and 3D printed scaffolds. (**a**) PEDOT, PSS, k-Carrageenan, methyl cellulose molecules; (**b**) colloidal structures combined with cells; (**c**) 3D printing; (**d**) various scaffold geometries are possible; and (**e**) final ionic crosslinking carried out in KCl bath to provide a rigid scaffold. (Reprinted from Rastin et al. (2020a), copyright © 2020 Royal Society of Chemistry, used with permission from Royal Society of Chemistry)

porosity of 84–92%, and GO was dispersed uniformly in the alginate matrix (Fig. 14.4). *In vitro* biomineralisation in SBF showed excellent hydroxyapatite formation on their surface over 30 days, with Alg-GO-Dex being better than Alg-GO, and *in vitro* studies with MG-63 osteoblast cells revealed good biocompatibility in all cases, but also increased cell proliferation with Alg-GO-Dex. The combination of added conductive GO and Dex improved the biological function of the alginate.

k-Carrageenan (κCA) is a negatively charged sulfonated polysaccharide obtained from red algae, also consisting of sheet and chain sections, and it is widely used as a thickening agent in the pharmaceutical and food industries, exhibiting a stiff gel behaviour after gelation with $K^+$ ions. As such, it has sufficient mechanical strength to support cells, and its rheology is suitable for 3D bioprinting applications (Liu et al. 2015). k-carrageenan (kCA) was combined with methyl cellulose to prepare MC/κCA hydrogels with added PEDOT:PSS to make them conductive (Rastin et al. 2020a). The thixotropy of the printing ink could be tuned by changing MC: κCA ratios. 4–8 wt% MC and 0.1 or 0.3 wt% PEDOT:PSS were added. MC-8/κCA already had a low conductivity of 0.0012 S/cm due to the ionic nature of κCA hydrogels, and this increased to 0.002 and 0.003 S/cm with 0.1 and 0.3 wt% PEDOT:PSS. Encapsulated human embryonic kidney 293 (HEK-293) cells in bulk hydrogels and 3D bioprinted structures with 0.1 wt% PDOT:PSS maintained a high cell viability of >96% over a week. Furthermore, to demonstrate the capability of the MC/κCA ink, multilayered 3D lattice, hexagonal and cylindrical structures were printed on the cm scale without the need for a secondary support bath, showing robust fabrication of physiological-scale structures by the 3D printing process.

A 3D printed "bionic ear" has also been produced using an alginate hydrogel with conductive Ag NPs. A chondrocyte (cartilage cell) seeded alginate hydrogel matrix was 3D printed with an electrically conductive Ag NP infused inductive coil antenna, connecting to cochlea-shaped electrodes supported on silicone (Mannoor et al. 2013). This resulted in the 3D integration of functional electronic components which had the precise anatomy and complex geometry of the human ear. The alginate hydrogel matrix had viable chondrocytes added at a density of ~$6 \times 10^{+7}$ cells/mL, and it was cross-linked and 3D printed along with a non-conducting silicone solution containing conducting Ag NPs; this allowed the fabrication of the biological, electronic and structural components of the bionic organ in a single process. The printed ear was immersed in chondrocyte culture media, maintaining good structural integrity and with shape retention under culture, and maintained its form after 10 weeks of *in vitro* culture. It had a cell viability of 91% with homogeneous chondrocyte distribution, and chondrocytes were alive and metabolically active after 10 weeks. The resistivity of the printed coil was found to be $1.31 \times 10^{-6}$ Ωm, which is only 2 orders of magnitude higher than pure silver, and it was capable of receiving electromagnetic signals over an extended frequency range from Hz to GHz—human hearing only operates in the 20 HZ to 20 kHz range.

### 14.2.6 Silk

Silk is a natural fibre made of proteins, 70% being fibroin that forms the filaments of the structural centre of the silk, and 30% sericin that is the sticky material that surrounds it. Fibroin consists of the sequence of long-chain amino acids Gly-Ser-Gly-Ala-Gly-Ala, and hydrogen bonds between N and H atoms form between fibroin chains, giving silk its high tensile strength. The fibroin consists of two phases, heavy chains containing aligned crystalline b-sheets and connecting light chains connected by a disulfide bond, as well as a glycoprotein. These link by further hydrogen bonds to form fibroin microfibrils, which in turn bundle together to form silk filaments, and two of these are combined side by side and covered by the adhesive and hydrophilic protein sericin to form a single silk fibre 10–13 mm in diameter (Fig. 14.1). Silk is produced by several insects, but generally only that of moth caterpillars has been used for textile manufacturing, especially the cocoons of the larvae of the mulberry silkworm *Bombyx mori*, produced by one of the worm's salivary glands during spinning.

Silk fabric has been used for at least 5500 years; moreover, silk has also been employed for centuries as a suture material in medicine, as is biocompatible, highly elastic and has a slow degradation rate, giving support for cells during tissue regeneration (Gungor-Ozkerim et al. 2018). More recently, silk has also been used for bioprinting and to make scaffolds. Silk was used to produce non-conductive small-diameter fibres which were shown to be favourable for the maturation of astrocytes (Qu et al. 2013). Astrocytes are glial cells, and they contribute the majority of cells that make up the human central nervous system. Performing metabolic, structural, homeostatic and neuroprotective tasks, clearing excess neurotransmitters, stabilising/regulating the blood-brain barrier and promoting synapse formation in the brain.

Silk-based fibres are a suitable basis for novel conductive fibre scaffolds, and if electrospun from fibroin can be much smaller in diameter than natural silk fibres. However, in the majority of cases, silk fibroin has been used as only an additive to organic polymers in relatively small quantities to add spinnability; as mentioned above, these scaffolds are outside the scope of this work and they will not be discussed. There are, however few studies in which silk fibroin is the major polymer component to make conductive biomaterials.

Aznar-Cervantes et al. (2012) enhanced the electrical conductivity of electrospun silk fibre (2.3 mm average diameter) by coating a layer of PPy through the chemical oxidation polymerisation method (coated fibres average diameter 2.6 mm). On examination, it was found that a large quantity if PPy was coated, between 46–53 wt%. Mechanical properties were improved, possibly due to the greater diameter, and a mesh of the silk-PPy could support the flow of a large range of anodic and cathodic currents up to ±8 mA at 10 mV/s in a cell proliferation bath, with greater electroactivity permitting anion storage and delivery during oxidation/reduction reactions in aqueous solutions. However, after 21 days the growth of adult human mesenchymal stem cells (ahMSCs, undifferentiated from bone marrow) and primary human skin fibroblast (hFb) cells on the silk-PPY was worse than uncoated silk; these data indicate that the PPy coating reduced the bioactivity, adhesion and proliferation of cells in comparison to the fibroin itself, which is highly bioactive.

Conductive silk fibre foams were fabricated using a salt leaching method with an interpenetrating network of pyrrole and 2-hydroxy-5-sulfonic aniline for bone tissue engineering, for electrical stimulation of stem cells for differentiation into different cell types (Hardy et al. 2015). 2.5 mmol of each conducting polymer were added to silk fibroin foam scaffolds 4 mm in height and diameter, and porosity was moderately reduced due to the interpenetrating network of the conducting polymers within the hydrogel-like matrix of cross-linked silk proteins, but with no significant differences in mechanical properties. The conductivity was $6.1 \times 10^{-4}$ S/ cm, stated to be of a similar order of magnitude to mammalian tissues, typically $\geq 10^{-4}$ S/cm. Cell viability tests showed no cytotoxicity, but cell viability decreased when cells were exposed to high concentrations of 0.2 mg/mL of the conductive silk fibres. Bone marrow-derived HMSCs adhered to both conductive and non-conductive silk scaffolds with and without electrical stimulation, and remained active in osteogenic medium for up to 30 days. Differentiation studies showed markers of osteogenicity were increased on conductive foams with or without electrical stimulation compared with non-conductive silk scaffolds. Although the non-conductive silk scaffolds supported differentiation of HMSCs to osteoblasts, electrical stimulation of HMSCs on the conductive silk scaffolds gave enhanced differentiation.

Unmodified silk fibroin was extracted and coated with PANI and PPy through *in situ* polymerisation, the fibre diameter growing from 29.5 μm in pure silk to 45.6 μm in PPY/SF, and 34.0 μm in PANI/SF, the surface of the conducting polymer-coated fibres also having increased surface roughness (Darshan et al. 2017). Electrical conductivity showed that the pure silk fibroin was non-conductive at $1 \times 10^{-11}$ S/cm, but the PPy and PANI coated fibres were $2.2 \times 10^{-5}$ and $1.6 \times 10^{-4}$ S/cm, respectively. The coated silk fibres were cytocompatible, and adhesion of human keratinocyte (HaCaT) epidermal cells on the conducting silk fibres was promoted compared to pure silk, especially for PANI/SF.

## 14.3 Conclusion and Future Perspectives

Many studies have focused on developing electrically conductive nanofibrous scaffolds through incorporation of conductive materials into polymeric nanofibers; as we have seen, however, there have been relatively few so far focussing on natural biopolymer-based fibre materials. The results described above clearly indicate the potential of natural polymers in this field; this is, therefore, a topic deserving more attention.

In addition to the higher sustainability in their preparation, already mentioned above, another important factor in the use of natural biopolymers is their inherent biodegradability, an increasingly important feature. Biopolymers such as chitosan, cellulose, alginate, collagen, etc. are easily biodegradable by enzymatic and/or hydrolytic reactions, despite their stability during use as biomaterials.

It is important to highlight that, although several biopolymers have been used as main component for the scaffolds, the conductive additive has been always a synthetic material; the only exception to this is the work performed by Scalera et al. (2021), where carbon derived from cork was employed. This is a research field which needs to be investigated more; indeed, carbon of natural origin should be preferred to graphene or CNTs as their synthesis could require toxic chemicals and have a relatively high impact on the environment. Moreover, contrasting reports have been published regarding the toxicity of both CNTs and graphene; it is known that the cytotoxicity of CNTs can cause disruption of intracellular metabolic pathways, membrane damage and they can also generate ROS, while the cytotoxic effects of graphene and its derivatives are not yet well established, and depends on concentration, size, charge and functional groups on the graphene surface. Therefore, their substitution with natural and potentially less toxic alternatives should be investigated. However, Shin et al. (2016) have shown that using deoxyribonucleic acid (DNA) as a natural surfactant to disperse CNTs in various biomaterials such as GelMA and HA greatly reduced their cytotoxicity.

So-called 4D printing employing conductive biopolymers is another avenue that requires more exploration. Smart materials with the ability to change their shape in response to external stimuli, such as stress/pressure, pH, and magnetic or electrical fields, have been studied in the field of tissue engineering for some time, and fabricated as 3D structures with the ability to mimic natural tissues. If the stimulus-responsive properties of such 3D smart materials can also transform their geometry over time, this is then known as 4D printing (where time is the fourth dimension). This goes beyond 3D printing, as 4D printing allows the real-time full mimicry of the dynamics of natural tissues, while 3D printed structures are static. In this case, the idea is to use the conductive properties of the scaffold material both during and after the scaffold fabrication process, as 4D printed structures can change their morphology after exposure to external stimuli, such as an electrical field. This can produce complex architectures, and enhance the structural and biological functionality of the printed scaffold (Gladman et al. 2016). An example using sustainable biopolymers is the methacrylated gelatin conductive scaffold made by Yang et al. (2021) that mimics the natural structure of skeletal muscle tissues, but there is much more future potential in this area exploiting sustainable natural biopolymers.

As well as requiring a rheology suitable for fibre production, usually via electrospinning, it has also been shown that fibre diameter is a key aspect that can regulate and direct cell behaviour. For example, the nanofibrous structure of myocardium extracellular matrix has a diameter in the order of 500 nm, and added conductivity would promote electrical propagation and functional coupling of cardiac muscle cells (cardiomyocytes), which are electroactive cells and can spontaneously contract under electrical stimulation (Noorman et al. 2009). Such conductive fibrous scaffolds have been developed and demonstrated on cardiac cells using non-natural or sustainable polymer fibres (Wang et al. 2017). Therefore, it would be interesting to study some of the conductive biopolymer nanofibres discussed above for such applications. Although natural silk fibres at ~10 mm are an order of magnitude larger than this, electrospun fibres from fibroin can be submicron in diameter.

Flexible and stretchable electronics have generated great interest for biomedical applications such as implantable medical devices and smart sutures, epidermal sensors, soft contact lenses, neuron-to-machine interfaces, artificial electronic muscles and artificial skin (Shin et al. 2016). Therefore, the 2D and 3D printing of sustainable conductive biopolymers on such flexible substrates will be a growing area of research in the future. Furthermore, as shown by the "bionic ear" discussed above, the concept of 3D printing living cells together with electronic components and growing them into functional organs has already arrived, and represents a new direction in merging electronics with biological systems (Mannoor et al. 2013). This "bionic ear" even had enhanced auditory sensing for radio frequency reception, extending beyond the normal human range.

Lastly, the potential for 3D printable structures based on sustainable conductive materials goes beyond just biomedical applications and biolelectronics, with fields such as "soft robotics" (Li et al. 2023), flexible electronics and wearable conducting fabrics (Shin et al. 2016), sensors and energy harvesting devices being of great potential interest, and commercial value, in many sectors and emerging technologies.

**Acknowledgements** Clara Piccirillo would like to acknowledge the Italian Ministry of Research (MUR) under the complementary actions to the NRRP (PNC0000007) "Fit4MedRob- Fit for Medical Robotics" Grant (contract number CUP B53C22006960001).

## References


Adedi A, Hasanzadeh M, Tayebi L (2019) Conductive nanofibrous Chitosan/PEDOT:PSS tissue engineering scaffolds. Mater Chem Phys 237:121882

Agarwal A, Kumar N, Srivastava A (2021) Highly elastic, electroconductive, immunomodulatory graphene crosslinked collagen cryogel for spinal cord regeneration. Mater Sci Eng C 118:111518

Aznar-Cervantes S, Roca MI, Martinez JG et al (2012) Fabrication of conductive electrospun silk fibroin scaffolds by coating with polypyrrole for biomedical applications. Bioelectrochemistry 85:36–43

Baei P, Jalili-Firoozinezhad S, Rajabi-Zeleti S et al (2016) Electrically conductive gold nanoparticle-chitosan thermosensitive hydrogels for cardiac tissue engineering. Mater Sci Eng C 63:131–141

Batool F, Kurnianwan TA, Mohyddin A et al (2024) Environmental impacts of food waste management technologies: a critical review of life cycle assessment (LCA) studies. Trends Food Sci Tech 143:104287

Bordoni M, Karabulut E, Kuzmenko V et al (2020) 3D Printed conductive nanocellulose scaffolds for the differentiation of human neuroblastoma cells. Cells 9:682

Cameron J, Skabara PJ (2020) The damaging effects of the acidity in PEDOT:PSS on semiconductor device performance and solutions based on non-acidic alternatives. Mater Horiz 7:1759–1772

Chung JY, Song M, Ha CW et al (2014) Comparison of articular cartilage repair with different hydrogel-human umbilical cord blood-derived mesenchymal stem cell composites in a rat model. Stem Cell Res Ther 5:39

Darshan GH, Kong D, Gautrot J et al (2017) Fabrication and characterization of conductive conjugated polymer-coated Antheraea mylitta silk fibroin fibers for biomedical applications. Macromol Biosci 17:1600443

Devi YGV, Prabhu A, Anil S et al (2021) Preparation and characterization of dexamethasone loaded sodium alginate-graphene oxide microspheres for bone tissue engineering. J Drug Deliv Sci Technol 64:102624

Dixton DT, Gormillon CT (2022) Conductive scaffolds for bone tissue engineering: current state and future outlook. J Funct Biomater 13:1

Fantini V, Bordoni M, Scocozza F (2019) Bioink composition and printing parameters for 3D modeling neural tissue. Cells 8:830

Gholizadeh S, Moztarzadeh F, Haghighipour N et al (2017) Preparation and characterization of novel funtionalized multiwalled carbon nanotubes/chitosan/b-glycerophosphate scaffolds for bone tissue engineering. Int J Biol Macromol 97:365–372

Gladman AS, Matsumoto EA, Nuzzo RG et al (2016) Biomimetic 4D printing. Nat Mater 15:413–418

Gungor-Ozkerim PS, Inci I, Zhang YS et al (2018) Bioinks for 3D bioprinting: an overview. Biomater Sci 6:915–946

Guo R, Li J, Chen C et al (2021) Biomimetic 3D bacterial cellulose-graphene foam hybrid scaffold regulates neural stem cell proliferation and differentiation. Colloids Surf B 200:111590

Hardy JG, Geissler SA, Aguilar D et al (2015) Instructive conductive 3D silk foam-based bone tissue scaffolds enable electrical stimulation of stem cells for enhanced osteogenic differentiation. Macromol Biosci 15:1490–1496

He T, Li B, Colombani T et al (2021) Hyaluronic acid-based shape-memeory cryogel scaffolds for focal cartilage defect repair. Tissue Eng A 27:748–760

Huang J, Hu X, Lu L et al (2010) Electrical regulation of Schwann cells using conductive polypyrrole/chitosan polymers. J Biomed Mater Res 93:164–174

Jiang L, Chen D, Wang Z et al (2019) Preparation of electrical conductive graphene oxide / chitosan scaffold for cardiac tissue engineering. Appl Biochem Biotech 188:952–964

Jisoo S, Choi EJ, Cho JH et al (2017) Three-dimensional electroconductive hyaluronic acid hydrogels incorporated with carbon nanotubes and polypyrrole by catechol-mediated dispersion enhance neurogenesis of human neural stem cells. Biomacromolecules 18:3060–3072

Józwiack T, Filipowska U, Szymczyk P et al (2017) Effect of ionic and covalent crosslinking agents on properties of chitosan beads and sorption effectiveness of Reactive Black 5 dye. React Funct Polym 114:58–74

Kanaan AF, Piedade AP (2022) Electro-responsive polymer-based platforms for electrostimulation of cells. Mater Adv 3:2337–2353

Kashi M, Baghbani F, Moztarzadeh F et al (2018) Green synthesis of degradable conductive thermosensitive oligopyrrole/chitosan hydrogel intended for cartilage tissue engineering. Int J Biol Macromol 107:1567–1575

Klontzas ME, Reakasame S, Silva R et al (2019) Oxidized alginate hydrogels with the GHK peptide enhance cord blood mesenchymal stem cell osteogenesis: a paradigm for metabolomics-based evaluation of biomaterial design. Acta Biomater 88:224–240

Kuzmenko V, Kalogeropoulos T, Thunberg J et al (2016) Enhanced growth of neural networks on conductive cellulose-derived nanofibrous scaffolds. Mater Sci Eng C 58:14–23

Lahiri D, Nag M, Dutta B et al (2021) Bacterial cellulose: production, characterization, and application as microbial agent. Int J Mol Sci 22:12984

Levengood SL, Zhang M (2014) Chitosan-based scaffolds for bone tissue engineering. J Mater Chem B 2:3161–3184

Li J, Lu B, Gu G (2023) 3D-printed PEDOT:PSS for soft robotics. Nat Rev Mater 8:604–622

Liang X, Qu B, Li J et al (2015) Preparation of cellulose-based conductive hydrogels with ionic liquid. React Funct Polym 86:1–6

Liu S, Chan WL, Li L (2015) Rheological properties and scaling laws of κ-carrageenan in aqueous solution. Macromolecules 48:7649–7657

Loh QL, Choong C (2013) Three-dimensional scaffolds for tissue engineering applications: role of porosity and pore size. Tissue Eng B 19:485–502

Mannoor MS, Jiang Z, James T et al (2013) 3D printed bionic ears. Nano Lett 13:2634–2639

Mantione D, Del Agua I, Sanchez-Sanchez A (2017) Poly(3,4-ethylenedioxythiophene) (PEDOT) derivatives: innovative conductive polymers for bioelectronics. Polymers 9:354

Martins AM, Eng G, Caridade SG et al (2014) Electrically conductive chitosan/carbon scaffolds for cardiac tissue engineering. Biomacromolecules 15:635–643

Mehrotra S, Dey S, Sachdeva K et al (2023) Recent advances in tailoring stimuli-responsive hybrid scaffolds for cardiac tissue engineering and allied applications. J Mater Chem B 11:10297–10331

Morello G, Quarta A, Gaballo A et al (2021) A thermo-sensitive chitosan/pectin hydrogel for long-term tumor spheroid culture. Carbohydr Polym 274:118633

Mueller D, Silva JP, Rambo CR et al (2013) Neuronal cells behavior on polypyrrole coated bacterial nanocellulose three-dimensional (3D) scaffolds. J Biomater Sci Polym Ed 24:1368–1377

Naguib M, Kurtoglu M, Presser V et al (2011) Two-dimensional nanocrystals produced by exfoliation of Ti3AlC2. Adv Mater 23:4248–4253

Nikolova MP, Chavali MS (2019) Recent advances in biomaterials for 3D scaffolds: a review. Bioact Mater 4:271–292

Noorman M, van der Heyden MA, van Veen TA et al (2009) Cardiac cell–cell junctions in health and disease: electrical versus mechanical coupling. J Mol Cell Cardiol 47:23–31

Oladzadabbasabadi N, Nafchi AM, Ariffin F (2022) Recent advances in extraction, modification, and application of chitosan in packaging industry. Carbohydr Polym 277:118876

Pourjavadi A, Doroudian M, Ahadpour A et al (2019) Injectable chitosan/k-carrageenan hydrogel designed with au nanoparticles: a conductive scaffold for tissue engineering demands. Int J Biol Macromol 126:310–317

Qu J, Wang D, Wang H et al (2013) Electrospun silk fibroin nanofibers in different diameters support neurite outgrowth and promote astrocyte migration. J Biomed Mater Res A 101:2667–2678

Radwan-Pragłowska J, Janus Ł, Piątkowski M et al (2020) 3D Hierarchical, nanostructured chitosan/PLA/HA scaffolds doped with $TiO_2$/Au/Pt NPs with tunable properties for guided bone tissue engineering. Polymers 12:792

Rastin H, Zhang B, Bi J, Hassan K, Tung TT, Losic D (2020a) 3D printing of cell-laden electroconductive bioinks for tissue engineering applications. J Mater Chem B 8:5862–5876

Rastin H, Zhang B, Mazinani A et al (2020b) 3D bioprinting of cell-laden electroconductive MXene nanocomposite bioinks. Nanoscale 12:16069–16080

Roshanbinfar K, Vogt L, Greber B et al (2018) Electroconductive biohybrid hydrogel for enhanced maturation and beating properties of engineered cardiac tissues. Adv Mater 28:1803951

Sachar S, Saxena RK (2011) Cytotoxic effect of poly-dispersed single walled carbon nanotubes on erythrocytes in vitro and in vivo. PLoS One 6:e22032

Scalera F, Monteduro AG, Maruccio G et al (2021) Sustainable chitosan-based electrical responsive scaffolds for tissue engineering applications. Sustain Mater Technol 28:e00260

Schöbel L, Boccaccini AR (2023) A review of glycosaminoglycan-modified electrically conductive polymers for biomedical applications. Acta Biomater 169:45–65

Shaharudin A, Aziz Z (2016) Effectiveness of hyaluronic acid and its derivatives on chronic wounds: a systematic review. J Wound Care 25:585–592

Shin SR, Farzad R, Tamayol A et al (2016) A bioactive carbon nanotube-based ink for printing 2D and 3D flexible electronics. Adv Mater 28:3280–3289

Singh N, Chen J, Koziol KK et al (2016) Chitin and carbon nanotube composites as biocompatible scaffolds for neuron growth. Nanoscale 8:8288–8299

Spencer AR, Primbetova A, Koppes AN et al (2018) Electroconductive gelatin methacryloyl-PEDOT:PSS composite hydrogels: design, synthesis, and properties. ACS Biomater Sci Eng 4:1558–1567

Spencer AR, Sani ES, Soucy JR et al (2019) Bioprinting of a cell-laden conductive hydrogel composite. ACS Appl Mater Interfaces 11:30518–30533

Srinivasan SY, Cler M, Zapata-Arteaga O et al (2023) Conductive bacterial nanocellulose-polypyrrole patches promote cardiomyocyte differentiation. ACS Appl Bio Mater 6:2860–2874

Stinstra JG, Hopenfeld B, MacLeod RS (2005) On the passive cardiac conductivity. Ann Biomed Eng 33:1743–1751

Truong TH, Musilová L, Kaspárová V et al (2022) New approach to prepare cytocompatible 3D scaffolds via the combination of sodium hyaluronate and colloidal particles of conductive polymers. Sci Rep 12:8065

Turnbull G, Clarke J, Picard F et al (2018) 3D bioactive composite scaffolds for bone tissue engineering. Bioact Mater 3:278–314

Wang L, Wu Y, Hu T et al (2017) Electrospun conductive nanofibrous scaffolds for engineering cardiac tissue and 3D bioactuators. Acta Biomater 59:68–81

Wang L, Hu S, Ullah MW et al (2020) Enhanced cell proliferation by electrical stimulation based on electroactive regenerated bacterial cellulose hydrogels. Carbohydr Polym 249:116829

Wessels MR, Moses AE, Goldberg JB et al (1991) Hyaluronic acid capsule is a virulence factor for mucoid group A streptococci. Proc Nat Acad Sci 88:8317–8321

Xue Y, Chen H, Xu C et al (2020) Synthesis of hyaluronic acid hydrogels by cross-linkng the mixture of high-molecular-weight hyaluronic acid and low-molecular-weight hyaluronic acid with 1,4-butanediol diglycidyl ether. RSC Adv 10:7206–7213

Yang J, Choe G, Yang S et al (2016) Polypyrrole-incorporated conductive hyaluronic acid hydrogels. Biomater Res 20:31

Yang GH, Kim W, Kim J et al (2021) A skeleton muscle model using GelMA-based cell-aligned bioink processed with an electric-field assisted 3D/4D bioprinting. Theranostics 11:48–63

Yi H, Patel R, Patel KD et al (2023) Conducting polymer-based scaffolds for neuronal tissue engineering. J Mater Chem B 11:11006–11023

# Biofabrication of Hierarchical Scaffolds for Tissue Engineering and *In Vitro* Models

# 15

César R. Casanova, Marta R. Casanova, Rui L. Reis, and J. Miguel Oliveira


**Abstract**

The field of tissue engineering, regenerative medicine, and *in vitro* models is undergoing transformative advancements, driven by the urgent need for structures that mimic the complex architecture and functionality of native tissues. Hierarchical scaffolds, characterized by macro-, micro-, and nanoscale features, have emerged as pivotal tools for efficiently regenerate different tissues/organs. These scaffolds can replicate the extracellular matrix with remarkable precision, offering structural support, enhanced cell alignment, and biocompatibility essential for effective tissue regeneration and *in vitro* models. Herein, the current advances in the biofabrication of hierarchical scaffolds, emphasizing the integration of advanced techniques such as additive manufacturing, electrospinning, freeze-drying, sol-gel, and self-assembly, are discussed. These methods enable the precise design of scaffolds tailored for diverse applications, from bone regeneration to disease modeling. The discussion extends to material innovation and selection criteria, balancing natural and synthetic polymers for optimized biocompatibility, mechanical stability, and controlled degradation. Additionally, sustainability is highlighted, showcasing renewable materials, energy-efficient processes, and circular economy principles that align with environmental responsibility. Hierarchical scaffolds offer transformative potential in regenerative


César R. Casanova and Marta R. Casanova are co-first authors and contributed equally to this work.


C. R. Casanova (✉) · M. R. Casanova · R. L. Reis · J. M. Oliveira (✉)
3B's Research Group, I3Bs – Research Institute on Biomaterials, Biodegradables and Biomimetics, University of Minho, Headquarters of the European Institute of Excellence on Tissue Engineering and Regenerative Medicine, AvePark, Parque de Ciência e Tecnologia, Rua Ave 1, Edifício 1 (Sede), Barco, Guimarães, Portugal

ICVS/3B's – PT Government Associate Laboratory, Braga/Guimarães, Portugal
e-mail: cesar.casanova@i3bs.uminho.pt; miguel.oliveira@i3bs.uminho.pt





J. M. Oliveira et al. (eds.), *Sustainable Scaffolds-based Strategies in Tissue Engineering and Regenerative Medicine*, Biomaterials, Bioengineering and Sustainability 8, https://doi.org/10.1007/978-3-031-96274-5_15

medicine and *in vitro* three-dimensional (3D) models, overcoming challenges like vascularization and scalability. By harmonizing biofabrication, functionality, and sustainability, they pave the way for next-generation diagnostics and advanced *in vitro* 3D testing platforms to revolutionize healthcare and biomedical research.

**Keywords**

Biofabrication · Hierarchical scaffolds · *In vitro* models · Sustainability · Tissue engineering

## 15.1 Introduction

The field of tissue engineering and regenerative medicine has experienced remarkable advancements over the past few decades, driven by the urgent need to develop advanced products and therapies that can restore, maintain, or enhance tissue and organ functions (Lewns et al. 2023; Xia et al. 2024). These innovations have been propelled by a deeper understanding of cellular behaviors—such as gene expression, signaling pathways, and cell-cell interactions, often studied using traditional two-dimensional (2D) culture models (Dou et al. 2021; Sousa et al. 2025). While these models have greatly advanced cellular biology, they fall short of replicating the dynamic three-dimensional (3D) environment of living tissues. This has led to a growing demand for 3D models that better mimic *in vivo* conditions, providing a more realistic platform for studying cellular responses, cell-materials and developing regenerative solutions (Hu et al. 2024; Dalton et al. 2020).

The development of effective 3D cell culture models relies heavily on the use of scaffolds that not only provide structural support but also enable nutrient transport and maintain biocompatibility. Among the various structural features of scaffolds, hierarchical and heterogeneous architectures are particularly significant (Hu et al. 2024; Zhang et al. 2021). Hierarchical scaffolds, characterized by macro-, micro-, and nanoscale features, closely replicate the complex structure of native tissues (Dalton et al. 2020; Liu et al. 2024). The incorporation of micropores facilitates the efficient transport of oxygen and nutrients and supports the removal of metabolic waste (Wang et al. 2021a, 2022). For example, porous scaffolds composed of polycaprolactone (PCL) and nano-hydroxyapatite (nHAp) have demonstrated improved osteogenic properties and cellular adhesion (Wang et al. 2022; Anthon and Valente 2022). Additionally, scaffolds with interconnected porous networks, such as those combining gelatin and synthetic polymers, have been shown to support cell proliferation and mimic extracellular matrix (ECM)-like environments (Xia et al. 2024; Dou et al. 2021; Wang et al. 2022; Anthon and Valente 2022). However, challenges such as low cell survival rates and inadequate oxygen supply persist, emphasizing the need for improved scaffold designs that can enhance tissue regeneration (Zhang et al. 2021; Anthon and Valente 2022; Lemos et al. 2021).

Cell alignment is a critical factor in tissue engineering, particularly for tissues where anisotropic organization is essential for function, such as blood vessels, cardiac muscle, neurons, and tendons (Wang et al. 2021a). In such types of tissues, aligned vascular endothelial cells regulate anti-thrombogenic reactions and biological signaling. By its turn, cardiomyocytes' alignment ensures synchronous contraction and efficient blood pumping (Lang et al. 2024; Inbody et al. 2021; DePalma et al. 2024). Advances in scaffold fabrication technologies, such as melt electrospinning writing (MEW) and solution electrospinning, have enabled the creation of scaffolds with topographical features that promote cell alignment, enhancing their integration and functionality (Wang et al. 2021a; DePalma et al. 2024). To achieve successful tissue engineering outcomes, scaffolds must act as temporary ECMs, supporting cell attachment, proliferation, and differentiation while being biocompatible, biodegradable, and mechanically suitable for the target tissue (Sousa et al. 2025; Anthon and Valente 2022; Inbody et al. 2021; Hedegaard and Mata 2020). Hierarchical scaffolds meet these requirements with their multiscale architecture, combining macroscale shape and strength, microscale features for vascularization, and nanoscale topographies for modulating protein adsorption and cell signaling, offering superior biological functionality over traditional designs (Zhang et al. 2021; Wang et al. 2021a; Lemos et al. 2021; Inbody et al. 2021).

Biofabrication has revolutionized the creation of hierarchical scaffolds, offering precise control over their architecture and composition through advanced techniques such as 3D printing, electrospinning, bioprinting, and self-assembly (Sousa et al. 2025; Dalton et al. 2020; Liu et al. 2024; Hedegaard and Mata 2020). These methods enable the fabrication of complex structures that closely mimic native tissue environments and allow customization for specific clinical needs, including patient-specific implants and tailored degradation profiles (Lewns et al. 2023; Sousa et al. 2025; Dalton et al. 2020; Hedegaard and Mata 2020). Simultaneously, there is an increasing focus on sustainability in scaffold design, with researchers incorporating renewable resources, biodegradable polymers, and energy-efficient methods to minimize ecological impact while addressing biomedical demands (Heiden et al. 2024; Pandian et al. 2024; Mancuso et al. 2021). These advancements not only enhance the potential of tissue engineering in regenerative medicine but also align with regulatory and public expectations for environmentally responsible innovation, ensuring progress benefits both healthcare and environmental sustainability.

The biofabrication of hierarchical scaffolds and their applications in tissue engineering and *in vitro* models are discussed herein. The design principles, benefits, and fabrication techniques that enable the creation of advanced scaffolds, as well as the materials used to balance biocompatibility, mechanical properties, and sustainability are overviewed. Their diverse applications, emphasizing their role in promoting cell survival, alignment, and tissue-specific functionality, are highlighted. Lastly, the challenges and future opportunities in scaffold design and biofabrication are discussed in depth. By integrating biofabrication, sustainability, and regenerative medicine, hierarchical scaffolds have the potential to revolutionize tissue engineering, advancing both scientific research and clinical applications.

## 15.2 Fundamentals of Hierarchical Scaffolds

Hierarchical scaffolds, characterized by their multiscale structural organization inspired by the ECM, have revolutionized tissue engineering and regenerative medicine by replicating the natural architecture of tissues to enhance biological functionality, mechanical stability, and biocompatibility (Dou et al. 2021; Hu et al. 2024; Liu et al. 2024).

### 15.2.1 Hierarchical Design Principles

The design of hierarchical scaffolds integrates macro-, micro-, and nano-scale features:

#### 15.2.1.1 Macroscale Features

These features provide the overall geometry and structural integrity of scaffolds, essential for load-bearing applications. Large interconnected pores enhance mechanical stability and allow efficient nutrient transport, which is critical for applications like bone and cartilage tissue regeneration (Lewns et al. 2023; Xia et al. 2024; Sousa et al. 2025; Wang et al. 2022). Scaffolds fabricated using fused deposition modeling (FDM) demonstrate the ability to replicate anatomical complexity while supporting vascularization and cellular integration (Xia et al. 2024). Biomimetic scaffolds with hierarchical porous structures composed of PCL and bioactive ceramics, such as nHAp, offer excellent mechanical properties while promoting osteogenic activity (Xia et al. 2024; Sousa et al. 2025; Wang et al. 2022).

#### 15.2.1.2 Microscale Features

Microscale porosity plays a key role in enhancing cellular integration by facilitating cell adhesion, migration, and proliferation (Zhang et al. 2021). These features also allow for the ingrowth of blood vessels, enabling efficient nutrient delivery to cells. Techniques like melt MEW are used to produce aligned microfibers that improve endothelial cell behavior and promote angiogenesis (Dalton et al. 2020; Wang et al. 2021a). The combination of MEW with solution electrospinning enables the production of scaffolds with well-defined micro-architectures and improved mechanical properties, tailored for tissue-specific applications (Wang et al. 2021a).

#### 15.2.1.3 Nanoscale Features

Nanoscale elements are critical for biochemical signaling and cellular interactions (Lemos et al. 2021; Casanova et al. 2020a, 2023). By mimicking the ECM's nanoscale fibrous network, scaffold surfaces enhance protein adsorption, cell adhesion, and differentiation. Electrospun nanofibers made from biopolymers like collagen and PCL closely replicate the ECM and support osteogenesis (Wang et al. 2021a, 2022; Casanova et al. 2020b). Nanoscale porosity also facilitates interactions with signaling molecules and growth factors, further enhancing tissue regeneration potential (Wang et al. 2021a, 2022; Casanova et al. 2020a, b, 2021a, b, 2023).

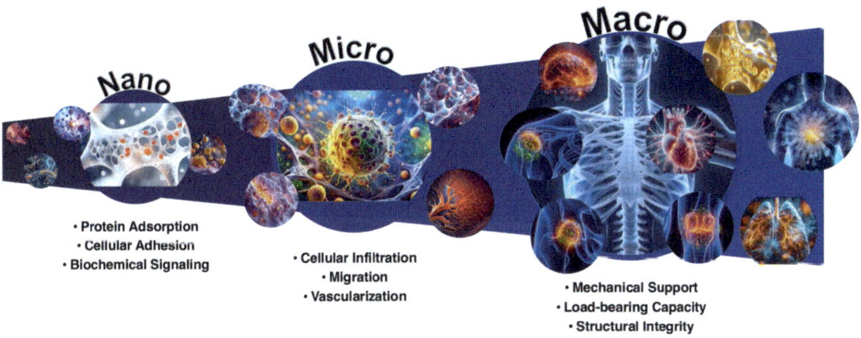


**Fig. 15.1** Schematic representation of hierarchical scaffold features at nano-, micro-, and macro-scale levels. Nano features promote protein adsorption, cellular adhesion, and signaling; micro features facilitate cellular infiltration, migration, and vascularization; and macro features ensure structural support, load-bearing capacity, and integrity. These multiscale attributes collectively enable effective tissue regeneration. (Selected graphical elements were sourced from Adobe Stock to enhance clarity and scientific visualization)

The integration of these multiscale features ensures that hierarchical scaffolds emulate the ECM's structural and functional complexity. Figure 15.1 illustrates how macroscale features provide mechanical stability and structural frameworks; microscale porosity supports cellular integration and vascularization; and nanoscale elements enable biochemical signaling and cell adhesion. Aligning scaffold designs with the natural architecture of the ECM optimizes their application in diverse tissue engineering and regenerative medicine contexts.

### 15.2.2 Advantages

Hierarchical scaffolds provide versatile platforms for tissue regeneration and advanced *in vitro* modeling. By mimicking the ECM through multiscale porosity, these scaffolds enhance cell attachment, proliferation, and differentiation while enabling the creation of biomimetic environments for studying diseases like cancer (Maity et al. 2024; Luo et al. 2024a). For instance, scaffolds with nanofibers and possessing micropore interconnectivity can replicate tumor stroma, supporting investigations into cancer invasion and drug resistance (Mittapelly et al. 2024; Chen et al. 2025; Kim et al. 2025).

To further illustrate the benefits of hierarchical scaffolds, Table 15.1 summarizes the advantages of macro-, micro-, and nanoscale features in enhancing tissue engineering applications (Lewns et al. 2023; Xia et al. 2024; Sousa et al. 2025; Dalton et al. 2020; Zhang et al. 2021; Wang et al. 2021a, 2022; Lemos et al. 2021; Casanova et al. 2020a, b, 2021a, b, 2023). These features collectively contribute to the structural, mechanical, and biological functionality required for complex tissue regeneration and disease modeling.

**Table 15.1** Advantages of multiscale features in hierarchical scaffolds

| Feature | Advantages | | Examples and applications | References |
|---|---|---|---|---|
| Macroscale | Structural integrity | Provide overall geometry, stability, and load-bearing capacity, essential for bone and cartilage regeneration | PCL/nHAp scaffolds with interconnected pores for bone regeneration. 3D-printed constructs mimicking large bone defects. | Lewns et al. (2023); Xia et al. (2024); Sousa et al. (2025); Wang et al. (2022) |
| | Interconnected porosity | Large pores (100–500 μm) improve vascularization and enable efficient nutrient and oxygen transport | | |
| | Anatomical mimicry | Enable replication of complex tissue shapes for patient-specific implants | | |
| Microscale | Cellular integration | Facilitate cell adhesion, migration, and proliferation through microscale pores (10–100 μm) | Melt electrospun scaffolds with aligned microfibers to support endothelial cell alignment. PLGA frameworks with microvascular networks for angiogenesis. | Dalton et al. (2020); Zhang et al. (2021); Wang et al. (2021a) |
| | Vascular network support | Promote blood vessel ingrowth and nutrient diffusion | | |
| | Tissue-specific design | Create aligned microfibers to guide cells in anisotropic tissues (e.g., tendons, blood vessels) | | |
| Nanoscale | Biochemical signaling | Mimic the ECM's nanoscale fibrous architecture to enhance cell signaling, adhesion, and differentiation | Electrospun collagen nanofibers for osteogenesis. nHAp particles for bone tissue engineering. | Wang et al. (2021a, 2022); Lemos et al. (2021); Casanova et al. (2020a, b, 2021a, b, 2023) |
| | Protein adsorption | Facilitate interaction with growth factors and signaling molecules, enhancing tissue regeneration | | |
| | Surface area | High surface area supports drug loading and controlled release | | |

*PCL* polycaprolactone, *nHAp* nano-hydroxyapatite, *PLGA* poly(lactic-co-glycolic acid)

In regenerative applications, hierarchical scaffolds promote vascularization by means of facilitating endothelial migration and capillary network formation. This capability supports both tissue viability and dynamic *in vitro* studies of angiogenesis (Liu et al. 2020; Sun et al. 2021; Wu et al. 2021). Their mechanical stability, achieved through materials like nHAp, ensures suitability for load-bearing applications while enabling the modeling of bone defects and metastases (Wang et al. 2022; Liu et al. 2021a; Yang et al. 2024; Aveic et al. 2024).

Additionally, hierarchical scaffolds support localized drug delivery and controlled release, enabling precise therapeutic testing (Hu et al. 2024; Wang et al. 2022; Maity et al. 2024; Wu et al. 2021; Liu et al. 2021a; Yang et al. 2024). Their adaptability across regenerative and disease modeling contexts highlights their transformative role in advancing tissue engineering and biomedical research.

## 15.3 Biofabrication Techniques for Hierarchical Scaffolds

### 15.3.1 Additive Manufacturing (3D Printing)

3D printing technology is referred to by various terms, including solid freeform fabrication (SFFF), rapid prototyping (RP), rapid manufacturing (RM), layered manufacturing (LM), additive manufacturing (AM), and 3D printing, reflecting its versatility and transformative impact across industries (Chaudhary et al. 2022; Germaini et al. 2022; Rahman et al. 2023). While each term emphasizes specific aspects of the process, they all represent the same fundamental concept, i.e. constructing objects layer by layer from digital designs (Germaini et al. 2022; Rahman et al. 2023; Monzón et al. 2014).

This technology has had a profound impact on the field of tissue engineering (Dou et al. 2021; Casanova et al. 2024; Ratheesh et al. 2021). AM offers unparalleled precision in controlling geometric architectures and material distribution, enabling the fabrication of highly accurate scaffolds (Dou et al. 2021; Ratheesh et al. 2021). These meticulously designed structures are essential for advancing *in vitro* models, as they closely replicate the architecture of native ECM. By creating a more biomimetic environment, AM significantly enhances the reliability and accuracy of *in vitro* models for both research and practical applications (Germaini et al. 2022; van Doorn et al. 2024; Bowes et al. 2024).

Although 3D bioprinting remains an innovative and rapidly evolving field, its origins can be traced back to the late 1980s (Germaini et al. 2022; Rahman et al. 2023; Song et al. 2024). It has since gained significant attention with advancements in materials and technology. The American Society for Testing and Materials (ASTM) F42 committee classifies AM processes based on parameters such as material type, solidification mechanisms, and deposition methods (Germaini et al. 2022; Monzón et al. 2014; Kawalkar et al. 2022). Vat-photopolymerization techniques, including stereolithography (SLA) and digital light processing (DLP), use photopolymer resins cured layer by layer with ultraviolet light (Chaudhary et al. 2022; Germaini et al. 2022; Monzón et al. 2014; Kawalkar et al. 2022; Li et al. 2023).

Material jetting involves depositing liquid materials layer by layer, which is then cured or solidified. Binder jetting relies on bonding powder-based materials with a liquid binding agent, enabling the production of intricate geometries (Germaini et al. 2022; Monzón et al. 2014; Kawalkar et al. 2022). In material extrusion, methods such as FDM use nozzles to extrude thermoplastic or bioink materials, forming structures layer by layer (Germaini et al. 2022; Monzón et al. 2014; Ratheesh et al. 2021; Kawalkar et al. 2022; Rajan et al. 2022). Powder bed fusion employs heat sources, such as selective laser sintering (SLS), electron beam melting, and direct metal laser sintering, to fuse powdered materials at precise locations (Germaini et al. 2022; Monzón et al. 2014; Song et al. 2024; Kawalkar et al. 2022). Sheet lamination processes, like ultrasonic additive manufacturing and laminated object manufacturing, bond thin material sheets to form structures. Directed energy deposition combines focused energy sources to melt and deposit materials simultaneously, as exemplified by techniques like laser cladding (Chaudhary et al. 2022; Germaini et al. 2022; Monzón et al. 2014; Song et al. 2024; Kawalkar et al. 2022; Li et al. 2023; Rajan et al. 2022).

While there are numerous subcategories within AM, the workflow is generally standardized. Key steps include the design phase in CAD software, exporting the file in STL file format (or additive manufacturing file format, AMF), importing the model into the machine, defining the processing parameters, building the structure, removing the printed components, post-processing as required, and finally applying the product to its intended use (Monzón et al. 2014; Kawalkar et al. 2022; Rajan et al. 2022).

Certain techniques within the various AM subcategories are particularly prominent in tissue engineering and *in vitro* models due to their ability to create complex biomimetic structures. Specific technological advancements such as SLA, FDM, SLS, and DLP have greatly improved the quality of scaffolds created through 3D printing (Chaudhary et al. 2022; Germaini et al. 2022; Song et al. 2024; Li et al. 2023).

SLA is favored for its high resolution, making it ideal for fabricating microscale features and achieving the highest fabrication accuracy. However, it is expensive, supports limited materials, and requires time-consuming post-processing (Germaini et al. 2022). FDM is one of the most economical, efficient, and high-speed AM techniques, with a wide range of available materials. This technique, however, presents lower resolution than SLA, is incompatible with cell printing, and produces structures with weak mechanical properties (Germaini et al. 2022; Rajan et al. 2022). SLS uses a laser as its power and heat source to weld solid powders at specific points, allowing the creation of 3D objects without the need for support structures. This capability enables complex geometries to be achieved. Nevertheless, the main disadvantages are the high-costs involved, slow processing, and incompatibility with hydrogels (Germaini et al. 2022; Song et al. 2024). DLP, also a Vat-photopolymerization technique similar to SLA, uses a digital light projector for faster curing. Like SLA, it offers high resolution and precision, but it is significantly faster. Additionally, DLP uses only the resin required for the build, minimizing waste as excess resin can often be reused. Nonetheless, it also presents limitations, such as restricted material options and the need for post-processing (Chaudhary et al. 2022; Xiao et al. 2022).

### 15.3.2 Electrospinning Fabrication

Electrospinning is a versatile technique for fabricating nanofibrous scaffolds, utilizing a high-voltage electric field to produce continuous micro- to nano-scale fibers from polymer solutions or melts (Ji et al. 2024; Elveren et al. 2023; Xing et al. 2023; Lian et al. 2024; Casanova et al. 2018). These electrospun structures have garnered significant attention in tissue engineering due to their ability to replicate the fibrous architecture of the ECM, providing high porosity, large surface area for cell adhesion, and customizable mechanical, biochemical, and biodegradation properties (Elveren et al. 2023; Xing et al. 2023; Casanova et al. 2018).

There are various electrospinning set-ups, each tailored for specific applications. Conventional electrospinning, the most commonly used method, utilizes a single nozzle to produce solid, single-component nanofibers, making it ideal for general tissue scaffolds (Elveren et al. 2023; Lian et al. 2024; Keirouz et al. 2023). Coaxial electrospinning advances this technique by enabling the fabrication of core-shell fibers using concentric nozzles, particularly suited for drug delivery systems and multifunctional materials (Keirouz et al. 2023; Rahmati et al. 2021). MEW eliminates the use of solvents, relying on heat to process polymers, which is advantageous for environmentally sensitive and medical applications (Xing et al. 2023). Additionally, multi-jet and needleless electrospinning techniques have been developed to enhance scalability, with multiple nozzles or free-surface jets significantly increasing fiber production for industrial applications (Wen and Wang 2024).

Despite these variations, the fundamental principle of fiber fabrication remains consistent. The process begins with the formation of the Taylor's cone at the needle tip under a high-voltage electric field. Electrostatic forces stretch the polymer solution into a fine jet, which undergoes elongation and solvent evaporation before depositing onto a collector (Lian et al. 2024; Keirouz et al. 2023; Rahmati et al. 2021). The collector plays a pivotal role in determining the arrangement and properties of the deposited fibers (Ji et al. 2024; Robinson et al. 2021; Rezvani Ghomi et al. 2022). The most commonly used types are static flat collectors and rotating collectors. A static flat collector produces randomly oriented, nonwoven fiber mats suitable for applications where isotropic properties suffice (Ji et al. 2024; Robinson et al. 2021; Rezvani Ghomi et al. 2022). In contrast, rotating collectors align fibers along a single axis by spinning at controlled speeds. This alignment enhances the mechanical properties of fibers in the direction of orientation, making them ideal for applications such as tendons, ligaments, or nerve regeneration (Ji et al. 2024; Robinson et al. 2021; Rezvani Ghomi et al. 2022).

The morphology of electrospun fibers can be finely tuned through the interplay of solution properties, process parameters, and environmental conditions (Elveren et al. 2023; Lian et al. 2024; Rahmati et al. 2021). For instance, higher solution viscosity increases fiber thickness but may disrupt jet stability if excessive (Elveren et al. 2023; Lian et al. 2024; Rahmati et al. 2021). Similarly, increasing the flow rate without adjusting voltage can produce thicker fibers with beads, while lower flow rates yield finer, uniform fibers but risk jet instability (Elveren et al. 2023; Lian et al. 2024; Rahmati et al. 2021). Voltage controls jet stretching, with higher voltages creating thinner fibers; however, excessive voltage may destabilize Taylor's cone (Keirouz et al. 2023; Rahmati et al. 2021). Environmental factors, such as humidity and temperature, also play critical roles. High humidity promotes porous fibers by slowing solvent evaporation, whereas low humidity results in smooth, dense fibers (Keirouz et al. 2023; Rahmati et al. 2021). Optimizing these interdependent parameters is essential for achieving precise control over fiber architecture and bioactivity.

Electrospinning plays a pivotal role in the biofabrication of hierarchical scaffolds, offering a versatile approach to replicate the fibrous architecture of native tissues (Ji et al. 2024; Lian et al. 2024; Rahmati et al. 2021; Liu et al. 2023). Its capacity to produce nano- and micro-fibrous layers with tunable properties supports the development of scaffolds for skin regeneration, neural interfaces, vascular grafts, and bone cartilage systems (Lian et al. 2024; Keirouz et al. 2023; Patel and Gundloori 2022). These scaffolds guide cell behavior, enhance nutrient diffusion, and facilitate biomolecule delivery, making them integral to tissue engineering and *in vitro* models (Lian et al. 2024; Keirouz et al. 2023; Rahmati et al. 2021; Patel and Gundloori 2022). Despite challenges in scaling, fiber alignment, and porosity control, advancements such as hybrid fabrication techniques and green electrospinning methods are overcoming these barriers. As part of a broader biofabrication tool-kit, electrospinning contributes to the creation of multi-scale, biomimetic scaffolds essential for advancing regenerative medicine and complex *in vitro* models (Elveren et al. 2023; Lian et al. 2024; Wen and Wang 2024).

### 15.3.3 Freeze-Drying, Sol-Gel, and Self-Assembly Methods

Biofabrication techniques such as freeze-drying, sol-gel processing, and self-assembly provide versatile approaches to designing porous, biomimetic scaffolds for tissue engineering and *in vitro* models.

Emulsion freeze-drying is widely used for fabricating 3D scaffolds with hierarchical structures and exceptional porosity (Aldemir Dikici and Claeyssens 2020; Jang et al. 2023). This technique enables the creation of scaffolds with porosity exceeding 90% and pore sizes ranging from 20 to 200 μm, tailored for applications requiring enhanced cell infiltration and nutrient diffusion. The process involves homogenizing a polymer solution with an organic solvent or water to form an emulsion, followed by rapid cooling to stabilize the structure (Silvestro et al. 2021). The frozen emulsion is then subjected to vacuum sublimation under low pressure to remove the solvent, resulting in a dry, porous scaffold. Critical parameters such as freezing rate, polymer concentration, and solution pH can be fine-tuned to achieve the desired pore size, interconnectivity, and mechanical properties. While this method is highly versatile, its energy demands and time-intensive nature pose scalability challenges, particularly for industrial production (Jakubowska and Lulek 2021; Karimi et al. 2021).

The sol-gel process is another interesting technique for producing hierarchical scaffolds with customized properties, including porosity, surface area, and bioactivity (Shoushtari et al. 2024). This method typically employs acid- or base-catalyzed reactions, microemulsion-based synthesis, and chemical modifications to generate complex, interconnected pore structures (Shoushtari et al. 2024). In tissue engineering applications, these features are critical for improving mechanical stability and promoting biological performance, such as cell adhesion and proliferation. Additionally, sol-gel techniques allow for the incorporation of bioactive components to enhance scaffold functionality (Shoushtari et al. 2024).

Self-assembly, a bottom-up fabrication approach, relies on molecular interactions such as hydrogen bonding, electrostatic forces, and van der Waals interactions to organize building blocks into hierarchical structures (Badekila et al. 2021). This technique enables the creation of biocompatible nano-structures without the need for external energy input. Self-assembly has been successfully applied to fabricate injectable nanofibrous peptide scaffolds, which hold great promise for tissue regeneration (Badekila et al. 2021). Despite its prevalence in nature, self-assembly approaches remain underexplored in tissue engineering. However, it offers significant potential when integrated with other fabrication methods, enabling the development of multifunctional and biomimetic (Badekila et al. 2021).

By leveraging the unique strengths of these biofabrication techniques, researchers can create scaffolds with tailored properties to meet the complex demands of tissue engineering and *in vitro* model development, Fig. 15.2 illustrates the advanced biofabrication methods for achieving hierarchical scaffold architectures. These techniques enable the precise integration of macro-, micro-, and nano-scale features, enhancing structural integrity, cellular interaction, and biochemical functionality.

## 15.4 Materials for Hierarchical Scaffolds

Material selection underpins the performance of hierarchical scaffolds in tissue engineering and *in vitro* models, demanding a precise balance among biocompatibility, mechanical integrity, and bioactivity (Lewns et al. 2023; Zhang et al. 2021; Hedegaard and Mata 2020; Koushik et al. 2023). Achieving this balance requires thoughtful integration of natural biopolymers, synthetic polymers, and composite or hybrid systems, optimized through controlled pore architectures, bioactive fillers, and growth factor incorporation (Casanova et al. 2019, 2020a, b, c, 2021a, b, 2023, 2024; Mendanha et al. 2023, 2024). Figure 15.3 illustrates the classification of materials used in the biofabrication of hierarchical scaffolds, categorizing them as: natural, synthetic, and composite types. It highlights how these materials contribute to the design of advanced, bioinspired structures that support tissue regeneration.

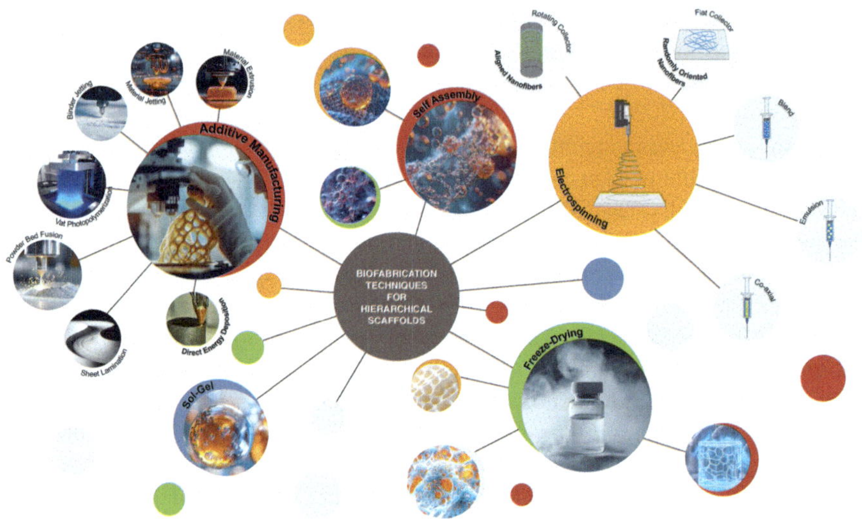


**Fig. 15.2** Schematic representation of advanced biofabrication techniques for hierarchical scaffold architectures. The diagram highlights key methods: additive manufacturing, electrospinning, freeze-drying, sol-gel processing, and self-assembly. (To ensure clarity and enhance scientific visualization, selected graphical elements were sourced from BioRender and Adobe Stock)

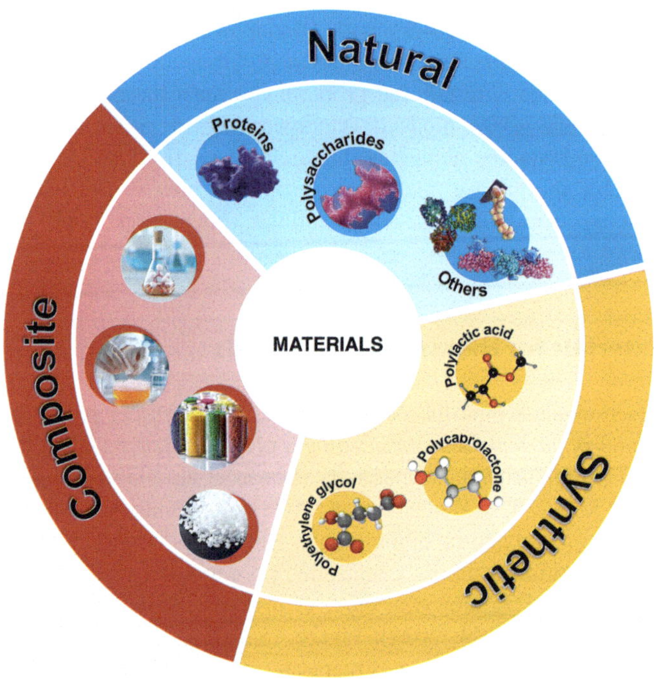


**Fig. 15.3** Classification of materials for the biofabrication of hierarchical scaffolds. Materials are classified as: natural, synthetic, and composite. (Visual elements were sourced from Adobe Stock)

Equally critical is the emphasis on sustainable sourcing, environmentally responsible processing, and scalability to ensure that advanced, bioinspired scaffolds not only enhance tissue regeneration but also adhere to principles of resource stewardship (Sousa et al. 2025; Wang et al. 2022; Heiden et al. 2024).

### 15.4.1 Natural Polymers

Natural polymers closely resemble native ECM components, providing inherent bioactivity and favorable cell-material interactions. When engineered into hierarchical scaffolds, these materials can incorporate multiscale porosity and architecture that support nutrient diffusion, cell infiltration, and tissue maturation (Dalton et al. 2020; Lewis et al. 2024; Tang et al. 2023).

**Collagen's**, a primary ECM protein, can be assembled into scaffolds that closely replicate the fibrillar architecture of native tissues. For example, collagen nanoyarns, as described by Chikelu et al. (2023), are 3D multiscale fibrous scaffolds produced via nanoyarn electrospinning (Chikelu et al. 2023). These scaffolds preserve collagen's triple helical structure, demonstrate improved mechanical properties, and support cell adhesion *in vitro*, with cells adopting spindle-like morphology typical of aligned fibrous surfaces.

**Chitosan's** cationic nature and enzymatic degradability allow for supportive cell-material interactions and controlled remodeling. Studies have demonstrated that chitosan, particularly when combined with other natural polymers or reinforced with hydroxyapatite, can be fabricated into hierarchical scaffolds that improve osteoconduction, promote antibacterial activity, and enhance nutrient exchange (Gu et al. 2024; Wang et al. 2021b; Liu et al. 2021b). For example, the incorporation of small intestinal submucosa and chitosan into a hierarchical sponge/hydrogel scaffold has been shown to mimic the natural micro-/nano-structures of the dura mater, promoting cell adhesion, proliferation, and tissue regeneration (Wang et al. 2021b). Similarly, chitosan combined with **silk fibroin** nanofibers has been utilized to create sponges capable of self-generating antimicrobial hypochlorous acid, effectively accelerating the healing of infected diabetic wounds. These examples underscore the versatility of chitosan-based biomaterials in addressing diverse biomedical challenges.

**Silk fibroin** provides exceptional mechanical strength and adjustable degradation rates, making it particularly well-suited for forming fibrous, load-bearing structures (Lemos et al. 2021; Yang et al. 2024). For example, silk fibroin/hydroxyapatite (SF/HAp) scaffolds have been engineered to possess a hierarchical macro- and microporous architecture that optimizes both mechanical stability and the 3D environment necessary for cell migration and proliferation (Shi et al. 2024). These scaffolds can also be functionally modified to enhance endogenous cell recruitment, ultimately facilitating robust tissue integration and promoting effective cartilage regeneration.

Beyond their biological advantages, natural polymers can be sustainably sourced. Collagen is often derived from food industry byproducts, while chitosan is extracted from marine waste, contributing to circular bioeconomy models (Vieira et al. 2023). Employing green processing techniques, minimizing solvent use, and implementing efficient purification methods further improve the environmental footprint of natural polymer scaffolds, ensuring that their fabrication aligns with both regenerative medicine goals and responsible resource stewardship (Heiden et al. 2024; Pandian et al. 2024).

### 15.4.2 Synthetic Polymers

Synthetic polymers are key in tissue engineering and *in vitro* models, offering tunable properties, controlled degradation, and scalability for creating tailored, biomimetic scaffolds (Lewns et al. 2023; Lang et al. 2024; Koushik et al. 2023).

**Polycaprolactone (PCL)**, one of the most widely utilized synthetic polymers, is notable for its mechanical stability and slow degradation, making it ideal for long-term scaffolds in applications such as bone and cartilage regeneration (Xia et al. 2024; Sousa et al. 2025; Wang et al. 2022; Yang et al. 2024). PCL scaffolds with hierarchical porous architectures have been shown to support vascularization and osteogenesis (Xia et al. 2024; Yang et al. 2024; Lu et al. 2023). For example, dual-pore PCL scaffolds fabricated using MEW and FDM promote nutrient transport and cell adhesion, combining macropores for vascularization and microscale features for osteogenic differentiation (Ratheesh et al. 2021).

**Polylactic acid (PLA)**, a biodegradable polymer derived from renewable resources, is another prominent material in tissue engineering (Dou et al. 2021; Mittapelly et al. 2024; Lu et al. 2023). Its ability to degrade into lactic acid, a naturally occurring metabolite, makes it both biocompatible and environmentally friendly. PLA scaffolds are often used in vascularized tissue engineering applications due to their tunable properties and compatibility with advanced biofabrication techniques (Dou et al. 2021).

**Poly(lactic-co-glycolic acid) (PLGA)**, a copolymer of PLA and poly(glycolic acid) (PGA), is widely employed for controlled drug delivery systems and temporary scaffolds (Dou et al. 2021; Mittapelly et al. 2024; Lu et al. 2023). A validated mechanistic model has been developed for PLGA-based implants, accounting for hydrolysis, diffusion, and erosion to optimize drug release profiles (Mittapelly et al. 2024).

**Polyethylene glycol (PEG)** is another widely used synthetic polymer, valued for its hydrophilicity, non-toxicity, and biodegradability (Xia et al. 2024; Guo et al. 2023; Jiao et al. 2024). When incorporated into PCL-based scaffolds, PEG creates multilevel porosity, improving cell infiltration and bone formation while allowing controlled degradation rates conducive to osteogenesis (Xia et al. 2024).

**Poly(3-hydroxybutyrate-co-3-hydroxyhexanoate) (PHBHHx)**, a biodegradable polymer, has been used to fabricate scaffolds with interconnected porous structures, supporting cell adhesion, proliferation, and differentiation (Liu et al. 2020).

In myocardial tissue engineering, synthetic polymers such as **dextran vinyl sulfone (DVS)** are utilized to fabricate fibrous matrices that mimic the mechanical anisotropy of cardiac ECM (DePalma et al. 2024). These matrices enhance cardiomyocyte alignment, contractility, and maturation, providing insights into tissue-specific mechano-sensing and maturation (DePalma et al. 2024).

Despite their advantages, achieving a balance between mechanical stability and biodegradability remains a key challenge in synthetic polymer design (Lang et al. 2024; Maity et al. 2024; Koushik et al. 2023). Slow-degrading polymers, such as PCL, can hinder cell infiltration and nutrient transport, while rapidly degrading materials may lose structural integrity prematurely. Strategies to address these challenges include tailoring polymer crystallinity, molecular weight, and processing methods to meet application-specific requirements. Synthetic polymers remain a cornerstone in the development of innovative scaffolds for regenerative medicine, thus offering tunability, reproducibility, and scalability essential for advanced tissue engineering applications.

### 15.4.3 Composite and Hybrid Materials

Composite and hybrid scaffolds combine natural and synthetic materials with inorganic components to address the need for hierarchical structures that closely mimic native tissues (Sousa et al. 2025; Wu et al. 2021; Ghorai et al. 2022). This approach leverages the unique properties of each material type, allowing the alignment of features at macro-, micro-, and nano-scale levels to guide cellular behavior, drive specific tissue responses, and improve overall scaffold biofunctionality. Such types of constructs are carefully designed to modulate key properties such as mechanical strength, degradation rates, and cellular interactions through spatially organized architecture and finely tuned biochemical cues (Zhang et al. 2021; Koushik et al. 2023; Liu et al. 2021b).

As an example, a modularized bioceramic scaffold integrated with a hydrogel membrane architecture has demonstrated hierarchical organization for periodontal regeneration (Wei et al. 2022). Custom-designed macropores in the bioceramic framework enhance cellular infiltration, while hydrogel membranes formed from gelatin methacrylate (GelMA), photoinitiator LAP, and hydroxypropyl methylcellulose (HPMC) support cell migration and nutrient exchange at the micro-scale. Nanoscale nonstoichiometric wollastonite particles release osteogenic ions, promoting tissue repair through hierarchical functionality (Wei et al. 2022).

Similarly, in bone regeneration, MEW PCL scaffolds with macroporous grids (~400 μm) provide structural support while interwoven gelatin nanofibers enhance hydrophilicity and mimic ECM-like cues at the microscale, fostering osteogenesis (Wang et al. 2021c). Another example is a PCL hybrid scaffold combined with nHAp, alginate, and type I collagen, where macropores (~380 μm) enable cell infiltration, the hydrogel matrix supports microscale cell encapsulation, and nHAp particles drive osteoconductivity, resulting in improved biocompatibility and bone formation (Sousa et al. 2025).

For modeling the blood-brain barrier, synthetic PEGDA and GelMA-based hydrogels have been used to create perfusable 3D systems with macroscale channels for dynamic fluid flow, microscale features for co-culture of endothelial and neuronal cells, and nanoscale tunability to enhance endothelial adhesion (Royse et al. 2024). This multiscale design captures the complexity of vascular networks and supports advanced neuropharmaceutical testing.

In cancer research, alginate-based systems combining oxidized and RGD-modified alginates have replicated the tumor microenvironment through hierarchical organization (Barros da Silva et al. 2024). Macroporous architectures support epithelial-stromal interactions, microscale hydrogels ensure uniform cell distribution, and nanoscale RGD motifs mimic ECM-like conditions enabling the study of tumor progression, metastasis, and therapeutic responses *in vitro* (Barros da Silva et al. 2024).

A polysaccharide-based composite hydrogel incorporating PLGA, collagen, and nHAp demonstrates the impact of hierarchical design in addressing critical-size bone defects (Lu et al. 2023). Macroporous networks foster vascularization, PLGA–HAp micro-spheres (10–100 μm) and collagen structures enhance cell infiltration and early tissue formation, and nHAp provides osteoinductive signals to guide differentiation and mineralization (Lu et al. 2023).

These examples highlight how the strategic integration of materials across scales enables scaffolds to emulate the complexity of native tissues. Table 15.2 summarizes key materials used in the biofabrication of hierarchical scaffolds, along with representative examples and their applications (Xia et al. 2024; Dou et al. 2021; Sousa et al. 2025; Liu et al. 2021b, 2024; Wang et al. 2021b, c, 2022, 2024; Lemos et al. 2021; Mittapelly et al. 2024; Yang et al. 2024; Chikelu et al. 2023; Gu et al. 2024; Shi et al. 2024; Vieira et al. 2023; Lu et al. 2023; Guo et al. 2023; Jiao et al. 2024; Barros da Silva et al. 2024; Nogueira et al. 2023; Soleymani Eil Bakhtiari and Karbasi 2024; Suamte et al. 2023; Xu et al. 2023a, 2024; Elomaa et al. 2020; Li et al. 2024; Kang et al. 2020; Zhang et al. 2022; Guzman-Soria et al. 2023; Ghomi et al. 2022; Ilomuanya et al. 2023; Huang et al. 2023; Chahsetareh et al. 2024; Kavitha Sri et al. 2023; Iswarya et al. 2023; Yu et al. 2023; Pradeep et al. 2022; Cai et al. 2022; Materón et al. 2021). The material selection in tailoring scaffolds to specific tissue engineering and *in vitro* model needs is a very important parameter to be considered. As the field progresses, prioritizing sustainably sourced materials, eco-friendly fabrication processes, and the responsible disposal of scaffolds will ensure that next-generation constructs meet both clinical and environmental demands.

## 15.5 Sustainability and Green Engineering in Scaffold Design

The integration of sustainability and green engineering principles into scaffold design reflects a critical evolution in tissue engineering and regenerative medicine, ensuring that innovations meet biomedical needs while aligning with ecological responsibility (Heiden et al. 2024; Vieira et al. 2023). The emphasis on using biodegradable materials, energy-efficient fabrication techniques, and lifecycle-focused

**Table 15.2** Materials used in hierarchical scaffold fabrication with examples and applications

| Type | Materials | | Examples/applications | References |
|---|---|---|---|---|
| Natural polymers | Proteins | Collagen, gelatin, elastin, silk fibroin, fibrin, keratin | ECM mimicry<br>Soft tissue scaffolds<br>Wound healing | Dou et al. (2021); Lemos et al. (2021); Chikelu et al. (2023); Shi et al. (2024); Vieira et al. (2023); Wang et al. (2021c); Nogueira et al. (2023); Soleymani Eil Bakhtiari and Karbasi (2024); Suamte et al. (2023) |
| | Polysaccharides | Chitosan, alginate, hyaluronic acid, dextran, starch, cellulose | Hydrogel formation<br>Drug delivery systems<br>Tissue scaffolds | Liu et al. (2024); Gu et al. (2024); Wang et al. (2021b); Liu et al. (2021b); Shi et al. (2024); Vieira et al. (2023); Barros da Silva et al. (2024); Suamte et al. (2023) |
| | Others | dECM | Tissue-specific scaffolds<br>Bioinks for 3D bioprinting | Xu et al. (2024); Elomaa et al. (2020); Li et al. (2024); Kang et al. (2020) |
| Synthetic polymers | PLA, PGA, PLGA, PCL, PEG, PU, PDMS, PVA, PNIPAAm; polyanhydrides, poly(β-amino esters) | | Controlled drug delivery<br>Structural scaffolds<br>Organ-on-a-chip devices | Xia et al. (2024); Dou et al. (2021); Sousa et al. (2025); Wang et al. (2022); Mittapelly et al. (2024); Yang et al. (2024; Lu et al. (2023); Guo et al. (2023); Jiao et al. (2024); Suamte et al. (2023); Zhang et al. (2022) |

(continued)

**Table 15.2** (continued)

| Type | Materials | | Examples/ applications | References |
|---|---|---|---|---|
| Composite materials | Natural-synthetic hybrids | Collagen-PLGA, gelatin-PCL, silk fibroin-PLA, hyaluronic acid-PVA, chitosan-PEG, alginate-PCL | Enhancing mechanical properties Tunable biodegradation | Guzman-Soria et al. (2023); Ghomi et al. (2022); Wang et al. (2024); Ilomuanya et al. (2023); Huang et al. (2023); Chahsetareh et al. (2024) |
| | Natural/polymer-biomaterials composites | HAp-collagen, bioactive glass-silk fibroin, alginate-hydroxyapatite | Bone regeneration Cartilage scaffolds | Kavitha Sri et al. (2023); Xu et al. (2023a); Iswarya et al. (2023) |
| | Nanomaterial hybrids | Graphene oxide-gelatin, nanocellulose-PVA, polymeric micelles (e.g., PEG-PLA), magnetic nanoparticles | Drug delivery systems Tissue engineering with enhanced electrical/ mechanical properties | Yu et al. (2023); Pradeep et al. (2022); Cai et al. (2022); Materón et al. (2021) |
| | Decellularized hybrid scaffolds | dECM-PCL, dECM-PVA | Organ-specific scaffolds Hybrid bioinks for complex tissue engineering applications | Elomaa et al. (2020); Li et al. (2024); Kang et al. (2020) |

*dECM* decellularized ECM, *PLA* polylactic acid, *PGA* poly(glycolic acid), *PLGA* poly(lactic-co-glycolic acid), *PCL* polycaprolactone, *PEG* polyethylene glycol, *PU* polyurethane, *PDMS* polydimethylsiloxane, *PVA* polyvinyl alcohol, *PNIPAAm* poly(N-isopropylacrylamide)

designs supports the development of advanced scaffolds that minimize environmental impact while enhancing clinical outcomes (Cheng et al. 2021).

Biodegradable polymers, such as poly(L-lactide) (PLLA) and PCL, along with natural polymers like gelatin and chitosan, form the foundation of sustainable scaffold design (Jiao et al. 2024; Cheng et al. 2021; Kim et al. 2021; Xu et al. 2023b). These materials, derived from renewable sources, are biocompatible and exhibit controlled degradation properties, making them ideal for minimizing ecological footprints while supporting tissue regeneration. For instance, PCL-based scaffolds treated with plasma-assisted techniques enhance hydrophilicity without introducing toxic agents, adhering to green engineering principles by reducing environmental harm during production (Kim et al. 2021). Similarly, the incorporation of surfactant-treated PLLA/PCL sponges demonstrates how renewable polymers can be optimized for biocompatibility while supporting scaffold functionality (Xu et al. 2023b).

Natural polymers like gelatin and chitosan provide inherent bioactivity and effectively mimic ECM environments, contributing to scaffold functionality (Jiao et al. 2024; Cheng et al. 2021).

Bio-based innovations in scaffold fabrication further highlight the role of green engineering. Mesoporous bioactive glass composites combined with biodegradable polymers demonstrate how bioinspired materials can balance mechanical strength and bioactivity while reducing ecological impact (Lian et al. 2021). Multifunctional designs, such as piezoelectric nanofiber-aerogel scaffolds, integrate mechanical and electrical functionalities to support tissue regeneration while minimizing additional material and energy demands (Chen et al. 2023).

Energy-efficient fabrication methods are vital for sustainable scaffold production. AM, for example, reduces material waste and enables the precise, patient-specific design of scaffolds (Hu et al. 2025). Plasma-assisted fabrication enhances properties such as hydrophilicity and surface functionality without high-energy inputs or harmful solvents, while solvothermal synthesis offers resource-efficient methods to produce advanced scaffold materials (Xu et al. 2023b; Martins et al. 2009; Zhang et al. 2023). Additionally, triboelectric and piezoelectric scaffolds harvest biomechanical energy from the body, providing renewable energy sources for long-term tissue regeneration without reliance on external power (Chen et al. 2023; Luo et al. 2024b).

Lifecycle-focused approaches and circular economy principles are also pivotal in scaffold development. For example, reactive oxygen species-responsive nanostructures degrade into non-toxic components after use, ensuring minimal environmental impact (Choi et al. 2022). Similarly, the reuse of bioelectric scaffolds with integrated energy recovery systems demonstrates how circularity can reduce waste and support sustainable healthcare practices (Luo et al. 2024b). The use of renewable materials like gelatin, collagen, and chitosan—often derived as byproducts of other industries—further reinforces a circular bioeconomy by transforming waste into high-value biomedical resources (Chikelu et al. 2023; Vieira et al. 2023; Nogueira et al. 2023). These advancements underscore the transformative potential of sustainability and green engineering in scaffold design.

## 15.6 Conclusions and Future Perspectives

Hierarchical scaffolds represent a pivotal advancement in tissue engineering and regenerative medicine, providing biomimetic structures that replicate the complexity of native tissues. By integrating macro-, micro-, and nano-scale features, these scaffolds enhance cell attachment, proliferation, and differentiation, addressing key challenges in tissue regeneration and *in vitro* modeling. Biofabrication techniques such as advanced manufacturing and electrospinning have enabled precise control over scaffold architecture, while sustainable materials and energy-efficient methods underscore the growing focus on environmental responsibility. Despite remarkable progress, challenges such as scaling biofabrication processes, achieving efficient vascularization, and balancing mechanical stability with degradation remain.

Regulatory hurdles and the need for standardized testing also slow clinical translation. Addressing these challenges requires interdisciplinary collaboration and innovative approaches. By integrating renewable materials, energy-efficient methods, and lifecycle-focused strategies, scaffold development is paving the way for tissue engineering solutions that address modern biomedical challenges while promoting ecological responsibility. Looking forward, emerging technologies like 4D printing, machine learning-aided design, and self-healing biomaterials offer exciting opportunities to create adaptive and intelligent scaffolds. Additionally, integrating hierarchical scaffolds with organ-on-a-chip systems can revolutionize *in vitro* models, advancing drug discovery and personalized medicine. Hierarchical scaffolds are poised to transform tissue engineering by harmonizing biofabrication, functionality, and sustainability. By tackling current limitations and embracing innovation, they promise to drive impactful advancements in healthcare while supporting sustainable development.

**Acknowledgments** This research acknowledges the funding provided by the UID/50026: 3B's-Biomaterials, Biodegradables and Biomimetics Research Group, University of Minho (3B's Res. Group/UMinho), CAN-TARGET (ref. NORTE2030-FEDER-02705300, CCDR-N), and the European Commission through the ONCOSCREEN (ID: 101097036) and EngVIPO (ID: 101183041) projects.

## References

Aldemir Dikici B, Claeyssens F (2020) Basic principles of emulsion templating and its use as an emerging manufacturing method of tissue engineering scaffolds. Front Bioeng Biotechnol 8:875. https://doi.org/10.3389/fbioe.2020.00875

Anthon SG, Valente KP (2022) Vascularization strategies in 3D cell culture models: from scaffold-free models to 3D bioprinting. Int J Mol Sci 23:14582. https://doi.org/10.3390/ijms232314582

Aveic S et al (2024) Three-dimensional in vitro model of bone metastases of neuroblastoma as a tool for pharmacological evaluations. Nanotheranostics 8:1–11. https://doi.org/10.7150/ntno.85439

Badekila AK, Kini S, Jaiswal AK (2021) Fabrication techniques of biomimetic scaffolds in three-dimensional cell culture: a review. J Cell Physiol 236:741–762. https://doi.org/10.1002/jcp.29935

Barros da Silva P et al (2024) An integrative alginate-based 3D in vitro model to explore epithelial-stromal cell dynamics in the breast tumor microenvironment. Carbohydr Polym 342:122363. https://doi.org/10.1016/j.carbpol.2024.122363

Bowes A et al (2024) Bioprinted high-cell-density laminar scaffolds stimulate extracellular matrix production in osteochondral co-cultures. Int J Mol Sci 25:11131. https://doi.org/10.3390/ijms252011131

Cai Y, Qi J, Lu Y, He H, Wu W (2022) The in vivo fate of polymeric micelles. Adv Drug Deliv Rev 188:114463. https://doi.org/10.1016/j.addr.2022.114463

Casanova MR, Reis RL, Martins A, Neves NM (2018) The use of electrospinning technique on osteochondral tissue engineering. Adv Exp Med Biol 1058:247–263. https://doi.org/10.1007/978-3-319-76711-6_11

Casanova MR et al (2019) Chondrogenesis-inductive nanofibrous substrate using both biological fluids and mesenchymal stem cells from an autologous source. Mater Sci Eng C Mater Biol Appl 98:1169–1178. https://doi.org/10.1016/j.msec.2019.01.069

Casanova MR, Reis RL, Martins A, Neves NM (2020a) Fibronectin bound to a fibrous substrate has chondrogenic induction properties. Biomacromolecules 21:1368–1378. https://doi.org/10.1021/acs.biomac.9b01546

Casanova MR et al (2020b) Spatial immobilization of endogenous growth factors to control vascularization in bone tissue engineering. Biomater Sci 8:2577–2589. https://doi.org/10.1039/d0bm00087f

Casanova MR, Reis RL, Martins A, Neves NM (2020c) Surface biofunctionalization to improve the efficacy of biomaterial substrates to be used in regenerative medicine. Mater Horiz 7:2258–2275. https://doi.org/10.1039/d0mh00542h

Casanova MR, Osorio H, Reis RL, Martins A, Neves NM (2021a) Chondrogenic differentiation induced by extracellular vesicles bound to a nanofibrous substrate. NPJ Regen Med 6:79. https://doi.org/10.1038/s41536-021-00190-8

Casanova MR, Reis RL, Martins A, Neves NM (2021b) Stimulation of neurite outgrowth using autologous NGF bound at the surface of a fibrous substrate. Biomolecules 12:25. https://doi.org/10.3390/biom12010025

Casanova MR et al (2023) Functional recovery of injured cavernous nerves achieved through endogenous nerve growth factor-containing bioactive fibrous membrane. Acta Biomater 168:416–428. https://doi.org/10.1016/j.actbio.2023.07.015

Casanova CR, Casanova MR, Reis RL, Oliveira JM (2024) Advancing diagnostics and disease modeling: current concepts in biofabrication of soft microfluidic systems. In Vitro Model 3:139–150. https://doi.org/10.1007/s44164-024-00072-5

Chahsetareh H et al (2024) Alginate hydrogel-PCL/gelatin nanofibers composite scaffold containing mesenchymal stem cells-derived exosomes sustain release for regeneration of tympanic membrane perforation. Int J Biol Macromol 262:130141. https://doi.org/10.1016/j.ijbiomac.2024.130141

Chaudhary R et al (2022) Additive manufacturing by digital light processing: a review. Prog Addit Manuf 8:331–351. https://doi.org/10.1007/s40964-022-00336-0

Chen Z et al (2023) Ultrasound-driven electrical stimulation based on 3D hierarchical porous piezoelectric nanofiber-aerogel scaffold promotes bone defect repair. Chem Eng J 470:144305. https://doi.org/10.1016/j.cej.2023.144305

Chen G et al (2025) Establishment of a human 3D in vitro liver-bone model as a potential system for drug toxicity screening. Arch Toxicol 99:333–356. https://doi.org/10.1007/s00204-024-03899-9

Cheng H et al (2021) Facile and green approach towards biomass-derived hydrogel powders with hierarchical micro-nanostructures for ultrafast hemostasis. J Mater Chem B 9:6678–6690. https://doi.org/10.1039/d1tb01477c

Chikelu CW et al (2023) Collagen nanoyarns: hierarchical three-dimensional biomaterial constructs. Biomacromolecules 24:1155–1163. https://doi.org/10.1021/acs.biomac.2c01204

Choi H et al (2022) Reactive oxygen species responsive cleavable hierarchical metallic supra-nanostructure. Small 18:e2202694. https://doi.org/10.1002/smll.202202694

Dalton PD, Woodfield TBF, Mironov V, Groll J (2020) Advances in hybrid fabrication toward hierarchical tissue constructs. Adv Sci (Weinh) 7:1902953. https://doi.org/10.1002/advs.201902953

DePalma SJ et al (2024) Matrix architecture and mechanics regulate myofibril organization, costamere assembly, and contractility in engineered myocardial microtissues. Adv Sci (Weinh) 11:e2309740. https://doi.org/10.1002/advs.202309740

Dou Y et al (2021) A hierarchical scaffold with a highly pore-interconnective 3D printed PLGA/n-HA framework and an extracellular matrix like gelatin network filler for bone regeneration. J Mater Chem B 9:4488–4501. https://doi.org/10.1039/d1tb00662b

Elomaa L, Keshi E, Sauer IM, Weinhart M (2020) Development of GelMA/PCL and dECM/PCL resins for 3D printing of acellular in vitro tissue scaffolds by stereolithography. Mater Sci Eng C Mater Biol Appl 112:110958. https://doi.org/10.1016/j.msec.2020.110958

Elveren B, Kurecic M, Maver T, Maver U (2023) Cell electrospinning: a mini-review of the critical processing parameters and its use in biomedical applications. Adv Biol (Weinh) 7:e2300057. https://doi.org/10.1002/adbi.202300057

Germaini M-M et al (2022) Additive manufacturing of biomaterials for bone tissue engineering – a critical review of the state of the art and new concepts. Prog Mater Sci 130:100963. https://doi.org/10.1016/j.pmatsci.2022.100963

Ghomi ER et al (2022) Electrospun aligned PCL/gelatin scaffolds mimicking the skin ECM for effective antimicrobial wound dressings. Adv Fiber Mater 5:235–251. https://doi.org/10.1007/s42765-022-00216-w

Ghorai SK et al (2022) A judicious approach of exploiting polyurethane-urea based electrospun nanofibrous scaffold for stimulated bone tissue regeneration through functionally nobbled nanohydroxyapatite. Chem Eng J 429:132179. https://doi.org/10.1016/j.cej.2021.132179

Gu M et al (2024) High-sensitivity, ultrawide linear range, antibacterial textile pressure sensor based on chitosan/MXene hierarchical architecture. iScience 27:109481. https://doi.org/10.1016/j.isci.2024.109481

Guo M et al (2023) Fabrication and validation of a 3D portable PEGDA microfluidic chip for visual colorimetric detection of captured breast cancer cells. Polymers (Basel) 15:3183. https://doi.org/10.3390/polym15153183

Guzman-Soria A et al (2023) Effect of electrospun PLGA/collagen scaffolds on cell adhesion, viability, and collagen release: potential applications in tissue engineering. Polymers (Basel) 15:1079. https://doi.org/10.3390/polym15051079

Hedegaard CL, Mata A (2020) Integrating self-assembly and biofabrication for the development of structures with enhanced complexity and hierarchical control. Biofabrication 12:032002. https://doi.org/10.1088/1758-5090/ab84cb

Heiden A et al (2024) Organic ink multi-material 3D printing of sustainable soft systems. Adv Mater 37:e2409403. https://doi.org/10.1002/adma.202409403

Hu X-B et al (2024) A hierarchical porous structure integrated three-dimensional electrochemical biosensor for cell culture and real-time monitoring. Microchem J 204:111061. https://doi.org/10.1016/j.microc.2024.111061

Hu Z et al (2025) 3D printing hierarchical porous nanofibrous scaffold for bone regeneration. Small 21:e2405406. https://doi.org/10.1002/smll.202405406

Huang L et al (2023) Rapidly in situ forming an injectable chitosan/PEG hydrogel for intervertebral disc repair. Mater Today Bio 22:100752. https://doi.org/10.1016/j.mtbio.2023.100752

Ilomuanya MO et al (2023) Evaluation of far-field electrospun polyvinyl alcohol/hyaluronic acid nanofibrous membranes for skin tissue engineering applications. Macromol Mater Eng 309:2300299. https://doi.org/10.1002/mame.202300299

Inbody SC, Sinquefield BE, Lewis JP, Horton RE (2021) Biomimetic microsystems for cardiovascular studies. Am J Physiol Cell Physiol 320:C850–C872. https://doi.org/10.1152/ajpcell.00026.2020

Iswarya S, Theivasanthi T, Gopinath SCB (2023) Sodium alginate/hydroxyapatite/nanocellulose composites: synthesis and potentials for bone tissue engineering. J Mech Behav Biomed Mater 148:106189. https://doi.org/10.1016/j.jmbbm.2023.106189

Jakubowska E, Lulek J (2021) The application of freeze-drying as a production method of drug nanocrystals and solid dispersions – a review. J Drug Deliv Sci Technol 62:102357. https://doi.org/10.1016/j.jddst.2021.102357

Jang J-W et al (2023) Review: scaffold characteristics, fabrication methods, and biomaterials for the bone tissue engineering. Int J Precis Eng Manuf 24:511–529. https://doi.org/10.1007/s12541-022-00755-7

Ji D et al (2024) Electrospinning of nanofibres. Nat Rev Methods Primers 4:1. https://doi.org/10.1038/s43586-023-00278-z

Jiao J et al (2024) The promotion of vascular reconstruction by hierarchical structures in biodegradable small-diameter vascular scaffolds. Biomater Adv 162:213926. https://doi.org/10.1016/j.bioadv.2024.213926

Kang MS et al (2020) Advanced techniques for skeletal muscle tissue engineering and regeneration. Bioengineering (Basel) 7:99. https://doi.org/10.3390/bioengineering7030099

Karimi M et al (2021) Fabrication of shapeless scaffolds reinforced with baghdadite-magnetite nanoparticles using a 3D printer and freeze-drying technique. J Mater Res Technol 14:3070–3079. https://doi.org/10.1016/j.jmrt.2021.08.084

Kavitha Sri A, Arthi C, Neya NR, Hikku GS (2023) Nano-hydroxyapatite/collagen composite as scaffold material for bone regeneration. Biomed Mater 18:032002. https://doi.org/10.1088/1748-605X/acc99e

Kawalkar R, Dubey HK, Lokhande SP (2022) A review for advancements in standardization for additive manufacturing. Mater Today Proc 50:1983–1990. https://doi.org/10.1016/j.matpr.2021.09.333

Keirouz A et al (2023) The history of electrospinning: past, present, and future developments. Adv Mater Technol 8:2201723. https://doi.org/10.1002/admt.202201723

Kim W et al (2021) Plasma-assisted multiscale topographic scaffolds for soft and hard tissue regeneration. NPJ Regen Med 6:52. https://doi.org/10.1038/s41536-021-00162-y

Kim J et al (2025) Assembly of glioblastoma tumoroids and cerebral organoids: a 3D in vitro model for tumor cell invasion. Mol Oncol 19:698–715. https://doi.org/10.1002/1878-0261.13740

Koushik TM, Miller CM, Antunes E (2023) Bone tissue engineering scaffolds: function of multi-material hierarchically structured scaffolds. Adv Healthc Mater 12:e2202766. https://doi.org/10.1002/adhm.202202766

Lang Z et al (2024) Construction of vascular grafts based on tissue-engineered scaffolds. Mater Today Bio 29:101336. https://doi.org/10.1016/j.mtbio.2024.101336

Lemos R et al (2021) Carbon nanotube-reinforced cell-derived matrix-silk fibroin hierarchical scaffolds for bone tissue engineering applications. J Mater Chem B 9:9561–9574. https://doi.org/10.1039/d1tb01972d

Lewis JA et al (2024) A bioactive supramolecular and covalent polymer scaffold for cartilage repair in a sheep model. Proc Natl Acad Sci USA 121:e2405454121. https://doi.org/10.1073/pnas.2405454121

Lewns FK et al (2023) Hydrogels and bioprinting in bone tissue engineering: creating artificial stem-cell niches for in vitro models. Adv Mater 35:e2301670. https://doi.org/10.1002/adma.202301670

Li W et al (2023) Stereolithography apparatus and digital light processing-based 3D bioprinting for tissue fabrication. iScience 26:106039. https://doi.org/10.1016/j.isci.2023.106039

Li B et al (2024) Fabrication and characterization of dECM/PVA/GelMA tissue-engineered small-diameter blood vessels. In: Zhou K (ed) Computational and experimental simulations in engineering. Springer, Cham, pp 1035–1045

Lian R et al (2021) Rational design and fabrication of biomimetic hierarchical scaffolds with bone-matchable strength for bone regeneration. Front Mater 7:622669. https://doi.org/10.3389/fmats.2020.622669

Lian S, Lamprou D, Zhao M (2024) Electrospinning technologies for the delivery of biopharmaceuticals: current status and future trends. Int J Pharm 651:123641. https://doi.org/10.1016/j.ijpharm.2023.123641

Liu Y et al (2020) Facilitated vascularization and enhanced bone regeneration by manipulation hierarchical pore structure of scaffolds. Mater Sci Eng C Mater Biol Appl 110:110622. https://doi.org/10.1016/j.msec.2019.110622

Liu A et al (2021a) Optimized BMSC-derived osteoinductive exosomes immobilized in hierarchical scaffold via lyophilization for bone repair through Bmpr2/Acvr2b competitive receptor-activated Smad pathway. Biomaterials 272:120718. https://doi.org/10.1016/j.biomaterials.2021.120718

Liu X, Wu Y, Zhao X, Wang Z (2021b) Fabrication and applications of bioactive chitosan-based organic-inorganic hybrid materials: a review. Carbohydr Polym 267:118179. https://doi.org/10.1016/j.carbpol.2021.118179

Liu P, Gao T, He W, Liu P (2023) Electrospinning of hierarchical carbon fibers with multi-dimensional magnetic configurations toward prominent microwave absorption. Carbon 202:244–253. https://doi.org/10.1016/j.carbon.2022.10.089

Liu S, Yang M, Xu W (2024) Three-dimensional hierarchical cellulose structures based on microbial synthesis and advanced biofabrication. Chem Bio Eng 1:876–886. https://doi.org/10.1021/cbe.4c00143

Lu G et al (2023) Polysaccharide-based composite hydrogel with hierarchical microstructure for enhanced vascularization and skull regeneration. Biomacromolecules 24:4970–4988. https://doi.org/10.1021/acs.biomac.3c00655

Luo X et al (2024a) Bioengineered human arterial equivalent and its applications from vascular graft to in vitro disease modeling. iScience 27:111215. https://doi.org/10.1016/j.isci.2024.111215

Luo B et al (2024b) An encapsulation-free and hierarchical porous triboelectric scaffold with dynamic hydrophilicity for efficient cartilage regeneration. Adv Mater 36:e2401009. https://doi.org/10.1002/adma.202401009

Maity S et al (2024) Recent developments in glioblastoma-on-a-chip for advanced drug screening applications. Small 21:e2405511. https://doi.org/10.1002/smll.202405511

Mancuso E et al (2021) Additively manufactured BaTiO(3) composite scaffolds: a novel strategy for load bearing bone tissue engineering applications. Mater Sci Eng C Mater Biol Appl 126:112192. https://doi.org/10.1016/j.msec.2021.112192

Martins A et al (2009) Surface modification of electrospun polycaprolactone nanofiber meshes by plasma treatment to enhance biological performance. Small 5:1195–1206. https://doi.org/10.1002/smll.200801648

Materón EM et al (2021) Magnetic nanoparticles in biomedical applications: a review. Appl Surf Sci Adv 6:100163. https://doi.org/10.1016/j.apsadv.2021.100163

Mendanha D et al (2023) Macrophage cell membrane infused biomimetic liposomes for glioblastoma targeted therapy. Nanomedicine 49:102663. https://doi.org/10.1016/j.nano.2023.102663

Mendanha D, Casanova MR, Gimondi S, Ferreira H, Neves NM (2024) Microfluidic-derived docosahexaenoic acid liposomes for targeting glioblastoma and its inflammatory microenvironment. ACS Appl Mater Interfaces 16:40543–40554. https://doi.org/10.1021/acsami.4c01368

Mittapelly N et al (2024) Mechanistic model for drug release from PLGA-based biodegradable implants for in vitro release testing: development and validation. ACS Appl Bio Mater 7:7453–7465. https://doi.org/10.1021/acsabm.4c01054

Monzón MD, Ortega Z, Martínez A, Ortega F (2014) Standardization in additive manufacturing: activities carried out by international organizations and projects. Int J Adv Manuf Technol 76:1111–1121. https://doi.org/10.1007/s00170-014-6334-1

Nogueira LFB et al (2023) Collagen/kappa-carrageenan-based scaffolds as biomimetic constructs for in vitro bone mineralization studies. Biomacromolecules 24:1258–1266. https://doi.org/10.1021/acs.biomac.2c01313

Pandian N et al (2024) Roadmap of algal autotrophic tissue engineering in the avenue of regenerative wound therapy. Mater Adv 5:7516–7533. https://doi.org/10.1039/d4ma00492b

Patel PR, Gundloori RVN (2022) A review on electrospun nanofibers for multiple biomedical applications. Polym Adv Technol 34:44–63. https://doi.org/10.1002/pat.5896

Pradeep HK, Patel DH, Onkarappa HS, Pratiksha CC, Prasanna GD (2022) Role of nanocellulose in industrial and pharmaceutical sectors – a review. Int J Biol Macromol 207:1038–1047. https://doi.org/10.1016/j.ijbiomac.2022.03.171

Rahman M et al (2023) A review on nanomaterial-based additive manufacturing: dynamics in properties, prospects, and challenges. Prog Addit Manuf 9:1197–1224. https://doi.org/10.1007/s40964-023-00514-8

Rahmati M et al (2021) Electrospinning for tissue engineering applications. Prog Mater Sci 117:100721. https://doi.org/10.1016/j.pmatsci.2020.100721

Rajan K, Samykano M, Kadirgama K, Harun WSW, Rahman MM (2022) Fused deposition modeling: process, materials, parameters, properties, and applications. Int J Adv Manuf Technol 120:1531–1570. https://doi.org/10.1007/s00170-022-08860-7

Ratheesh G, Shi M, Lau P, Xiao Y, Vaquette C (2021) Effect of dual pore size architecture on in vitro osteogenic differentiation in additively manufactured hierarchical scaffolds. ACS Biomater Sci Eng 7:2615–2626. https://doi.org/10.1021/acsbiomaterials.0c01719

Rezvani Ghomi E et al (2022) Advances in electrospinning of aligned nanofiber scaffolds used for wound dressings. Curr Opin Biomed Eng 22:100393. https://doi.org/10.1016/j.cobme.2022.100393

Robinson AJ et al (2021) Comparative analysis of fiber alignment methods in electrospinning. Matter 4:821–844. https://doi.org/10.1016/j.matt.2020.12.022

Royse MK et al (2024) Development of a 3D printed perfusable in vitro blood-brain barrier model for use as a scalable screening tool. Biomater Sci 12:4363–4375. https://doi.org/10.1039/d4bm00663a

Shi W et al (2024) Three-dimensional printed silk fibroin/hyaluronic acid scaffold with functionalized modification results in excellent mechanical strength and efficient endogenous cell recruitment for articular cartilage regeneration. Int J Mol Sci 25:10523. https://doi.org/10.3390/ijms251910523

Shoushtari MS et al (2024) Sol–gel-templated bioactive glass scaffold: a review. Res Biomed Eng 40:281–296. https://doi.org/10.1007/s42600-024-00342-x

Silvestro I et al (2021) Chitosan scaffolds with enhanced mechanical strength and elastic response by combination of freeze gelation, photo-crosslinking and freeze-drying. Carbohydr Polym 267:118156. https://doi.org/10.1016/j.carbpol.2021.118156

Soleymani Eil Bakhtiari S, Karbasi S (2024) Keratin-containing scaffolds for tissue engineering applications: a review. J Biomater Sci Polym Ed 35:916–965. https://doi.org/10.1080/09205063.2024.2311450

Song Y, Ghafari Y, Asefnejad A, Toghraie D (2024) An overview of selective laser sintering 3D printing technology for biomedical and sports device applications: processes, materials, and applications. Opt Laser Technol 171:110459. https://doi.org/10.1016/j.optlastec.2023.110459

Sousa AC et al (2025) Hybrid scaffolds for bone tissue engineering: integration of composites and bioactive hydrogels loaded with hDPSCs. Biomater Adv 166:214042. https://doi.org/10.1016/j.bioadv.2024.214042

Suamte L, Tirkey A, Babu PJ (2023) Design of 3D smart scaffolds using natural, synthetic and hybrid derived polymers for skin regenerative applications. Smart Mater Med 4:243–256. https://doi.org/10.1016/j.smaim.2022.09.005

Sun F et al (2021) Directly construct microvascularization of tissue engineering trachea in orthotopic transplantation. Mater Sci Eng C Mater Biol Appl 128:112201. https://doi.org/10.1016/j.msec.2021.112201

Tang L et al (2023) A biomimetic in situ mineralization ECM composite scaffold to promote endogenous bone regeneration. Colloids Surf B Biointerfaces 232:113587. https://doi.org/10.1016/j.colsurfb.2023.113587

van Doorn ECH et al (2024) Advancing 3D engineered in vitro models for heart failure research: key features and considerations. Bioengineering 11:1220. https://doi.org/10.3390/bioengineering11121220

Vieira H et al (2023) Current and expected trends for the marine chitin/chitosan and collagen value chains. Mar Drugs 21:605. https://doi.org/10.3390/md21120605

Wang C et al (2021a) Multi-scale hierarchical scaffolds with aligned micro-fibers for promoting cell alignment. Biomed Mater 16:045047. https://doi.org/10.1088/1748-605X/ac0a90

Wang J et al (2021b) A biomimetic hierarchical small intestinal submucosa-chitosan sponge/chitosan hydrogel scaffold with a micro/nano structure for dural repair. J Mater Chem B 9:7821–7834. https://doi.org/10.1039/d1tb00948f

Wang Z et al (2021c) Fabrication and in vitro evaluation of PCL/gelatin hierarchical scaffolds based on melt electrospinning writing and solution electrospinning for bone regeneration. Mater Sci Eng C Mater Biol Appl 128:112287. https://doi.org/10.1016/j.msec.2021.112287

Wang Y et al (2022) 3D printed integrated bionic oxygenated scaffold for bone regeneration. ACS Appl Mater Interfaces 14:29506–29520. https://doi.org/10.1021/acsami.2c04378

Wang X et al (2024) A novel electrospun polylactic acid silkworm fibroin mesh for abdominal wall hernia repair. Mater Today Bio 24:100915. https://doi.org/10.1016/j.mtbio.2023.100915

Wei Y et al (2022) Modularized bioceramic scaffold/hydrogel membrane hierarchical architecture beneficial for periodontal tissue regeneration in dogs. Biomater Res 26:68. https://doi.org/10.1186/s40824-022-00315-0

Wen X, Wang L (2024) Mass production of electrospun nanofibers. In: Qin X, Wang L (eds) Electrospinning: fundamentals, methods, and applications. Wiley-VCH, Weinheim, pp 29–43

Wu M et al (2021) Bioinspired redwood-like scaffolds coordinated by in situ-generated silica-containing hybrid nanocoatings promote angiogenesis and osteogenesis both in vitro and in vivo. Adv Healthc Mater 10:e2101591. https://doi.org/10.1002/adhm.202101591

Xia D et al (2024) Robust hierarchical porous polycaprolactone/nano-hydroxyapatite/polyethylene glycol scaffolds with boosted in vitro osteogenic ability. Colloids Surf A Physicochem Eng Asp 681:132740. https://doi.org/10.1016/j.colsurfa.2023.132740

Xiao T et al (2022) All digital light processing-3D printing of flexible sensor. Adv Mater Technol 8:2201376. https://doi.org/10.1002/admt.202201376

Xing J et al (2023) Multi-material electrospinning: from methods to biomedical applications. Mater Today Bio 21:100710. https://doi.org/10.1016/j.mtbio.2023.100710

Xu J et al (2023a) Electrospun the oriented silk fibroin/bioactive glass @ silk fibroin/polycaprolactone composite bi-layered membranes for guided bone regeneration. Colloids Surf A Physicochem Eng Asp 676:132224. https://doi.org/10.1016/j.colsurfa.2023.132224

Xu C et al (2023b) Thoroughly hydrophilized electrospun poly(L-lactide)/poly(epsilon-caprolactone) sponges for tissue engineering application. Macromol Biosci 23:e2300143. https://doi.org/10.1002/mabi.202300143

Xu P, Cao J, Duan Y, Kankala RK, Chen A (2024) Recent advances in fabrication of dECM-based composite materials for skin tissue engineering. Front Bioeng Biotechnol 12:1348856. https://doi.org/10.3389/fbioe.2024.1348856

Yang J, Fatima K, Zhou X, He C (2024) Meticulously engineered three-dimensional-printed scaffold with microarchitecture and controlled peptide release for enhanced bone regeneration. Biomater Transl 5:69–83. https://doi.org/10.12336/biomatertransl.2024.01.007

Yu Q et al (2023) Graphene oxide/gelatin nanofibrous scaffolds loaded with N-acetyl cysteine for promoting wound healing. Int J Nanomedicine 18:563–578. https://doi.org/10.2147/IJN.S392782

Zhang L et al (2021) Hierarchical and heterogeneous hydrogel system as a promising strategy for diversified interfacial tissue regeneration. Biomater Sci 9:1547–1573. https://doi.org/10.1039/d0bm01595d

Zhang Z et al (2022) Sustainable hierarchical-pored PAAS-PNIPAAm hydrogel with core-shell structure tailored for highly efficient atmospheric water harvesting. ACS Appl Mater Interfaces 14:55295–55306. https://doi.org/10.1021/acsami.2c19840

Zhang Y et al (2023) Sulfur-doped NiFe hydroxide nanobowls with wrinkling patterns for photothermal cancer therapy. ACS Nano 17:20024–20033. https://doi.org/10.1021/acsnano.3c05244

The manufacturer's authorised representative in the EU is Springer Nature Customer Service Centre GmbH, Europaplatz 3, 69115 Heidelberg, Germany. If you have any concerns regarding our products, please contact ProductSafety@springernature.com

Printed and bound by CPI Group (UK) Ltd, Croydon, CR0 4YY
05/10/2026
02228133-0007